# PUTNAM'S GEOLOGY

Professor William Clement Putnam, the original author of this book, died on March 16, 1963, at the age of fifty-four. He had taught geology for thirty-two years. This book is the result of his experience as a teacher, and of his love of teaching beginning students.

# PUTNAM'S GEOLOGY

FIFTH EDITION

## Peter W. Birkeland
## Edwin E. Larson

UNIVERSITY OF COLORADO

NEW YORK   OXFORD
OXFORD UNIVERSITY PRESS
1989

Oxford University Press

Oxford   New York   Toronto
Delhi   Bombay   Calcutta   Madras   Karachi
Petaling Jaya   Singapore   Hong Kong   Tokyo
Nairobi   Dar es Salaam   Cape Town
Melbourne   Auckland

and associated companies in
Berlin   Ibadan

Published by Oxford University Press, Inc.,
200 Madison Avenue, New York, New York 10016

Oxford is a registered trademark of Oxford University Press

Library of Congress Cataloging-in-Publication Data
Putnam, William Clement, 1908–1963.
[Geology]
Putnam's geology. — 5th ed. / Peter W. Birkeland, Edwin E. Larson.
p.   cm.   Includes bibliographies and index.
ISBN 0-19-505630-2.   ISBN 0-19-505517-9 (pbk.)
1. Physical geology. I. Birkeland, Peter W., 1934–   II. Larson, Edwin E., 1931–
III. Title. IV. Title: Geology.
QE28.2.P87 1989
550—dc19       88-4007   CIP

2 4 6 8 9 7 5 3 1

Printed in the United States of America
on acid-free paper

# PREFACE

The order of presentation of material in this edition is somewhat different from earlier editions. We have divided the book into four parts, each containing related subject matter.

In Part I we present general concepts and information about the planetary system, geology, geological time, and the earth's physical features and materials.

In Part II we have grouped six chapters that discuss deformation of crustal rocks and the building of the earth's surface. These processes, which are primarily constructive, are internally generated; that is, the energy that drives them comes from within the earth.

Part III contains nine chapters that relate to the leveling of the earth's surface; that is, the erosion of the uplands and filling in of the lowlands. By and large, these processes are externally generated by energy derived from the sun and work in many cases under the influence of gravity.

Part IV, a single chapter, discusses geological and social aspects of resources and energy, with emphasis on the future of scarce resources and energy options.

This four-part structure is not limiting, since each chapter can stand on its own. Instructors thus can assign reading in any order that fits their schedules.

# ACKNOWLEDGMENTS

We are grateful for the help of instructors in universities all over the country who read and commented on individual chapters from the previous edition and those who criticized the final manuscript. We would especially like to thank the following: William Atkinson, Victor Baker, Subir Banerjee, Raymond Burke, Arthur Bloom, William Bradley, William Brennan, Scott Burns, Anthony Gangi, Michael Fix, Bert Nordlie, William Rose, William White, and Duane Wohlford. Finally, we would like to thank Fred Luiszer for his help in preparation of the manuscript, and the staff at Oxford: our editor, Joyce Berry, and Karen Lundeen, for her work on the illustrations.

# CONTENTS

# PART I

# General Concepts

# 1

# THE PLANETARY SYSTEM

We live on earth, the third planet from our central star, the sun. With the aid of telescopes and other astronomical equipment, we can examine the sun and eight other planets and various moons, asteroids, and comets in our solar system. Beyond that the sky is literally filled with stars, bright and faint, near and far. Most of the stars we see are part of a large, rotating pinwheel-shaped body of about 10 billion stars called the **Milky Way galaxy** (Fig. 1-1) that takes about 250 million years to make one turn around its center. Our galaxy derives its name from the white, hazy band traversing the sky that is visible on dark nights away from city lights. As you gaze at the milky path, you are looking along the plane of the galaxy.

As large and bright as we think the sun to be, it is but an average-sized star in our galaxy. Along with its planetary system, the sun is located not in the center of the galaxy, but relatively far out on one of the pinwheel arms.

During the previous spin of our galaxy, 250 million years ago, when our planetary system was similarly positioned, most of today's continents were part of one giant landmass. Reptiles were becoming the dominant life form, and mammal-like animals (our ancestral line) had just come into existence. Two spins ago (500 million years), continental landmasses had been nearly leveled by erosion, and life, dominated by invertebrate forms, was virtually restricted to the seas. We can only wonder what will transpire in the next spin of the galactic wheel.

It is truly humbling that our galaxy is but one small part of a visible universe that includes about 1 billion individual galaxies—islands of stars in the vastness of space (Fig. 1-2). The size of the universe and the amount of material in it are all but incomprehensible. It is so large that light traveling at a speed of 300,000 km/sec (186,000 mi/sec) takes many millions, even billions, of years to reach the earth from the edge of deep space.

So, as we look progressively farther out into the universe, we are looking back in time. This causes our picture of the universe to be somewhat distorted because we see it as it *was*, not as it *is*. For example, Chinese, Arab, and American Indian ob-

**Fig. 1-1.** Spiral galaxy in the Ursa Major constellation, as seen through a telescope, closely resembles our Milky Way galaxy. Individual stars in the galaxy are not discernible. The bright large and small spots are relatively nearby stars in the Milky Way. (Hale Observatories)

**Fig. 1-2.** The Virgo cluster of galaxies, which is about 70 million light years away from the earth, contains about 250 large galaxies, including a relatively large number of spiral galaxies. In this photograph, the galaxies show up as eliptical to nearly circular fuzzy white spots. (Kitt Peak National Observatory, National Optical Astronomy Observatories)

servers saw the catastrophic explosion of a large star (now called a supernova event) in A.D. July 1054. For several days during its death rattle, the star shone so brightly that it was visible with the naked eye during the day. Light from this star, a relatively close neighbor of our solar system, took 5000 years to travel to earth. The early astronomers had been treated to celestial fireworks that had actually occurred 50 centuries earlier, in 3946 B.C. Modern astronomers document the occurrence of extraordinary astral events—the births

and deaths of stars, even whole galaxies. Space is a vast and fascinating arena, of which our solar system is but a miniscule part.

This is not to say, however, that our solar system and our earth are inconspicuous. In fact, to most of us, they are probably the only "real" part of the universe. Our lives spin out, from day to day, year to year, according to the rhythms dictated by the sun, earth, and moon as they rotate on their own axes and around each other. The sun provides warmth and light for life and energy to oceanic and atmospheric processes; the moon lights our way at night and is responsible for tidal ebbs and flows, and the other planets, many of which can be seen with the naked eye, are our solar-system companions.

Before 1960 knowledge of our solar system came primarily from direct observations of the earth, the study of meteorites that fell to earth, and the telescopic observations of other planets that began in the early 1600s.

Earthbound observations, even by means of huge, sophisticated telescopes of the day, were limited by the atmospheric layer through which incoming light had to pass. To progress, astronomy had to move beyond the envelope of air around the earth. In the mid-1960s balloonists carried instruments to more than 15 km (9.3 mi) above the earth, but the telescopes they used were necessarily small. The obvious solution was, first, to establish relatively large artificial satellites that would orbit the earth beyond the atmosphere and, second, to send space probes to the reaches of the solar system. In the last 20 years both objectives have been realized.

Today scores of satellites exist, collecting such data as the commonly seen images of surface weather conditions. Some of those in operation (from about 1978), such as the sea satellite (SEASAT) and the magnetic satellite (MAGSAT), have provided us with indispensable data on the surface elevations of the sea, the earth's gravity and magnetic fields, polar wobble, and so on. In the late 1980s a space telescope will be placed in orbit high above the earth's atmosphere. From that vantage point it will even be possible to determine visually if Barnard's star, a single-star solar system 6 light years from earth, possesses planetary bodies similar to the larger ones in our system.

Space travel really began with the Apollo missions of the 1960s. In the initial phases astronauts circled the earth testing both their equipment and their physiological responses to weightlessness in preparation for the first manned mission to the moon in 1969.

But space exploration is not a simple matter. To place a human being on the moon was probably the most awesome engineering feat ever achieved.

Twenty-four men traveled to the moon from 1969 to 1972, and 12 of them actually walked on its surface. Since 1972 much of the exploration has been aimed beyond the moon. Spacecraft have now visited Mercury, Venus, Mars, Jupiter, and Saturn. Voyager 2 passed Saturn early in 1981, Uranus early in 1986, and is now on its way to Neptune for a rendezvous in the summer of 1989.

The unmanned spacecraft of today are sophisticated vehicles with special radar units that scan the surface features below clouds. Some spacecraft are designed to take high-quality photographs and to sense and measure the magnetic field and chemical composition of a planet as well as a wide spectrum of electromagnetic radiation. Others are designed to land gently on a planetary surface and determine the presence of extraterrestrial life.

Since the 1970s the Soviet Union has placed a number of space stations in orbit, each occupied by teams of astronauts who lived in space for extended periods of time. In the early 1980s the United States inaugurated space shuttle flights. Our astronauts performed experiments in weightlessness high above the earth and returned in a stubby-winged free-flying glider. This phase of exploration was curtailed after the catastrophic explosion of the space shuttle Challenger and the death of all seven of its crewmembers on January 28, 1986.

## THE SUN, PLANETS, ASTEROIDS, AND COMETS

The principal components of our solar system (Table 1-1) are nine planets, differing in size and density, that orbit the sun in near-circular elliptical paths, all roughly in the same plane (Fig. 1-3). Most have their own satellite bodies, or moons, in orbit. The inner four, the so-called **terrestrial planets,** are composed chiefly of dense rocky material like that of the earth, whereas the outer five, with the exception of Pluto, are large bodies composed chiefly of gases—particularly in the liquid and gaseous states.

Between the two groups is the **asteroid belt,** a zone in which quantities of "rocks" of various sizes orbit the sun. Individual and irregularly shaped, these bodies range in size from particles the size of dust to asteroids with diameters of several kilometers. The largest body in this zone, Ceres, is

**Table 1-1** Features of the Solar System

| Principal Statistics | Mercury | Venus | Earth | Mars | Jupiter | Saturn | Uranus | Neptune | Pluto |
|---|---|---|---|---|---|---|---|---|---|
| Mean distance from sun (millions of kilometers) | 57.9 | 108.2 | 149.6 | 227.9 | 778.3 | 1,427 | 2,869 | 4,496 | 5,900 |
| Mean distance from sun (astronomical units) | 0.387 | 0.723 | 1 | 1.524 | 5.203 | 9.539 | 19.18 | 30.06 | 39.44 |
| Period of revolution around sun | 88 days | 224.7 days | 365.26 days | 687 days | 11.86 yr | 29.46 yr | 84.01 yr | 164.8 yr | 247.7 yr |
| Rotation period on axis | 59 days | 243 days retrograde | 23 hr 56 min 4 sec | 24 hr 37 min 23 sec | 9 hr 50 min 30 sec | 10 hr 14 min | 15 hr retrograde | 22 hr or less | 6 days 9 hr |
| Inclination of axis | <28° | 3° | 23°27' | 23°59' | 3°5' | 26°44' | 82°5' | 28°48' | ? |
| Inclination of orbit to ecliptic | 7° | 3.4° | 0° | 1.9° | 1.3° | 2.5° | 0.8° | 1.8° | 17.2° |
| Eccentricity of orbit | 0.206 | 0.007 | 0.017 | 0.093 | 0.048 | 0.056 | 0.047 | 0.009 | 0.25 |
| Equatorial diameter (kilometers) | 4,880 | 12,104 | 12,756 | 6,787 | 142,800 | 120,000 | 51,800 | 49,500 | 6,000(?) |
| Mass (earth=1) | 0.055 | 0.815 | 1 | 0.018 | 317.9 | 95.2 | 14.6 | 17.2 | 0.1(?) |
| Volume (earth=1) | 0.06 | 0.88 | 1 | 0.15 | 1,316 | 755 | 67 | 57 | 0.1(?) |
| Density (water=1) | 5.4 | 5.2 | 5.5 | 3.9 | 1.3 | 0.7 | 1.3 | 1.7 | ? |
| Atmosphere (main components) | None | Carbon dioxide | Nitrogen, oxygen | Carbon dioxide | Hydrogen, helium | Hydrogen, helium | Helium, hydrogen, methane, ammonia | Hydrogen, helium, methane | None detected |
| Atmospheric pressure at surface, bars | $10^{-6}$ | 90 | 1 | 0.006 | ? | ? | ? | ? | ? |
| Known satellites | 0 | 0 | 1 | 2 | 15 | 17 | 14 | 2 | 1 |
| Ring structure | No | No | No | No | Yes | Yes | Yes | ? | ? |

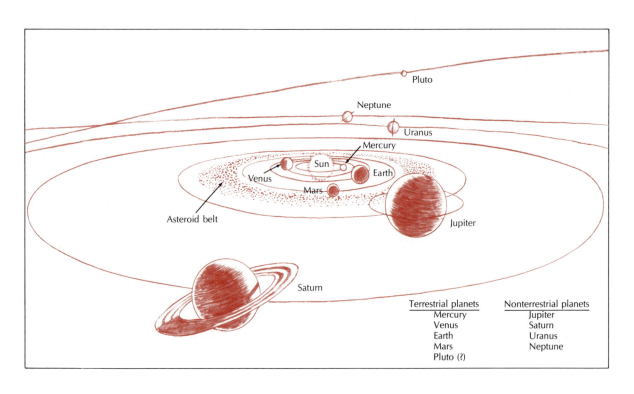

Terrestrial planets
Mercury
Venus
Earth
Mars
Pluto (?)

Nonterrestrial planets
Jupiter
Saturn
Uranus
Neptune

about 300-km (187-mi) across. A few of the meteorite showers observed on earth appear to originate in this zone. Most meteorite showers, however, particularly the most spectacular ones, are associated with debris trails liberated during the disintegration of comets. Composed of rock, dust, and ices, comets traverse the solar system in elliptical paths. Sometimes in their journey they are far beyond the outermost planet; other times they cut across the paths of all the planets to make a close swing around the sun. In 1985 and 1986 skywatchers were treated to the return of the famous Halley's comet, which had last made a spectacular visit in 1910. Two earth-launched probes sailed very close to the solid head of the comet, providing astronomers with some of the best data yet about the nature of a comet's core.

We will now describe the more significant features of the sun and the planets, starting from the planet closest to the sun on outward. Following the description of the earth, we will discuss the moon, our closest companion in the solar system.

## The Sun

The center of our solar system is a huge rotating sphere of white-hot gases—330,000 times more massive than the earth and 109 times as large in diameter. More than 99.8 percent of all the mass in the solar system is centered in the sun. The sun's mean density is 1.4 g/cm$^3$—slightly more dense than liquid water.

What we normally see from the earth is the sun's reddish, low-density **chromosphere** and beneath that, the outermost part of the **photosphere**—both composed largely of noncharged gas atoms at temperatures near 5400° C (9750° F). Much of the photosphere is in constant motion as heat is transported from below. Above the chromosphere there is a changing zone of ionized gases (the **corona**), which is clearly visible only when the chromosphere and the photosphere are blocked out, during an eclipse of the sun by the moon (Fig. 1-4). The gas particles acquire their energy from

magnetic lines of force and shock waves from deep within the sun and appear to have temperatures near 2.5 million degrees centigrade (4.5 × 10$^{6°}$ F). The corona has no outer limit, no boundary. The ionized particles move out in space to form the **solar wind,** which sweeps outward at speeds of 300 to 600 km/sec (675,000 to 1,350,000 mi/hr). The solar wind causes a comet's tail to be blown away from the sun, and it is this stream of particles that is trapped in the magnetic field of the earth, resulting in auroral displays and radio interference.

Physicists estimate that temperatures near the sun's center are close to 15 million degrees centigrade (27 × 10$^{6°}$ F). Particles at that temperature move so fast that they tend to fly off from the sun. However, the sun's huge mass creates such a large gravitational pull that most particles are held toward the center. The size of the sun, then, is the result of an equilibrium between expansional and contractional forces. Pressures near its center are 0.5 million million kg/cm$^2$ (7 trillion lb/in.$^2$), which is equivalent to the weight of forty 100,000-ton aircraft carriers concentrated in an area the size of a single perforation on a postage stamp. Of the 92 naturally occurring elements found on earth, more than 60 have been identified through spectroscopic analysis of the sun. Hydrogen (60 to 80 percent by weight) and helium make up about 95 percent of the total.

The sun releases an immense amount of energy into space—in all directions. The amount has been estimated to be roughly 5 × 10$^{23}$ hp. Today most solar physicists believe that this incredible amount of energy can only be continuously supplied through thermonuclear reactions deep inside the sun. In particular, it is thought that reactions involving the joining (fusion) of atoms are the basic source of the sun's energy. If only 0.45 kg (1 lb) of hydrogen protons and neutrons combined to form helium nuclei (the type of reaction that takes place in the hydrogen bomb), 1000 kw of power would be released continuously for 135 years.

So, the sun and apparently all other stars are nuclear breeding pots in which different elements with lighter nuclei fuse to create more complex nuclei and thus give off heat and light.

Many, perhaps most, stars in the heavens do not occur singly as the sun does; they exist in groups

**Fig. 1-3.** Nine planets circle the sun, more or less in the same plane. The four inner planets are small and composed mostly of rocky material; four of the outer five are large and composed mostly of hydrogen and helium.

**Fig. 1-4.** The sun's corona, photographed through a telescope during an eclipse, is composed of ionized gases that radiate outward to form the solar wind. (Yerkes Observatory)

of two or three. Apparently, material sufficient to initiate and continue nuclear reactions was concentrated in two or three bodies. In our solar system Jupiter, which is composed largely of hydrogen and helium, appears to have fallen just short of initially accumulating enough material to start the fires of its own nuclear furnace.

### Mercury

A small planet, Mercury is not much bigger than earth's moon. It is closest to the sun and receives about nine times as much sunlight as does the earth. As a result its surface is extremely hot—so hot, in fact, that lead would melt there. Because of its size, the planet lacks an appreciable atmo-

sphere; thus the huge fireball sun blazes down on it from a jet-black sky.

Mercury possesses a magnetic field one hundred times weaker than that on earth and a moonlike landscape pocked by meteorite-impact craters of all sizes (Fig. 1-5). The strength of the magnetic field supports the idea that Mercury possesses a relatively large iron core (nearly as big as our moon) beneath a relatively thin 640-km (400-mi) outer shell of rock.

### Venus

Venus, completely covered by thick, reflective yellow-white clouds, is the brightest body in our sky—next to the sun and moon. The planet rotates

8

slowly on its axis once every 243 earth days in a direction opposite (retrograde) to the earth. Temperature increases steadily downward from −50° C (−60° F) at cloud-top elevations to about 475° C (890° F) at the surface. The atmosphere is composed almost entirely of carbon dioxide (97 percent) and nitrogen (2 percent). The gas cover is so dense that the atmospheric pressure on Venus is about 90 times that on earth at an equivalent elevation. It is thought that the extremely hot Venusian surface temperatures are the result of a supergreenhouse effect whereby long wavelength radiation from solar-heated surface rocks is absorbed by the abundant carbon dioxide in the atmosphere and cannot escape.

## The Earth

The earth, the third planet out from the sun, is relatively dense and rocky and roughly spherical in shape. Almost three-fourths of the surface is covered by ocean waters (Fig. 1-6). One complete orbit of the earth around the sun takes approximately 365 1/4 days at a mean distance of 140

**Fig. 1-5.** The cratered surface of Mercury photographed from a spacecraft at a distance of 77,800 km (48,400 mi). The distance across the bottom of the photograph is about 600 km (375 mi). (NASA, Jet Propulsion Laboratory)

**Fig. 1-6.** View of the earth from the moon (foreground) about 384,000 km (240,000 mi) away. The photo was taken by members of the 1969 Apollo 11 mission just after arriving at, and going into orbit around, the moon. (NASA)

million km (994 million mi). At this distance from the sun, there has been sufficient radiant energy for life to evolve in abundance, and it now is fairly certain that the planet earth is the only haven for life within our solar system.

The earth revolves once per day on its axis, which is inclined at an angle of 23° 27' to the plane of its orbit around the sun. In response to centrifugal forces associated with that rotation, which are greatest at the rotational equator, the earth possesses a slightly flattened spherical shape (ellipsoid of rotation) like an underinflated beach-ball. The discrepancy is barely noticeable; the distance from the earth's center to the pole is about 6356.9 km (3950.2 mi), whereas that from the center to the equator is 6378.4 km (3963.5 mi).

An improved timepiece—based on the electronic measurement of atomic vibrations in the element cesium—is capable of recording variations in the length of a day to billionths and trillionths of a second. With it scientists can demonstrate that the spinning earth is slowing down, a phenomenon recorded as a progressive, very slight lengthening of each successive day. Most scientists agree that the slowing of rotation is primarily the result of gravitational interaction—a sort of drag—between the earth and its moon.

Apparently, this slowing has been going on for as long as the earth has had its moon. It is documented by Babylonian records of solar eclipses and, before recorded time, by studies of the number of growth lines deposited per lunar month on ancient seashells, particularly corals. Scientists now know that 300 million years ago, a year would have been 440 days long.

The earth possesses a magnetic field, the strongest of the terrestrial planets, which is inclined about 11 1/2° to its rotational axis. The overall density of the earth is about 5.5 g/cm$^3$. This represents its total mass ($6 \times 10^{27}$ g) divided by its total volume ($1.084 \times 10^{27}$ cm$^3$). For comparison, the density of the earth is about five and one-half times that of water. As the average density of rocks close to the earth's surface is only about 2.7 to 3.0 g/cm$^3$, it is certain that a very dense material must exist at depth. In fact, quantitative data from other sources (see chap. 5) indicate that, near the earth's center, the material is four-to-five-times more dense than the average surface rock.

Among the planets the earth's atmosphere is unique, consisting mostly of nitrogen (78 percent by volume) and oxygen (21 percent by volume). The remaining 1 percent is constituted primarily of argon, carbon dioxide, and water vapor. The air close to the earth's surface, compressed by the weight of the air above it, is much more dense than that higher up. Nearly all the atmosphere lies within 96 km (60 mi) of the earth; above that there is less air than in the best vacuum we can produce in the laboratory. Above 960 km (600 mi) the atmosphere consists mostly of hydrogen and helium; above 2400 km (1500 mi) only small, fast-moving molecules of hydrogen are found. As the earth orbits the sun, some of the light, active gases (mostly helium and hydrogen) are lost into space, left behind to become part of the interplanetary gas mixture.

It is the presence of oxygen, carbon dioxide, and water vapor on earth that is of prime interest to geologists. Most organisms cannot survive without oxygen, and carbon dioxide is vital to plants. Water, which occurs as a vapor in relatively small amounts in the atmosphere, plays several roles: it is essential to most plants and animals; it is the prime absorber of radiant heat in the atmosphere; and it is the principal substance involved in weathering and erosion.

The earth's atmosphere appears to have evolved to its present state through geological time by the slow action of many processes. Among these are (1) loss into space of many of the less dense gases, (2) the addition of gases from organic processes and of gaseous emanations associated with volcanic activity, and (3) the accumulation of such gases as argon and helium, which are radioactive-isotope decay products.

The atmosphere is in continuous agitation and flow, as is quite evident from global weather patterns. The driving force for the circulation of the air is, in one way or another, related to the energy received from the sun.

Water in the liquid state is present predominantly in the world's oceans, which cover about 71 percent of the earth's surface, and subordinately in lakes, streams, and within pores in the layers of rock near its surface. The oceans supply water to the atmosphere through evaporation by the sun's rays. Water vapor is carried by prevailing wind systems far inland, where it precipitates to recharge the surface waters and groundwaters and to nourish plants and animals. Chemical weathering of rocks is greatly aided by the moisture that lingers at the earth's surface after the water has flowed away. And runoff of excess water is the prime agent in erosion of land. Eventually, of course, much of the water that falls on the land returns to the oceans and the atmosphere. This continual circulatory replenishment is termed the *hydrologic cycle* and is of utmost importance to the well-being of the planet (chap. 12).

The waters of the ocean are constantly in motion. Some currents are wide and feeble; others are narrow and strong. Currents are not restricted to surface layers. Flow patterns also exist at deeper levels and bring about vertical as well as lateral movement of the ocean waters. The circulation pattern of the deeper currents generally differs greatly from that of the surface currents.

Surface flow is dominated by the prevailing winds, which push at the surface water, eventually creating a slow-moving ocean drift. Like atmospheric currents, water currents are affected by the earth's rotation; they, too, continue to curve as they flow until large rotating cells develop.

The earth is composed predominantly of high-density, mostly solid matter. Our best information about this material comes from rocks at the surface. Drilling on land and at sea has allowed us to directly observe the earth at depths of more than 12,200 m (40,000 ft), and analysis of volcanic products has given us some idea of the temperature gradient and rock material to 200 to 250 km (125 to 155 mi) below the surface. For knowledge of the earth beyond those depths, we must rely on geophysical studies—most of them involving interpretations of the waves that emanate in all directions during an earthquake (see chap. 5). Some useful data, however, have also been obtained from studies on the magnetic and gravitational fields at the earth's surface and on the regional loss of heat from the earth's interior.

The data suggest that the earth has a pronounced density gradient, with rocks near the center being four-to-five-times more dense than surface rocks (Fig. 1-7). The greatest variety of rocks, which have an average density of only 2.7 g/cm³, are found on the continents and extend beneath them to maximum depths of around 70 km (44 mi). Rocks just below the oceans have an average density of 3.0 g/cm³ and project downward only 5 to 6 km (3 to 4 mi). These two zones are termed the *continental crust* and the *oceanic*

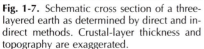

**Fig. 1-7.** Schematic cross section of a three-layered earth as determined by direct and indirect methods. Crustal-layer thickness and topography are exaggerated.

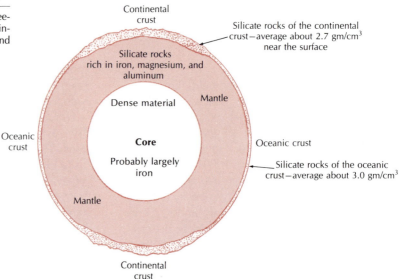

*crust,* respectively. Because of the greater thickness of the lower-density continental crust, the continents stand higher than do the rocks of the ocean basins (see chap. 6). The rocks beneath the crust, down to about 2900 km (1800 mi), are rich in silicon and oxygen, with lesser amounts of magnesium, iron, and aluminum; they constitute the zone called the *mantle.* Below that depth, the earth's *core* is thought to be nearly pure iron (possibly with small amounts of silicon, nickel, and sulfur)—both in liquid and solid state. The magnetic field of the earth appears to be generated in the liquid part of the iron core (see chap. 5).

Exactly how the earth came to be layered into the core, mantle, and crust will probably forever remain a mystery. Some geophysicists believe that **density layering** occurred quickly, as a result of the complete melting of the entire protoearth shortly after it came into being. In the molten state the heavier materials could, under the influence of gravity, easily make their way toward the center of the earth and the lighter ones would tend to move outward. Some of the heat needed to produce melting could have come from decay of radioactive isotopes, which were more abundant initially; some may have been released during gravitational compression of an initially larger, low-density protoearth into the smaller, more dense earth we know. Yet many earth scientists feel that such heat sources would have been insufficient to completely melt the earth. Alternatively, if temperatures became high enough to melt only the iron, then metal droplets could slowly make their way to the center of the earth and density stratification would be slow. Whatever the process, there is evidence that by about 3 billion years ago their existed a molten core large enough to produce a magnetic field.

Raymond Siever and Frank Press, two prominent American geoscientists, have recently called the process of density layering and the movement of iron to the center of the earth the *iron catastrophe.* They believe that it may have been the most significant geological event that has ever happened. When it occurred, the earth not only became density stratified, but volatile substances, particularly water, were brought from the depths to form our oceans directly and, through a partial dissociation of the water, to produce indirectly some of the oxygen needed for life development. Moreover, a natural subsurface heat engine was set in motion that has kept the earth geologically active through its first 5 billion years. As of now this activity shows no signs of abatement.

Many of the earth's phenomena, including volcanism, mountain building, lateral movement of continental blocks, and earthquakes, result from processes related to the deep-seated heat engine. Most of them appear to be particularly active at depths of less than 250 km (155 mi), but some activity occurs as far down as 700 km (435 mi), possibly even deeper. Many of the internal processes can be considered **geologically constructional,** that is, mountains form, or the earth's surface grows upward or is uplifted, as a result of such processes. The internally generated constructional processes are discussed in Part II.

Earth scientists have long searched for a unified theory or model that would neatly tie together all the internally generated processes. Today many geologists believe that the recently formulated theory of *plate tectonics* (from the Greek word *tektonikos,* "carpenter" or "builder") fulfills the requirement. According to the model the outer shell of the earth down to about 100 km (62 mi) is rigid and composed of a number of segments or plates— somewhat like the tiles on a bathroom floor—that can move with respect to each other in response to internal sources of power (Fig. 1-8). At some plate boundaries the plates push together, at some they pull apart, and at still others they slide laterally past one another. It has been postulated that all the internal processes and the attendant surface features and phenomena result from plate motion. Present-day plate boundaries include most of the world's earthquake belts, volcanically active regions, deep-sea trenches, island arcs, and some mountain chains.

Plate tectonics is undoubtedly the dominant theory for the internally generated processes of the earth—a theory that has revolutionized earth science. Virtually all geologists now accept the idea that continents drift and change in surface position and have done so throughout much of the geological past.

The earth's surface is also subject to such **land-**

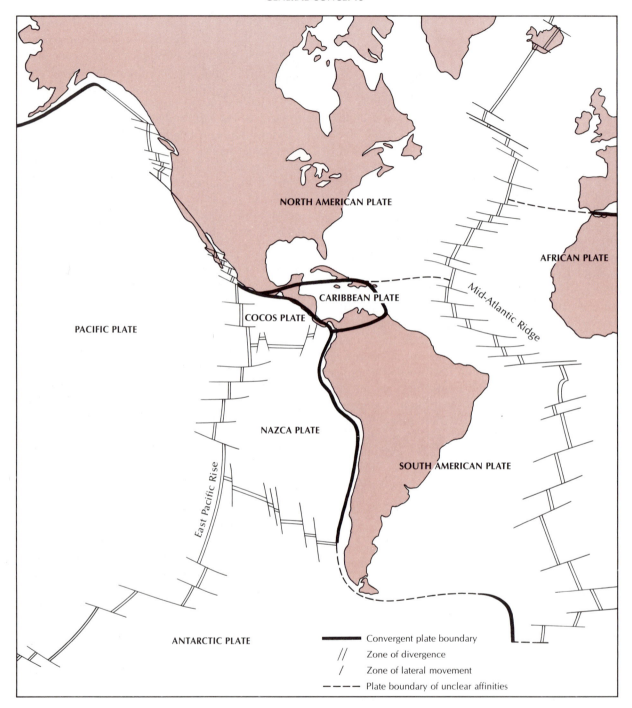

**Fig. 1-8.** The Western Hemisphere, showing the formally named rigid plates, bounded by zones of divergence, lateral movement, and convergence. According to the plate-tectonic model, virtually all internally generated phenomena are brought about by plate-margin interactions.

*leveling processes* as weathering, erosion, and deposition of sediments—processes that tend to reduce the high spots and to fill in the low ones. The various surface processes and how they level the earth will be discussed in Part III.

As noted earlier, the principal energy source for the surface processes on earth, and for life itself, is the sun. There seems little likelihood that the solar power source will fail for several billion years to come.

The face of the earth has changed continuously down through the geological ages by the interaction of constructional and land-leveling processes. When the latter dominates for long periods—as it did, for example, between 1200 and 500 million years ago—the continents are reduced to broad, low-relief features that barely rise above sea level. When the former is dominant, however—as it has been in the recent geological past—the continental masses are built up until they stand, studded with lofty mountain peaks, high above the oceans.

**The Moon**   The earth is the closest planet to the sun that possesses a moon—an orbiting satellite body. During its orbit around the earth, at an average distance of about 400,000 km (240,000 mi), the moon revolves once on its axis. Therefore, it always shows but one face to earthbound observers; the far side of the moon was never seen until the Apollo missions in the 1960s. Because of its relatively small mass (1/81 that of the earth), the moon retains but little atmosphere—in fact, the density of the lunar atmosphere is about a thousand times less than that achieved in the best laboratory-created vacuum on earth. Perhaps the best-known feature of the moon is its cratered surface, which is easily observed through a telescope (Fig. 1-9). Early observers, however, were divided as to whether the craters resulted from meteorite impacts or volcanic eruptions in the airless surface environment. Finally, on July 20, 1969, U.S. astronauts were the first to step onto the lunar surface. In six Apollo missions our astronauts carried out a battery of scientific experiments and brought back 382 kg (840 lb) of lunar rocks. The data gathered from these missions have increased our knowledge about the moon immeasurably.

Instruments were set out at each station the astronauts established to monitor tremors from moonquakes and the impact of meteorites. Moonquakes are both less common and smaller in magnitude than are quakes on the earth. Scientists have conjectured that, like the earth, the moon consists of concentric layers. The outermost layer is about 1-km (0.6-mi) thick and composed of fragmented material, the result of meteorite bombardment. Beneath, down to about 60 km (37 mi), is a layer of solid rock probably composed of dark-colored lava (basalt) or the lighter-colored rock that is typically exposed in the lunar highlands. The next zone appears to extend to a depth of 1000 km (620 mi) and is a rocky layer rich in magnesium and iron. There is evidence of a core—extending from 1000 km (620 mi) to the center at 1800 km (1120 mi)—that differs from the zone above only in being partially molten; no clear-cut evidence suggests the existence of an iron core at the center.

Scientists are eager to know if the moon's core is metallic because of conflicting magnetic data obtained during the Apollo missions. Magnetometers aboard the spacecraft had revealed an extremely weak lunar field. The permanent magnetism of samples returned to earth, however, is relatively intense—seemingly greater than what could have been acquired in the moon's present weak field. This prompts questions: Did the moon at one time possess a stronger field, generated (as was the earth's) in a molten iron core? Was the moon once elsewhere in the solar system, a place where the magnetic field was stronger when the rocks were magnetized? The contradictory magnetic data are perhaps the most puzzling enigma of the lunar investigation. They have led to several explanations concerning the origin of the moon, none of which is entirely satisfactory.

On the moon's far side, large meteorite craters are common; however, the large, dark, circular maria (plural of mare) lowlands, so common on the near side and visible to the earth, are rare here.

Rocks from the lunar maria are predominantly basalt, which is also common on earth. Lunar basalts, however, contain more titanium and iron than do their terrestrial counterparts. Most scien-

**Fig. 1-9.** Large and small meteorite-impact craters on the near side of the moon (illumination from left to right). The largest crater, Goclenius, is about 60-km (37-mi) across. The linear depressions in and near Goclenius appear to have resulted from fracturing. (NASA)

tists agree that the dark maria rocks were formed by large-scale eruptions of molten lava. Radiometric age dating of the rocks shows that basaltic floods occurred over large parts of the moon between 3.8 and 3.2 billion years ago.

Rocks collected in the lighter-colored highland regions contain more calcium and aluminum and less iron, magnesium, and titanium than do maria rocks. They appear to have originated deep beneath the surface. After formation, they have been uplifted, perhaps by the impact of large meteorites. All highland rocks have been dated at 4.2 to 4.0 billion years—ages that confirm the previously held idea that the highland rocks are older than the maria rocks. The oldest age, 4.2 billion years, is greater than that for earth rocks (3.8 billion years), yet it falls short of the estimated time of the origin of the solar system—about 4.5 billion years

ago. Perhaps future missions to the moon will uncover rocks of that age.

The astronauts also brought back many samples of lunar breccia, a shock-welded jumble of angular fragments of preexisting maria and highland rocks shattered during meteorite impacts. The largest meteorites appear to have bombarded the moon about 4 to 3 billion years ago, producing huge cratered depressions. Later flooding by basalt flows filled these depressions and produced the flat-bottomed maria.

## Mars

Known as the red planet, Mars is not much larger than our moon. Because of its nearness to earth, Mars has long been the object of study. Features visible through the telescope convinced some of

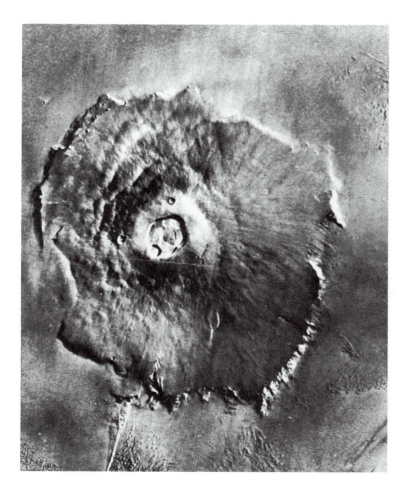

**Fig. 1-10.** Olympus Mons, the largest volcano on Mars seen directly from above. Notice the high, steep escarpment around the base. This cone is about 500 km (310 mi) in diameter, a truly huge volcanic structure. (NASA)

the early observers that Mars was inhabited by intelligent beings. Black lines first observed in 1877 by the Italian astronomer Giovanni Schiaparelli (1835–1910) were interpreted as canals, dug to bring water from the polar regions to the arid equatorial region. This view was obliterated in 1965 when Mariner 4 flew within 10,000 km (6220 mi) of Mars and sent back pictures showing a bleak, cratered landscape more reminiscent of the moon than the earth. Later views sent back by Mariner 6 and Mariner 7 showed no signs of the canals or any other evidence of intelligent life. The probes also revealed that the planet's atmosphere, which is largely carbon dioxide, is so sparse that the air pressure is less than one-hundredth that at sea level on the earth. In the frigid polar regions, the carbon dioxide freezes to form dry ice, which falls to the ground like our snow. The Martian polar caps, which shrink and swell seasonally, are composed largely of dry ice.

When Mariner 9 arrived at Mars in 1972, a planetwide dust storm completely obscured the surface for weeks. As the dust finally began to settle, the views of Mars became spectacular. First to protrude through the dust shroud was the big-

gest volcano ever seen (Fig. 1-10), since named Olympus Mons. It rises 24 km (15 mi), nearly three times as high as Mount Everest. And it is about 500-km (310-mi) across at its base, which is terminated by a steep, high cliff. Were Olympus Mons in California, it would reach from Los Angeles to San Francisco. This volcano is one of a group of four huge cones.

The cameras also revealed a canyon unequaled on earth. This chasm, called Valles Marineris, is 4000-km (2500-mi) long, up to 200-km (125 mi) wide, and up to 6-km (4-mi) deep. Remember that Mars is a relatively small planet with only one-seventh the volume of the earth. This makes the scale of the volcanoes and canyons on Mars truly awesome.

Perhaps the most dramatic aspects of the Martian exploration came when Viking spacecraft were sent to the planet to search for life. Lander craft with television cameras were detached from orbiting vehicles and directed toward the surface. The first pictures showed a barren surface strewn with angular rocks of all sizes. There were no signs of life. Soon after touchdown, mechanical scoops retrieved some of the Martian soil, and three sep-

**Fig. 1-11.** The turbulent atmosphere of Jupiter as viewed from Voyager 1, March 1, 1979. The well-known Great Red Spot is in the upper right. (NASA, Jet Propulsion Laboratory)

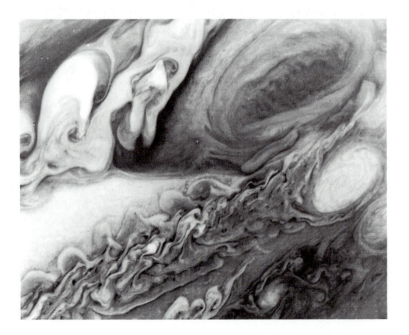

**Fig. 1-12.** A spectacular eruption on Io, one of Jupiter's moons. The volcanic plume, composed largely of sulfur and sulfur dioxide, extends about 320 km (200 mi) into space. (NASA)

arate experiments designed to detect microorganisms were performed. The results were ambiguous, but most researchers have concluded that not even microorganic life exists on Mars.

## Jupiter

Jupiter, the largest planet in our solar system, is easily visible from earth with the naked eye. Its most conspicuous features are a large elongate red spot on its surface (first viewed by telescope in 1665) and a retinue of 15 moons, including the 4 large ones discovered by Galileo (1564–1642) in the early 1600s shortly after he acquired a telescope. Voyager 1 and Voyager 2, after two years and a 645-million km (400-million mi) space journey, closely approached Jupiter and obtained measurements and photographs of the planet and 5 of its moons.

The Great Red Spot was shown to be a vast atmospheric storm, twice as wide as the earth, rotating counterclockwise every six earth days (Fig.

1-11). Huge lightning flashes were seen crackling through the cloud tops.

One of the startling facts revealed by the Voyager missions was an equatorial ring-structure resembling that around Saturn. Jupiter's rings, however, are thin, diffuse, and invisible from the earth. Probes verified that the planet's atmosphere is largely composed of hydrogen (~75 percent) and helium (~20 percent). Small amounts of sulfur compounds appear to produce the reds, oranges, and yellows of the planet's cloud layers. Other sensors indicated that Jupiter possesses the strongest dipolar magnetic field in the solar system, tipped about 10° from its rotational axis, and that the planet gives off about twice as much heat as it receives from the sun. This latter fact supports the idea that heat trapped inside Jupiter during its early days of formation is leaking away.

One of the main tasks of the Voyager missions was to photograph the Galilean satellites, which carry the names of friends and lovers of the Roman god Jupiter. What the imaging team expected to see when the cameras focused on Io was the crater-pocked face of a dead moon. Instead, they saw a sensational spectacle—eight different volcanic vents simultaneously erupting, some spewing plumes as high as 320 km (200 mi) above the surface (Fig. 1-12). Besides the earth, Io is the only body in the solar system that is volcanically active. Later images from Voyager 2 showed six of the volcanoes still erupting. As far as we can tell, these volcanoes are unlike their earthly counterparts in that the material being ejected is primarily elemental sulfur and sulfur dioxide—which accounts for the red-to-orange hue of Io's surface.

## Saturn

Next beyond Jupiter is Saturn, also one of the large, low-density planets. Its most characteristic feature is a well-developed set of equatorial rings (Fig. 1-13). Along with the craters on the earth's moon and the red spot on Jupiter, the rings of Saturn are among the favorites of backyard astron-

omers. Its atmosphere is, like that on Jupiter, composed mostly of hydrogen and helium.

Space probes indicate that Saturn's dipolar magnetic field is aligned exactly along its rotational axis, but it is much smaller than that of Jupiter. At the cloud-top level, the field was found to be nearly equal to that at the earth's surface. Saturn also emits about three times as much thermal energy as it receives from the sun and must, therefore, possess a source of internal heat.

Late in 1980 Voyager 1 made its closest approach to Saturn. Its photographs showed that the atmosphere is highly turbulent and possesses a reddish spot similar to Jupiter's, but smaller. The ring-structure is larger and more complex than was first thought, consisting of many dark and light zones. Additional moons were observed, bringing the total to 17 encircling the planet.

## Uranus

Although Uranus is 67 times larger than the earth, it is so far from the earth that it escaped detection until 1781. It is possible to see Uranus with the

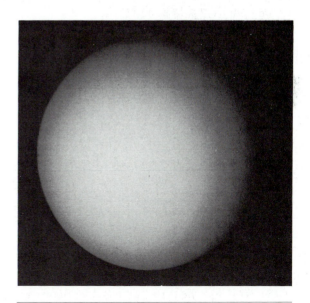

**Fig. 1-14.** Uranus, as viewed from Voyager 2, January 17, 1986, at a distance of 9.1 million km (5.7 million mi), seven days prior to closest approach. At this point in its passage, Voyager 2 is about 3 billion km (1.87 billion mi) from earth. (NASA, Jet Propulsion Laboratory)

**Fig. 1-13.** The complexity of Saturn's ring-structure is evident in this awesome Voyager 1 computer image. (NASA)

naked eye, but conditions must be optimal. The telescope revealed Uranus as a greenish-blue orb surrounded by 5 moons.

In January 1986 Voyager 2 reached Uranus and began to relay data acquired from its sensors. Because of the planet's great distance from earth, the transmission of data traveling at the speed of light required about 2 3/4 hours. Photos revealed 14 moons and 10 separate narrow rings. Uranus, as shown by Voyager 2, also possesses a magnetic field. Curiously, this field is tilted about 55° to its rotational axis. Some scientists have speculated that the odd tilt is associated with a transitional stage in the 180° shift of magnetic poles on the planet; perhaps, however, it is related to the unusual orientation of the rotational axis—unlike the other planets, it is virtually in the plane of its orbit around the sun.

As viewed from Voyager 2, the upper, visible surface of Uranus is the top of the atmospheric cloud cover (Fig. 1-14). Most of the atmosphere, which shows evidence of large-scale lateral and turbulent flow, consists of hydrogen and helium and lesser amounts of ammonia and methane.

## Neptune

Few facts about Neptune have been established. Viewing it through a telescope from earth is like examining a dime with the naked eye at a distance of about 1.6 km (1 mi). We hope to learn more about this planet when Voyager 2 reaches it in the summer of 1989.

## Pluto

Pluto, the smallest planet, is farthest from the sun most of the time. Because of its small size and great distance from the earth it was not discovered until 1930. Its orbit is highly elliptical, so much so that at times it is closer than Neptune to the sun.

Pluto is unlike most of the outer planets—it is small and dense. Because of this and its unusual orbit, it has been suggested that the planet was at one time a moon of Neptune, dislodged during a catastrophic solar event and subsequently sent into orbit around the sun.

## ORIGIN OF THE SOLAR SYSTEM

Our solar system is about 5 billion years old—so ancient that the details of its origin will probably remain hazy for a long time to come. Moreover, planets are dark bodies that weakly transmit illumination received from companion stars. This makes other planetary systems—thought to be common within our galaxy and beyond—invisible to observations from our solar system. As a result we lack other examples for comparison with our own system.

Virtually all astronomers agree that a solar system such as ours is a natural consequence of star formation, which has occurred and is occurring throughout the universe. The basic theory of star formation is that in response to some initial perturbation a nebulous cloud of gases (mostly hydrogen and helium), dust, and ice in a local portion of a galaxy begins to condense. At this time temperatures are near absolute zero and no warming sun exists. Gravitational contraction of the condensed region continues but so slowly that turbulent motions are smoothed out as the vast cloud begins to slowly rotate. In time the nebular cloud becomes a rotating disc-shaped body in which most of the mass is concentrated in the central bulge. Clumps of more densely packed material in the less massive peripheral-disc zone occasionally collide with one another, eventually coalescing into larger clumps (protoplanets), which, like the material in the central zone, are composed largely of hydrogen and helium along with smaller amounts of the heavier elements. In the process of colliding and coalescing, some of the smaller clumps begin to circle the larger clumps.

At about this time contraction of the central mass into a protosun creates enough heat for nuclear reactions to take place. The mass begins to glow and then to radiate intensely. The resultant rapid heating vaporizes the lighter elements, especially hydrogen and helium, on those clumps near the sun, and the radiation sweeps the material outward. Bodies far from the developing sun, however, remain sufficiently cool to retain most of the lighter matter. The inner planets, therefore, tend to be rocky and relatively dense; the outer planets tend to be composed largely of low-density materials: hydrogen, helium, and water.

Because of differences in size and initial composition, each newly formed planet experiences a somewhat different life cycle. Small planets (and moons) go through their planetary stages of development relatively rapidly, within a few billion years, and subsequently become dead. In contrast others (like our earth) remain active throughout the life of the solar system. Distance from the sun also affects planetary behavior. For instance, Venus and earth are nearly identical (it is believed) in composition, density, and size. Yet Venus, only 41 million km (25.75 million mi) closer to the sun, has markedly different atmospheric and near-surface conditions than has the earth.

In some cases a planet within a particular solar system satisfies all the conditions necessary for the initiation and sustenance of life, for which we may express our humble gratitude.

## SUMMARY

1. The earth is one of nine planets in our solar system that orbit our central star, the sun—all in the same direction and roughly in the same plane. The sun is but one of about 10 billion stars that make up the large spiraling *Milky Way galaxy*. It is estimated that the visible universe consists of about 1 billion galaxies.

2. The sun, 109 times larger in diameter than the earth, possesses more than 99.8 percent of all the mass in the solar system. The *chromosphere* and outermost *photosphere,* composed of noncharged gas atoms at a temperature near 5400° C, (9750° F), are what we normally see from the earth. Beyond these is a tenuous zone, the *corona,* which consists of ionized particles that move rapidly away from the sun. Near the center of the sun, temperatures climb to about 15 million degrees centigrade ($27 \times 10^6$ ° F). The sun's energy appears to be produced by the fusion of lighter atomic nuclei into heavier ones and the destruction, in the process, of a relatively small amount of mass.

3. Mercury, the closest planet to the sun, is small, hot and lacks an appreciable atmosphere. Its surface is pocked by numerous meteorite-impact craters.

4. Venus, similar is size to the earth, is completely surrounded by a dense hot atmosphere composed largely of carbon dioxide.

5. The earth, the third planet from the sun, possesses an average density of 5.5 g/cm³, about twice that of its surface rocks. The obvious inference is that some very dense material must be present below the earth's surface.

6. The earth's slightly flattened ellipsoidal shape is the result of the balance between its inward pull on all its particles of matter and its rotation on its axis.

7. The earth's atmosphere is principally nitrogen (78 percent) and oxygen (21 percent). The atmosphere also contains small amounts of other gases, many of which are important, including carbon dioxide and water vapor. The composition of the atmosphere has undoubtedly changed with time. Circulation of the atmosphere is driven by energy from the sun.

8. The earth's oceans cover 71 percent of its surface area. These waters constitute the largest reservoir of moisture and are the wellspring of the *hydrologic cycle.*

9. The earth is layered. The outer layer of relatively light silicate rock extends down to about 6 km (4 mi) beneath the ocean basins *(oceanic crust)* and up to about 70 km (44 mi) under the continents *(continental crust).* The next layer is *the mantle.* It is largely silicon and oxygen with lesser amounts of magnesium, iron, and aluminum in a solid state; it extends down to 2900 km (1800 mi). *The core,* the innermost zone, is mostly iron and is partly liquid and partly solid.

10. *Volcanism, mountain building, earthquake activity,* and *continental drift* rebuild the earth's surface. They are generated internally by process that occurs to a depth of 700 km (435 mi) and possibly deeper. Other processes, such as *weathering, erosion,* and *sedimentation,* act to level the surface and are powered by the sun's energy. The face of the earth has changed continuously through time in response to the interaction of *constructional* and *land-leveling* processes.

11. The theory of *plate tectonics* forms the basis for understanding and interrelating the internally generated phenomena. The outer shell of the earth appears to be composed of a number of rigid plates that move: pushing together, diverging, or sliding

laterally. Most of the constructional phenomena are developed at the plate margins.

12. The moon contains only 1/81 of the mass of the earth; therefore it virtually lacks an atmosphere. Its surface typically has been pockmarked by meteorite impacts. Like the earth, it is layered and apparently composed entirely of silicate material; an iron core seems to be lacking. The dark *maria* regions are composed of basalt that erupted between 3.8 and 3.2 billion years ago. The upland regions are underlain by rocks richer in calcium and aluminum but poorer in iron, magnesium, and titanium than are lunar basalts. The highland rocks have been dated at 4.2 to 4.0 billion years.

13. Mars, a small planet, possesses only a thin atmosphere that is relatively rich in carbon dioxide. In the frigid polar regions, this gas freezes to form dry-ice polar caps. Spectacular surface features include huge volcanoes and deep canyons. Space probes thus far have shown no signs of life.

14. *Jupiter*, the largest planet, is largely composed of hydrogen and helium, with perhaps a small rocky or molten iron core at its center. We see only the tops of the atmospheric clouds, which are in constant motion. The Great Red Spot is a large circulating storm system. Jupiter has a diffuse ring-structure, the strongest magnetic field in the solar system, and 15 moons. On its moon, Io, active volcanism has been detected.

15. *Saturn* resembles Jupiter in many ways. Its ring-structure, however, is much more complex. It possesses 17 moons.

16. *Uranus* is also a large nonterrestrial planet that possesses 14 moons and a very diffuse ring-structure. Its moderate magnetic field is tipped at 55° to its axis of rotation, which is nearly in the solar plane of rotation.

17. *Neptune* is a large nonterrestrial planet about which we know very little. In overall aspect it resembles the other nonterrestrial planets. Tiny *Pluto* appears to be a rocky planet and probably at one time was one of the moons of Neptune.

18. Our solar system probably started as a diffuse cold cloud of gas, dust, and ice (nebula) that slowly contracted. Most of the mass concentrated in a central bulge, whereas lumps and clumps of the remaining material circulated in an equatorial plane. In time the central body became hot enough to allow nuclear fusion reactions to occur, and eventually became our central star, whereas the clumps formed planets and satellite moons.

# QUESTIONS

1. Describe the various parts of our solar system and the earth's position in the Milky Way galaxy.
2. What accounts for both the similarities and differences among the different planets?
3. Why do you think that some but not all of the planets possess magnetic fields of variable intensity?
4. What are the principal physical features of the earth?
5. What planet resembles our moon most closely? Why?
6. How does our moon differ from the earth?
7. Venus, similar to the earth in many ways, rotates very slowly on its axis. Would you expect Venus to be ellipsoidal in shape?
8. How did the layering of the earth occur? How does it differ from that of other planets? Our moon?
9. Describe the theory of plate tectonics?

# SELECTED REFERENCES

### General

Berry, R., 1986, Uranus: the voyage continues, Astronomy, vol. 14, pp. 6–22.

Gore, R., 1980, Voyager views Jupiter's dazzling realm, National Geographic, vol. 157, no. 1, pp. 2–29.

Hartman, W.K., 1980, A climactic year in solar-system exploration clears up mysteries about the planets' satellites, Smithsonian, vol. 10, pp. 36–46.

Hodge, P.W., 1969, Concepts of the universe, McGraw-Hill, New York.

Kaufman, W.J. III, 1979, Planets and moons, W.H. Freeman, San Francisco.

Kopal, Z., 1979, The realm of the terrestrial planets, Wiley, New York.

Mission to Jupiter and its satellites, 1979, Special Science

Reprint, American Association for the Advancement of Science, Washington, D.C.

Pioneer Saturn encounter, 1979, NASA, Ames Research Center, Moffett Field, Calif.

Voyager encounters Jupiter, 1979; NASA, Jet Propulsion Laboratory, Calif. Institute of Technology, Pasadena.

Wood, J.A., 1979, The solar system: Prentice-Hall, Englewood Cliffs, N.J.

### Earth and Moon

Beiser, A., and the editors of Life, 1962, The earth, Time Inc., New York.

Donn, W.L., 1972, The earth: our physical environment, Wiley, New York.

French, B.M., 1977, The moon book, Penguin, New York.

King-Hele, D., 1967, The shape of the earth, Scientific American, vol. 217, no. 4, pp. 67–76.

Wilson, J.T., comp, 1970, Continents adrift: readings from Scientific American, W.H. Freeman, San Francisco.

Wyllie, P.J., 1976, The way the earth works: an introduction to the new global geology and its revolutionary development, Wiley, New York.

**Fig. 2-1.** The Grand Canyon, Arizona, was carved into the Colorado Plateau by the Colorado River over millions of years. (USGS)

# 2

# THE NATURE OF GEOLOGY AND GEOLOGICAL TIME

Geology—the study of the earth—is a broad-based, multidisciplinary science, encompassing the applications of the fundamental sciences, such as physics, chemistry, biology, and mathematics as well as other multidisciplinary sciences, such as astronomy. Geology is also an observational science in which the scientists—called *geologists*—spend a great deal of time looking at and recording the earth's features—large and small—and contemplating their meaning.

In science, the *scientific method* can be applied to its greatest extent. First, experiments are performed and the various physical and chemical parameters are measured. Based on the results, a plausible hypothesis (explanation) is advanced. subsequent experiments test the hypothesis; if it is shown to be in error, it is discarded and another is advanced. Actually, several equally plausible hypotheses may be advanced to fit the observations—this approach is called *multiple working hypotheses.* Then, the search continues for evidence that will disprove one or another of the several hypotheses. As new evidence turns up, all existing hypotheses may be discarded and a new, seemingly more-plausible one advanced. The best hypothesis that emerges is called a *theory,* on

which predictions of expected experimental results are based. A theory that continues to meet the tests of experiments over a long period of time is declared a *natural law*—the law of gravity is an example.

Regrettably, in geology the establishment of theories and laws is not so clear-cut for a variety of reasons. The "experiment"—commonly on an immense scale and of extreme complexity—has already been performed in nature and all that is observable is the result. Even the scope of observation is restricted because rocks at the earth's surface usually are not well exposed, being masked by soil and other deposits.

Attempts to duplicate natural processes and geological formations in the laboratory are not always successful because of uncertainties about natural conditions present long ago—particularly those relating to the composition of materials, the rates of physical and chemical change, and the total length of time involved in geological processes. Earth scientists know, for example, that the formation of a mountain range hundreds-of-kilometers long is a complex process requiring millions of years. Obviously, such a process cannot be duplicated in a laboratory. What can be gained, however, is a clearer understanding of the process—knowledge that may help other earth scientists to solve a particular complex problem.

Even the monitoring of processes going on today is not simple. On Hawaii, for example, earth scientists have thoroughly studied the timing and nature of intermittent volcanism, but they still know very little about what is actually going on deep beneath the island's surface where the lava originates.

In spite of such difficulties, earth scientists, using careful observation integrated with their knowledge of physical, chemical, and biological processes, are able to draw conclusions and to advance hypotheses and theories.

Although geological processes are occurring at present, most of the landforms and rocks (Fig. 2-1) we find today at the earth's surface are relics of past ages when the world and its living organisms were different. All we have left to tell us of those distant times is the fragmentary rock record or perhaps the landscape itself. Through careful scrutiny of rocks and landforms as well as knowledge of the way geological events occur on earth today, geologists attempt to recreate events of the past. Underlying their ability to draw conclusions and to advance hypotheses about bygone ages is the assumption that the present is the key to the past, a concept called *uniformitarianism.* The term means only that the processes in evidence in the world today probably existed in the past—that is, rivers under the influence of gravity probably flowed in earlier days much as they do today; ancient winds probably blew in response to atmospheric disturbances; volcanoes erupted; and earthquakes shook the land, much as they do now.

The concept of uniformitarianism reaffirms that the natural laws have prevailed on our planet continuously and unchangingly through the ages. The idea is basic not only to geology, but to all the other sciences as well. Materials attract each other through gravitational, electrostatic, and magnetic forces in the same way now as they always did; hydrogen, for example, joins with oxygen to form water now as it did in the past.

The alternative view is that all that surrounds us has no connection with the past. Such a view would lead to utter confusion—a world in which the physical, chemical, and biological laws fail to

hold; in which there is no continuity; and in which each geological age is characterized by processes and events never seen before and probably never to be seen again. Fortunately, all evidence seems to indicate that such is not the case.

Before about 1700 the concept of uniformitarianism was not readily accepted. The dominant doctrine—*catastrophism*—held that the earth had been the scene of repeated large-scale catastrophes of global proportion, each of which had wrought extensive changes both in the physical features of the earth and in all living things. Yet by the mid-1700s it became apparent to some observers that geological features, even those on a huge scale, such as the Alps and Himalayas, could result from natural processes if they acted over a sufficiently long time span. The validity of this view, which was championed first by a Scotsman, James Hutton (1726–97), and subsequently by his friend and countryman, John Playfair (1748–1819), was still being discussed almost a century later. But it was not until about 1830 that Charles Lyell (1797–1875), an Englishman sympathetic to the views of Hutton, documented the reality of uniformitarianism through many careful observations of rocks and landforms in western Europe.

In 1859 Charles Darwin (1809–82) eloquently formulated the basis of biological uniformitarianism. Greatly influenced by the work of Lyell, Darwin reportedly read one of that geologist's texts during his voyage on HMS *Beagle.*

Once the ideas of geological uniformitarianism were accepted, earth scientists could start to unravel the tangled web of the geological past.

Although modern geoscientists have adopted the general concept of uniformitarianism, they also accept the notion that the rates of geological processes varied in the past. These processes usually occur at a reduced level of activity; from time to time, however, the pace quickens. Commonly, most geological changes result from short bursts of increased intensity. For example, in cutting a deep canyon, such as the Grand Canyon, a river is most erosive during the relatively short-lived flood stages associated with violent storms. The canyon thus represents the long-term results of a succession of relatively small catastrophic floods of local extent.

Similarly, the movement of two blocks of the

earth's crust in opposite directions along a fault might constitute several tens of kilometers over a time span of 10 million years. The total displacement, however, is the result of a succession of small, rapid displacements. Beach erosion, too, is primarily accomplished during those catastrophically short episodes when ocean-current and wave action are at peak intensity, such as during cyclonic storms.

In the past few years geologists have come full circle, accepting the possibility that some of the catastrophic events in our geological past may have had more than local significance. In the early 1980s Nobel prizewinner Luis W. Alvarez and his son Walter reached the dramatic conclusion that about 65 million years ago many forms of life—including the dinosaurs—were forced to the brink of extinction by a sequence of events that involved a large asteroidal body (perhaps a comet) hitting the earth. The main piece of data underlying their assessment was the discovery of relatively large amounts of the element iridium in very thin 65-million-year-old strata in Italy and in Denmark. The metal iridium, closely akin to platinum, is rare in rocks at the earth's surface but fairly common in metallic meteorites. The Alvarezes concluded that the widespread iridium-rich zone indicated the presence of global fallout—specifically dust and debris—generated by a large explosion created by the impact of an extraterrestrial asteroid about 10-km (6-mi) across.

Such an impact, they calculated, would have produced a crater 160-km (100-mi) across. Pulverized rock and asteroidal material would have been blown into the atmosphere and carried around the world to produce darkness and chilling cold over much of the earth's surface for about 10 years. Photosynthetic plants would have been the first to succumb to the cold and lack of sunlight. Soon thereafter, in domino fashion, more extinctions would have followed—first among the plant-eating animals, then among the meat-eating animals. The venerable dinosaurs, whose lineage spanned nearly 150 million years, may have been snuffed out almost instantaneously. The Alvarezes further concluded that the other six or seven large-scale episodes of extinction in the geological record may also have been related to asteroid impacts.

Currently, the asteroid-impact hypothesis is hotly debated; moreover, the debate has created a worldwide flurry of activity among earth scientists to find geological evidence to test the Alvarezes' hypotheses.

Some concerned scientists have, on the basis of the asteroid-impact model, concluded that explosions and fires related to an all-out nuclear war would likewise result in the injection of large quantities of dust and smoke into the atmosphere, bringing about an end to earth's biota.

## GEOLOGICAL TIME

As is now apparent, the history of the earth involves the passage of eons of time. Yet to most of us a few hundred years of history encompass a great deal of time. We look at fragmentary accounts from the days of early Greek, Roman, and Egyptian civilizations and refer to them as ancient history. The gulf between our own culture and those civilizations seems so great that we find it difficult to relate to them. It has only been through the study of the earth that we have come to realize the true vastness of the fourth dimension, time, which extends back to the beginning of our world and beyond. In fact, as stated by Adolph Knopf (1882–1966), a well-known U.S. geologist, it is the development of the concept of the immensity of geological time that has been perhaps the most significant scientific contribution made by geology. Whereas in the historical world we deal with hours, days, and years, in the geological one we deal with thousands, millions, and billions of years—spans so great that they cannot be realistically conceived by the human mind.

Within the three dimensions of space and the fourth dimension of time, geological processes have been—and still are—at work. Given enough time, it is possible for even the slowest process to bring about large-scale changes. Time, then, is perhaps the most essential ingredient in the uniformitarian concept.

To estimate the passage of time, it is necessary to use some device that records the occurrence of equally spaced events. We now use mechanical or electronic clocks as monitors; in older societies, people used sundials, sand clocks, or water clocks.

Conventional clocks are fine for conventional

**Table 2-1** Early Estimates of the Earth's Age

| Scientist | Date | Method | Age Estimate |
|---|---|---|---|
| Archbishop Ussher | Mid–1600s | Biblical chronology | Nearly 6000 yr |
| William Sollas | 1883 | Stratigraphic data | 26 million yr |
| John Joly | 1899 | Sea salinity | 100 million yr |
| Lord Kelvin | Over span beginning in mid 1860s, extending to 1880s | Cooling of earth and sun | Initially 100 million yr, refined to about 20 million yr |

purposes, but unconventional methods are required in recording geological time. An unusual estimate of the earth's age was made in the mid-1600s by Archbishop James Ussher (1581–1656) of the Irish Protestant church. The archbishop firmly believed that the Bible contained a complete record of the world's events since its inception. Therefore, by adding up all the life spans of successive generations recorded in the Old Testament, he thought that he could arrive at a first-rate approximation of the time of the earth's origin. According to his painstaking calculations, the date of inception was the evening of October 22, 4004 B.C. (Table 2-1). Ussher's estimate was a refinement of a figure similarly determined by John Lightfoot, a distinguished Greek scholar and vice chancellor of Cambridge University. The date 4004 B.C. was even referred to in the Great Edition of the English Bible published in 1701. For a century thereafter, any Christian who maintained that the earth's development took more than about 6000 years could be charged with heresy.

Some geologists attempted to approximate the age of the earth by first estimating the combined total thickness of sediments laid down since the earth's beginning, and then dividing that amount by the rate at which sediments accumulate.

This method, however, is potentially inadequate. For example, no one knew then—or knows now for that matter—how to determine the rate at which sediments are deposited. Obviously, the rate for a conglomerate made of boulders 1-m (3 ft) across is different from that of a limestone deposit consisting of the remains of microscopic marine organisms. Furthermore, few if any sediments are deposited without interruption; there may be times of accumulation separated by inter-

vals in which no deposition occurs or during which erosion removes previously deposited sediments. To add to the difficulties, no place is known on the earth's surface where strata representing the total age of the earth were laid down continuously. Nonetheless, in 1883 the European geologist William Sollas (1849–1936), on the basis of stratigraphic thickness, estimated that the earth was a minimum of 26 million years old, a figure nearly three orders of magnitude greater than that given by Archbishop Ussher (Table 2-1).

Another attempt to relate present-day processes to the age of the earth was made in 1899 by the Irish geophysicist John Joly (1857–1933). The amount of salt dissolved in the oceans had been determined fairly accurately by that time. Joly reasoned that if a close estimate could be made of the amount of salt that rivers contribute each year to the oceans, he could establish a value for the age of the sea and, from this, a rough estimate of the age of the earth. The method employed by Joly had been suggested by Edmund Halley (1656–1742), of Halley's comet fame, almost 200 years earlier. Halley, then, deserves recognition as the first person to suggest a reasonable, geologically based method of determining the earth's age.

There are many pitfalls and problems associated with the sea-salinity method. Immense quantities of salt now present in salt beds have been removed from the oceans by evaporation of seawater in ages past, some of which is now being recycled back to the sea as the salt beds erode. The estimate also does not take into account the probability that the amount of salt added to the sea each year through erosion has not remained constant throughout geological time. Furthermore, there now is good evidence to suggest that the salinity is buffered, such that after initially rising to a value close to the present level, it has remained so ever since.

Both Joly and Halley realized some of the problems associated with the sea-salinity method. They concluded, however, that the calculated value would at least provide a first-order approximation—a minimum value of the earth's age. By this method Joly estimated the seas to be 100 million years old (Table 2-1).

One of the most interesting episodes concerning early attempts at estimating the earth's total age centers on the renowned British physicist William Thomson (1824–1907), later dubbed Lord Kelvin. In 1866 he published a paper of less than 250 words in which he mathematically demonstrated, on the assumption that the earth was initially molten, that it could be no older than about 100 million years (Table 2-1). Subsequently, he added two other arguments—one based on the cooling of our sun from a molten state, the other on the shape of the earth—to substantiate his earlier estimate of 100 million years.

By the late 1800s the uniformitarian doctrine had been accepted wholeheartedly by earth scientists. Although no one knew exactly how long the earth had been spinning, most geologists advocated thousands of millions of years to account for all of the many known geological and biological events. Geologists, however, were unable to counter Kelvin's arguments and most, grudgingly, came to accept Kelvin's curtailed time estimate. To accommodate the many events, they increased the rates at which the processes proceeded.

By the 1880s Lord Kelvin was convinced that some of the assumptions in his calculation had been flawed. During subsequent refinements, he eventually whittled the estimated age down to 20 million years—too short a span to be acceptable to many geologists (Table 2-1). A philosophical feud then erupted between the physicist and geologist camps that lasted until the end of the century when radioactivity was discovered. Soon after came the knowledge that radioactive decay provided significant quantities of additional heat to the earth. It then became apparent that Lord Kelvin's arguments—based only on continued loss of nonrenewable heat from the sun and earth—had been in error. Intriguingly, the recognition of radioactivity not only removed the constraints Kelvin had placed on the age of the earth, but ultimately provided a means for obtaining the actual age of the earth. We return to this interesting topic later in this chapter.

## Relative Time

During the 1700s and 1800s, although a few earth scientists were trying to establish total ages for the earth and certain geological features, most were involved in deciphering geological records in their local areas of study. They placed all the events recorded in the rocks into a **relative time sequence**—by determining whether an event occurred before or after another one. In our daily lives we do much the same thing, stacking the events of the day into a succession based on the order of their occurrence.

Only a few methods are used to establish the relative time sequence. Some of them are so straightforward that they seem obvious; yet in the formative days of geology, when little was known about the earth, such methods were extraordinary. Each represents a discovery made by an individual only after painstaking observation. For instance, Nicolaus Steno (1638–86), a Danish physician and theologian, curious about the earth's origin, arrived in 1669 at the conclusion that in sequences of layered rocks a layer will be older than that above it, younger than that below it (Fig. 2-2). His reasoning was that before a stratum could be laid down, the layer below must have already been deposited. This simple rule is called the **law of superposition.** It is used primarily to determine the relative ages of sedimentary rocks or lava flows in sequences that have not been structurally inverted, that is, turned upside down.

Another rule established by Steno is that sediments were originally deposited horizontally or nearly horizontally (Fig. 2-3). Therefore, if one finds a sequence of strata tipped up at an angle, the event that produced the tilting must have occurred after the strata were laid down. Exceptions to this rule, the **law of original horizontality,** do exist, but they are rare.

James Hutton established the **law of cross-cutting relationships** when he demonstrated that if one body of rock cuts across another body of rock, the

**Fig. 2-2.** Sedimentary rocks exposed in Coal Canyon, Arizona. Tens of millions of years were required for the deposition of this thick section. Erosion rates are commonly measured in centimeters per thousand years, so you can estimate that millions of years were required to form a canyon of such grandiose proportions. (Tad Nichols)

latter must be older than the former, that is, one rock must already be in place for another, a younger one, to cut across it. For example, sedimentary rock layers commonly contain younger, once-molten tabular bodies, called dikes, that cut across the preexisting strata. Similarly, faults (fractures along which movement has occurred) and most cracks that cut across a rock body must be younger than the rock in which they are found.

Another rule, the **law of inclusions,** states that if fragments of one rock body are included (i.e., contained) in another, then the rock body from which the inclusions came is the older. For example, foreign rock fragments that occur in a lava flow must have been derived from a rock body in existence prior to the time that flowage occurred.

Finally, weathering and removal of material by erosion can occur only near the surface of the earth and can only affect rocks already on the surface. In some areas, because of these processes, a gap may exist in the geological rock record—a place where the sequence of rock layers or units is interrupted. This break is called an **unconformity** because part of the rock record is missing and

**Fig. 2-3.** Sedimentary rocks exposed near Bluff, Utan: looking north over the San Juan River. The originally horizontal rock layers were tilted to their present positions. (John S. Shelton)

**Fig. 2-4.** The wall of the Grand Canyon is composed of almost 1.5 km (1 mi) of strata in which several major geologic events can be dated. The oldest rocks—the Grand Canyon Series—were deposited in upward succession over a long period of time, tilted, and eroded to a nearly flat surface. Subsequently, the sequence of flat-lying sediments above was deposited. The old surface of erosion separating the tilted strata below from the flat-lying layers above is a large-scale unconformity representing a time gap of nearly 1 billion years. The most recent phase of geologic activity is erosion, particularly by the cutting of the canyon. (L.F. Noble, USGS)

the sequence does not "conform." An uncomformity results from a period of erosion in the geological past during which previously formed rocks were lost, followed by a period of deposition of additional layers of rock. In the walls of the Grand Canyon, for example, several unconformities can be seen. One of the most significant is located near the bottom of the canyon; rocks below are older than the flat-lying sediments above by nearly three quarters of a billion years (Fig. 2-4).

The features used in relative age determination are shown in Figure 2-5; by means of these criteria, it is possible to establish a temporal sequence of geological events for most areas on earth (see Fig. 2-4).

Although it is comparatively easy to establish a relative time sequence locally, it is usually much more difficult to correlate rocks in one area with rocks in another, even when the other area is relatively close by. The best method devised thus far is to compare the physical features of the rocks themselves. If a sequence of strata made up of identifiable layers can be traced or mapped into another area, then correlation between the areas is easily accomplished. If each layer of a sequence of strata also has diagnostic features that can be identified in the same sequence as those in a nearby area, then correlation can be made with some assurance.

Sometimes, however, strata change laterally in physical character or the strata present in one area do not show up in another, nearby area (Fig. 2-6). Correlation in these cases becomes much more difficult. If one area is quite far removed from another, correlation is even more difficult.

Fortunately, many sedimentary rocks contain *fossils,* the preserved remains or traces of prehistoric life (Figs. 2-7, 2-8, and 2-9). In most cases the hard parts of an organism (e.g., bone or exterior shell) are fossilized. But soft parts can also be preserved. Tracks, trails, burrow structures, and even feces may be preserved and are, therefore, included under the broad definition of a fossil.

Fossils have proved to be of great value in correlating strata and in determining the relative age of the layers in which they are found. They are especially useful in correlations between widely separated areas, even on different continents. Fossils also form the basis for the *relative geological time scale:* A means of placing successions of fossil forms (and the strata in which they are found) throughout the world in their correct order of occurrence. The scale clearly shows that life has changed continuously in form and kind throughout geological time.

Fossils have excited interest since prehistoric times. They appear in ancient drawings and coats of arms; for centuries they have been cherished because of the interesting decorative effect they give to polished stone. The word itself is from the Latin *fossilis,* ''something dug up.''

Fossils have not always been so well understood as they are today, To some in the remote past, fossils were the handiwork of Satan, placed in the world to confuse mortals. To others—for example, Avicenna (980–1037), the Arab philosopher responsible for the revival of interest in the works of Aristotle—they were inorganic petrifications that grew within rocks: Only by chance did they come to resemble bones or shells of living creatures. Ignorance extended even to the remains of prehis-

**Fig. 2-5.** Features that aid geologists in the relative dating of rocks.

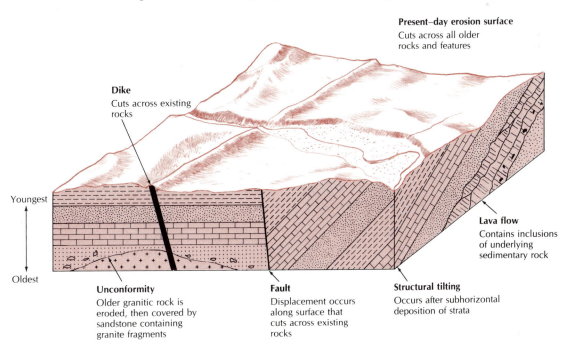

Present–day erosion surface
Cuts across all older rocks and features

Dike
Cuts across existing rocks

Youngest

Oldest

Lava flow
Contains inclusions of underlying sedimentary rock

Unconformity
Older granitic rock is eroded, then covered by sandstone containing granite fragments

Fault
Displacement occurs along surface that cuts across existing rocks

Structural tilting
Occurs after subhorizontal deposition of strata

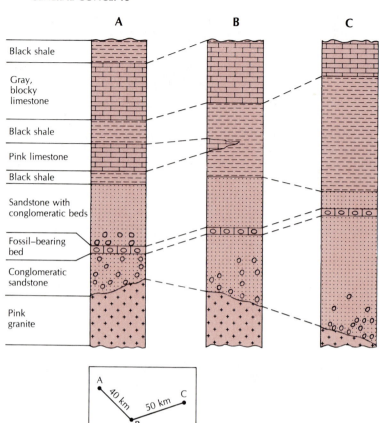

A B C

Black shale

Gray,
blocky
limestone

Black shale

Pink limestone

Black shale

Sandstone with
conglomeratic beds

Fossil–bearing
bed

Conglomeratic
sandstone

Pink
granite

A
40 km
B
50 km
C

**Fig. 2-6.** Correlation of hypothetical rock units in three partial stratigraphic sequences. Notice lateral changes.

toric people, such as the Cro-Magnons and Neanderthals, whose bones were sometimes carted to village churchyards for proper interment.

It remained for an untutored and eminently practical canal builder, William Smith (1769–1839), to establish the relationship between stratified rocks and the fossils they contain. One reason for his success is that he lived and worked at the right time and place—during a period when many canals were being built in England. The advent of the Industrial Revolution coupled with the post-Napoleonic prosperity of Britain placed a premium on cheap and efficient transportation; canals proved to be the answer.

In the construction of canals, the recognition of various rocks and an understanding of their physical properties are of prime importance. If the rocks are harder than expected, if they slump and cave readily, or if they allow water to drain away, then the canal may be expensive to build or un-

successful, or both. Smith found that the rock structure of midland and eastern Britain was relatively simple and that the various rock units were distinctive. He carefully noted the nature of all the kinds of rocks in which the canals were cut. All told, he tramped up and down the English countryside for 24 years making observations. When he was done, he had made some fundamental discoveries: (1) that the strata in southeastern England occurred in the same order—for example, chalk beds were always found above coal layers (unless disturbed structurally), never vice versa; (2) that different layers contained distinctive assemblages of fossils; and (3) that the distribution of sedimentary rocks in southeastern England could be represented on a map.

The geological map of England and Wales that Smith published in 1815 was the world's first. Since Smith's time, geological mapping has become an effective means of depicting the geolog-

36

**Fig. 2-7.** A 50-million-year-old spider [about 4-mm (0.16-in.) across] preserved in amber, Hot Springs, Arkansas. Notice the fine detail of preservation. (W.B. Saunders)

**Fig. 2-8.** *Olenellus fremonti,* a trilobite from the Marble Mountains, California. Its length is about 11.5 cm (4.5 in.). This animal, related to the hermit crab, lived in the oceans between 600 and 225 million years ago; then it disappeared from the face of the earth. (Takeo Susuki)

**Fig. 2-9.** A fossil *Ichthyosaurus,* a reptile that swam the seas about 175 million years ago. The modern porpoise resembles it in form. With a large mouth full of sharp teeth, an ichthyosaur, no doubt, presented a formidable appearance as it hunted the ancient seas. (Smithsonian Institution)

ical features of any given area; today it represents one of the most important tools of the geologist.

Smith was not aware of the evolutionary significance of the fossils he collected so patiently over the years. Yet he had found the key by which strata could be correlated with one another, not only locally but continents apart.

Such a method of correlation is no simple matter, however, because the organisms that became fossilized lived in different types of environments. Furthermore, during the long span of geological time, some groups of animals and plants gradually evolved into more complex creatures, whereas others, (e.g., the horseshoe crab and cockroach) remained essentially unchanged. Still others, including dinosaurs and flying reptiles, became extinct.

This brings us to an important contribution that *paleontology*—the study of ancient life—makes to the whole realm of contemporary thought. From fossils preserved in thousands of feet of stratified rocks found in diverse places, paleontologists have demonstrated that many forms of life have evolved from relatively simple to complex hierarchies of plants and animals. In fact, the fossil record was to Darwin, and remains for us today, the strongest body of evidence to support the theory of *organic evolution.* It is the tangible historical record of the stream of life on earth. To be sure, it is fragmentary for a number of reasons: the difficulty of preserving the soft parts of flora and fauna; the minimal likelihood of preserving fossils in general and, particularly, in some natural environments; and the minimal chance of a fossil being formed, then being freed from entombment in the sediments by erosion, and still later being discovered by a geologist. With new discoveries turning up almost daily, however, the record is becoming more complete. Year after year we are finding fewer missing links in the chain of life.

The record eloquently depicts how life began simply in the early stages of the earth's history with the development of primitive single-cell bacteria and blue-green algae. Slowly, with time, more complex, free-moving invertebrate forms, ever-increasing in diversity, developed in the shallow waters of the world's oceans. Then a new form appeared, providing evidence of the first vertebrate animals on earth—the protofish and sharks.

The record of plant life similarly documents a trend from simple beginnings to an ever-increasing diversity. Some plants eventually colonized on land areas—the first life to do so—followed "soon after" (from a geological standpoint) by the land-dwelling vertebrates. The more recent part of the story demonstrates a continuing trend of plant and animal complexity and diversity right up to today.

The fragmentary fossil record is replete with false and multiple starts of some life forms, failed experiments (extinctions), rapid changes in some groups of organisms, and little or no changes in others. There have been short episodes when a large portion of the biota died away, as if wiped out by a scourge, and other times when a variety of life burgeoned. Certainly, the account is clear on one point: the pattern of life has never repeated itself. Any organism that has disappeared from the face of the earth has never reappeared.

The fossil record provides a giant family tree, with most of the roots and links of the biological world since it began, and it documents an ever-changing transformation (evolution) of the life forms through time. The tree itself, formed by fragmentary fossil evidence, is indisputable; the *meaning* of the evolutionary pattern is, however, open to philosophical debate.

## Geological Time Scale

As earth scientists accumulated knowledge of strata in different parts of the world as well as in a single region, they were able to compare rock layers on the basis of their fossil content. It was first necessary to determine the order of deposition in one region. When that was done, the succession of rocks there might be compared with a different succession of rocks—perhaps those a whole continent away and not necessarily the same kinds of rocks at all, but ones containing fossils of the same geological age.

The most valuable fossils for correlating rocks are those with comparatively short histories (geologically speaking) yet which, in their limited time on earth, achieved a wide geographical distribution and underwent rapid evolutionary changes. Such ideal examples are called *index fossils.* In some cases we take our cue from an association of fossils rather than from a single species.

| EON | ERA | PERIOD | EPOCH | ABSOLUTE TIME | LIFE FORMS |
|---|---|---|---|---|---|
| **Phanerozoic (evident life)** | **Cenozoic** (recent life) | Quaternary | Holocene | | Rise of mammals and appearance of modern marine animals  |
| | | | Pleistocene | 2 | |
| | | Tertiary | Pliocene | 5 | |
| | | | Miocene | 24 | |
| | | | Oligocene | 37 | |
| | | | Eocene | 58 | |
| | | | Paleocene | 66 | |
| | **Mesozoic** (middle life) | Cretaceous | Late | | Abundant reptiles (including dinosaurs); more advanced marine invertebrates  |
| | | | Early | 144 | |
| | | Jurassic | Late | | |
| | | | Middle | | |
| | | | Early | 208 | |
| | | Triassic | Late | | |
| | | | Middle | | |
| | | | Early | 245 | |
| | **Paleozoic** (ancient life) | Permian | Late | | |
| | | | Early | 286 | First reptiles |
| | | Carboniferous — Pennsylvanian | Late | | |
| | | | Middle | | |
| | | | Early | | |
| | | Carboniferous — Mississippian | Late | | |
| | | | Early | 360 | First terrestrial vertebrates—amphibia |
| | | Devonian | Late | | |
| | | | Middle | | |
| | | | Early | 408 | |
| | | Silurian | Late | | |
| | | | Middle | | |
| | | | Early | 438 | First vertebrates—fish |
| | | Ordovician | Late | | |
| | | | Middle | | |
| | | | Early | 505 | |
| | | Cambrian | Late | | Primitive invertebrate fossils  |
| | | | Middle | | |
| | | | Early | 570 | |
| **Precambrian** | | | | 3800+ | Meager evidence of life |

**Fig. 2-10.** The composite geological time scale now in use is a combination of the relative time scale—based on the superposition of beds and the fossil character in the strata—and absolute ages—determined primarily by radiometric dating of igneous and metamorphic rocks.

39

Through the application of the law of superposition, it is possible to ascertain, at least locally, which fossils are older than others. Eventually, as more and more stratigraphic sections were correlated, it was possible to establish a chronological order on the basis of the fossils in the rock.

The geological time scale devised in this manner (Fig. 2-10) was a great step forward; it enabled geologists around the world to compare the timing of geological processes and events. The two major divisions of the scale, the Precambrian and Phanerozoic eons, primarily reflect a division between older rocks in which fossils are extremely rare, even nonexistent, and younger rocks in which they are relatively common.

The Phanerozoic is, in turn, divided into three eras—Paleozoic (ancient life), Mesozoic (middle life), and Cenozoic (recent life)—each successively characterized by more modern life forms.

With geological history, as with human history, events closer to us leave more complete and decipherable evidence than distant events for which the records have grown more and more fragmentary. The result is that more is known about later happenings in earth's history than about earlier episodes; that knowledge is reflected in the larger number of subdivisions in the last part of the geological time scale.

Each of the eras is divided into a number of smaller divisions termed **periods;** in turn, these are divided into even smaller units called **epochs.** Most periods are named for regions in which rocks containing fossils characteristic of their segment of geological time were found. Strata deposited in the Mississipppi Valley and in Pennsylvania provide two of the period names for the late Paleozoic Era. Neither of these period terms, however, is used in Europe; there, the two periods are treated as a single unit, the Carboniferous, which acquired its name from the coal-bearing strata of England.

Other period names with origins that can be recognized readily are the Devonian, named for rocks that crop out along the southwestern tip of Great Britain in Devon, and the Jurassic, named for rock exposures in the Jura Mountains along the western border of Switzerland. A less familiar period is the Permian—named after the ancient kingdom of Permia in Russia. The ancient name of Wales, Cambria, is the source for the name of the

Cambrian Period, whereas the Ordovician and Silurian perpetuate the memory of ancient tribes whose homes had been in Wales—the Ordovices in the north, the Silures in the south.

Some of the other periods are named not for places, but for the physical characteristics of their rocks. The Cretaceous, derived from the Latin word *creta* (''chalk''), refers to the exposures of chalky rock along the cliffed coast of southern England. The Triassic (from the Greek, meaning ''threefold'') takes its name from the fact that in Germany the rocks of that period are divided into three distinctive layers—a limestone in the middle, with reddish sandstones and shales above and below.

The most accepted way of dividing the Cenozoic Era is into the Tertiary (meaning ''third'') and the Quarternary (meaning ''fourth''). The division represents a sort of cultural lag, because we no longer speak of the Primary and Secondary rocks as our geologist forebears did. With the first and second of a series gone, it seems strange to speak of a third and fourth, but you will see these forms employed consistently in most geological writing today; such is the force of habit.

The epochal subdivisions of the Cenozoic are for the most part a special case. Most of them were established by Charles Lyell according to the percentage of now-extinct marine molluscan fossils that are present in the sediments. Thus in Eocene (from the Greek words for ''dawn'' and ''recent'') strata, about 95 to 99 percent of the fossil species are extinct. In the Miocene (Greek words for ''less'' and ''recent''), 60 to 80 percent of the fossil species no longer exist. In the Pleistocene (Greek words for ''most'' and ''recent''), only 10 percent or less of the fossil shells are those of species no longer living in the seaways of the world.

The Pleistocene Epoch is regarded by many geologists as coinciding with the Ice Age, and although the length of time represented by the multiple advances and retreats of the ice sheets appears to vary from place to place, it probably is close to 2 million years.

## Absolute Time

The relative geological time scale provides a framework within which to view the events that shaped the earth. It contains no information, however, about when a specific event actually took place. We know, for example, that the dinosaurs lived after the plants found in the coal beds of Pennsylvania. But how long afterward and for how long? Or at what specific time or over what time span was the Sierra Nevada range formed? Or when did mammals supplant reptiles as the dominant vertebrate group? A few earth scientists made ballpark estimates about the total age of the earth, but up to the end of the nineteenth century no method existed that could provide reliable estimates of the absolute age or duration of individual events.

In 1895, however, a door to the past was opened by the discovery of radioactivity—a discovery that led to techniques that have since enabled us to obtain some cherished goals—not only the total age of the earth, but the dates of events that occurred throughout its history.

Few people alive in 1895 were aware of the eventual significance to the world that a simple experiment by French physicist Antoine-Henri Becquerel (1852–1908) was to have. He had been intrigued by the discovery of Wilhelm Roentgen (1845–1923), within the same year, of the X-ray and wondered whether the phenomenon of phosphorescence bore any relationship to the strange rays. To test the possibility that various substances might be able to pick up energy from the sun, he exposed a number of them to the sun's rays—with no discernible effect. By chance he observed that when some uranium-bearing salts were placed on photographic paper, the paper darkened as though subjected to some sort of radiation, and that this was true whether or not the uranium salts had first been exposed to the sun's rays. Becquerel's discovery was the first demonstration of *radioactivity.*

Soon, others began experimenting with radioactive substances, among them the distinguished physicist Lord Rutherford (1871–1937), born Ernest Rutherford in New Zealand. In 1902 he discovered that when unstable radioactive atoms disintegrated, they formed completely different atoms and radiation was given off. The initial work by several people, including the French chemists Marie (1867–1934) and Pierre Curie (1859–1906), indicated that both uranium and thorium were radioactive elements. During the early 1900s it was also found that several forms of these substances could exist with the same elemental properties, differing only in atomic weight. To these variants of the same element the radiochemist Frederick Soddy (1877–1956) gave the name *isotopes* (from the Greek, meaning "same place"). Eventually, it was concluded that only certain isotopes are radioactively unstable; in these the decay is generally restricted to the nucleus of the atom.[1] It has been demonstrated that each radioactive element breaks down at its own rate. For some isotopes the rate of decay is very fast, nearly complete decay occurring within a fraction of a second; for others the rate is extremely slow, such that many millions or even billions of years are required for complete decay.

In simple radioactive decay, atoms of a radioactive isotope (called the *parent isotope*) will slowly and spontaneously disintegrate over a period of time to form an equivalent number of atoms of the decay isotope (called the *daughter isotope*). The ratio of daughter to parent changes continuously with time, thereby providing a measure of the total elapsed time of the decay process. The time that it takes for a radioactive isotope to decay to one-half its initial quantity is called its *half-life.* Each radioactive isotope has a specific half-life. The process of radioactive breakdown is illustrated in Figure 2-11. After one half-life, one-half the parent remains; after the next half-life, only one quarter; after the next, only one-eighth; and so on. After about seven half-lives, very little of the parent remains.

Rutherford recognized the importance of naturally occurring isotopes and in 1905 proposed the use of radioactive decay as a means of dating rocks. He pointed out that there was not one clock, but as many as there were usable radioactive isotopes. In 1907 an analytical chemist, Bertram B. Boltwood (1870–1927), announced the first *radiometric date* ever obtained—410 million years for a uranium-bearing mineral contained in a rock from Connecticut. Soon after, dates were obtained

---

[1] An atom is the smallest unit of an element that can exist and still display all of its elemental properties. Simply, an atom consists of a dense, tiny kernal at its center—the nucleus—surrounded by a diffuse cloud of small negatively charged particles—the electrons—whose number depends on the particular element.

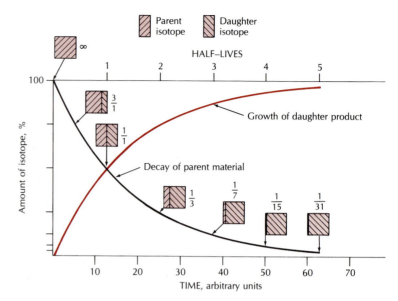

**Fig. 2-11.** Decay curve for a hypothetical radioactive isotope (parent) and growth curve for daughter isotope. The boxes indicate the ratio of parent to daughter isotopes at specific points along the curve. The continuous change of this ratio is the basis for radiometric dating. The time required for the parent to decay to one-half its initial amount, called the half-life, is unique for each radioactive isotope.

for other rocks, ranging from 400 million to 2200 million years. A great many factors were not taken into account in the early dating methods, and many of the first dates were in error. The important thing, however, is that a method had been discovered that held great promise of providing absolute ages for minerals and rocks and that even the first dates determined by the method indicated the earth to be hundreds of millions of years old. By the 1930s the earth's age was already tentatively established as 2000 million years.

The minerals initially used in absolute dating were restricted to those containing three radioactive isotopes—uranium-238, uranium-235, and thorium-232. (The numbers indicate the atomic weight of the isotope.) The process of nuclear disintegration is long and involved, but each of the three reaction chains eventually produces one of the isotopes of lead as a stable end product. The gas helium is also produced in many intermediate steps of the breakdown. In fact, many of the first dates were based on the ratio of radioactive uranium or thorium to helium; but later dates, which proved to be much more reliable, were based on the ratio of uranium or thorium to lead. Besides uranium and thorium isotopes, three others—potassium-40, rubidium-87, and carbon-14—have proved very useful in the absolute dating of rocks (Table 2-2). Recently, with advances in tech-

nology, a relatively rare isotope, samarium-147, is also becoming useful in dating very old rocks.

With the exception of carbon-14, all the commonly used isotopes are similar in that they possess extremely long half lives—generally greater than 1 billion years. It is essential in the dating of events that occurred millions or billions of years ago to employ isotopes whose half-lives are similarly long. Carbon-14, because of its relatively short half-life, can only be used to date events back to about 80,000 years.

The useful isotopes of uranium, thorium, rubidium, and potassium occur only in very small quantities in most rocks—and more often in granitic than in other types. Samarium-147, which is even rarer, occurs preferentially in basalt. Because these isotopes typically are present in only trace amounts,

**Table 2-2** Commonly Used Radioactive Isotopes, with Their Breakdown Products and Rates of Decay

| Radioactive Isotope | Daughter Product | Half-life |
|---|---|---|
| Uranium-235 | Lead-207 | 713.0 million yr |
| Uranium-238 | Lead-206 | 4.5 billion yr |
| Thorium-232 | Lead-208 | 13.9 billion yr |
| Potassium-40 | Argon-40 | 1.3 billion yr |
| Rubidium-87 | Strontium-87 | 47.0 billion yr |
| Samarium-147 | Neodymium-143 | 106.0 billion yr |
| Carbon-14 | Nitrogen-14 | 5730 yr |

they do not form their own minerals, but exist as impurities in the more common rock-forming minerals. Moreover, because of the rarity of the radioactive isotopes, the procedure of obtaining reliable dates from rocks and minerals is a very complicated one. It requires elaborate laboratory methods and sophisticated equipment to measure precisely the small amounts of parent and daughter isotopes. The work is carried out in few places, and a level of performance has now been reached so that little error is introduced by laboratory procedures.

The principal source of error comes from the minerals and rocks themselves. If a mineral is to produce a reliable date, parent or daughter materials cannot have entered or left the mineral during its entire existence. It is not always easy to tell how well this restriction has been met. One simple precaution that eliminates some of the uncertainty is to collect only unweathered rocks, avoiding those that appear chemically altered. Rocks dated by the potassium/argon method are also troublesome because the daughter product, argon-40, is a gas. As such, it is extremely difficult to keep trapped within the atomic latticework of the host mineral and tends to diffuse slowly out of the mineral structure. If a rock is subjected to increased temperatures, the likelihood of argon loss increases. Reduction in the amount of argon-40 leads to an underestimation of the age of the rock. If the temperature remains sufficiently high for a long enough period, all the previously trapped argon will escape. In effect, the radioactive clock is reset. The dating of such a rock will give the time of thermal resetting—in some circumstances this can still be useful.

The best safeguard against hidden errors in radiometric dating comes through the use of consistency checks. One obvious check is to date a mineral or rock by more than one radiometric method. For example, if both rubidium–strontium and potassium–argon methods—methods based on two different decay processes and rates—are applied to the same rock and if the results agree, the likelihood is high that the date is reliable. If the two dates differ appreciably, then additional dating methods must be used to determine which, if either, is the correct date. Another consistency test may be made by comparing the absolute and relative ages of rocks. For example, rocks that are known by physical correlation to be the same age should produce nearly identical dates.

Not all rocks, however, can be dated by radiometric methods. First, the minerals of the rock must have formed within a short time and must have remained a closed system thereafter. Datable rocks, then, are for the most part those that have formed as a result of igneous or metamorphic processes. The time of the deposition of sedimentary rocks, which commonly are made up of fragments of igneous and metamorphic rocks as well as other sedimentary rocks, usually cannot be radiometrically dated because any dating of the fragments only reveals the age of the source rock—not the time of deposition.

Second, the rocks and minerals to be dated must initially have contained sufficient radioactive material to be measured. Third, enough time must have elapsed since the origin of the rock to allow measurable quantities of daughter products to accumulate. As a general rule, most rocks younger than about 250,000 years cannot be dated with a high degree of confidence. Because of the relatively long half-life of rubidium-87 (47 billion years), this isotope cannot in most cases be used with rocks younger than about 30 to 40 million years old. For rocks younger than that, the daughter product would be essentially unmeasurable. Because the content of samarium-147 is even smaller and the half-life is about twice that of rubidium-87, in most instances the samarium–neodymium method cannot be applied to rocks younger than about 1 billion years.

Interweaving the fossil-based time scale with the absolute time scale is not as simple as it might seem at first. Virtually all radiometric dates have been obtained from igneous and metamorphic rocks, whereas fossils—the basis for the relative time scale—are almost exclusively restricted to sedimentary rocks, which usually cannot be radiometrically dated. However, it is possible to bracket the age of the associated fossil-bearing strata and to place absolute ages on the relative time scale by dating igneous rocks interlayered with sedimentary rocks, such as lava flows and ash layers; or by dating dikes and other rock bodies that cut across sedimentary sequences; or by dating igneous and metamorphic rocks that unconformably underlie sedimentary rock sections. In this way, particularly during the last 25 years, thousands of radiometric dates from igneous and metamorphic rocks have been used to construct an absolute-age framework for the fossil-based relative time scale—resulting in the modern composite geologic time scale (Fig. 2-10).

A glance at this scale demonstrates that the Precambrian Eon includes most of geological time. Although there are few fossils in this eon, radiometric dating now enables the geologist to unravel its record. And with absolute dating methods, geologists can even determine such things as rates of evolution and the time spans over which mountains were formed. Rocks in the ocean basins have been found to be relatively young (Table 2-3). In fact, none appear to be older than 150 to 200 million years. On the continents and subcontinents the rocks display a wide range in age. Most continents contain some rocks as old as 3. to 3.2 billion years.[2]

Rocks about 3.5 billion years old, however, have only been found in a few places, notably in Minnesota; Swaziland, Africa; and the western part of Australia (Table 2-3). Holding the distinction of being the very oldest—3.8 billion years—is a pebble of volcanic rock from a conglomerate in western Greenland. Recently, there have been reports from Australia of a radiometric age of 4.2 billion years obtained from individual tiny crystals of the mineral zircon. A sample of the zircon crystals has since been sent to a French laboratory for verification, but as yet French scientists have not confirmed such an ancient date.

Meteorites are the most ancient materials ever found. Dated at about 4.6 billion years, geologists consider them the best indicators of how old our earth is and even when the solar system came into being—estimates partially corroborated by dates of up to 4.2 billion years obtained from moon rocks gathered during the Apollo missions. As old as our solar system appears to be, its age falls far short of that of the Milky Way galaxy. On the basis of astronomical considerations, that galaxy

**Table 2-3** Significant Maximum Radiometric Dates

|  | Age (billions of years) |
|---|---|
| On continents and subcontinents |  |
| Minnesota, U.S. | ≅3.5 |
| Swaziland, Africa | ≅3.5 |
| Western Australia | ≅3.5 |
| Western Greenland (oldest known age) | 3.8 |
| Western part of Australia | 4.2(?) |
| In ocean basins | ≅0.15 |
| On the moon |  |
| Maria | 3.2–3.8 |
| Highlands | 4.0–4.2 |
| Meteorites | ≅4.6 |

is estimated to be between 9 and 19 billion years old.

To comprehend better the immensity of geological time, imagine that the age of the earth represents a single year. Then the record of abundant fossils would extend back a scant 40 days or so; humans have been on the earth a matter of hours; and all of recorded history amounts to about a minute. Nearly 10 2/3 months of the year are taken up by the Precambrian Eon. Yet because the rock record is so poor and fossils are essentially nonexistent for this long span of time, we know the least about it. Without radiometric dating, we would know almost nothing. Figure 2-12 is an attempt to pictorially describe the long spiral of geological time and the significant changes wrought over that expanse of time.

## CARBON-14 DATING AND DENDROCHRONOLOGY

A discussion of radiometric dating would be incomplete without a few words about carbon-14. This isotope has been left to last because its role in establishing the overall framework of the geological time scale has been minimal. However, in placing archaeological traces of earlier cultures in historical perspective and in fleshing out details of geological history during the later phases of the Ice Age, the use of this isotope has been significant.

Carbon-14 decays to nitrogen-14 with a half-

[2]The huge discrepancy between the maximum ages of the continents and those of the ocean basins requires an explanation, and we will return to this puzzle in chapter 4.

**Fig. 2-12.** Depiction of the long span (~4.5 billion years) of geological time, punctuated by important physical and biological events. ( After W.L. Newman, 1983, "Geologic Time," USGS booklet)

life of only 5730 years. With conventional dating methods, it can only be used effectively for obtaining dates that go back to 40,000 years. In a few laboratories the employment of technologically advanced equipment has nearly doubled the range over which carbon-14 can be used.

The discovery of carbon-14 is linked to cosmic-ray research of the upper atmosphere which was undertaken in 1939 by geophysicist Serge Korff. He noted that cosmic rays produce electrically neutral secondary particles (neutrons) in their initial collision with nitrogen gas molecules. During collision with the abundant isotope nitrogen-14, those neutrons, he predicted, would react to free the positively charged particles (protons), thus forming the radioactive isotope carbon-14.

Carbon-14 combines with oxygen to form carbon dioxide, which circulates in the atmosphere and dissolves in the ocean waters (Fig. 2-13); it is absorbed by plants and animals at the earth's surface. Carbon-14 radioactively decays to nitrogen-14 in time, but new carbon-14 is continuously being formed in the atmosphere by cosmic-ray bombardment, so that an equilibrium level is

Interaction of cosmic rays
with $N_2$ in outer atmosphere
produces $^{14}C$, which combines with oxygen
to form $^{14}CO_2$

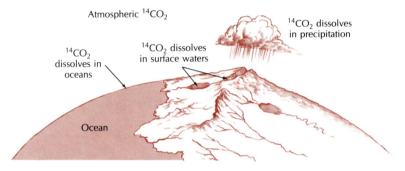

Atmospheric $^{14}CO_2$

$^{14}CO_2$ dissolves
in precipitation

$^{14}CO_2$ dissolves
in surface waters

$^{14}CO_2$
dissolves in
oceans

Ocean

**Fig. 2-13.** Generation of carbon-14 ($^{14}C$) in the outer atmosphere and its subsequent infiltration into oceans, surface waters, and the atmosphere.

maintained. As long as an organism lives, it continuously replenishes its carbon-14 content; once it dies, however, the isotope that has accumulated in the body continues to decay, but it is no longer replenished (Fig. 2-14). Through measurement of the carbon-14 level, it is possible to determine how long an organism has been dead—up to a maximum of about 40,000 to 80,000 years, depending on the analytical methods employed.

The basic assumption made in carbon-14 dating is that after an organism dies, it no longer absorbs carbon-14 from its surroundings. Yet some circumstances render that assumption invalid. For example, if the organic matter to be dated is buried,

**Fig. 2-14.** Carbon-14 cycle in living and dead organisms.

| Living organisms | Dead organisms |
|---|---|

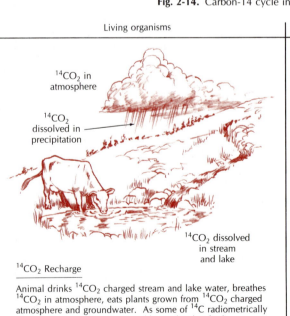

$^{14}CO_2$ in
atmosphere

$^{14}CO_2$
dissolved in
precipitation

$^{14}CO_2$ dissolved
in stream
and lake

$^{14}CO_2$ Recharge

Animal drinks $^{14}CO_2$ charged stream and lake water, breathes $^{14}CO_2$ in atmosphere, eats plants grown from $^{14}CO_2$ charged atmosphere and groundwater. As some of $^{14}C$ radiometrically decays, it is replenished to equilibrium level in both animals and plants.

At time of death, animal no longer breathes, drinks, or eats; therefore, there is no exchange of $^{14}C$ to replace that lost through radioactive decay. Similarly, plants cease to take in atmospheric and groundwater $^{14}C$ when they die. From that time on decay from previously established equilibrium level occurs.

circulating groundwater may provide a ready source of carbon-14, which then may enter into partial equilibrium with the carbon-14 already present. Another problem in dating involves contamination of the ancient organic-rich sample by entanglement with roots of modern plants that can extend downward from the surface for 2 m (6.5 ft) or more. Because of such uncertainties and difficulties with the laboratory procedure, any carbon-14 dates, particularly those older than 30,000 years, should be viewed with some reservation.

Carbon-14 dating is also based on the assumption that the amount of carbon-14 produced in the atmosphere has remained relatively constant over the last 40,000 to 80,000 years. If variations in the production of the isotope did occur and lasted several hundreds or thousands of years, they would lead to systematic errors in the radiocarbon dates obtained from organisms that lived during those times.

In the early days of carbon-14 dating, there was only a nagging suspicion that the production of carbon-14 varied; there seemed to be no way to obtain valid data to indicate the magnitude of possible errors. However, with the advent of **tree-ring dating** (also termed **dendrochronology**), a

**Fig. 2-15.** Tree rings of a Douglas fir that began growing in A.D. 113 and was cut down in 1240. This specimen was used as a construction timber in a prehistoric structure south of Mesa Verde in southwestern Colorado. (Laboratory of Tree-Ring Research, University of Arizona)

means for comparing carbon-14 dates with actual calendar dates became available. Each year that a tree lives, it normally adds one growth layer to its circumference. In a cross section from the tree, the growth layers appear as rings, and they can be counted (Fig. 2-15). In the ring structure of very old trees, some successions of rings are close together; in others they are wider apart. The spacing apparently reflects the rate of growth in response to the climatic conditions prevailing at the time. The spacing of the tree rings provides a "signature," and all trees that grew in the same locality during the same time span show the same pattern of variation. Thus a key is provided for matching rings within a local geographical area of similar climate. By counting and describing patterns in successively older trees, a tree-ring master pattern can be compiled for a particular geographical area and specific time span. Today through the use of tree rings from the long-lived bristlecone pine, one such tree-ring master pattern has been pieced together for the southwestern United States, extending back in time, year by year, for about 7000 years. Dendrochronology has provided an alternative method of dating archaeological sites whenever logs and tree trunks can be found in association with cultural material. In addition, by carbon-14 dating the wood itself, it has become possible to determine systematic variations in the radiocarbon dates (Fig. 2-16). The results indicate that from the present to about 500 years ago and from 2100 to 7100 years ago,

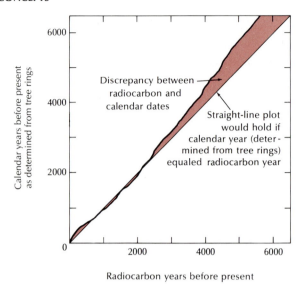

**Fig. 2-16.** Relationship of radiocarbon and calendar age as determined from carbon-14 dating of wood of a known calendar age. (After W.M. Wendland and D.C. Donley, 1971, "Radiocarbon and Calendar Age Relationships," *Earth and Planetary Science Letter*, vol. 11, pp. 135–39)

radiocarbon ages are younger than calendar ages, whereas from 500 to 2100 years ago the two chronologies are essentially the same. For dates near 7100 years ago, the error in the radiocarbon dates can amount to as much as 700 years. Most laboratories working on radiocarbon chronology now use that information to correct their carbon-14 dates.

## SUMMARY

1. Geology, the multidisciplinary study of the earth, is based on other sciences, such as physics, chemistry, biology, astronomy, and mathematics. As in other sciences, explanations of geological observations are expressed as hypotheses and theories. Those that stand the test of time are called *natural laws.*

2. One of the basic concepts that permits the unraveling of geological history is *uniformitarianism,* which maintains that the processes we see in action today are probably the same as those of the past. There is good evidence, however, that the rates of these processes are variable.

3. Within a local area of study, it is normally possible to place the events recorded in the rocks into a *relative time sequence* on the basis of *original horizontality* and *superposition* of beds, *cross-cutting relationships,* the *law of inclusions,* and determination of *erosion surfaces.*

4. Through the observation of *fossils* (the remains and traces of prehistoric life), which form an evolutionary succession, a *relative geological time scale* has been developed. From fossil content, it is possible to date and correlate strata in widely separated areas in relationship to one another—even on different continents.

5. The discovery of radioactivity in 1895 eventually led

to methods for the *absolute dating* of minerals and rocks, thus enabling the establishment of an absolute time frame. The oldest rocks found on earth so far are 3.8 billion years old, whereas meteorites are about 4.6 billion years old, an age that apparently marks the beginning of our solar system.

6. *Carbon-14* and *tree-ring dating* provide additional means of obtaining dates for very recent geological materials.

# QUESTIONS

1. What is the difference between relative time and absolute time?
2. How does the composite geological time scale differ from the relative time scale?
3. What are the various precautions, problems, and safeguards that are involved in obtaining valid radiometric dates?
4. List all the possible methods that can be used to establish a relative time sequence.
5. How are fossils used to establish the relative time scale?
6. Define the following: radioactivity, isotope, half-life, parent/daughter ratio.
7. Explain how radioactive isotopes are used to establish the composite time scale.
8. Without using the book, draw a reasonable facsimile of the composite geological time scale. List all the eras and periods and insert absolute dates at each break between the eras and at the base.
9. How is carbon-14 formed in the atmosphere?
10. How and why is carbon-14 dating restricted in its use?

# SELECTED REFERENCES

Adams, F.D., 1954, The birth and development of the geological sciences, Williams & Wilkins, Baltimore. (Repr. 1954, Dover, New York.)

Cohee, G.V., Glassner, M.F., and Hedberg, A.D., eds., 1978, The geologic time scale, American Association of Petroleum Geologists, Studies in Geology No. 6, Tulsa, Okla.

Dalrymple, G.B., and Lamphere, M.A., 1969, Potassium-argon dating. W.H. Freeman, San Francisco.

Eicher, D.L., and McAlester, A.L., 1980, History of the earth, Prentice-Hall, Englewood Cliffs, N.J.

Faure, G., 1986, Principles of isotope geology, Wiley, New York.

Hubbert, M.K., 1967, Critique of the principle of uniformity, *in* Uniformity and simplicity, C.C. Albritton, Jr., ed., Geological Society of America Special Paper 89.

Knopf, A., 1949, Time and its mysteries, series 3, New York University Press, New York.

Libby, W.F., 1961, Radiocarbon dating, Science, vol. 133, pp. 621–29.

McIntyre, D.B., 1963, James Hutton and the philosophy of geology, *in* The fabric of geology, C.C. Albritton, Jr., ed., Addison-Wesley, Reading, Mass.

Newman, W.L., Geologic time, U.S. Geological Survey booklet.

Palmer, A.R., 1983, Geologic time scale, the decade of North American geology, Geology, Vol. II, p. 503.

Thackray, J., 1986, The age of the earth, Cambridge University Press, New Rochelle, N.Y.

Wendland W.M., and Donley, D.C., 1971, Radiocarbon and calendar age relationships: earth and planetary science letters, vol. 11, pp. 135–39.

York, D., and Farquhar, R.M., 1972, The earth's age and geochronology, Pergamon, New York.

**Fig. 3-1.** Eye-catching group of amethystine quartz crystals from Vera Cruz, Mexico, approximately 18 cm (7 in.) across in the long direction. (M. Halberstadt)

# 3

# MATTER, MINERALS, AND ROCKS

Down through the ages philosophers, alchemists, and chemists have considered the composition of matter. Ancient Greeks had no means other than their own senses for investigating the physical world; their knowledge about that domain, therefore, was rudimentary. Even so, by about 400 B.C., the philosopher Democritus (c. 460–c. 370 B.C.) concluded, entirely on the basis of logic, that matter must be made up of small indivisible bits, which he called *atoms,* all of which are similar and eternal. About 50 years after this, the esteemed philosopher Aristotle (384–322 B.C.) proposed that the principal *elements* of terrestrial matter were fire, water, air, and earth, the last one providing the essence that gave objects the property of solidity.

In succeeding centuries, as alchemists strived to attain the elusive goal of turning base metals (e.g., lead, zinc, or copper) into gold, it became apparent that the physical world was not as simple as Aristotle supposed. Fire, water, and air were still thought to be elemental, but earth came to be recognized as a mixture of many things.

By the end of the eighteenth century, chemists had accumulated a fairly extensive knowledge of the basic building blocks of matter. The French chemist Antoine Lavoisier (1743–94) correctly identified 23 elements, which he defined as pure substances, and indicated that other elements probably existed. He was also one of the first to state that a chemical compound is a pure substance consisting of two or more elements in combination.

In the early 1800s, the English physical chemist John Dalton (1766–1844), drawing on the previously formulated atomic theory of the Greeks, hypothesized that all matter is composed of tiny invisible atoms that have different weights and chemical properties. An element, he concluded, is composed of only one kind of atom. He even went so far as to say that when two or more elements combine, their atoms form identical groups of atoms called *molecules.*

Today, there are 106 known elements, 92 of which are found naturally at the earth's surface. Spectral analysis of the sun shows that most of the naturally occurring ones also exist there, although not in the same proportions as on earth.

The next hurdles were cleared through experiments carried out in the late 1800s and early 1900s, particularly through those involving radioactive substances. Almost all atoms were found to be composed of three kinds of particles: negatively charged **electrons,** positively charged **protons,** and **neutrons,** which carry no charge whatever. In any atom there are as many negative particles (electrons) as there are positive ones (protons); therefore, the electrical charge is balanced. In atoms of a particular element, the number of protons or electrons is always a fixed number (the **atomic number**). But the number of neutrons can vary slightly from atom to atom of that element. Species of an element that differ only in the number of neutrons in the atoms (and, therefore, in **atomic weight**) are termed isotopes of that element. Both isotopes, uranium-235 and uranium-232, possess the chemical properties unique to uranium—they differ only in the number of neutrons in the atoms.

It was also found that electrons, protons, and neutrons are roughly the same size—about $10^{-12}$ cm (one-millionth of one-millionth of a centimeter) in diameter. An electron, however, has much less mass than the other two particles; it is only 1/1836 and 1/1837 as massive as a proton and a neutron, respectively. From its mass and neutral charge, a neutron can be considered as essentially a proton and electron combined.

An ingenious experiment by the English physicist Lord Rutherford in the early twentieth century demonstrated that in a single atom all the protons and the neutrons exist in a very dense central kernal, or **nucleus,** that contains more than 99.9 percent of the atom's mass, but represents only about one-trillionth of its volume. The nucleus is at the center of the essentially spherical atom, having a diameter of about $10^{-8}$ cm (one-hundredth of one-millionth of a centimeter) in which one or more electrons—the number depending on the particular element—move rapidly in continual agitation: now closer to the nucleus, now farther away (Fig. 3-2). Atoms that contain many electrons resemble, to a degree, a diffuse swarm of gnats. When many electrons are present, some are constrained to a path close to the nucleus, whereas others tend to circle near the outer margin of the electron "cloud." These outer electrons enter into chemical reactions, as when two elements combine to form a compound.

Because atoms are so tiny, large numbers of them are required to make them visible to the eye. A one-half pound lead brick, for example, contains nearly $1 \times 10^{24}$ atoms (1 followed by 24 zeroes).

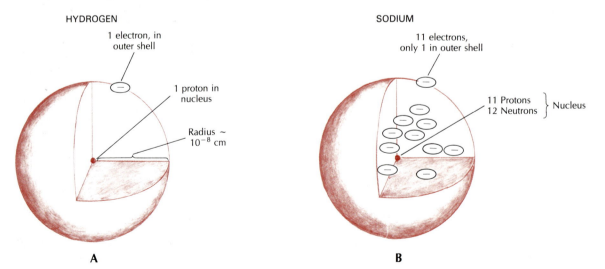

**Fig. 3-2.** Atoms of hydrogen and sodium. In both cases the atoms are essentially spherical and have radii of about $10^{-8}$ cm. **A.** The hydrogen atom, the simplest one of all, contains only 1 proton in the nucleus and 1 electron, the latter restricted to the outer shell. **B.** The sodium atom is much more complex, containing a nucleus made up of 11 protons and 12 neutrons as well as an electron cloud consisting of 11 electrons. Only 1 electron, however, moves at the outer limit of the electron shell.

HYDROGEN

1 electron, in
outer shell

1 proton in
nucleus

Radius ~
$10^{-8}$ cm

SODIUM

11 electrons,
only 1 in outer shell

11 Protons
12 Neutrons
} Nucleus

A

B

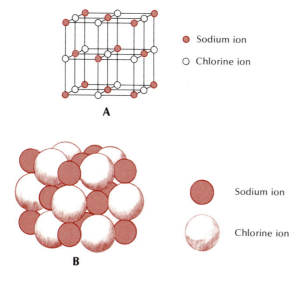

● Sodium ion

○ Chlorine ion

**A**

Sodium ion

Chlorine ion

**B**

**Fig. 3-3.** The structural arrangements of sodium and chlorine ions in the mineral halite. **A.** An exploded view depicts the relative positions of the ions. All ions are shown as points in order to stress their spatial configuration. **B.** This representation more closely approximates the actual stacking of the spherical positive and negative ions in the mineral. Notice that the chlorine anions are considerably larger than the sodium cations and that the electron shells just touch and merge with each other. **C.** (Below) Halite cubes consist of billions of sodium and chlorine ions bonded together ionically; the cubic geometric shapes of the crystals reflect the regular internal ionic arrangement shown in **A** and **B.**

Another way of demonstrating the large number of atoms (and molecules) present in a small volume of matter is to imagine the following experiment: If you pour two quarts of water into the ocean and "stir well," any cup of water subsequently taken from the ocean will contain *one* molecule from the two quarts you poured in.

Atoms normally combine, or **bond,** with other atoms. In some cases the atoms that bond together are all of the same element, as in the metal gold and the gas hydrogen. In other cases, however, atoms of two or more elements bond together in a definite proportion, to form a **chemical compound.**

If one atom of chlorine, a toxic gas commonly used in swimming pools, bonds with one atom of sodium, a highly reactive metal, the resultant compound is sodium chloride (table salt). Naturally occurring sodium chloride is called **halite** (Fig. 3-3). Or if two atoms of chlorine bond with one atom of calcium, the resultant compound is calcium chloride, a salt used as a drying and dehumidifying agent.

## BONDING

Two or more atoms bond through electrical attraction that results from the interaction of one or more electrons from the outer parts of their elec-

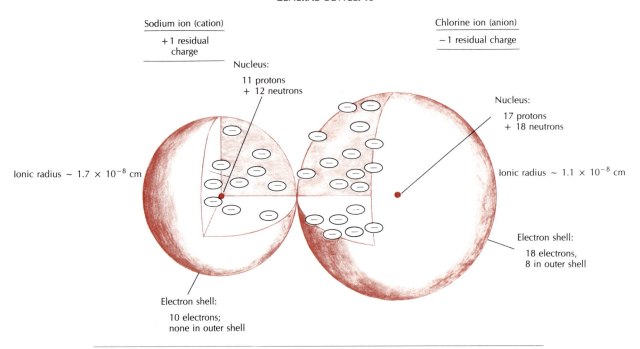

Sodium ion (cation)

+1 residual charge

Chlorine ion (anion)

−1 residual charge

Nucleus:

11 protons
+ 12 neutrons

Nucleus:

17 protons
+ 18 neutrons

Ionic radius ~ 1.7 × 10⁻⁸ cm

Ionic radius ~ 1.1 × 10⁻⁸ cm

Electron shell:

18 electrons,
8 in outer shell

Electron shell:

10 electrons;
none in outer shell

**Fig. 3-4.** A sodium atom has 1 loosely bound electron in its outer shell, whereas a chlorine atom has 7 electrons, missing only 1 to make it completely full. When the 2 atoms come together, the outer sodium electron moves to the shell of the chlorine atom, producing a positively charged sodium ion (cation) and a negatively charged chlorine ion (anion). Electrostatic attraction holds the 2 ions together, and an ionic bond is formed. Loss of the outer-shell electron in the sodium ion is accompanied by a reduction in radius size, whereas the addition of 1 electron to the outer shell in the chlorine results in an increase in radius.

tron clouds. Chemists have described several types of bonding. The particular type that occurs in a compound is determined by the nature of the outer bonding shells of the various atoms involved.

In one type of bonding, there is an actual transfer of one or more electrons from the electron cloud of one atom to that of another—one atom loses electrons, the other gains them. The resulting positively and negatively charged particles are called *ions* (Fig. 3-4). The ion carrying a positive charge is called a **cation,** the ion carrying a negative charge an **anion.** Usually the positively charged cations are formed by the loss of one or more electrons from atoms of the metallic elements, such as calcium, sodium, or potassium. **Ionic bonds** result from the electrostatic attraction between positively and negatively charged particles and normally are quite strong. Halite is an example of a compound in which the bonding is primarily ionic.

In another type of bonding, the electron, rather than transferring completely from the donor atom to the acceptor, remains halfway between the atoms and is shared by both. This is called **covalent bonding;** it, too, is quite strong. In this way two atoms of hydrogen will join to one atom of oxygen to form a water molecule (Fig. 3-5).

In many compounds the bonding is neither solely ionic nor covalent, but alternates between the two. At times the electron moves into the electron cloud of the other atom and the bonding momentarily is ionic; at other times the electron is shared by both atoms and the bonding is covalent. The atoms in most solid geological substances are held together by this alternating **ionic-covalent** bonding.

A third type of atomic bond, less common than the previous two, is the **metallic bond.** As you might expect from its name, such bonding is found principally among the metals in their uncombined state. In a metal all the atoms are exactly the same

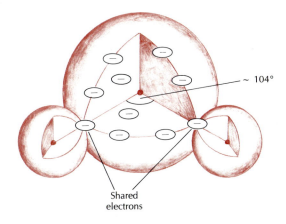

**Fig. 3-5.** Covalent bonding of two smaller hydrogen atoms with one larger oxygen atom to form a water molecule. The single electron from each hydrogen atom is shared with the outer electron shell of oxygen at a relatively constant angle between the centers of the three atoms of 104°.

**METALLIC BONDING**

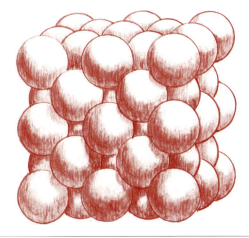

**Fig. 3-6.** Close packing arrangement of atoms of the same size in metallic bonding; each sphere is surrounded by 12 others. The bond between any 2 atoms is covalent; however, because any 1 atom can share electrons with any of its 12 neighbors, extra electrons are available for bonding. As a result, the electrons continuously drift and bonding electrons come from different neighbors at different times.

size, and they pack around each other like marbles. The tight packing of the equally sized spheres results in each atom being surrounded by 12 other atoms (Fig. 3-6), so that the electron cloud of each atom must react simultaneously with the clouds of its 12 closest neighbors. The bonding between spheres tends to be covalent. However, because more than enough electrons are available from all 12 surrounding atoms to complete bonding, those not in use at any moment are relatively unattached and tend to drift through the metal. Because all of the electrons in the metal are similar, an electron forming a covalent bond may at any moment be replaced by an electron that has been drifting. The replaced electron is then free to drift on to another atom. Because of the musical chairs played by the electrons, metallic bonding has also been called *time-shared covalent bonding*. This abundance of moving, loosely held electrons gives metals their characteristic abilities to conduct heat and electricity and to reflect light.

Molecules—that is, integral groups of atoms—can also be bonded together. The molecule already has fairly strong internal bonding, which is primarily covalent. Thus those electrons used in the *internal bonding* are generally unavailable for bonding with other molecules. In most cases, however, the molecules do not exhibit electrical neutrality in all directions; because of the positioning of the internal bonds, one or more locations on the margin of the molecular group will display a slightly greater positive or negative charge. Bonding between molecular groups results from the electrostatic attraction between the marginal positive and negative charge anomalies. Accordingly, such bonds are relatively weak—generally only about 1/100 of the strength of an ionic or covalent bond. Ice is characterized by this type of relatively weak bonding between water molecules (see Fig. 3-5). Both hydrogen atoms in the molecule tend to be slightly more positively charged than the average, whereas the relatively large oxygen atom tends to be more negatively charged in the region diametrically opposite the two hydrogen atoms.

In many compounds, particularly those involving atoms of more than two different elements, several types of bonds, or levels of bonding, coexist.

The shapes of ions and atoms, as noted earlier,

are essentially spherical. But not all are the same size. First, the more complicated atoms—those with more protons, neutrons, and electrons—tend to be of larger diameter. Second, the act of losing, gaining, or sharing an electron can affect the size of the atom or ion. Metals, which readily give up one or two electrons during bonding, tend to decrease in radius as a result. Atoms of iron are about 1 1/4 angstrom (Å) in diameter, whereas ionically bonded iron has a diameter only about half that size. Conversely, atoms that acquire electrons during bonding tend to increase in size. An oxygen atom in the unbonded state is about 3/4 Å across; in the ionic-covalent state, in which each atom acquires two electrons, its diameter is nearly twice as large. Because of these processes, cations tend to be smaller than anions. To a lesser degree, cation and atomic size are also dependent on the number of nearby ions of opposite charge that stack around a particular ion during bonding in the formation of a compound.

## STATES OF MATTER AND THE STRUCTURE OF MINERALS

Compounds exist in one of three different states—*solid, liquid,* or *gas*—depending on the strength of bond between the various atoms and on the temperature and pressure. Water, for example, can exist at the earth's surface in all three states. If liquid water is cooled sufficiently, it freezes to a solid; if heated sufficiently, it becomes a gas. It is apparent, then, that the principal difference between the three states is one of heat content, of which temperature is a measure. From experiments on gases in the 1800s, it became clear that temperature was also a measure of the average translational speed of the gas molecules: The higher the temperature, the faster the molecules move. When you blow up a balloon, the molecules move in all directions in a random fashion, hither and yon—an orderly pattern of the molecules' positions is totally lacking. At any one time many molecules strike the balloon wall, creating an outward pressure. If the balloon is heated, the molecules move faster and strike the wall harder and more frequently, thereby resulting in an increase in pressure and an associated expansion of the balloon.

In a gas, such as nitrogen or oxygen, the molecules move too quickly for any type of molecular bonding to occur. If, however, a gas is sufficiently cooled, the atoms (or molecules) slow down until they no longer are independent of one another. Then attractive, or bonding, forces become strong enough to cause the atoms or molecules to begin to stick to one another, and the liquid state is attained. Although relatively slow moving, each atom in a liquid still tends to move independently of every other one. There are occasions, however, when two or more atoms bond together momentarily. In the next moment when those atoms resume their aimless wandering, the bonds are ripped apart. Thus, around a particular molecule (or molecules) or within a small volume of liquid, there will exist from time to time (when bonds exist) an ordered assemblage of molecules. **Short-range ordering,** as that state is called, is characteristic of all liquids.

Because the atoms or molecules continuously change partners, a liquid lacks rigidity and is able to flow. During flow some portions move more rapidly than others, requiring that additional interatomic bonds are broken. It follows that the stronger the bonds, the greater the resistance to flowage. The resistance to flowage is called **viscosity.** Liquids that flow with difficulty are said to be of high viscosity; those that flow easily are of low viscosity. Increased temperatures in a liquid are associated with greater speeds in the molecules and atoms as well as decreases in bonding among them. Accordingly, an increase in temperature is almost always paralleled by a decrease in viscosity, and vice versa.

Continued cooling of a liquid progressively slows down the atoms or molecules. Eventually, they become so slow that bonds between them can form and remain firm. At that point the change from liquid to solid will have occurred. In the formation of a solid, the spherical atoms and ions fit together in the closest way possible for both their size and the type of bonding that is prevalent. Thus a solid is not a jumble of atoms or molecules; instead, each atom or molecule occupies a particular, set position with regard to those surrounding it. An ordered latticework of atoms is created—a little like the repetitive pattern on wallpaper, but in three dimensions. In this state a **long-range** *order* exists, and the material is said to be crystalline, or to possess **crystallinity**. (The word *crystalline* comes from crystal, the Anglicized Greek word for ice.) With few exceptions, only one particular atomic ordering is possible for each compound.

*Minerals* are crystalline compounds and elements that occur naturally. To be a mineral the material must also be inorganic (i.e., formed through nonbiological processes) and have a definite chemical composition or a restricted range of compositions.

It was the study of geological minerals that eventually led to our knowledge of long-range ordering, or crystallinity, which is exhibited by most compounds in the solid state. Today **mineralogists** (those who study minerals) have described

and named about 2500 minerals. Many centuries ago, however, only a few were known—generally those that were showy in appearance or had utility. For example, the common mineral quartz (Fig. 3-7), which characteristically occurs in clusters of shiny, icelike crystal spires, was well known in the Middle Ages; certain ore minerals that yield iron, copper, lead, and silver have been recognized for thousands of years. Minerals were used by our early ancestors just as they were found. Clay was molded into bricks and formed into pottery; flint and jade were fashioned into weapons and jewelry; the oxides of manganese and iron were made into pigments and paints; turquoise and amethyst were set into jewelry and other ornamentation; and gold, silver, and copper were used for ornaments or utensils. In time it was

**Fig. 3-7.** Cluster of quartz crystals from Crystal Springs, Arkansas. These transparent crystals possess a characteristic external form bounded by a number of relatively flat surfaces. (Smithsonian Institution)

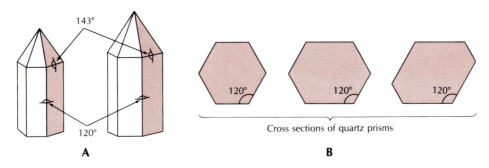

**Fig. 3-8. A.** Commonly occurring external forms of quartz crystals; the six-sided prism (straight-sided form) and pyramid (pointed form). The angle between two adjacent prism faces is always 120° regardless of the difference in size between crystals **(A)** or apparent distortions in shape **(B).** The fact that crystals of a particular mineral have similar external forms led early mineralogists to suspect an orderly atomic arrangement.

found that usable metals could also be extracted from certain minerals by heating or smelting.

By the eighteenth century the identification and understanding of minerals increased rapidly, largely because chemists were trying to understand the nature of matter and were using naturally occurring minerals in their experiments. Although chemists and mineralogists began to understand the chemical properties of the materials they worked with, neither group could explain why many minerals were bounded by planar crystal faces.

By 1669 Nicolaus Steno, the Danish physician who made contributions to our understanding of stratigraphic principles, demonstrated that similar faces on quartz crystals always meet at the same angle regardless of crystal size and shape (Fig. 3-8). Later it was shown that the surface structure of crystals is always the same in any given mineral species. As the field of crystallography grew, the crystal forms of many kinds of minerals were studied. Regardless of the large number studied, however, it became apparent that all the many crystal forms could be classified into just seven basic groups of systems, a fact that seemed indicative of an underlying order of some sort.

As early as 1611 Johannes Kepler (1571–1630), the famous German astronomer, aware that snowflakes always have a hexagonal pattern (Fig. 3-9), conjectured that their regularity in form was probably due to the geometrical arrangement of their minute building blocks. In 1748 René Haüy (1743–

1822), a French crystallographer, concluded that any mineral exhibiting an observable crystal form must be made up of small polyhedral units, each of which must have the same symmetry as the whole crystal (Fig. 3-10).

By the early 1880s, following acceptance of Dalton's atomic theory, mineralogists concluded that minerals were chemical compounds with definite compositions and that each compound must consist of a large number of atoms uniquely fitted together in a regular three-dimensional pattern.

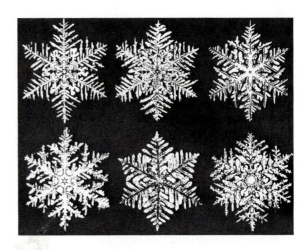

**Fig. 3-9.** Naturally occurring ice crystals (snowflakes) possess a six-sided symmetry—as do all ice crystals—suggesting a regular internal ordering. (Moody Institute of Science)

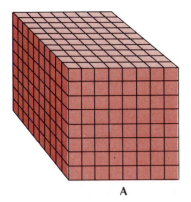

**A**

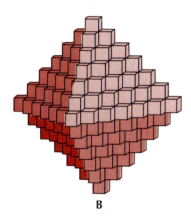

**B**

**Fig. 3-10.** Different crystal forms are made up of many small units having the same symmetry as the crystal itself. **A.** Cube; **B.** Octahedron.

a glass plate engraved with fine, closely spaced parallel lines. When light passes through the plate, the grooves interfere with, or diffract, the light waves in such a way that they are dispersed into a spectrum of colors, each color corresponding to light of a different wavelength. For a diffraction grating to be effective, the lengths between wave crests of the light and the distance between diffraction grooves should roughly be equal. What was needed for the investigation of the very-short-wavelength X-rays was a grating several orders of

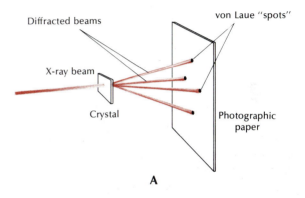

**A**

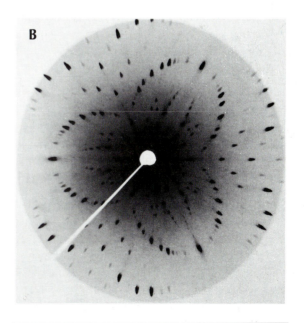

**Fig. 3-11.** Von Laue X-ray pattern. **A.** The X-ray beam is broken into a number of smaller beams by the internal crystal structure. **B.** Von Laue spots obtained by X-ray analysis of the mineral calcite. (Bernhardt J. Wuensch)

Crystallographers finally came to realize that the planar faces so characteristic of each particular mineral must be the outward reflection of the internal ordering—but they had no way of proving it for some time to come.

The proof came in 1912 as the result of an intuitive experiment performed by German physicist Max von Laue (1879–1960) and his associates. The X-ray was discovered by Wilhelm Roentgen in 1895, but little was known about it until 1912. Von Laue and his colleagues were interested in understanding X-rays and conjectured that, like light rays, they might be wavelike in character, but of exceedingly short wavelengths. To demonstrate those wave properties, they needed a better analytical device than the available diffraction grating. A standard **diffraction grating** consists of

magnitude finer than the standard grating. In a stroke of insight von Laue hit on the idea that the tiny atoms in a crystalline substance might be systematically arranged in layers spaced closely enough together to serve as a diffraction grating.

After the usual false starts and mishaps, he was successful in sending an X-ray beam through a crystal of copper sulfate and onto a photographic plate behind the crystal (Fig. 3-11A). When the plate was developed on it, there appeared a pattern of dots that von Laue interpreted as evidence of X-rays that had been diffracted through interactions with regularly arranged layers of atoms (Fig. 3-11B). Besides showing beyond a doubt that X-rays are wavelike in nature, the experiment revealed that the planes of atoms in a crystal are regularly spaced. X-rays had proved to be the key to the door of an unseen, ordered world in miniature, its very existence previously only inferred from surface measurements of interfacial angles and the regular geometry of crystal faces.

Later experiments with X-rays demonstrated that all minerals, whether they exhibit crystal faces or not, have a crystalline, long-range order. Through X-ray analysis it is now possible to determine the distances between layers in a crystal lattice and thereby, the exact radii of the constituent atoms and ions.

## PHYSICAL PROPERTIES OF MINERALS

### Crystallographic Structure

The crystallographic structure of a particular mineral is also a natural consequence of the way in which the atoms of the elements making up the compound bond together. Accordingly, the crystallographic structure of a particular mineral is the same, no matter where the mineral comes from.

For example, carbon can exist as either of two minerals—diamond or graphite. Diamond is very hard, in fact, so hard that it is used as an abrasive; it shines brilliantly, can take and hold a polish, and breaks smoothly along certain planes. Graphite, on the other hand, is relatively dull and soft and separates into small flakes that easily slide over one another.

We can explain the profound differences between minerals that are chemically the same by understanding their very different crystallographic arrangement of atoms. In a diamond crystal each atom is in direct contact with four other atoms and covalently shares the four electrons in its outer shell with them. The bonds are exceptionally strong, resulting in the extreme hardness of the mineral. However, the geometrical arrangement of atoms is such that there exist four sets of parallel planes that mark zones of weaker bonding. Because of these planes, the diamond cutter can cleave larger stones into smaller ones with precision.

The carbon atoms in graphite, on the other hand, occur in regularly spaced, thin, parallel sheets. The bonds uniting the atoms within each sheet are many times stronger than those between one atomic layer and its neighboring sheet above or below. Graphite, therefore, splits easily, parallel to the sheets, and the sheets slide easily over one another.

The crystallographic structure is, obviously, a definitive property of a mineral and can be used to aid in its identification. But it is only one of several distinct properties that each mineral displays.

### Color

Color is the most obvious property that minerals possess, but it is not always the most definitive. When white light falls on a mineral, some of the wavelengths are absorbed and some are reflected. The color of the mineral corresponds to the wavelengths of reflected light. Black minerals, for example, absorb essentially all the light that falls on them. What causes a mineral to absorb only certain wavelengths is varied and complex. It can be a fundamental property that is directly related to chemical composition, as in the blues or greens of some copper minerals. Or it may be unrelated to composition, depending instead on crystal structure and the type of bonding (as in diamonds and graphite). In some cases it is caused by foreign atoms contained in the crystal structure. For example, pure quartz is colorless, whereas the colored varieties of that mineral (so sought after by collectors) contain small quantities of impurities. Therefore, we cannot use color with the same confidence in identifying minerals as in naming birds. Considerable experience is needed to determine whether the color of a mineral is significant.

One way to avoid ambiguities concerning mineral color is to grind the mineral into a fine powder. In the powdered form, the intrinsic color is shown, whereas color resulting from impurities are lost. A simple way to make the test is to rub the mineral on a piece of unglazed porcelain *(streak plate)*, leaving a thin film of powder, or *streak*, of true color on the white porcelain background. Two iron-containing minerals—hematite and magnetite—superficially resemble each other in many physical properties. However, the streak of hematite is red-brown, whereas that of magnetite is black-brown. In this case the streak is a most definitive property.

## Cleavage

*Cleavage* is the ability of a mineral to split along closely spaced parallel planes (Fig. 3-12). Not all minerals have that ability; many fracture along widely spaced surfaces that can be relatively smooth but that are often curved or irregular. One notable (and familiar) type of fracture surface, commonly produced when glass is broken, roughly resembles the valve of a seashell and, appropriately, is called

Fig. 3-12. Cleavage fragment of the common mineral calcite. Because of the characteristic regular arrangement of the atoms in the mineral, it always breaks (cleaves) along three smooth plane surfaces, any two of which meet at an angle of either 74°55' or 105°05'. The resulting shape is rhombohedral.

Fig. 3-13. Conchoidal fracture on a quartz fragment. Notice the nearly concentric arrangement of rounded grooves and ridges resembling the markings on a seashell. (Janet Robertson)

a *conchoidal fracture* (from the Greek *konkolides,* "like a shell") (Fig. 3-13).

Some minerals are characterized by only one set of cleavage planes (i.e., they only break in one direction), others by two, three, four, or (rarely) six. Because of crystal symmetry, no minerals possess five or more than six sets of cleavage planes. Not only the number of cleavages, but the angles between them as well are characteristic for each mineral. Some cleavages meet at right angles; many do not. If splitting occurs easily and produces a nearly unblemished flat surface, the cleavage is said to be *perfect.* In some minerals the cleavage is less perfect and can be called *good, distinct,* or *indistinct,* depending on the roughness of the resultant surface. The geometrically repetitive nature of cleavage, its planar character, and the distinctive orientation of the planes are strong evidence that cleavage, like the crystal form of minerals, is a property determined by the regular geometric ordering of the atoms in the mineral. In fact, for some small crystals that are difficult to observe either with the naked eye or with the aid of a hand lens, shiny cleavage surfaces can commonly be confused with crystal faces. It takes a keen eye and experience to make the proper distinction. Cleavage directions generally run parallel to those crystallographic planes in which the packing of ions is greatest.

## Hardness

The hardness of a mineral—its ability to withstand being scratched—is also related to atomic structure. Even a small scratch on the surface of a mineral requires the breaking of bonds and the separation of atoms; the ease of separation depends on the kinds of atoms and bonding present. Fortunately, hardness is an easy property to determine—a harder mineral will scratch a softer one.

The scale we use today to compare hardness was devised more than a century ago by an Austrian mineralogist, Friedrich Mohs (1773–1839), and so bears his name. In this numerical scale 10 minerals that essentially cover the range of hardness variability are arranged in order from the softest (No. 1, talc) to the hardest (No. 10, diamond). By means of the scale (Table 3-1), the

**Table 3-1** Mohs Hardness Scale

| Mineral | Hardness | Equivalent in Hardness |
|---|---|---|
| Diamond | 10 | |
| Corundum | 9 | |
| Topaz | 8 | |
| Quartz | 7 | |
| Potassium feldspar | 6 | glass, knife blade |
| Apatite | 5 | |
| Fluorite | 4 | |
| Calcite | 3 | penny |
| Gypsum | 2 | fingernail |
| Talc | 1 | |

hardness of all minerals can be determined. The hardness numbers, although in sequential order, are not equally spaced apart. For example, the actual interval between diamond—to which Mohs assigned a value of 10—and corundum—assigned a value of 9—is greater than the rest of the scale combined. If absolute values were assigned to the various minerals used in this scale, diamond would have a value of about 42.

## Luster

When a mineral is viewed in ordinary light, the amount of light reflected from its surface and the way it is reflected determine its **luster.** Essentially all lusters fall into two groups: *metallic* and *non-metallic.* The first term is applied to minerals that reflect light in about the same way that polished metals, such as iron and copper, do. In general, metallic luster is characteristic of minerals in which metallic bonding is prevalent. Minerals with a metallic luster commonly are opaque (they will not allow light to pass through), even along thin edges held up against the light. A metallic luster is common to native (uncombined) metals and to minerals composed of a metal combined with sulfur.

Most common minerals exhibit nonmetallic lusters, of which certain common ones have been singled out. If a mineral reflects light to about the same degree as glass, it has a **vitreous (glassy) luster.** Other nonmetallic lusters are essentially self-explanatory: earthy, greasy, waxy, dull, resinous, pearly, silky.

## Specific Gravity

The specific gravity of a mineral may be defined as the weight of a specified volume of the mineral divided by the weight of an equal volume of water at 4° C (39° F)—the temperature at which liquid water is most dense. It expresses how much more dense a mineral is than a common substance, water. For example, quartz has a specific gravity of 2.7, which means that its density is 2.7 times that of water. The specific gravity of any mineral is, within limits, characteristic of that mineral, and it is determined by the atomic weights of the elements that make up the compound and how tightly the atoms or ions are packed in the crystallographic structure.

A rough estimate of the specific gravity of a mineral can be made (after some experience has been gained) simply by hefting the mineral by hand. Some minerals feel light, whereas others, particularly those with metallic lusters, feel heavy.

## Magnetic Properties

A few minerals, particularly those composed principally of iron and oxygen or iron and sulfur, will possess fairly strong magnetic properties; however, most minerals will not. A simple test of this property can be made with the use of a small hand magnet.

## MINERAL DESCRIPTIONS

A mineral, to review briefly, is a naturally occurring inorganic compound (or element) that is crystalline in nature and that possesses a definite chemical composition or a restricted range of compositions. A few materials are still considered minerals even though they do not strictly meet all these requirements. Two of the notable exceptions are opal, a noncrystalline solid, and mercury, a liquid. More than 2500 minerals have been recognized on the face of the earth. Some of their names—derived from Greek, Latin, Old English, or other tongues—describe such properties as color, crystal form, density, or cleavage. Other names are derived from the geographical locality in which they were found. Most minerals, however, bear the names of people—mineralogists, crystallographers, scientists, explorers, mine owners, mining engineers, public officials.

Minerals do not occur in equal abundance; some are relatively common, most are rare. The abundance of a mineral more or less reflects the quantities of the component elements that are available for its formation at or near the earth's surface (Table 3-2). Because of the nonuniform distribution of elements on the earth, minerals that are rare on a worldwide basis can be quite abundant locally.

Only 10 elements, surprisingly, occur in sizable amounts. These 10 account for about 99 percent of the total mass of minerals at the earth's surface; all the others—many important to the life and prosperity of humans—account for less than 1 percent of the total. Of the elements making up continental rocks, oxygen and silicon together total nearly two-thirds of the mass. The seven metals—aluminum, iron, calcium, sodium, potassium, magnesium, and titanium—make up most of the rest. The tenth element is hydrogen. The average composition of the crust beneath the oceans is only slightly different from that of the continents (Table 3-2).

In comparison to the other common elements, oxygen atoms (ions) are large and make up about 94 percent of the volume of the continental crust. When you step on a rock, you tread mostly on oxygen ions.

**Table 3-2** Average Chemical Composition of the Continental and the Oceanic Crust

| Element | Weight (mass) | |
| --- | --- | --- |
| | Continental Crust | Oceanic Crust |
| Oxygen (O) | 46.6% | 45.4% |
| Silicon (Si) | 27.2 | 22.8 |
| Aluminum (Al) | 8.1 | 8.7 |
| Iron (Fe) | 5.0 | 6.4 |
| Calcium (Ca) | 3.6 | 8.8 |
| Sodium (Na) | 2.8 | 1.9 |
| Potassium (K) | 2.6 | 0.3 |
| Magnesium (Mg) | 2.1 | 4.1 |
| Titanium (Ti) | 0.4 | 0.8 |
| Hydrogen (H) | 0.1 | 0.1 |

[a] The symbols for the elements are in parentheses.

Tetrahedron, with oxygen ion at each corner; back, hidden line is dashed

Oxygen ion

Silicon ion, hidden

**Fig. 3-14. A.** The SiO₄ tetrahedron, consisting of one silicon ion surrounded by four oxygen ions. **B.** The same pattern with lines that connect the centers of the oxygen ions emphasizes the tetrahedral arrangement.

A

B

Because of the overwhelming abundance of oxygen and silicon, it is only natural that *silicates*—those minerals containing silicon and oxygen—are the most plentiful on earth. There are many silicates; most contain varying amounts of one or more of the seven relatively abundant metals and in some instances small amounts of hydrogen..Silicates are thought to predominate all layers of the earth, right down to its core. The core itself appears to be largely made of iron.

In relation to most other cations the silicon cation is very small, and it bonds covalently to the four large oxygen anions (radii ≅ 1 1/3 Å) that fit around it to produce a four-sided tent-shaped form called a *tetrahedron (pl. tetrahedra)* (Fig. 3-14).

The centers of the four oxygen cations are at the points of the tetrahedron, each side of which is an equilateral triangle. The form can exist singly or it can be joined to other tetrahedra by sharing a common oxygen anion to produce a varied array of single- and double-chain structures, rings, or three-dimensional networks (Fig. 3-15). Because there are many more oxygen anions relative to the central silicon cation, the tetrahedra possess a residual negative charge. Although these tetrahedra are the basic structural building blocks, to form a mineral they have to be bonded together by positive ions—generally the common metal ions, such as those of iron, calcium, sodium, potassium, magnesium, and titanium. The cations fit

**Fig. 3-15.** Some of the diversity of silicate structures—increasing in complexity—that can be formed by various combinations of the basic SiO₄ tetrahedron.

**SILICATE STRUCTURES**

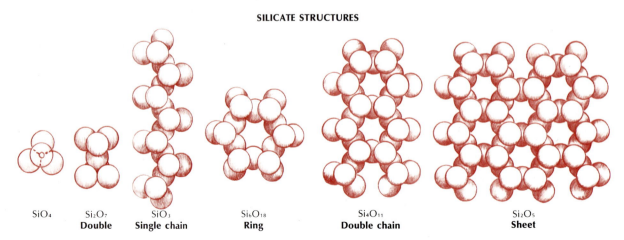

| SiO₄ | Si₂O₇ | SiO₃ | Si₆O₁₈ | Si₄O₁₁ | Si₂O₅ |
|------|-------|------|--------|--------|-------|
|  | **Double** | **Single chain** | **Ring** | **Double chain** | **Sheet** |

**Table 3-3** Geologically Important Minerals

| Silicates | | Nonsilicates | |
|---|---|---|---|
| | Quartz<br>Potassium feldspar<br>Plagioclase<br>(feldspar)<br>Muscovite (mica) | **Native elements,** including gold, silver, platinum, diamond, copper, iron, nickel, sulfur, graphite<br>**Sulfides,** such as argentite, chalcocite, bornite, galena, sphalerite, chalcopyrite, cinnabar, molybdenite, pyrite | **Oxides,** including magnetite, hematite, goethite<br>**Halides,** such as halite, fluorite<br>**Carbonates,** particularly calcite, dolomite<br>**Sulfates,** including anhydrite, gypsum |
| **Ferromagnesian minerals** | Biotite (mica)<br>Hornblende<br>(amphibole)<br>Augite (pyroxene)<br>Olivine | | |
| | Garnet<br>Chlorite<br>Clay | | |

into available spaces in the crystal lattice, in between the tetrahedral units.

Although silicates make up the bulk of the common minerals, some ***nonsilicates***—particularly native elements, oxides, halides, carbonates, and sulfates—are volumetrically abundant.

Of the large number of minerals described, those making up most of the solid earth number only about 40 (Table 3-3). The minerals fall into two groups: silicates and nonsilicates. There is also a subgroup of four silicates, the ***ferromagnesian minerals,*** that contain cations of iron and magnesium and tend to be dark in color.

We shall now explain the shorthand notation used to indicate the chemical makeup of a mineral compound. Rather than spelling out each element composing a mineral, one- or two-letter symbols are used to represent each element. Silicon is Si, oxygen is O, and $SiO_2$ is the formula for silicon dioxide, the mineral known as quartz. The subscript 2 indicates that quartz is made up of two parts oxygen to one part silicon. With the chemical formula, one can quickly see what the constituent elements of a mineral are and in what proportion they occur. The symbols for many of the common elements are given in Table 3-2. In addition, sulfur is represented by S, carbon by C, copper by Cu, lead by Pb, and zinc by Zn.

## Silicates

**Quartz** ($SiO_2$) Quartz has a vitreous luster, a hardness of 7 (on the Mohs scale), and when pure,

it is clear and colorless. In fact, the Greeks thought it was a kind of frozen water. The name may be derived from a Saxon word meaning "cross-veined." Quartz, which is particularly resistant to surface weathering, lacks cleavage, but it commonly fractures conchoidally (see Fig. 3-13). If quartz grows freely, it crystallizes customarily in a six-sided crystal form, which is terminated by a sharp-pointed pyramid at each end. If quartz grows into cavities, as it commonly does, it will possess only one pyramid at the end of the crystal that extends into the opening (see Fig. 3-7). Crystals such as these, often 12 cm (5 in.) or more in length, are commonly sold in gem and mineral shops. In most rocks quartz occurs in association with other minerals as tiny grains, 2- to 3-mm (0.08- to 0.12-in.) across, that generally lack crystal faces. When they are fresh the disseminated grains may sparkle like tiny bits of glass. Quartz is perhaps the most widely recognized mineral of all, being familiar to most people. Common varieties, mostly differentiated on the basis of color, include clear, amethyst (purple), rose, and smoky. A finely crystalline variety is called *chalcedony;* its darker form is called *flint.*

**Feldspar** The feldspars are by far the most abundant of the common minerals, probably making up at least 50 percent of the rocks at the earth's surface. The name comes from the Swedish word *feld* ("field") and from *spar,* a mineral commonly found in fields overlying granite. The two most common varieties are *potassium feldspar,* which

65

**Fig. 3-16.** Aggregate of blocky crystals of potassium feldspar, collected from the Front Range, Colorado. (Department of Library Services, American Museum of Natural History. Photo by Lee Boltin, Neg. No. 123996)

is rich in potassium, and *plagioclase,* which is rich in sodium and calcium. The $SiO_4$ tetrahedra in these minerals are joined in a strong three-dimensional network that possesses planes of weakness in two directions at, or nearly at, right angles to each other (Fig. 3-16). Cleavage along the two planes of weakness and a hardness of 6 provide two of the most characteristic properties of the feldspars. In many rocks they occur in well-formed crystals with a tabular form, so that on weathered or broken surfaces they often resemble small rectangular gravestones or fence laths.

***Potassium Feldspar*** ($KAlSi_3O_8$)    There are several common species of potassium feldspar (among

them orthoclase); these vary in the way in which their ions are arranged. For our purposes, little will be gained by differentiating among them. Potassium feldspar often has a vitreous luster; however, it also commonly resembles glazed porcelain, like the surface of a dinner plate. The mineral is usually milky white or flesh pink.

***Plagioclase***    ($NaAlSi_3O_8 \cdot CaAl_2Si_2O_8$)    This mineral, like potassium feldspar, has a vitreous luster. Its color is most likely to be white or pale gray, although some varieties show a beautiful iridescence (play of colors) much like those of a peacock's feathers. At times it is almost as clear as glass. Plagioclase is probably best distinguished from potassium feldspar or quartz by examining the crystal or cleavage surfaces of the mineral for *striations*—a multitude of closely spaced, parallel straight lines that look as if they had been engraved on the surfaces. Striations, well developed in plagioclase, are not found in potassium feldspar.

If we analyze the chemical composition of plagioclase taken from several different rocks, we find that the mineral contains calcium, sodium, aluminum, and silicon as the principal cations, but the proportions of each vary in each rock. The reason for this is easily understood when we compare the relative sizes of the four cations. Those of sodium and calcium are relatively large ($\cong 1$ Å), whereas those of silicon and aluminum are relatively small ($\cong 1/3$ and $1/2$ Å, respectively). In a way we can think of the crystal lattice as blind. During crystal growth from a liquid containing a mixture of atoms, the lattice accepts any ion as long as it is of an appropriate size—for example, sodium and calcium cations are nearly equal in size and can simply interchange or substitute for one another, as can aluminum and silicon ions. Calcium, however, cannot substitute for silicon because the size difference is too great. This process of interchanging cations in mineral formation is called **solid solution.** Moreover, cations are positively charged particles, and during interchange the neutrality of the electrical charge in the lattice must be maintained. The cations of sodium, calcium, aluminum, and silicon carry electrical charges of $+1$, $+2$, $+3$, and $+4$, respectively. Therefore, a sodium ion ($+1$) cannot be substituted for a calcium ion ($+2$) unless at the same time a silicon ion ($+4$) is substituted for an aluminum ion ($+3$). This type of solid solution, in which pairs of cations simultaneously substitute, is called **coupled solid solution.** Temperature during the formation of plagioclase plays an important role in determining the degree of coupled solid solution. Crystals formed at fairly high temperatures tend to be rich in calcium, whereas those formed at low temperatures are generally rich in sodium. Crystals formed at intermediate temperatures possess calcium and sodium in intermediate proportions. A simple way of informing the reader that two components have the same charge, are nearly identical in size, and can substitute in any proportion for one another is to enclose the symbols in parentheses: For example, (Mg,Fe) indicates that magnesium and iron are freely interchangeable, as in the case of $(Mg,Fe)_2SiO_4$, olivine.

The term *solid solution* is usually restricted to those instances in which both the substitution is a common occurrence and the chemical composition of the mineral is significantly affected by the substitution—as in the case of plagioclase or olivine. Limited solid solution on a scale so small that the mineral composition is not greatly altered occurs commonly throughout the mineral world, depending on the limited availability of the "foreign" ions and on the appropriate ionic radii and charges. In any particular mineral those elements that occur in minute quantities are called **trace elements.**

**Mica** The micas include a number of closely related minerals, all of which possess one perfect, or nearly perfect, cleavage and have a hardness ranging from 2 to 3. X-ray analysis shows that the crystal of mica consists of parallel sheets, like the pages in a book, which are made up of $SiO_4$ tetrahedra strongly bonded together at their bases (see sheet structure in Fig. 3-15). The sheets themselves are bonded less strongly to layers of potassium ions along which splitting occurs relatively easily—thereby producing the characteristic perfect cleavage of mica (Fig. 3-17). The two most common rock-forming micas are *muscovite* and *biotite*.

**Muscovite** $(KAl_3Si_3O_{10}(OH)_2)$ The common name for muscovite is **white mica,** and it is generally colorless, gray, or transparent, especially when split into thin sheets. Muscovite was used for the tiny windows of medieval Europe before the widespread use of glass enabled more light to enter the gloomy interiors of houses. *Muskovy* was the name given to Old Russia. Muscovite has a pearly to vitreous luster, and in sunlight the cleavage plates of the tiny mineral grains, which are common in many rocks, shimmer and shine.

**Biotite** $(K(Mg,Fe)_3AlSi_3O_{10}(OH)_2)$ Named in honor of the French physicist Jean Baptiste Biot (1774–1862), biotite is commonly called **black mica.** Its chemical formula indicates that in addition to the ions common to muscovite, it contains iron and

## MUSCOVITE STRUCTURE

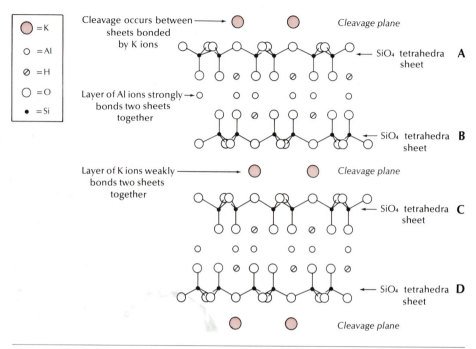

**Fig. 3-17.** The double-sandwich structure of muscovite. The points of the SiO₄ tetrahedra in the sheets (**A** and **B**) face each other and are strongly bonded by aluminum (Al) ions. Sheets (**C**) and (**D**) are similarly bonded. These double layers, however, are bonded by potassium (K) ions, which form a relatively weak bond; the one perfect cleavage in mica occurs along these planes of weakness.

magnesium (in solid solution). Thin cleavage sheets of it lack the degree of transparency of muscovite. Black mica ranges in color from dark brown to greenish black to black, and in many rocks it occurs as jet-black flakes that shine like satin in the sun. In some cases the small flakes appear to have an hexagonal outline.

**Hornblende**  (Ca₂Na(Mg,Fe)₄(Al,Fe,Ti)₃Si₆O₂₂(OH)₂) This ferromagnesian mineral is the most common of a large and complex group with similar physical properties, the **amphiboles.** Hornblende, a dark mineral, is commonly dark green or jet black, and it shines as brightly as a lacquered surface in the sun.

As a rule the crystals are long and narrow, and one of the most distinctive properties of hornblende is its cleavage pattern (Fig. 3-18). The SiO₄ tetrahedra are joined in long double chains that parallel the long axis of the crystal, and the two good cleavages, which intersect each other at angles of 56° and 124°, are parallel to planes of weakness (weaker bonds) that exist between groups of chains.

**Augite**  (Ca(Mg,Fe,Al)(Si,Al)₂O₆)   The ferromagnesian mineral augite takes its name from the Greek word for "luster." Superficially, it looks like hornblende in that it is dark and possesses two good cleavage planes and a vitreous luster. Augite crystals, however, generally are stubbier, to the point of being nearly equidimensional; their two cleavage planes meet nearly at right angles (87° and 93°), and viewed in cross section, the crystals are nearly square (see Fig. 3-19). The cleavage planes in augite also parallel planes of weaker bonds between silicate chains, but in this case the chains are single rather than double.

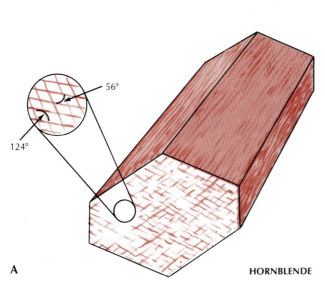

A HORNBLENDE

B

**Fig. 3-18. A.** Drawing of a hornblende crystal, showing the characteristic flattened six-sided face, elongate shape, and two good cleavages at 56° and 124°. **B.** The crystal in cross section (microscope photograph). (Edwin E. Larson)

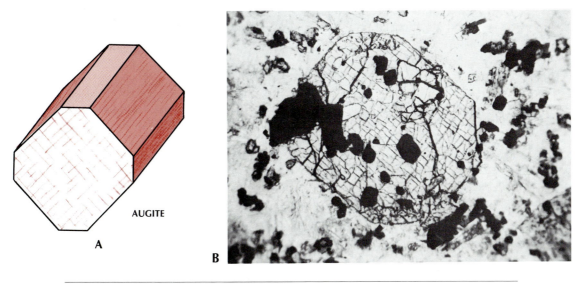

AUGITE

A

B

**Fig. 3-19. A.** Drawing of an augite crystal (pyroxene), showing the characteristic, nearly equant, eight-sided shape, stubby form, and two good cleavages that meet nearly at right angles (87° and 93°). **B.** The crystal in cross section (microscope photograph). (Edwin E. Larson)

**Fig. 3-20.** Garnet crystals displaying the common, 12-sided (dodecahedral) shape. This specimen comes from Austria; the largest crystals are about 2 cm (1 in.) across. (M. Halberstadt)

The principal distinctions between hornblende and augite are (1) hornblende crystals tend to be long and narrow, whereas augite crystals are short and stubby; (2) hornblende has two cleavages parallel to the long axis of the crystal that meet at 56° and 124°, whereas augite has two that intersect each other at approximately right angles; (3) hornblende crystals seen in cross section approach a rhombic pattern, whereas augite crystals are more nearly square.

**Olivine** $((Mg,Fe)_2SiO_4)$   Another ferromagnesian mineral, olivine occurs as rounded crystals that are green, granular, and glassy. It has a hardness of 6.5 to 7. The name, from the Latin, refers to its green (olive) color. Its transparent gem variety is called *peridot*.

Olivine lacks cleavage and fractures conchoidally. It is most common in basaltic lavas, in which it looks like tiny bits of dark-to-light-green bottle glass. Olivine is prone to alteration—generally transforming into a reddish or green-yellow fibrous material. Olivine is a solid solution mineral with variable amounts of iron and magnesium. As in the case of plagioclase, solid substitution is dependent on temperature. Olivine formed at relatively high temperatures possesses a preponderance of magnesium, whereas that formed at low temperatures is richer in iron. Most olivine has formed at higher temperatures and is, accordingly, richer in magnesium than in iron.

**Garnet**   Garnet is the name given to a group of minerals of similar silicate structure that possess great chemical diversity because of solid substitution; most varieties contain calcium, iron, and aluminum. Garnet has no cleavage and breaks with an uneven or conchoidal fracture. It possesses a resinous to a vitreous luster and a hardness of about 7. The colors of garnet are highly varied, but usually they appear red, brown, or yellow. The name refers to the resemblance of red garnets in color and equant shape to the seeds of pomegranates. Garnet almost always occurs in well-formed, equidimensional crystals—perhaps its most distinctive property (Fig. 3-20).

**Chlorite**   Chlorite is a complex group of hydrous silicates that contain Mg and Al and, to a lesser degree, Fe and other metals. It usually occurs in scaly or thinly banded masses. Its color is commonly grass green to blackish-green; its streak generally is also greenish. Its name, appropriately, derives from a Greek word for "light green." Chlorite exhibits a vitreous, or pearly, or sometimes dull luster. Other properties include a hardness of only 1 to 2.5, and a single perfect cleavage.

**Clay**   Clay is the common name for a group of hydrous aluminosilicate minerals that result from the weathering of rocks. Predominantly white, it can be easily stained by impurities and usually feels greasy to the touch. Clay—particularly when breathed on—often exhibits a distinctive odor, somewhat like the air just after the start of a summer rain. Some varieties adhere to the tongue or become plastic when moistened. Customarily, clay occurs in soft, compact earthy masses composed of aggregates of very tiny crystals, each of which exhibits one perfect cleavage. The clays are divided into three main classes: *kaolinite, illite,* and the *smectites.* The chemical formula for kaolinite is $Al_4Si_4O_{10}(OH)_8$. Illite in addition contains potassium ions; its composition $(KAl_5Si_7O_{20}(OH)_4)$ is similar to that of muscovite. The smectite group represents a hydrous aluminosilicate containing variable amounts of calcium, sodium, magnesium, and iron cations; its most notable property is its ability to absorb into its crystal structure large quantities of water, resulting in a significant volume increase. An alternative name for the smectites, appropriately, is the swelling clays. Soils that contain appreciable amounts of swelling clays are brick hard in the heat of summer but turn to goo during the wet season.

The best way to differentiate between the various fine-grained clays in a sample is by X-ray diffraction.

## Nonsilicates

**Native Elements**   Only a comparatively few elements occur in the uncombined (native) state. In total volume native elements represent a very small part of the mineral world. However, they more than make up for volume in monetary value, including such precious minerals as *gold* (Au), *silver* (Ag), *platinum* (Pt)—all metals—and diamond, the nonmetal element. Other strategic native metals include *copper* (Cu), *iron* (Fe), and *nickel* (Ni); important native nonmetals include *sulfur* (S) and *graphite*. All of the native elements share one chemical property—an aversion, at least under appropriate circumstances, to forming compounds.

All of the native metals are characterized by metallic bonding, whereas covalent bonding is the principal type exhibited by sulfur, diamond, and graphite. Diamond, unlike the other minerals, is an exotic mineral, formed under the high temperatures and pressures hundreds-of-kilometers deep in the earth; it is brought to the earth's surface during an unusual volcanic event.

**Sulfides**   The sulfide minerals are important primarily because they make up most of the ore minerals. The sulfides represent compounds consisting of sulfur in combination with one or more metal elements, including silver, copper, iron, lead (Pb), zinc (Zn), mercury (Hg), nickel, and molybdenum (Mo). Most sulfides are opaque, of high specific gravity, and exhibit metallic to submetallic luster. In most cases, they possess mixed metallic/covalent/ionic bonding, with metallic being predominant and ionic the least significant. Many are brightly colored, representing some of the showier minerals displayed in gem and mineral shops. Commonly encountered sulfides include *argentite* (Ag$_2$S), *chalcocite* (Cu$_2$S), *bornite* (Cu$_5$FeS$_4$), *galena* (PbS), *sphalerite* (ZnS), *chalcopyrite* (CuFeS$_2$), *cinnabar* (HgS), *molybdenite* (MoS$_2$), and *pyrite* (FeS$_2$)—the last is also known as fool's gold.

**Oxides** The oxide group of minerals includes those in which one or more of the metal elements combine—generally through covalent/metallic bonding—with oxygen. *Magnetite, hematite,* and *goethite* are the three most abundant minerals in this group. Magnetite (Fe$_3$O$_4$), the black oxide of iron (and an important ore of iron), occurs most often as disseminated grains in rocks rich in ferromagnesian minerals. Opaque, it has a metallic to submetallic luster, lacks cleavage, and exhibits a hardness of about 7. Its most distinctive property is its ability to be attracted to a strong magnet. Its name derives from the locality in which the mineral was first found, the ancient Mediterranean kingdom of Magnesia.

Hematite (Fe$_2$O$_3$), also an ore of iron, takes its named from the Greek word for "bloodlike stone." It looks like magnetite when it occurs in its granular form but is less magnetic, exhibits a hardness of between 5 and 6, and produces a red to red-brown streak. At times it occurs as a very fine pigment that colors a rock entirely red, often so intensely that it looks as if the rock were painted. .It is the mineral responsible for the red color of the sedimentary rocks so prevalent in the vicinity of Grand Canyon, Arizona. One variety possesses a metallic luster; when cut and polished, it makes fine semiprecious gemstones called *black diamonds*. Most hematite, however, has a dull, fibrous, or earthy luster. Some of the latter varieties are referred to as red ocher, a natural paint pigment.

Goethite (HFeO$_2$), named after the famous German poet and philosopher, Goethe (1749–1832), is a hydrated oxide of iron. The mineral, which exhibits one perfect cleavage and a hardness of 5, generally occurs in earthy, silky, or fibrous yellowish-to-dark-brown masses. Its streak, which is one of the principal means of differentiating it from hematite, is yellow-brown.

Frequently, hematite, goethite, and other less common hydrous iron oxides occur together. In this combination the name *limonite* is used.

**Fig. 3-21.** Cubic fluorite crystals adorn a mass of smaller calcite crystals. The specimen, which is from France, measures 14 cm (5.5 in.) across. (M. Halberstadt)

**Fig. 3-22.** Typical scalenohedral crystal form of calcite; an intergrown crystal. The form takes its name from the fact that each face has the outline of a scalene triangle. (M. Halberstadt)

**Halides** The halides are mineral compounds that form when cations combine with the halogen elements—chlorine (Cl), bromine (Br), fluorine (F), and iodine (I). The two most common minerals in this group are *halite* (NaCl) and *fluorite* ($CaF_2$).

Halite is transparent to translucent and possesses three perfect cleavages at 90° and a hardness of 2.5. Although usually white or colorless, it can occur in pastel shades of yellow, red, purple, and blue. Perhaps its most diagnostic property is its salty taste; in fact, its name stems from the Greek word for "salt."

Fluorite, also transparent to translucent, possesses a hardness of 4 and a vitreous luster (Fig. 3-21). Its color varies widely but often it displays a light green, yellow-green, or purple cast. Its name conveys the fact that many varieties of this mineral **fluoresce**—that is, glow when struck by ultraviolet light. Unlike nearly all other minerals, fluorite cleaves perfectly in four directions. If care is taken, fluorite can be shaped by careful cleaving into exquisite octrahedra (eight-sided forms).

**Carbonates** Carbon has a tendency to link covalently with three oxygen atoms to form the stable carbonate ion $CO_3$, which has a charge of $-2$. Carbonate ions can join, mostly ionically, with metal cations (including mainly calcium, magnesium, and iron) to form the various members of the carbonate group. Two of the most important minerals in this group are *calcite* and *dolomite*. Calcite ($CaCO_3$) is normally a light-colored (white, pale yellow) or colorless mineral, although the color may range from yellow and orange to brown and black, depending on the amount and nature of the impurities present. Its name is derived from the Greek word for limestone, *khálix*. Calcite has a vitreous luster, a hardness of 3, and occurs in a variety of crystal forms, which commonly are six-sided (Fig. 3-22). It possesses nearly perfect cleavage in three directions (see Fig. 3-12), and the intersections of the cleavage planes produce a rhombohedral pattern—that is, when the mineral breaks into fragments, each of the faces is rhombic or approximately diamond-shaped.

One type of calcite is called *onyx marble* or *Mexican onyx*. Because of its fine-grained texture and relative softness, this mineral is particularly suitable for sculpting.

Dolomite ($CaMg(CO_3)_2$) resembles calcite su- perficially. To distinguish one from the other, place the unknown mineral in cold, dilute hydrochloric acid. Calcite reacts readily, with a vigorous release of bubbles, whereas dolomite reacts much more quietly and slowly. Dolomite also has a slightly greater hardness (3.5), a higher specific gravity, a slight curve of its crystal faces, and most often a pearly luster. The mineral was named after Silvain Dolomieu (1750–1801), a French geologist and mineralogist.

**Sulfates** Surrounding itself with four covalently bonded oxygen atoms, sulfur can produce a stable configuration—the sulfate ion—that acts as the basic building block in the sulfate mineral group. All told, there are about 10 minerals in this group, of which *anhydrite* and *gypsum* are the most abundant.

Anhydrite ($CaSO_4$) is colorless to bluish white or violet as a rule, possesses a hardness of about 3, and has three cleavages at 90° that vary from perfect to good. It commonly occurs in crystalline masses. Anhydrite can be differentiated from calcite by its three 90° cleavages and by its higher specific gravity. Moreover, anhydrite does not fizz when immersed in dilute acid. It differs from gypsum particularly in its greater hardness and in those three cleavages at 90°.

Gypsum ($CaSO_4 \cdot 2H_2O$) represents the hydrated form of calcium sulfate. Its hardness is only 2; its luster is vitreous to pearly, at times silky; it ranges from colorless to white, to various shades of yellow, red, and brown. Gypsum possesses one perfect cleavage and two less well-developed ones that in some cases lead to a fibrous form with a silky luster called *satin spar*. Alabaster is a fine-grained massive variety of gypsum that is sometimes used in sculpture.

## ROCKS

Most rocks are composed of a number of different mineral grains differing in their properties, such as their color, luster, grain size, and grain shape (Fig. 3-23). These grains in most cases are relatively tiny, making many of the properties of the various minerals hard to assess. When the crystal in question is no larger than a grain of rice, it is difficult to determine its luster, hardness, number of cleavages and the number of angles at which these

**Fig. 3-23.** A rock composed of a number of different minerals, some black, some gray, some white. The individual grains of each mineral can be differentiated from those of other minerals on the basis of their physical properties. When the crystals are very small, this process can be extremely difficult and exasperating. (Edwin E. Larson)

meet. However, with persistence, nearly all the minerals in any common rock can be identified. Most rocks are composed of only a few minerals and differ not only in the kinds and abundance of their minerals, but also in **texture,** which pertains to grain size and how the grains fit together.

Geologists have found that most differences in mineralogy and texture are related to the way rocks have formed. By studying rocks and comparing their similarities and differences, it is possible to obtain clues about formational processes and events. In fact, the study of rocks has become our principal means of understanding the earth.

As an aid to comparing similarities and differ-

ences among rocks, geologists have devised a rock classification system. Biologists have done the same thing for the animal and plant kingdoms. To be sure, the scheme—a form of pigeonholing—is arbitrary, and inevitable exceptions, borderline cases, and overlaps exist. But for our purposes it works reasonably well for most rocks. Fundamentally, the classification is *genetic:* based on the origin of the rocks. Each major grouping, consisting of rock types that have formed in much the same way, is itself subdivided into successively smaller units. The smallest divisions are based on rock texture, mineralogy, and proportion of mineral constituents, each with a formal name—for example, *gran-*

ite, shale, slate. Once a rock has been classified, the name itself carries most of the genetic, mineralogical, and textural information concerning that rock. This classification enables rocks of all ages from around the world to be easily compared.

Here we describe only the major subdivisions of rock classification; later chapters give detailed descriptions of the rocks within each group and the processes that form them.

All rocks can be placed into one of three main groups: *sedimentary, igneous,* or *metamorphic;* these, in turn, are divisible into second-level categories (Table 3-4).

**Table 3-4** Principal Divisions of the Genetic Rock Classification

| Prime Divisions | Secondary Divisions | |
| --- | --- | --- |
| **Sedimentary** Formed near the earth's surface by depositional processes | Clastic | Fragments of variable sizes and shapes derived from the breakup of preexisting rocks |
| | Crystalline | Result of precipitation of carbonates, sulfates, and chlorides from ocean and lake waters |
| | Bioclastic | Fragments of hard parts (usually shells) of organisms |
| **Igneous** Formed by solidification of molten material (magma) | Volcanic | Formed of solidified magma resulting from a volcanic eruption |
| | Plutonic | Crystallized from magma beneath the earth's surface |
| **Metamorphic** Formed beneath the surface by growth of new minerals in response to heat, pressure, and volatiles | Foliated | Possess a layered appearance related to the formation of new minerals in bands generally perpendicular to directed pressure |
| | Nonfoliated | No obvious metamorphic banding of minerals |
| | | Generally indicative of environment free from directed pressure |

## Sedimentary Rocks

Of the three genetic rock families, **sedimentary rocks** are the most easily understood because most of the processes responsible for their formation are observable at or near the earth's surface. Sedimentary rocks cover about 75 percent of the earth's surface, forming a thin, discontinuous veneer over the more-abundant igneous and metamorphic rocks, which are the true foundations of our continents and ocean basins.

Many sedimentary rocks can be thought of as secondary, or derived, rocks in that they are composed of bits and pieces of preexisting rocks held firmly together by a cementing mineral. The resulting texture is called *clastic* (from the Greek word for "broken"). Examples of clastic sedimentary rocks are (1) *sandstone,* which consists of cemented sand grains; (2) *conglomerate,* which consists of larger (gravel-sized), rounded, cemented fragments; and (3) *shale,* which consists of very small indurated, laminated particles of clay. Another class of sedimentary rock results from chemical precipitation in lake or seawater (e.g., anhydrite and halite); some result from the accumulation of a variety of organic remains, particularly shell fragments (bioclastics).

Sedimentary rocks, which form on land and the floors of lakes and seas, are built up through the slow deposition of material, layer on layer. The layers are called *strata;* a single layer is a *stratum* (from the Latin, meaning "blanket" or "pavement"). Individual layers may range from paperthin sheets to beds tens-of-meters thick. Some beds are limited to small areas; others can be traced for many kilometers.

## Igneous Rocks

**Igneous** (Greek for "fire") **rocks** are those that have solidified from a molten, silica-rich liquid to which the name *magma (melt)* has been given. All magmas originate deep below the earth's surface where temperatures are relatively high. If magmas find a path to the surface, they erupt, cool, and solidify to form *volcanic rocks.* Magma, which never reaches the surface, ultimately cools and crystallizes below the surface to form *plutonic rocks.* The adjective is derived from the name for the Greek god of the lower world, Pluto.

In most igneous rocks, the various mineral grains form an interlocking network of crystals—some with crystal faces, most without. This network, called a **crystalline texture,** results from the progressive and simultaneous growth of many mineral grains during solidification, or crystallization, of the magma. The mineralogy, average grain size, and differences in grain size give clues to the conditions under which crystallization occurred. For example, beneath the surface, magma cools and crystallizes more slowly than it does at the surface. For that reason the mineral grains in a plutonic rock generally grow much larger than those that crystallize quickly during the rapid cooling of a lava flow. A common plutonic rock is a *granite,* whereas an equivalent volcanic rock is a *rhyolite.* These and other common igneous rocks are described in chapter 7.

## Metamorphic Rocks

**Metamorphic rocks** are products of heat, pressure, and fluids acting inside the earth on preexisting rock material. The word *metamorphic* is from the Greek, meaning "change in form." Beneath the surface a preexisting rock, such as shale, can undergo recrystallization, to form a metamorphic mineral aggregate, such as slate. In some cases, instead of being randomly oriented, as is true of many igneous rocks, the resultant minerals may become aligned in layers parallel to one another.

This layering is called **foliation** (from the Latin *folium,* meaning "leaf"). Foliation development can range from weak to strong. In some metamorphic rocks the foliation is so strongly developed that it superficially resembles the stratification of sedimentary rocks. These metamorphic layers, however, consist of interlocking crystals segregated into layers of dark minerals and layers of light-colored minerals. In some cases the long axes of elongate minerals (like hornblende) form in parallel alignment to produce a **lineation.**

In most types of metamorphism, the rock undergoes little or no change in chemical composition through mineral recrystallization. The elements originally present simply regroup themselves under conditions of higher temperatures and pressures to form new minerals that are stable in the new subsurface environment. In some cases, however, heated gases and fluids .circulating within the earth—often associated with plutonic igneous activity—introduce new material into the recrystallizing material.

Metamorphic rocks are complex because they can be made from any of the preexisting rock types: igneous, sedimentary, or even previously metamorphosed rocks. Moreover, they have no single mode of origin—in certain cases temperature is the important factor; in others, directed pressure; in some, the nature of the fluid phases; and in still others, all of these factors are significant.

## SUMMARY

1. An *element* is composed of the same kinds of atoms; a *compound* is composed of groups of different kinds of atoms.
2. Atoms consist of *electrons, protons,* and *neutrons.* In the atoms of a particular element, the number of electrons equals that of protons and is a fixed number *(atomic number).* But the number of neutrons can vary to a small degree.
3. Atoms bond together through *ionic, covalent,* and *metallic bonding* processes that involve interaction of the outer-shell electrons. Molecules can also bond together through electrostatic forces.
4. Matter can exist in three different states: *gas, liquid,* and *solid.* The particles that make up a gas are *randomly ordered;* those in a liquid possess *short-range ordering;* those in a solid exhibit *long-range ordering* or *crystallinity.*
5. *Minerals* are naturally occurring inorganic crystalline compounds and elements of a definite chemical composition or a restricted range of chemical compositions. The crystalline nature of minerals has been established through *X-ray diffraction* studies.
6. Each mineral has a definite set of properties that can be used to identify it. Those properties commonly used are *color, cleavage* (or lack of it), *hardness, luster, specific gravity,* and *magnetic properties.*
7. Oxygen and silicon make up almost 75 percent of the earth's solid outer layers. Accordingly, the most abundant minerals in the outer layers *(silicates)*

consist in large part of these two elements. The basic building block of silicates, the SiO₄ *tetrahedron,* can exist singly or in any array of chains, rings, sheets, and three-dimensional networks.

8. Important rock-forming minerals are relatively few in number and include the following silicates: *quartz, potassium* and *plagioclase feldspars, muscovite* and *biotite micas, hornblende, augite, olivine, garnet, chlorite,* and *clay.* Four of these, called ferromagnesian silicates, are relatively rich in iron and magnesium: biotite, hornblende, augite, and olivine. Nonsilicates, although less abundant than the silicates, are, nonetheless, geologically important. Significant nonsilicate mineral groups include the native elements, sulfides, oxides, halides, carbonates, and sulfates.

9. Rocks are composed of aggregates of one or more minerals, the grains of which fit together in various ways *(texture).* Most mineralogical and textural differences are related to the way in which different rocks form.

10. The system of rock classification is genetic. The smallest units in the classification are based on the rock's texture, mineralogy, and the proportions of its mineral constituents. All rocks fall into one of three main categories: *sedimentary, igneous,* and *metamorphic.*

11. Most *sedimentary rocks* are made of fragments *(clasts)* of preexisting rocks cemented together. Some are the result of chemical precipitation or accumulation of organic remains. The source of *igneous rocks* is a molten material called *magma.* Magma solidifying below ground forms *plutonic rocks;* on the surface it forms *volcanic rocks. Metamorphic rocks* form beneath the surface from the recrystallization of preexisting rocks under the influence of heat, pressure, and the chemical activity of fluids.

## QUESTIONS

1. Describe the parts of an atom and outline the various ways atoms bond together.
2. What is a crystalline substance? Are all minerals crystalline substances? Are all crystalline substances minerals?
3. Describe the various physical properties of minerals. Which are fundamental properties? Which properties are related to the crystal lattice?
4. How are the crystal form and cleavage of any mineral related? Why is there a lack of any type of cleavage in quartz?
5. What are the most common elements of the earth's crust? How do these elements bond together to produce the common minerals?
6. List 10 silicates and 10 nonsilicates; give a thumbnail sketch of the physical properties of each.
7. What are the distinguishing characteristics of each of the following mineral pairs: potassium feldspar/plagioclase; muscovite/biotite; hornblende/augite; calcite/dolomite; magnetite/hematite; olivine/quartz; quartz/potassium feldspar; calcite/anhydrite; anhydrite/gypsum; halite/anhydrite; anhydrite/fluorite; gold/pyrite.
8. Name the three classes of rock and give the characteristic features of each. How is each class formed?

## SELECTED REFERENCES

Berry, L.G., and Mason, B., 1959, Mineralogy, W.H. Freeman, San Francisco.

Bloss, F.D., 1971, Crystallography and crystal chemistry, Holt, Rinehart & Winston, New York.

Bragg, L., 1968, X-ray crystallography, Scientific American, vol. 219, no. 1, pp. 58–79.

Dana, E.S., and Ford, W.E., 1949, Dana's textbook of mineralogy, Wiley, New York.

Deer, W.A., Howie, R.A., Zussman, J., 1978, An introduction to the rock-forming minerals, Longman Group, London.

Desautels, P.E., 1968, The mineral kingdom, Grosset & Dunlop, Madison Square Press, New York.

Ehlers, E.G., and Blatt, H., 1982, Petrology: igneous, sedimentary, and metamorphic, W.H. Freeman, San Francisco.

Hurlbut, C.S., Jr., and Klein, C., 1977, Manual of mineralogy, 19th ed., Wiley, New York.

Mason, B., 1958, Principles of geochemistry, Wiley, New York.

Pough, F.H., 1976, Field guide to rocks and minerals, 4th ed., Houghton Mifflin, Boston.

Vanders, I., and Kerry, P.F., 1967, Mineral recognition, Wiley, New York.

# PART II

# Internal Processes

**Fig. 4-1.** Continental drift in action. The coastlines of the Arabian Peninsula and the Somali Republic in northeast Africa fit together so well that there is little doubt that at one time the two regions were joined and have since split apart from one another. (NASA)

# 4

# CONTINENTAL DRIFT AND PLATE TECTONICS

Earth science has been through a revolution as profound as that which rocked biology in the mid-1800s when Darwin presented his paper on evolution or that which rattled physics in the early 1900s when non-Newtonian quantum mechanics was introduced. The revolution has changed the way scientists view the earth and its internal processes. In the 1700s and 1800s geologists were concerned mainly with the continents and knew little about the oceans. Continents were viewed as fairly static features. It was thought that they might have grown slightly with time through marginal accretion and moved slowly up and down, but by and large they were locked in place on the face of the earth.

Except for the rare individual, geologists concentrated on understanding events and processes of local extent, not on those that might affect large portions of the globe or the earth as a whole. This approach changed drastically with the revolution. The new philosophy that emerged, the ***plate-tectonics theory,*** came about largely from an acceptance of continental drift (Fig. 4-1) and study of the ocean basins. Earth scientists completely reinterpreted geological history in terms of the new theory. And geological thought became unified.

Despite the theory's popularity, we should realize that it is only a more modern way of looking at the earth. The older fixed-continents theory was appropriate at the time of its acceptance. As new data and techniques become available, older ideas are succeeded by newer ones. Plate tectonics probably will be improved on. The theory has already been modified, and other changes will undoubtedly be necessary. The real test of the theory will come with time. Thus, what is perceived as revolutionary thought currently may well become tomorrow's dogma. Such is the way scientific thought advances—in fits and starts.

## CONTINENTAL DRIFT

The idea that there was once one great landmass that subsequently broke into drifting continental-size fragments is not new. Yet most of the early suggestions of continental drift were of little substance, having been made after a cursory examination of a world map.

In 1620 Sir Francis Bacon (1561–1626), in his book *Novum Organum,* commented that the west coast of Africa and the east cost of South America were too similar in shape to be the result of fortuitous processes. Later, in 1658 the Frenchman François Placet also remarked that the similarity of these coastlines suggested that the continents had been joined together at one time, and he put forth the idea that their separation may have occurred during the biblical flood. About 200 years later, Antonio Snider-Pelligrini wrote that the continents bordering the Atlantic must have once been contiguous, basing his conclusion on the similarity in fossil plants in North American and European coalbeds. He even made a graphical reconstruction of the supercontinent. By 1908 two scientists, F.B. Taylor and H.B. Baker, had reached the same conclusion from their studies of the distribution of mountain chains around the world.

But all of these ideas were presented in such a casual manner that they evoked little support and even less interest. Moreover, they ran contrary to classical geological thought. In 1900 most geologists believed that the continents were permanently fixed in position; they regarded as pure fantasy the notion that continents could slip and slide about on the earth's surface. But a storm of controversy regarding the positional constancy of continents was brewing, one that would break in the early 1900s and would last for nearly half a century.

Alfred Wegener (1880–1930) was an instructor of meteorology in Germany in 1910. He, too, had observed the coastlines of South America and Africa and was struck by their similarity. Yet it was only a fleeting idea. Then in 1911 he came across an article on the similarity of fossils on the two continents that rekindled his interest in continental drift. Wegener became obsessed with the idea and worked tirelessly to uncover evidence to indicate that the continents around the Atlantic Ocean were once one huge landmass. He finally published his findings, with supporting evidence, in 1915.

Wegener was convinced that this continental mass, which he called *Pangaea* (Fig. 4-2), had split into fragments that began to drift apart. The response to the publication was mild at first, but by 1924 a groundswell of opposition had developed, particularly among geologists in the United States.

R.T. Chamberlin, an American earth scientist, wrote, "Can we call geology a science when there exists such differences of opinion on fundamental matters as to make it possible for such a theory as this to run wild."[*] Even as late as 1944, the prominent American geologist Bailey Willis (1857–1949) held that Wegener's theory was nothing but a fairy tale that should be downplayed because of its deleterious effect on the developing minds of students.

Wegener, in his synthesis, included evidence from physical geology, geodesy, geophysics, paleontology, zoology, and paleoclimatology. He was not an expert in any of these fields and chose, unfortunately, some poor and even inaccurate examples. Seizing on Wegener's errors, the opposition ridiculed his hypothesis mercilessly. They railed against the evils of the drift theory with an almost religious fervor, most of them prefacing their remarks with pleas for open-mindedness. In reality, their outcry was a reactionary response against a change in scientific philosophy.

One of the opposition's main points was that there was no plausible mechanism for continental movement. Wegener had proposed two mechanisms: one related to gravitational attraction between the continents, the moon, and the sun; the other related to centrifugal pull on the spinning earth. But almost everyone agreed that the forces produced in both instances were woefully inadequate. Most earth scientists, therefore, discounted the hypothesis altogether.

A few scientifically liberal individuals, however, accepted the idea and tried to accumulate data to support it. Perhaps the most capable disciple was the South African geologist Alexander du Toit (1878–1948). He began studies in both Africa and Brazil to ferret out undeniable proofs of continental drift. Du Toit eventually proposed that not one but two supercontinents existed before drift took place. He named them *Gondwana* and *Laurasia.* Gondwana supposedly consisted of all the continents now in the Southern Hemisphere and Laurasia consisted of North America, Greenland, Europe, and Asia. An oceanic waterway he called the *Tethys Sea* separated the two landmasses.

The great Scottish geologist Arthur Holmes

---

[*]Chamberlin, R.T., 1928, "Some of the Objections to Wegener's Theory," *in* W.A.J.M. van Watershoot van der Gracht, ed., Theory of Continental Drift, *American Association of Petroleum Geologists,* Tulsa, Oklahoma, pp. 83–87.

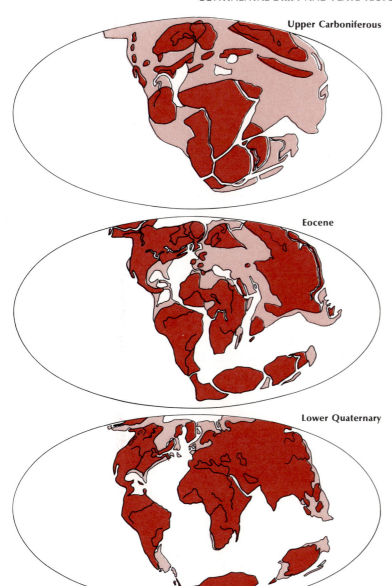

**Upper Carboniferous**

**Eocene**

**Lower Quaternary**

**Fig. 4-2.** Breaking up of the world's continents according to Wegener. He called the original giant continent *Pangaea*. The light areas are deep ocean, the gray areas land, and the stippled areas shallowly submerged continental shelf.

(1890–1965), also impressed by Wegener's idea, suggested in 1927 that **thermal convection** in the earth's mantle might be responsible for the movement of continents (Fig. 4-3). Heat can be transferred in three different ways: by radiation, conduction, and convection. Convection entails the physical transport of hot, less-dense substratum upward in a gravity field, such as that of the earth. This rise of material initiates a lateral inflow of matter to take the original material's place, thus initiating more or less circular flow lines called *convection cells* (Fig. 4-3). Within these cells, material rises, moves laterally, or sinks. Convection cells can readily be observed by heating water or soup. In liquids such as this, flow is relatively rapid. Flow can also occur in solids, but it is slow and not easily observed. Holmes proposed that lateral flow associated with slow convection in

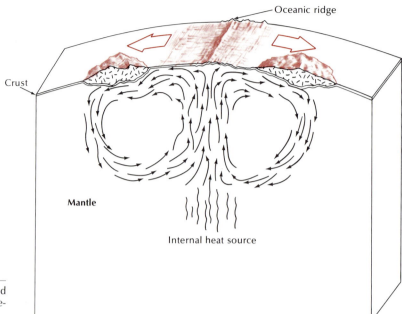

**Fig. 4-3.** Hypothetical thermal currents and convection cells in the mantle and their relation to continental drift.

the solid mantle accounted for lateral drift of the continents.

At last the advocates of the drift theory could point to a mechanism powerful enough to move continents. Most earth scientists presently support the idea to some degree, but no one knows the extent to which the solid mantle flows convectively.

As evidence increased to support Wegener's hypothesis so, too, did the number of his adherents. In time many influential geologists, such as Felix Vening Meinesz (1887–1966), S. Warren Carey, and Sir Edward Bullard (1907–80), were listed among its supporters; thus the theory became impossible to ignore.

## Evidence for Continental Drift

**Fit of Continental Coastlines**   One of the strongest pieces of evidence was the simple observation of how well continental coastlines fit together. The earliest reconstructions, however, were made on maps: two-dimensional representations of a three-dimensional globe that distort earth features to some degree. So, the first fits of continents were quite crude (Fig. 4-4), as the opposition was quick to point out. Some of the poor fit, of course,

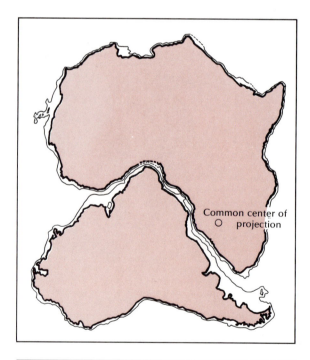

**Fig. 4-4.** Fit of South America and Africa 185 m (600 ft) below sea level, as proposed by Wegener. Sea-level configuration of continents is shown by heavy line. Although there are gaps between the continental margins, the fit is fairly good. (After S.W. Carey, 1958, "A Tectonic Approach to Continental Drift," in S.W. Carey, ed., *Continental Drift, A Symposium*, Geology Department, University of Tasmania Press, pp. 177–355)

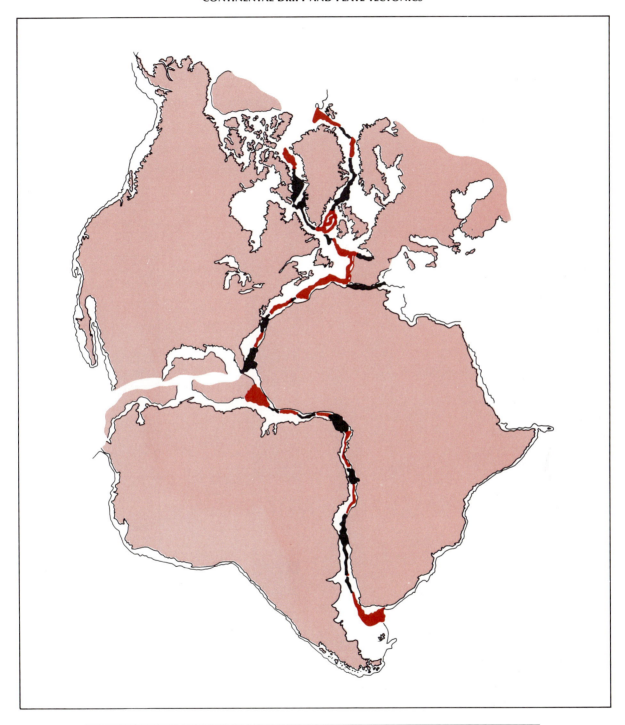

**Fig. 4-5.** Computer fit of all continents bordering the Atlantic. The thickness of the black line represents the amount of overlap or gap. Fitted at a level 1000 m (3280 ft) below sea level. (After E.C. Bullard et al., 1965, "The Fit of Continents Around the Atlantic," in P.M.S. Blackett et al., eds., *A Symposium on Continental Drift,* Philosophical Transactions Royal Society, London, vol. 100, pp. 41–51)

resulted from matching the landmasses at the coastlines. Soon it was realized that erosion can greatly modify coastal margins; subsequently, the continents were fitted at a line below sea level, a line that more truly represented the shape of the continent. In 1955 Carey showed that South America and Africa could be fit together neatly at a level 2000 m (6560 ft) below sea level.

In 1965 Bullard and two associates, using a computer, obtained an even better fit of the continents around the Atlantic. They chose 1000 m (3280 ft) below sea level as their depth-of-fit line and were amazed at how slight were the gaps and overlaps (Fig. 4-5).

**Stratigraphic and Structural Similarities** If the continents dovetailed, should not significant geological features then also match across the continental boundaries? First proposed by the Japanese geoscientists, Seiya Uyeda, such an idea could be

**Fig. 4-6.** Generalized columnar section of rocks from southeastern Brazil and southwestern Africa. Notice the great similarity in sections up to about 100 million years ago. (After D. Tarling and M. Tarling, 1971, *Continental Drift,* Doubleday, p. 34)

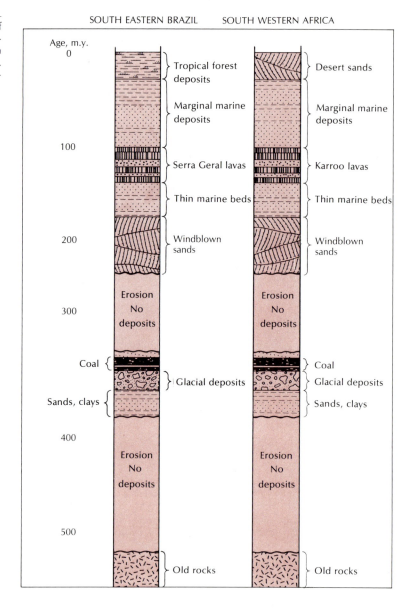

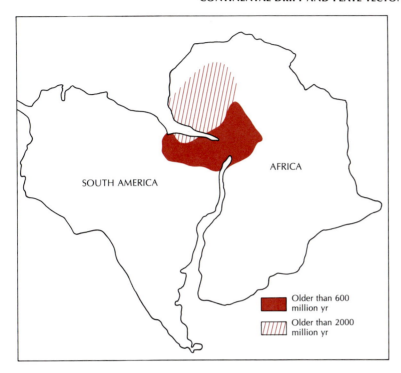

SOUTH AMERICA

AFRICA

■ Older than 600 million yr

▨ Older than 2000 million yr

**Fig. 4-7.** Matching the boundary between rocks of different ages in Africa and South America. The age boundary extends from the vicinity of Accra, Ghana, to that of São Luís, Brazil. (After P.M. Hurley, 1968, "The Confirmation of Continental Drift," *Scientific American*, vol. 229, no. 4, pp. 60–69)

termed the *torn newspaper concept:* For example, if a page of newsprint is ripped into several pieces and then reassembled, the pieces should not only fit back together, but the lines of print should continue across the torn boundaries as well.

The truncated rock record on two or more continents was one feature that could be matched. Figure 4-6 shows two generalized stratigraphic sections, one from southeast Brazil, the other from southwest Africa. The records should be very similar if the fit of the continents is valid—and they are nearly exact in detail. Evidence such as this provided strong support for the concept of continental drift.

In the late 1960s Patrick Hurley, an American geochronologist, found that in the vicinity of Ghana in West Africa a distinct boundary between rocks of two different ages exists: one more than 2000 million years, the other about 600 million years. The boundary trends southwestward to the west coastline of the continent. Hurley reasoned that the same boundary might be found in South America, on the east coast of Brazil. Indeed, a team of American and Brazilian scientists delineated a

boundary between 2000- and 600-million-year-old rocks that appeared to be an extension of the African trend (Fig. 4-7).

Segments of mountain chains in Europe, Greenland, and North America all are about the same age. One segment runs through Scandinavia, Scotland, and Northern Ireland; one through eastern Greenland; and one through eastern North America. If the continents are reassembled as shown in Figure 4-5, then most of the mountain segments fit together in a nearly continuous chain. Again, the evidence supports the idea of drift, but by itself is not conclusive.

**Paleoclimatology** Portions of all the continents in the present-day Southern Hemisphere were affected by glaciation during the Pennsylvanian and Permian periods (Fig. 4-8A). Many of the glacial deposits and the polished and scoured bedrock are located close to the equator, where today continental glaciation is most unlikely. Moreover, from the alignment of glacial grooves and the orientation of chattermarks left on the bedrock, it appears that the ice formed in the oceans and

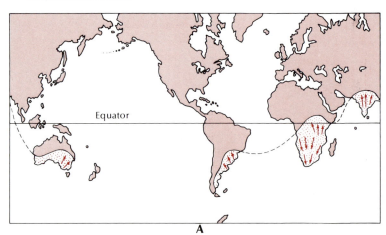

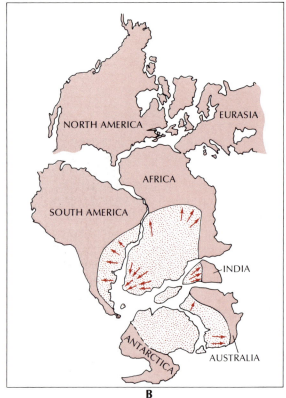

**Fig. 4-8.** Distribution of late Paleozoic glacial deposits as seen *(above)* **A.** on a present-day map and **B.** on a map in which continents have been reassembled according to Wegener. Directions of glacial flow are indicated by arrows. On the present-day map, the glaciers appear to originate in the oceans; in the reconstruction, glacial directions form reasonable patterns, indicating flow outward from a single large ice sheet. (After H. Takeuchi, S. Uyeda, and H. Kanamori, 1970, *Debate About the Earth,* Freeman, Cooper)

flowed toward land. Yet geologists know of no modern or relatively recent glaciers that behave in this way. Rather, ocean-margin glaciers form on the continental uplands and flow seaward.

If one reconstructs Pangaea (or Gondwana), however, all the glaciated regions fit together into a single large area unbroken by seaways (Fig. 4-8B), and the unusual glacial markings can be explained.

From a study of the rocks on the different continents, it is also possible to obtain an idea of the climates that existed when the rocks were laid down. The distribution of the mapped paleoclimates form a disjointed, checkerboard pattern on the continents as they now stand. When the continents are pieced back together, however, the patterns appear continuous.

**Fossils** If all the continents had been one in the past, animals and plants would have been able to migrate freely, restricted only by the boundaries of their natural habitats and such impassable features as mountain ranges. Fossils found on the various continents today, then, should show a similarity in ancestry, dating back to the time when the continents were joined. Wegener and his followers attempted to point out such similarities, noting, for example, that 64 percent of Carboniferous and 34 percent of Triassic reptile groups are the same on the southern continents. But data in the early 1900s were sketchy.

As more and better paleontological data have accumulated, similarities in fossil forms have been shown to exist between the continents. As recently as 1969 fossil land-restricted amphibia and reptiles of Triassic Age, discovered in Antarctica, were determined to be exactly like Triassic fossils from Africa, Madagascar, and Australia.

**Paleomagnetism** Before discussing paleomagnetic evidence, it is necessary to explain the earth's magnetic field—an invisible force field that everywhere permeates the earth and extends outward from its surface for hundreds to thousands of kilometers. The best way to visualize the field is to place a bar magnet, which has two magnetic poles, on a piece of paper at the center of a circle representing a cross section of the earth (Fig. 4-9). Invisible lines of force, the number being determined by the strength of the magnet, converge at one magnetic pole and diverge at the other. Each line of force is continuous, so that a line emerging from one pole will bend through space and eventually reenter the magnet at the other pole, as depicted in Figure 4-9. The illustration shows that the lines are essentially parallel to the earth's surface at the equator but that their angle of intersection with the surface increases toward the poles. At a spot above the earth's north magnetic pole, the field lines point vertically downward. A similar angular relationship is shown by the lines that emerge south of the equator. Because the earth is three-dimensional, lines of force like those shown in cross section completely encircle the earth. Such a field configuration is called a **dipole field.** The earth's dipolar field is generated within the metallic core. (We speculate on the likely means of generation in chap. 5.)

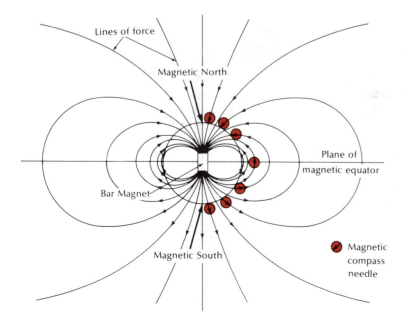

**Fig. 4-9.** Field lines around a bar magnet, shown in cross section through the bar's axis. Arrows on the field lines indicate the direction of the magnetic field; the magnetic compass needle is parallel to them. Notice that the inclination of the compass needle is horizontal at the magnetic equator and vertical at the magnetic poles.

Lines of force

Magnetic North

Bar Magnet

Plane of magnetic equator

Magnetic South

Magnetic compass needle

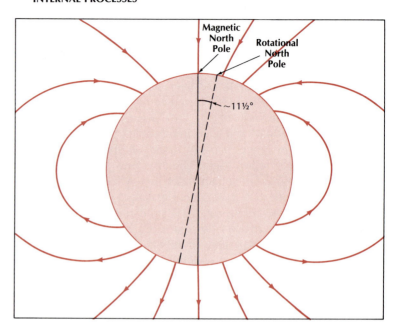

**Fig. 4-10.** Present-day angular offset of the dipolar axis of the earth's magnetic field with respect to the rotational axis.

To map the dipolar magnetic field, we can use a small magnetic compass and move it about from place to place along the heavy outline circumscribing the bar magnet. The compass needle will align parallel with the field lines at any particular point, and a unique orientation of the needle will result (Fig. 4-9). The north end of the needle will always point north, but the angle that the needle makes with the heavy line representing the earth's surface will change with location. If one can determine magnetic north and the magnetic inclination, it is possible to calculate the position of the magnetic pole.

It has long been known that the magnetic and rotational poles are not coincident. At present, the north magnetic pole lies at 78 1/2° N latitude, 69° W longitude (Fig. 4-10), a situation that has prevailed since at least the early 1830s when the field was first described in detail. This is thought to be a short-lived anomaly that will slowly change over the next several thousand years. This conclusion is based on the assumption that the earth's magnetic dipolar axis is coincident with the rotational axis when the field directions are averaged over sufficiently long periods of time (tens- to hundreds-of-thousands of years). It follows from this assumption that, if the average position of the

magnetic pole for an ancient geological time can be determined through paleomagnetic studies of rocks, the position of the ancient geographical pole can also be determined.

Almost all rocks contain some minerals that can become permanently magnetized. The most common is magnetite ($Fe_3O_4$), which in its pure state will retain its permanent magnetization *(remanence)* up to a critical temperature—580° C (1076° F)—known as its ***Curie temperature.*** Above 580° C, the mineral will be virtually nonmagnetic. Thus, when a lava starts to cool and solidify, it will remain unmagnetized until its temperature drops to 580° C, the Curie temperature of magnetite. As the temperature continues to decline, the lava will acquire a permanent magnetization parallel to the earth's magnetic field lines at that locality and at that time, which is called ***thermoremanent magnetization (TRM).*** The TRM is figuratively frozen in the lava, so that the lava will record the initial field direction acquired during cooling regardless of later changes in the earth's magnetic field.

Generally, the lava cools and acquires its remanence in only a few weeks, so that the magnetic directions frozen in are only for that geological moment. To determine average dipolar directions,

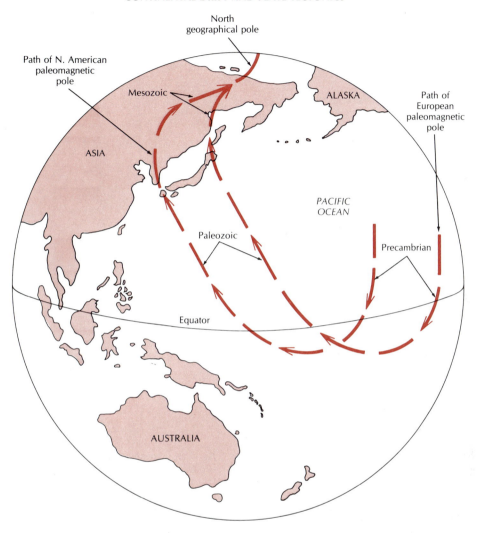

**Fig. 4-11.** Two apparent polar wander curves from the Precambrian Eon to the present, one based on data from North American rocks and one based on data from European rocks. The paths are similar in shape but not coincident in position, suggesting that North America and Europe have been drifting since the Triassic Period. (After A. Cox and R. Doell, 1960, "Review of Paleomagnetism," *Geological Society of America Bulletin*, vol. 71, p. 758)

paleomagnetic measurements must be made on many different flows from the same area. Only then can the best estimate of the average north paleomagnetic pole be made, which should be a close approximation of the paleogeographic pole. Such information can then be used to reconstruct the drift paths of the continents through time.

Sedimentary rocks can also acquire permanent magnetization. For example, it is possible for small magnetic clasts (generally magnetite) that accu-

mulate during sedimentation to orient themselves like tiny compass needles in the intergranular space between silicate grains just after deposition in lakes and oceans while the sediments are still slushy. In the paleomagnetic study of a sedimentary rock, many samples are taken from a section of strata, each sample representing a different time (as in the study of lava flows), and the average north paleomagnetic pole is calculated.

After initial rock formation, chemical processes

may also lead to the formation of magnetic iron oxides (particularly hematite) and the development of *chemical remanent magnetization* (CRM), which parallels the earth's field at the time the magnetic minerals were formed.

In the 1940s and 1950s paleomagnetism was a growing branch of geophysics. One of the leaders in the field was the English geophysicist Stanley K. Runcorn. He and his associates measured the fossil magnetism in many rock samples ranging in age from late Precambrian to Recent. From the averaged data, they calculated the position of the north rotational pole for the rocks of each geological age. When the poles for European rocks were plotted, it was found that they fell on a curved path that ran from south of the equator to the North Pole (Fig. 4-11). The younger the pole, the closer it was to the present-day geographical pole. The plot, called an *apparent polar wander curve,* seemed to indicate that, with regard to Europe, either the North Pole had moved or wandered or that the European continent had shifted and rotated in position with time. Today most earth scientists favor the latter explanation.

When Runcorn and his associates plotted the paleomagnetic poles for North American rocks covering the same time span, they obtained a similar apparent polar wander curve, but it was displaced about 40° toward the west from the European curve (Fig. 4-11). The similarity between the two curves suggested that both Europe and North America had undergone similar changes in latitude and rotation relative to the pole. Runcorn attributed the discrepancy between the two curves to continental drift. When North America was "moved" back next to Europe, the two curves were nearly coincident over most of their paths from the Precambrian through the Triassic. Then the paths diverged. Runcorn concluded that continental drift must have been initiated in the Triassic and that it had been going on ever since.

Runcorn published his results in 1962 when the theory of continental drift was still unacceptable to most earth scientists. The evidence appeared so incontestable that it swayed many scientists who had been fence-sitting until that time. Additional paleomagnetic studies of rocks of all ages from all the continents have confirmed the first results.

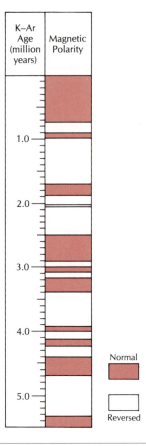

**Fig. 4-12.** Magnetic polarity time sequence back to about 5.4 million years ago, based on potassium/argon (K/Ar) dating of lava flows with known paleomagnetic directions. Notice that the field has repeatedly changed back and forth between normal and reverse polarity. (After A. Cox, 1969, "Geomagnetic Reversals," *Science,* vol. 163, p. 237)

**Linear Ocean-Floor Magnetic Anomalies** From time to time, the earth's magnetic field undergoes major episodes of fluctuation—called *magnetic reversals*—during which the dipolar field can die away to almost zero, followed by regeneration with the north and south magnetic poles switched in polarity. Such events occur about every 250,000 to 500,000 years.

The last major magnetic reversal ended about 730,000 years ago. Since then, the field has been "normal," (the same as today's field) except for minor fluctuations. Radiometric dating of lava samples from around the world has established

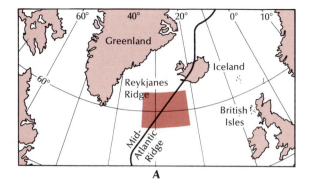

**A**

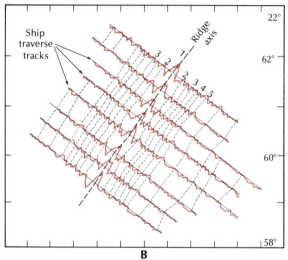

**B**

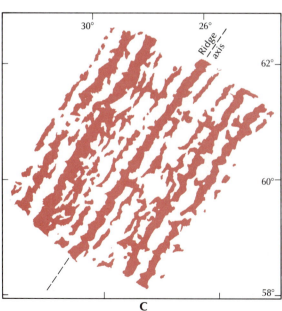

**C**

the sequence of polarity changes in detail for the last few million years (Fig. 4-12). The sequence is called the ***magnetic polarity time sequence.***

The transition from normal to reversed polarity, or vice versa, takes about 1000 to 5000 years. As the field changes over, it remains at a very low intensity.

During the last 20 years or so, oceanographic vessels have been sampling cores of the upper sedimentary layers of the ocean floors. These sediments, deposited during the last several million years, contain a record of magnetic polarity changes in their permanent magnetization, just as the rocks on land do. It has been possible to correlate these records with the magnetic polarity time sequence and thereby date the seafloor cores throughout their length. Sedimentation rates have thus been established for different parts of the ocean basins, and fossils contained in the cores have been accurately dated.

Beginning in the 1960s, many oceanographic research vessels towed magnetometers to measure the earth's magnetic field as they crisscrossed the oceans. After a few years, it became apparent that the intensity of the field over parts of the oceans varied. In some places it was slightly stronger than the average, in others it was slightly weaker. When plotted on a map the strong and weak regions formed alternating linear belts (called ***linear magnetic anomalies***) paralleling the axes of the oceanic ridges. Detailed surveys were then made to see how the pattern varied across the oceanic ridges and rises. Not only were magnetic anomalies linear, but the pattern on both sides of the ridges was

---

**Fig. 4-13. A.** Magnetic anomalies mapped in the area of Reykjanes Ridge, south of Iceland. **B.** The relative variation in the earth's magnetic field determined by ship traverses across the ridge. **C.** The inferred areal distribution of the variations. The black areas represent higher-than-average field intensity, the white areas lower-than-average. There is a bilateral symmetry to the anomalies, such that those on one side of the ridge are mirrored on the other—as indicated by the numbers in the center drawing. (**A.** After J.R. Heirtzler et al., 1965, "Magnetic Anomalies over the Reykjanes Ridge," *Deep-sea Research,* vol. 13, p. 247. **B.** After F.J. Vine, 1960, "Magnetic Anomalies Associated with Mid-Ocean Ridges," in R.A. Phinney, ed., *History of the Earth's Crust, A Symposium,* Princeton University, p. 73, Fig. 6. By permission of Princeton University Press)

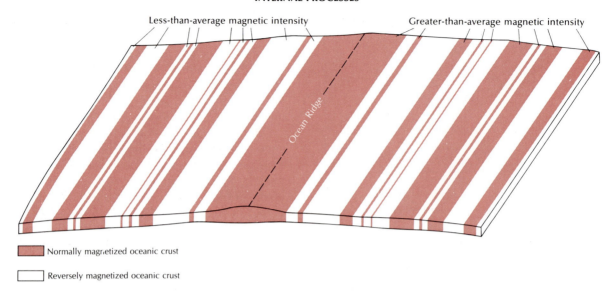

**Fig. 4-14.** An idealized magnetic anomaly pattern in the area of an ocean ridge and the magnetic polarity of oceanic crustal rocks that could account for the pattern.

**Fig. 4-15.** Comparison of the linear magnetic anomaly pattern with the magnetic polarity time sequence, from which it is inferred that the ocean floor is spreading apart. Apparently, lava flows and dikes are added to the region of the ridge axis, where they cool and become magnetized parallel to the field direction at that time. As divergence continues, the field changes polarity, new material is added, and older material is carried away from the axis in both directions.

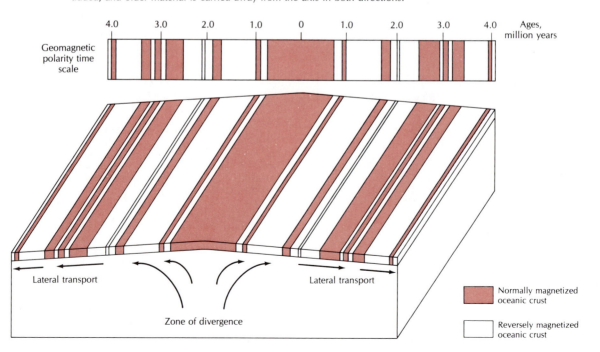

the same, outward from a higher-than-average linear anomaly that ran along the ridge axes (Fig. 4-13).

In 1963 two English geophysicists, F.J. Vine and D.H. Matthews, gave a plausible explanation of the mystery of the linear anomalies. They concluded that parts of the ocean floor beneath the sedimentary layers must be magnetized in a normal direction—that is, parallel to today's field, whereas other parts must be reversely magnetized. The higher-than-average field intensities, they reasoned, corresponded to the normally magnetized stripes, whereas the lower-than-average anomalies corresponded to the reversely magnetized stripes (Fig. 4-14). The explanation was based on the idea that the fossil magnetism that paralleled the present-day field *(normal direction)* would add to the field's total intensity and that the fossil magnetism opposite from today's field *(reversely magnetized)* would detract a little from it. The bilateral symmetry could only mean that for each normal or reversely magnetized stripe on one side of the ridge, there was a nearly identical counterpart on the other side. When the widths of the magnetic stripes on one side of a ridge axis were plotted against the land-based magnetic polarity time sequence over the last 5 million years—the span of accurate radiometric age dates—their relative positions matched nearly exactly, right up to the present (Fig. 4-15). The youngest subsediments were along the ridge axes and became progressively older outward in both directions. The discovery was exciting because it indicated that new material was continually being added to the crust along the ridge axis and subsequently was being split and moved laterally away as if on two outward-moving conveyor belts. Dating of the magnetic stripes provided evidence that the oceanic ridges and rises are zones of divergence and that the ocean basins are young features continually being formed.

By means of the magnetic polarity time sequence, it was possible to date the oceanic linear magnetic anomalies and to determine rates of spreading along the zones of divergence: this proved to be about 2 to 6 cm (1 to 2.3 in.) per year.

When linear magnetic anomalies in parts of the ocean basins were identified and correlated, it was possible to determine when continental fragmen-

tation began and the paths in which the continents drifted.

**Deep Drilling in the Oceans**   If the oceanic ridges and rises are zones of divergence along which spreading occurs, then the ocean floor should become progressively older outward from the ridge crests. With the launching of the oceanic drilling vessel *Glomar Challenger,* it was possible to test this hypothesis. One of the first lines drilled was in the South Atlantic Ocean, where eight sites were chosen, each progressively outward from the Mid-Atlantic Ridge. Drilling was halted when the drill bit penetrated the sedimentary layers and began to bite into the hard basaltic rocks beneath. The cores were dated by the fossils they contained, and the age of the sediments directly overlying the basaltic rocks was taken as a rough estimate of the age of the rocks beneath. Dating the basalts themselves was not reliable because they had

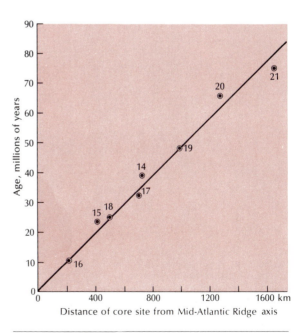

**Fig. 4-16.** The age (determined paleontologically) of ocean sediments immediately above basaltic rock plotted as a function of distance from the Mid-Atlantic Ridge axis. The solid line represents the best fit to the data. (After A.E. Maxwell et al., 1970, ''Deep-sea Drilling in the South Atlantic,'' *Science,* vol. 108, pp. 1047–59. Copyright © 1970 by the American Association for the Advancement of Science. By permission of A.E. Maxwell)

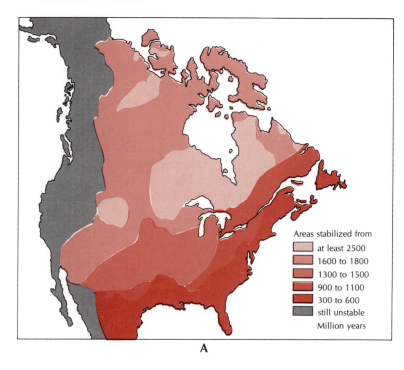

A

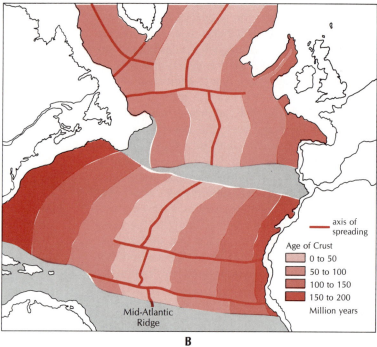

B

**Fig. 4-17. A.** Pattern of ages of continental basement rocks of North America. **B.** Age pattern of basement rocks in North Atlantic Ocean. Continents are of great antiquity, whereas ocean basins are geologically young. Notice in **A** that the oldest rocks are in the interior and that the youngest rocks are near the margin.

been extremely altered through chemical interaction with the ocean waters.

The results (Fig. 4-16), much to the elation of the advocates of drift, showed a continuous increase in age outward from the ridge axis. The slope of the curve, virtually a straight line, indicates uniform spreading from the Mid-Atlantic Ridge at a rate of about 2 cm (1 in.) per year—just about the rate determinable from the linear magnetic anomaly pattern in that part of the South Atlantic.

Continued deep-sea drilling turned up another fact, which at first seemed astonishing—in none of the oceans were the basaltic rocks older than about 150 million years. The ocean basins are all geologically young features! Yet radiometric dating of continental basement rocks shows that parts of most continents are 2500 to 3000 million years

old or more. Somehow the continents have lasted for billions of years, floating about like flotsam and jetsam, whereas the adjacent ocean basins extend no farther back in time than the Jurassic (Fig. 4-17). Why is there no old oceanic crust? We will ponder this question later in the chapter.

### The Theory of Continental Drift in Recent Times

During the 1960s and 1970s much effort was directed toward finding a theory that could account not only for the drifting of continents, but also for all the other internally generated phenomena, including mountain building, genesis of magma, and metamorphism.

Linear magnetic anomalies indicate that new

**Fig. 4-18.** Kamchatka Peninsula and the Kuril Islands arc, showing the relation of earthquakes, volcanoes, and deep-sea trench. (After B. Gutenberg and C.F. Richter, 1954, *Seismicity of the Earth, 2nd ed.*, Princeton University Press)

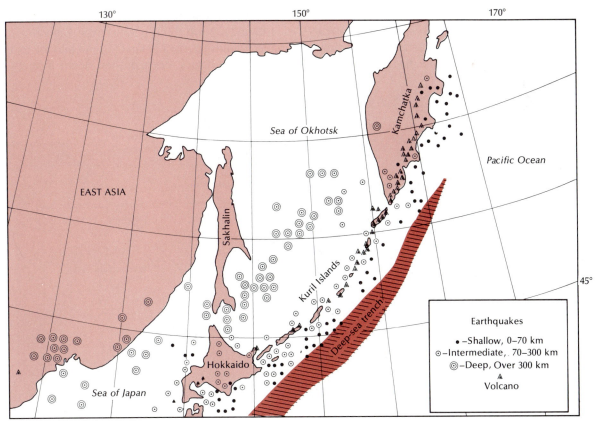

crust is continually being generated along the oceanic ridges and rises, but what happens to it after it is formed? Some geologists have suggested that the earth expands to accommodate the new crust. However, if the rate of expansion in the geological past was anywhere near the rate of recent seafloor spreading, the earth would have been an extremely small body initially, small enough to make that idea unrealistic.

Most geologists, therefore, have accepted the idea that crustal material is being destroyed. In the early 1960s the American geologist Harry Hess (1906–69) suggested that the crust was most likely consumed in the vicinity of the *island arcs* of the world, an idea that was also championed by the American oceanographer Robert Dietz. Island arcs are linear features and the sites of a variety of geological phenomena (Fig. 4-18). The islands themselves generally consist of a string of volcanoes, commonly steep sided and picturesque, like Fujiama. Oceanward from the island chains are linear trenches, the deepest places of the world. Earthquakes are common in island arcs, occurring down to depths of about 700 km (435 mi). When the points of origin of the quakes are plotted on a cross section at right angles to an arc system, they define an inclined seismic zone, the **Benioff zone,**

that begins at the surface near the trench and dips beneath the island arc at a steep angle, commonly about 45° (Fig. 4-19). The zone was named after Hugo Benioff (1899–1968), an American seismologist who pioneered studies on the distribution of earthquakes at continental margins. In most cases, the Benioff zone can be likened to a large thrust-fault zone along which the oceanward block pushes under (underthrusts) the continentalward block.

**New Global Tectonics: a Forerunner Hypothesis** In the late 1960s Jack Oliver and Bryan Isacks, American geophysicists, studied the seismic waves in the Tonga island arc system in the southwest Pacific to determine the possibility that crustal material disappears along island arcs. They recorded earthquake waves on islands both east and west of the deep Tonga trench and found that the apparent speeds of the waves varied in relation to the location of the recording stations. Oliver and Isacks reasoned that this must be the result of differences in the strength of the rocks beneath the island-arc system (Fig. 4-20). They concluded that a layer of high-strength (brittle) material about 100-km (62-mi) thick parallels the Benioff zone and extends nearly 500 km (310 mi) into the earth, seemingly having been pushed down through a

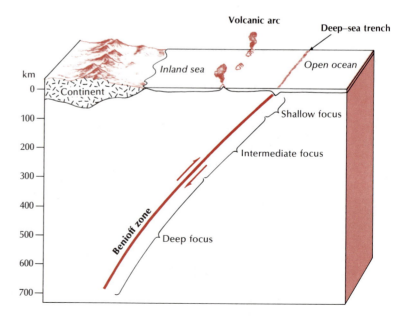

**Fig. 4-19.** The association of a deep-sea trench, a volcanic island arc, and the Benioff zone.

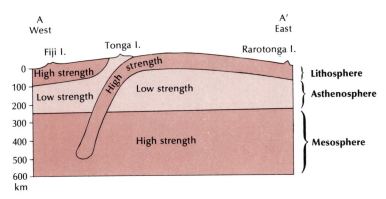

**Fig. 4-20.** A hypothetical cross section, A–A' extending from the Fiji Islands through the Tonga Islands to Rarotonga Island, illustrating from interpreted seismic waves the disposition and strength of the rock layers beneath the islands. (After J. Oliver and B. Isacks, 1967, "Deep Earthquake Zones, Anomalous Structures in the Upper Mantle and the Lithosphere," *Journal of Geophysical Research,* vol. 72, pp. 4272)

softer layer of lower-strength material. They called the outer, brittle layer the **lithosphere** (rock sphere) and the lower-strength zone, which extended from about 100 km (62 mi) down to 250 km (155 mi), the **asthenosphere** (soft sphere). Beneath the asthenosphere, the mantle, in response to increased depth, again becomes stronger, to form what they called the **mesosphere** (middle sphere).

In 1968 Isacks, Oliver, and another geophysicist, Lynn Sykes, incorporated their findings into a hypothesis they called **new global tectonics.** They postulated that new lithospheric material, formed at the ocean rises, moves laterally away from these zones toward the island arcs (Fig. 4-21). In the vicinity of an island-arc system, the lithosphere bends sharply downward and sinks or is pushed into the softer asthenosphere below. It is much like a giant treadmill, with the upwelling and generation of new lithospheric material along one

zone and the downwarping and consumption of the lithosphere along another zone. With this hypothesis these three geophysicists provided the answer to the riddle of the young basins and the old continents. Ostensibly, the ocean basins are continually being formed along the ocean rises and consumed along the island-arc/trench systems. Those portions of the lithosphere that contain relatively low-density continental blocks apparently were more buoyant and resisted downwarping, such that the continents continued to exist over long periods of geological time. Subsequently, the basic tenets of new global tectonics were incorporated into an even broader theory.

## PLATE TECTONICS—THE MODERN THEORY

Most earthquakes occur near the oceanic ridges and rises and along the Benioff zones of island

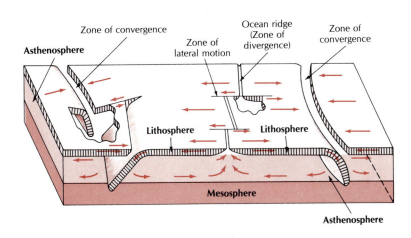

**Fig. 4-21.** The kinds of movement possible in the lithosphere and the relation of such movements to compensating flow in the asthenosphere. (After B. Isacks, J. Oliver, and L.R. Sykes, 1968, "Seismology and the New Global Tectonics," *Journal of Geophysical Research,* vol. 73, p. 5857)

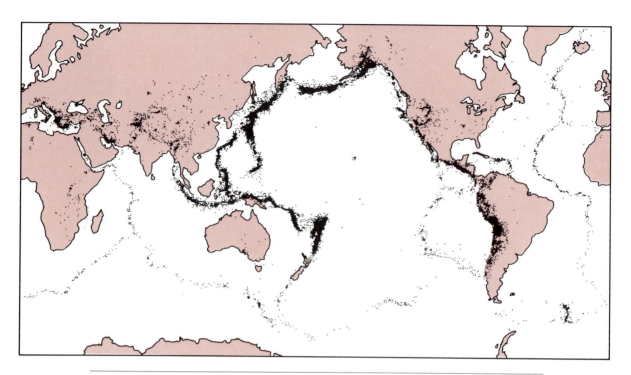

**Fig. 4-22.** Distribution of earthquake epicenters from 1961 to 1967. Most earthquakes occur in narrow belts that are now considered to mark the edges of moving lithospheric plates. Notice the belt of intense activity along the Pacific Ocean margin. (After M. Barazangi and J. Dorman, 1968, ''World seismicity maps compiled from ESSA, epicenter data, 1961–1967,'' *Bulletin of the Seismological Society of America,* vol. 59, pp. 309–80.)

**Fig. 4-23.** The earth's named lithospheric plates, zones of convergence, and zones of divergence. Arrows indicate the general movement of the plates.

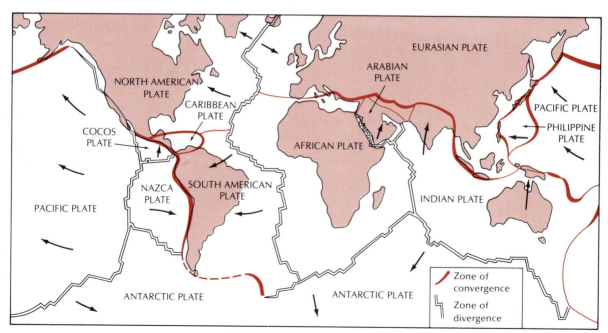

arcs. Figure 4-22 is a plot of the foci for a large proportion of the quakes that occurred between 1961 and 1967. The pattern confirms that most of the world's seismic activity is confined to narrow linear zones. Geologists have postulated from the pattern that the earth's crust is divided into a small number of rigid blocks, or **plates,** that move and jostle one against another (Fig. 4-23). Seismic activity takes place at the edges of the plates, but the plates themselves are nearly devoid of activity. The *plate-tectonics theory* is now used by most earth scientists to explain the internally generated processes and features of the earth.

### Divergence Zones

Where plates move apart—areas called **divergence zones** (or pull-apart zones)—new lithospheric material is formed (Fig. 4-23). The oceanic ridges and rises are divergence zones that are characteristically associated with basaltic volcanism and shallow-focus earthquakes. Rocks, when heated, tend to expand; therefore, the newly formed hotter rocks near the axis of the divergence zones occupy more volume, which results in an elongate welt—an oceanic ridge or rise. The outer part of the rise is stretched and commonly responds by forming a downdropped keystone block along the crest, that is, an axial valley. As the lithospheric material moves away from the ridge crest, it cools and contracts, and the elevation of the ridge sys-

tem is lowered. As a rule earthquakes near the divergence zones occur at depths of less than 20 km (12 mi) and are restricted to the ridge axes or the fracture zones that cross them at nearly right angles.

Since 1970 divergence zones have been studied by submersible craft in different places, including the Mid-Atlantic Ridge, the Galapagos off the coast of South America, and the East Pacific Rise south of Baja California. These zones have been found to be relatively narrow—in some places the line of demarcation is almost knife-sharp; in others, it is a few kilometers wide. Newly formed pillow lavas are common, indicating largely undetected submarine basaltic volcanism.

Sensors show that the temperature of the seawater near the divergence-zone axis is often higher than normal in response to heating from the recently erupted ridge basalts. Heat is also transferred, apparently, from the deeper, hotter parts of the ridge through convection of the oceanic waters that permeate the broken and fissured basalt flows. The interaction of the ocean waters with the lava commonly results in extensive rock alteration. Unusual marine organisms are concentrated in the heated waters at both the Galapagos center and the East Pacific Rise. Sulfide minerals rich in copper, iron, zinc, cobalt, lead, silver, and cadmium also precipitate from hot-water vents along the divergence zone axes. In some places, sulfide concentration is so great that the rising jets of hot

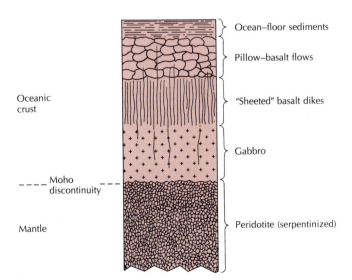

**Fig. 4-24.** Idealized section of an ophiolite approximately 6-km (4-mi) thick. The many "sheeted" dikes are thought to be emplaced along a divergent plate boundary that fed ocean-floor pillow-basalt flows above; the gabbro apparently represents the slower cooling magma bodies from which the dikes originated. Periodotite, consisting mostly of olivine, represents the oceanic upper mantle beneath the ophiolite.

water are dark in color—thus their name, *black smokers*. Certain rich mineral deposits, such as those copper ores on the Mediterranean island of Cyprus, are now thought to have formed in this way.

In a few places on the continents, traces of ancient oceanic crust are found. The explanation for these unusual occurrences is that after formation at the divergence zone, the seafloor moves progressively across the ocean basin and ultimately into a zone of plate convergence, where—during consumption of the ocean lithosphere—thin, highly deformed slices are thrust onto the continental plate margin. These unusual rocks, consisting of deep-sea sediments, serpentinized pillow basalts, and mafic igneous intrusions, are

called *ophiolites*—from the Greek *ophis* (serpent) and *lithos* (stone)—because of the abundance of serpentine (Fig. 4-24).

## Lateral-Movement Zones—Transform Faults

Throughout the world oceanic ridges are transected and sharply offset by a great number of narrow fracture zones. Because of the numerous offsets, the map traces of the rises and ridges (Fig. 4-25) zigzag across the oceans. In 1965 the Canadian geologist J. Tuzo Wilson postulated that sharp offsets in the divergence zones should lead to an unusual type of fault activity, which he called **transform faulting.** He reasoned that along the transverse fracture zone and between the ends of

**Fig. 4-25.** Earthquake epicenters near the Mid-Atlantic Ridge and fault motions along fracture zones. (After L. Sykes, 1967, "Mechanism of Earthquakes and Nature of Faulting in the Mid-oceanic Ridges," *Journal of Geophysical Research,* vol. 72, p. 2137)

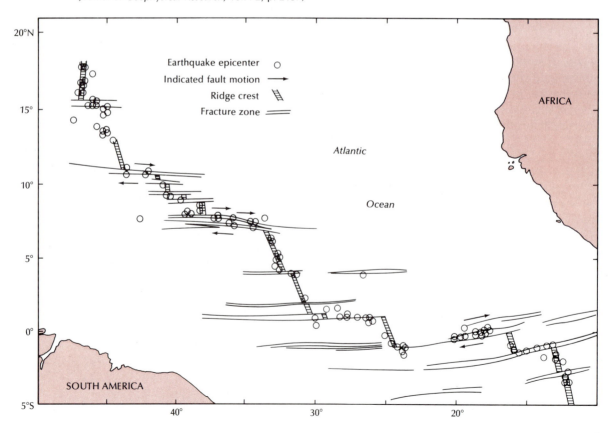

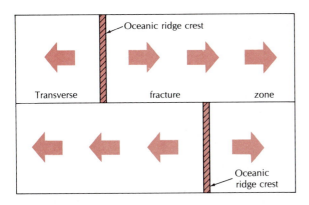

**Fig. 4-26.** A divergence zone offset along a transverse fracture zone. Because of the offset, the diverging plate material moves in opposite directions along the fracture zone, but only between the offset ridge ends. That part of the fracture zone between ridge ends (hatchured) is susceptible to transform faulting and earthquakes. Ridge crests also tend to be seismically active.

the displaced ocean rise (Fig. 4-26) strike-slip faulting (and earthquake generation) should prevail because the plates are moving laterally past one another. Beyond the offset ridge ends, however, the plates on either side of the fracture zone move in the same direction, thus the number of earthquakes should be minimal. Transform faulting, then, should be restricted to that part of the fracture zone between the ends of the offset ridges.

Several years later, Lynn Sykes confirmed the existence of transform faulting by plotting the epicenters of many recent earthquakes in the Atlantic Ocean (see Fig. 4-25). He found they were concentrated either over the ridge crest or along the fracture zones between the ends of the dislocated ridge-crest ends, just as Wilson postulated. Sykes also studied the nature of the earthquake waves and verified that the blocks slide laterally one past the other. The seismicity along transform faults is another proof that the oceanic ridges are active spreading zones.

Transform faulting does not cause the separation of ridge axes but occurs because they are already offset. The cause of the initial offset in the ridge lines is unclear, although it is undoubtedly related to the generation of divergence zones.

Faults along which oceanic ridges are offset are called **ridge-ridge transform faults.** One of the longest of these faults is the San Andreas Fault on the west coast of North America along which the northern end of the East Pacific Rise in the Gulf of California (Baja California) is offset from the southern end of a short segment of oceanic rise off the Oregon shore.

Other kinds of transform faults exist but are less common than the ridge-ridge type. Their names are self-explanatory: ridge-trench, trench-trench, and so on.

## Convergence Zones

The curvilinear island arcs of the world mark zones where lithospheric plates push together—**convergence zones**—and where lithospheric material is destroyed (Fig. 4-23). As a slab of lithosphere bends over and slowly descends into the asthenosphere, it is heated. Eventually, the descending slab heats up and becomes indistinguishable from the mantle rocks that surround it (Fig. 4-27). The depth to which a plate edge can penetrate depends on the rate of movement, but no plate extends below 700 km (435 mi).

Most convergence zones occur at the junction of an oceanic plate and a continent-bearing plate. In three places, however, the zone occurs between two oceanic plates: the South Sandwich Islands zone off the southern tip of South America, the Tonga–Kermadec Islands zone north of New Zealand, and the Izu–Mariana Island zone south of Japan.

As the relatively cold oceanic plate descends, frictional drag occurs. Sporadic release of the strain in the slab produces a belt of shallow-, intermediate-, and deep-focus quakes characteristic of island arcs (Fig. 4-27).

Sediments derived from the nearby continental landmasses and from erosion of the islands themselves can accumulate in the deep trench next to the arc. As the oceanic plate moves downward, these stratified rocks are dragged down with it, a process called **subduction.** At first they will be folded and faulted (Fig. 4-28); then, as they move to greater depths (and increasing temperatures), they will be heated and metamorphosed, as discussed in chapter 9. Friction along the slippage surface and deformation of the slab also produce thermal energy. Under these conditions sediments

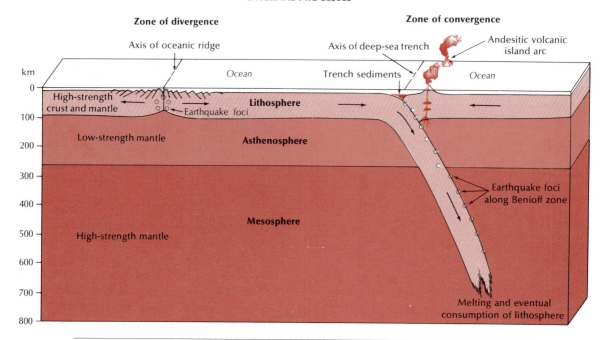

**Fig. 4-27.** Relation of a divergence zone to a convergence zone. New lithospheric material, formed along oceanic ridges, moves outward and bends sharply downward into a zone of convergence, where it is eventually consumed at depth. The axis of a deep-sea trench forms where the slab bends downward, and a line of andesitic volcanoes forms where heating adjacent to the descending slab is sufficient to melt rocks. Shallow-, intermediate-, and deep-focus earthquakes of the Benioff zone are thought to be generated along the upper surface of the slab.

and parts of the descending slab may melt. The magma formed tends to rise to the surface, feeding the arc volcanoes (Fig. 4-27).

Subduction slabs dip downward into the asthenosphere at varying angles, generally from 30° to 60°. Rarely, as along the Izu–Mariana Islands zone, is it near 90° or, as along some parts of the Peru–Chile convergent zone, is it as low as 10°. If the angle is low, the generation of magma is greatly curtailed.

Convergence-zone magmas come from a variety of source rocks: subducted sediments, the descending lithospheric slab just below the Benioff zone, the mantle just above the Benioff zone, and even the continental crust. This topic is discussed more fully in the section on igneous processes (chap. 7). Eventually, along a convergence zone, a mountain chain will be formed, a topic we will cover in chapter 6.

Only one convergence zone between two continental plates is known. It extends in an easterly direction from the northern shore of the Mediterranean Sea, through Iran, Pakistan, and into the Himalaya Range. Because *both* plates are buoyant, and therefore equally resistant to bending and descending, a wide zone between the plates has been crushed and crumpled. Shallow earthquakes are common in this zone. We will describe convergence of this type and the formation of the world's greatest mountain chain in chapter 6.

### Geometrical Description of Plate Motion

Ridge-ridge transform faults that cut across an oceanic ridge are commonly subparallel and curved. The tightness of curvature increases progressively along a ridge segment while the amount of lateral spread along the ridge decreases. These geometrical relations, illustrated in Figure 4-29, appear to fit best with the idea that the local divergence of two plates occurs by angular rotation of the plates about an axis.

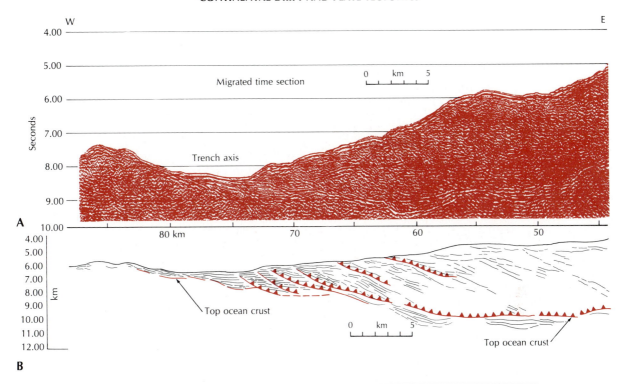

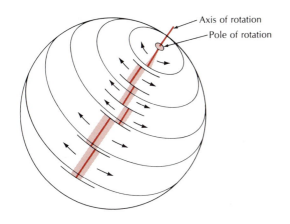

**Fig. 4-28.** Folding and faulting of sediments subducted in trench offshore of Peru. **A.** Seismic reflection record. **B.** Structural interpretation of the record. Light lines represent bedding planes; heavy lines with teeth represent thrust faults. (R. von Huene, L.D. Kulm, and J. Miller, 1985, "Structure of the Frontal Part of the Andean Convergent Margin," *Journal of Geophysical Research,* vol. 90, p. 5433)

**Fig. 4-29.** The relative displacement of two lithospheric plates can be described as an angular divergence about an axis of rotation. Notice that the tightness of curvature of the concentric circles-of-motion about the pole of rotation decreases and the absolute amount of opening along the divergence zone increases with greater angular distance from the pole of rotation.

## New Postulates

The basic concept of plate tectonics is simple and elegant: a few, brittle lithospheric plates diverge, converge, or slide laterally past one another and, in so doing, set in motion a host of constructional processes. Yet it has become apparent during the last several years that some observable processes and features are not covered under the initial statement of the theory. This has required some modifications of the basic theory.

**Microplates** The plate-tectonic model was initially meant to apply to the movements of a small number of large lithospheric plates. It has become apparent, however, that in some parts of the world, particularly along the north shore of the Mediterranean and in the Caribbean, the lithosphere is composed of numerous ***microplates***—small plates that can move in complicated ways.

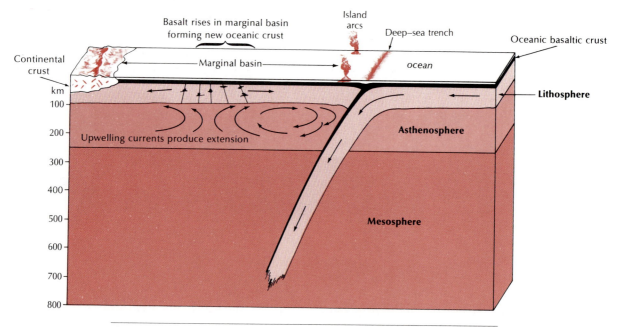

**Fig. 4-30.** Hypothetical model to account for the development of back-arc upwellings.

**Fig. 4-31.** Depicts J. Tuzo Wilson's theory that a hot rising plume melts rocks just below or within the lithosphere and creates intraplate basaltic volcanoes. Once formed, the cones move laterally from the hot spot and are carried piggyback on the moving lithospheric plate.

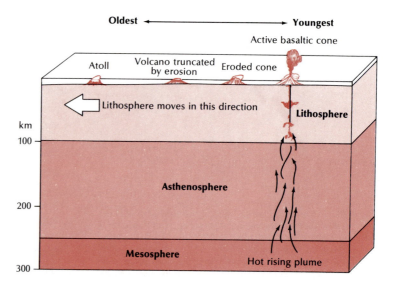

**Soft Plates** Paleomagnetic studies of rocks along plate margins have shown that some parts of the lithosphere are not particularly brittle and that they have undergone bending related to plate motions. The nearly 90° bend at the south end of the Izu–Mariana Islands zone is an example, as is the significant change in trend of the western edge of South America in Peru.

**Back-Arc Upwelling** Oceanographic studies, initiated by the American geologist Daniel Karig, have shown that secondary divergence zones commonly form between continental margins and island-arc systems. In time this leads to an oceanward migration of the arc and trench system—a process that has brought about a widening of the Philippine Sea between the Philippine and Mariana Islands, for example. Geophysicists now think that such **back-arc upwelling** is the result of a convectional eddy current behind the island arc that acts in response to the mantle's heating up, caused by the descending slab (Fig. 4-30).

**Mantle Hot Spots and Plumes** Not all volcanism is restricted to the zones around plate margins; some, such as that in Hawaii or in Yellowstone National Park (Wyoming), occurs in the middle of the plates (see chap. 7). Such volcanoes are not readily accounted for within the framework of the plate-tectonics model. The Canadian geophysicist J. Tuzo Wilson speculates that somewhere at depth in the mantle, perhaps as far down as the core-mantle boundary, localized **hot spots** form. Heating reduces the density of the mantle material, which then begins to rise toward the surface in a fingerlike **plume.** The rising heated rock undergoes partial melting; this leads to the eruption of magma that builds volcanoes in the centers of plates (Fig. 4-31).

Nearly two dozen volcanically active centers have been identified as the result of hot-spot activity. Half of these are located along, or close to, oceanic ridges and rises. This coincidence has prompted some geologists to speculate that the location of hot spots determines the ultimate positions of ridges and rises. And some have further theorized that all of the plumes bend over as they reach the base of the lithospheric plates and exert a push, like that of moving fingers, thus causing lateral movement of the plates.

Because many plumes supposedly are generated in the relatively static mantle beneath the lithospheric plate, the volcanoes, once formed, may be cut off from the magma source and carried away on the moving plate. The result is the formation of a line of volcanoes that increase in age away from the active hot spot (Fig. 4-31). The Hawaiian Islands form such a chain of volcanoes; the youngest ones are over active vents on the largest island, Hawaii.

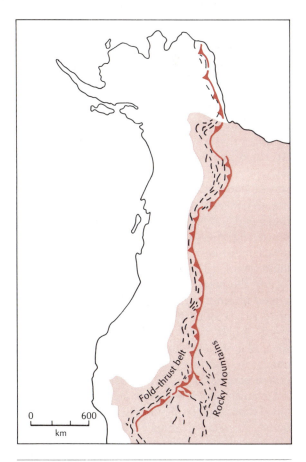

**Fig. 4-32.** Postulated extent of accreted terranes (including suspect terranes) along the western margin of North America are in white. Rocks indigenous to North America are in shaded area. (After P. Coney, 1981, "Accretionary Tectonics in Western North America," in W.R. Dickinson, et al., eds., *Relations of Tectonics to Ore Deposits in the Southern Cordillera, Arizona Geological Society Digest,* vol. 15, pp. 23–37)

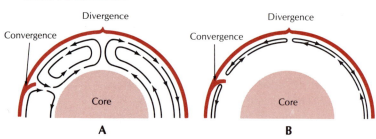

**Fig. 4-33.** Cross section of the earth with **(A)** deep convection and **(B)** shallow convection. (After E. Orowan, 1969, "The Origin of the Oceanic Ridges," *Scientific American*, vol. 221, no. 5, pp. 102–19)

Volcanic activity in the eastern arm of the Snake River plain, Idaho, which appears to have progressed steadily up the plain toward Yellowstone National Park during the last 15 million years, is thought to be the result of drift of the North American plate over the Yellowstone hot spot.

All geologists do not subscribe to the theory of hot spots. They point out that the same effect could be produced when magma rises through the lithosphere along the front edge of an advancing vertical crack.

**Accreted and Suspect Terranes** In the study of ancient convergence zones, geologists have observed that subduction of the oceanic plates is not always complete. Fragments and wedges of oceanic rocks and sediments may not have been carried down with the subducting plate or once carried down, they may have popped back up. Eventually, these "foreign" rock sections have been plastered against the nonsubducting plate margin. Such *accreted terranes* are common along the west coast of North America (Fig. 4-32). Many of the fragments found in accreted terranes are severely sheared and folded, but some remain intact. Commonly, they show evidence of having been transported parallel to the coast along faults.

Geologists think that some accreted blocks may travel thousands of kilometers from their place of origin. Accreted fragments that are suspected of having moved such great distances are called *suspect terranes.*

### Mechanism for Plate Tectonics

What makes the plates move? The simple truth is that we are not sure. The energy flow is internal, but its exact source remains a mystery.

One idea, introduced by Arthur Holmes, is that internal heating in the earth's mantle is uneven, so that it flows or convects much like slowly boiling oatmeal. Places where the hot currents rise and move apart correspond to zones of divergence; places where currents converge and descend correspond to zones of convergence (Fig. 4-33). Lateral transport of the plates corresponds to the horizontally moving portions of the convection cells.

Most geologists today favor some form of convection as the principal plate-driving mechanism. However, they do not agree on the details. Does convection occur throughout the mantle (down to about 600 to 700 km, [375 to 435 mi]) or is it restricted to the relatively soft asthenosphere (Fig. 4-33)?

A new computer technique that analyzes multitudes of seismic records now provides information on the thermal regime of the mantle. The technique, called *seismic tomography,* is similar to computerized axial tomography (CAT) scans used in medicine. The results indicate that heated material beneath some oceanic ridges and rises is restricted to the upper 200 km (125 mi); some segments, in fact, appear to be fed by lateral transport of material at shallow depth. The data also strongly suggest that the distribution pattern of hot and cold regions in the upper 200 km is unlike that at greater depths. Most oceanic hot spots appear to be fed by materials from the oceanic ridges and rises. The vertical pattern of hot and cold regions supports the idea of convection down to about 500 to 600 km (310 to 375 mi). As yet, there is no clearcut evidence of convection to greater depths.

Some geophysicists have proposed other mechanisms for some plate motion. One is based on

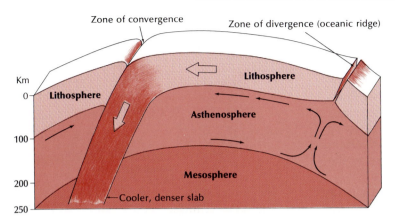

**Fig. 4-34.** The hypothetical sinking of the relatively cold and dense lithospheric slab into the asthenosphere and mesosphere. As the slab sinks, it pulls the oceanic lithosphere along behind it, causing an opening at the ocean ridge. This, in turn, causes passive convection.

the differences in density between the descending lithospheric slab and the mantle into which it sinks (Fig. 4-34). It is thought that the cold, sinking slab is more dense than the mantle beneath and that the slab sinks under its own weight, dragging the plate behind it like an immense flattened tail. However, some geophysicists question the idea that the lithosphere is more dense than the mantle.

They also point out that in the plates diverging from the Atlantic and Indian oceans, there are no convergence boundaries and no sinking slabs to provide for plate motion.

Another theory involves gravitational sliding of the plates away from the topographically high oceanic ridge and rise crests. Opponents to this view, however, cite several areas in the world

**Fig. 4-35. A.** Initial state of the earth. **B.** After expansion, the brittle lithosphere would crack apart and zones of divergence would form; as material flowed up to fill in the voids developing in the zones of divergence, mantle convection would be generated, creating convergent zones.

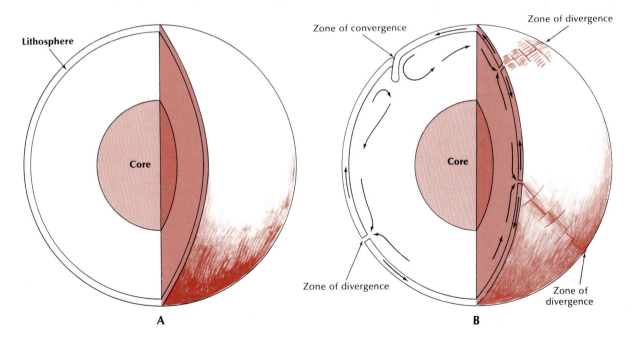

where the downslope gradient appears to be inadequate for sliding to take place.

One other mechanism—expansion—deserves mention as a means of triggering plate movement, as least during the last 150 million years (Fig. 4-35). If the brittle lithosphere split apart as a result of expansion of the earth, then new material would have to move upward to fill the space created by the splitting. Lateral inflow would then take place. Most geologists do not favor expansion as the mechanism for plate movement.

Almost all earth scientists subscribe to one or more of the theories just described. Yet there are questions and problems concerning all of them. It might be that the actual mechanism is not simple, but involves two or more processes. Or perhaps the real mechanism is still not known.

## Assessment of the Plate-Tectonics Theory

The theory of plate tectonics as it stands is still simple and elegant. In spite of the success of the theory, however, most proponents admit that it has deficiencies and inconsistencies and that it leaves some important questions unanswered.

The theory has been more successful in explaining phenomena in the oceans than in the continents. This is understandable because it is based largely on observations made in the ocean basins, which appear simpler in origin and relatively young (none older than Jurassic), whereas rock-forming cycles on the continents extend back 3.5 billion years and more.

There are also these unresolved questions: Why is there a bilateral symmetry to the ridge systems and bordering ocean basins in the Atlantic and Indian oceans, but none in the Pacific Ocean? Why does the formation of new lithosphere occur in all oceans, but its consumption take place almost exclusively around the margin of the Pacific? Why do many deep-sea trenches appear to be of recent geological origin and to have formed through extension rather than compression?

Several active strike-slip faults around the Pacific margin, such as the Alpine Fault in New Zealand, the Denalie Fault in Alaska, and the Atacama Fault in South America, are not well explained by the plate-tectonic theory. Motion on most of these faults parallels the linear zones of

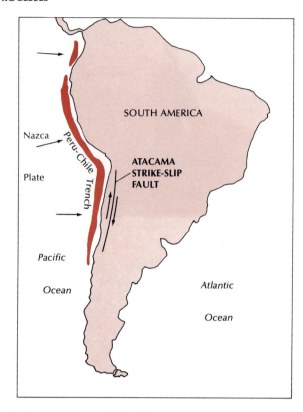

**Fig. 4-36.** South America, showing the Peru–Chile Trench and the Atacama strike-slip fault. The postulated movement of the Nazca plate is shown by the three separate arrows. Notice that the direction of the plate motion is nearly at right angles to movement along the fault.

convergence at right angles to the motion of the converging plates (Fig. 4-36).

The origin of the prominent east-west mountain belt that includes the Alps and the Himalayas is also enigmatic. This belt seems to have been formed by the north-south convergence of plates during the last 50 to 60 million years. Yet the evidence from the linear magnetic anomalies for that same time interval indicates that virtually all plate motion was toward the east or west.

Questions also arise about the African plate. Because it is bounded on the west and on the east by active divergence zones, one might suppose that it is being compressed and that a linear convergence zone, parallel to the divergence zones, might form within it. In fact, the plate is being pulled apart, as shown by deformation, volcanism, and earthquakes in the East African Rift area (Fig. 4-37).

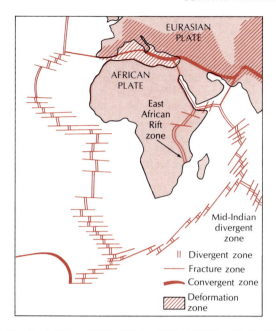

**Fig. 4-37.** The African plate and surrounding region, showing relations of divergent, fracture, convergent, and deformation zones.

Some aspects of volcanism also raise questions. For example, steep-sided volcanoes should typify convergence zones. Yet volcanoes in the Cascade Range of the Pacific Northwest, which resemble convergence-zone volcanoes in all other respects, are associated with an atypical subduction zone. The trench is poorly defined and earthquakes occur over a broad zone at depths of less than 70 km (44 mi). The spotty occurrence of igneous rocks throughout the western United States over the last 70 million years, up to 1600 km (1000 mi) inland from the present west coastal margin, is difficult to explain in terms of the plate-tectonics theory.

Such deficiencies, inconsistencies, and ques-

tions require further investigation of the nature of internal processes. And, as the theory is tested and reassessed, minor and perhaps even major revisions—a necessary part of the scientific process—will be required.

One of the most significant tests will depend on precise measurements between far-distant points to determine the interplate displacements. Since 1977 scientists in the United States, Western Europe, Japan, and South Africa have been measuring the distances between sites on the earth by astronomical and satellite methods. These techniques enable distances between observational sites separated by hundreds and thousands of kilometers to be determined within 1 to 2 cm (0.5 to 1 in). Repeated measurements over 5 to 10 years, therefore, have the potential of establishing the average drift rates between plates.

Measurements taken over 5 years between observatories in North America and Sweden indicate an opening of the Atlantic Ocean of between 1cm/yr and 2 cm/yr (0.5 and 1 in./yr)—just about the rate predicted from analysis of the linear magnetic anomalies along the Mid-Atlantic Ridge. Yet the data also indicate a great deal of inexplicable random fluctuation in the motion between the two plates.

Data from stations on the North American, Pacific, and Eurasian plates now show that drift between these plates is close to predicted values. This is heartening news, indicating that the plates do move as postulated. One cautionary note is sounded, however, by data from observatories on the North American plate that suggest the distance between Massachusetts and Texas has been shrinking by about 1cm/yr (0.5 in./yr). This intraplate motion is unexpected. It brings home the point that that interplate and intraplate motions may be more complicated than had previously been thought. Only time will tell.

## SUMMARY

1. Before the twentieth century, several theorists had suggested that some continents had split and drifted apart. In 1911 Alfred Wegener claimed that such drift did occur, and a great controversy over the theory of *continental drift* began.
2. The evidence that led to general acceptance of the

hypothesis included fit of continental coastlines, stratigraphic and structural similarities between continents, paleoclimatic reconstructions and distribution of the late Paleozoic glacial deposits, distribution of similar fossil groups, paleomagnetic pole wander curves from different continents, linear magnetic anomalies over the oceanic ridges and rises, and age of cores from ocean sediments.

3. By 1960 most geologists believed in continental drift, but a unified theory was needed—one that would explain drift along with other internally generated phenomena, such as volcanism, mountain building, metamorphism, and seismicity.

4. *New global tectonics* was an attempt to provide a unified theory. The outer 100 km (62 mi) of the earth was defined as the *lithosphere,* the zone between 100 and 250 km (62 and 155 mi) as the *asthenosphere,* and the zone beneath 250 km as the *mesophere.* The lithosphere is strong and brittle, the asthenosphere is relatively weak. New lithospheric material supposedly forms at the oceanic ridges and moves laterally away from the ridge systems toward the island arcs. At trench-island arc systems, lithospheric plates bend sharply over and plunge into the asthenosphere.

5. *Plate tectonic theory* is the most recent unifying concept. It is thought that the lithosphere is subdivided into a number of large plates that are moving. Some converge, some diverge, and some slide laterally past one another. Zones of divergence, where new lithospheric material is forming, are characterized by high heat flow, basaltic volcanism, and shallow-focus earthquakes. Zones of lateral movement *(transform faults),* which cross the ridges and rises nearly at right angles, are characterized by shallow-focus earthquakes. Zones of convergence are marked by deep oceanic trenches, andesitic volcanism, and a seismic zone, the *Benioff zone,* which extends down to 700 km (435 mi). Shallow-, intermediate-, and deep-focus earthquakes occur along that zone.

6. Such features as microplates, deformation of plates, back-arc upwelling, mantle hot spots and plumes, and accreted terranes are not directly addressed by the plate-tectonic theory.

7. The mechanism that causes plates to move is unclear. *Mantle convection,* a density difference between the lithosphere and asthenosphere, gravity sliding, and expansion of the earth have all been suggested as possible candidates.

8. Some data .do not fit into the plate-tectonic model convincingly.

## QUESTIONS

1. What was Alfred Wegener's revolutionary idea? What led him to his conclusion?
2. What were the ideas used in opposition to Wegener's hypothesis? How well were they founded?
3. List and describe briefly all the evidence used to support continental drift up to 1960.
4. Describe new global tectonics. What was it based on? How did it differ from Wegener's theory?
5. What is the theory of plate tectonics? What is its basis? How does it differ from new global tectonics?
6. Describe in detail, zones of convergence, divergence, and lateral movement.
7. What geological processes and features can be explained in terms of plate tectonics?

8. What are some specific features or phenomena that appear to be associated with, but that are not directly related to, plate tectonics?
9. Wegener's initial ideas about continental drift did not postulate a reasonable cause. Is this problem adequately addressed in the new plate-tectonics theory?
10. What are some of the obvious flaws or inadequacies in the theory of plate tectonics?
11. What do you think will be the ultimate fate of plate tectonics in the future?

## SELECTED REFERENCES

Beloussov, V.V., 1974, Sea-floor spreading and geologic reality *in* Plate tectonics: assessments and reassessments, C.F. Kahle, ed., American Association of Petroleum Geologists, Tulsa, Okla., pp. 155–66.

Bird, J.M., and Isacks, B., eds., 1972, Plate tectonics, selected papers from the Journal of Geophysical Research, American Geophysical Union, Washington, D.C.

Carter, W.E., and Robertson, D.S., 1986, Studying the earth by very-long-baseline interferometry, Scientific American, vol. 255, no. 5 pp. 46–54.

Cox, A., 1973, Plate tectonics and geomagnetic reversals, W.H. Freeman, San Francisco.

Cox, A., and Hart, R.B., 1986, Plate tectonics—how it works, Blackwell Scientific Publications, Palo Alto, Calif.

Hallam, A., 1973, A revolution in earth science, Oxford University Press, Oxford.

Heirtzler, J.R., 1968, Sea-floor spreading, Scientific American, vol. 219, no. 6, pp. 60–70.

Marvin, W.B., 1973, Continental drift, the evolution of a concept, Smithsonian Institution Press, Washington, D.C.

Maxwell, J.C., 1974, The new global tectonics: an assessment, C.F. Kahle, ed., American Association of Petroleum Geologists, Tulsa, Okla., pp. 24–42.

Orowan, E., 1969, The origin of the oceanic ridges, Scientific American, vol. 221, no. 5, pp. 102–19.

Sullivan, W., 1974, Continents in motion: the new earth debate, McGraw-Hill, New York.

Takeuchi, H., Uyeda, S., Kanamori, H., 1970, Debate about the earth, Freeman, Cooper, San Francisco.

Tarling, D., and Tarling, M., 1971, Continental drift, Doubleday, Garden City, N.Y.

Toksoz, M.N., 1975, The subduction of the lithosphere, Scientific American, vol. 233, no. 10, pp. 88–101.

Vine, F.J., and Matthews, D.H., 1963, Magnetic anomalies over oceanic ridges, Nature, vol. 199, pp. 947–9.

Wegener, A., 1966, The origin of continents and oceans, Dover, New York.

Wilson, J.T., ed., 1976, Continents adrift and continents aground, W.H. Freeman, San Francisco.

Windley, B.F., 1984, The evolving continents, Wiley, New York.

Wyllie, P.J., 1976, The way the earth works, Wiley, New York.

# 5

# EARTHQUAKES AND THE EARTH'S INTERIOR

It has been estimated that over the last 4000 years about 15 million people have died from earthquakes. Each year more than a million earth tremors occur around the world; of these, about 50 are intense enough to cause significant loss of life and property (Fig. 5-1) and about 10 are capable of causing many deaths and great destruction. The average cost of earthquake damage amounts to $7 billion yearly; in 1976, nearly 700,000 people died from earthquakes. On the other hand, earthquakes have provided the greatest amount of information about the interior of the earth through the waves associated with them.

We begin by recounting a few of the numerous case histories of earthquakes; then we discuss the occurrence, effects, and prediction of earthquakes; and finally we describe the nature of earthquake waves and the information they provide.

**Fig. 5-1.** Earthquake damage at the church of Monte di Buia in northern Italy. The Friuli earthquake of May 6, 1976, killed nearly one hundred people. (James Stratta)

## CASE HISTORIES OF EARTHQUAKES

### United States and Mexico

The United States is subject to earthquakes, particularly in the West. During its existence as a country, however, they have caused relatively little loss of life and property primarily because few major quakes have occurred in densely populated urban areas.

**San Francisco** In the spring of 1906 San Francisco was the queen city of the California Coast and the gateway to the Orient. Her business districts hummed by day while the Barbary Coast sang by night. At 5:12 A.M. on April 18, a region covering about 975,000 km² (376,000 mi²) surrounding San Francisco was severely shaken by an earthquake, leading to the nearly complete destruction of the city. Surprisingly, the death toll was low—about 700 persons.

The population of the United States is becoming increasingly urban, and many of the problems San Francisco faced at that time are exactly those that

**Fig. 5-2.** Sacramento Street, San Francisco, just after the 1906 earthquake. Notice how the brick fronts of the buildings spilled across the streets. The people are watching one of the great fires caused by the quake. (Arnold Genthe, California Palace of the Legion of Honor)

might confront civil defense agencies today, with the additional burden of immense traffic jams. Had the earthquake struck later when people were up and about, the casualty list would undoubtedly have been much larger. Actually, destruction related directly to the tremor was only moderate—perhaps only 25 percent of the total. Most was attributed to the fire that followed the tremor (Fig. 5-2).

The San Francisco earthquake emphasized that the kind of ground on which buildings are constructed is crucial in determining the extent and nature of structural damage. Buildings founded on solid rock showed slight damage when compared with virtually identical structures built on water-logged or unconsolidated ground. Destruction was especially severe, for example, in the downtown area where about 20 square blocks had been built on ground reclaimed from San Francisco Bay after the Gold Rush of 1849. There, on a sludgy foun-dation made up of water-soaked refuse buried under poorly consolidated mud and silt, the most severe damage occurred.

In that day the Victorian influence prevailed, and most buildings were encrusted with ornamentation and gingerbread—extreme earthquake hazards. San Francisco, to its sorrow, was unusual in one regard: It was one of the larger cities of the world with buildings constructed principally of wood. Typically, the residential section consisted of block-long rows of wooden multistoried houses or apartments on narrow lots. The ensuing fires, once they reached these buildings, were virtually unstoppable.

**Fig. 5-3.** The San Andreas Fault is one of the major faults in the United States, running nearly two-thirds the length of California. Almost without exception, the trace of the fault, shown here in the Carrizo Plains, is easily discernable. (William A. Garnett)

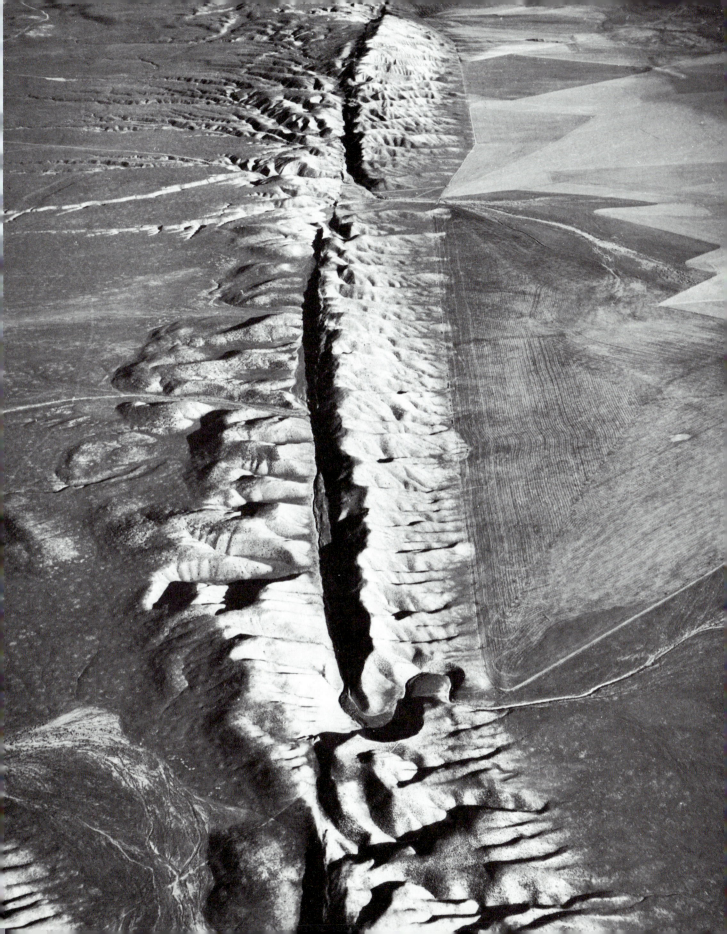

Fires from a variety of causes broke out at many points almost immediately after the strongest shocks. At first, there was little awareness that fire was the true enemy. People either gawked at the blazes or attacked them piecemeal—but not too successfully—because most water lines beneath the streets had ruptured and water flow to the fire hoses dropped to a trickle.

Fire started near the waterfront and swept inland across the broken city. Should you visit San Francisco, try to visualize the swath, 18-blocks deep, swept by fire—from the Embarcadero at the waterfront inland to Van Ness Avenue, the first wide street where a fireline could be held. Elsewhere vain attempts were made to check the advancing flames by dynamiting whole rows of buildings.

From a scientific point of view, the San Francisco quake was particularly significant. The California State Earthquake Commission, appointed by the governor, immediately conducted an exhaustive investigation. Hundreds of people were interviewed and evidence was collected about earthquake-related destruction.

The evidence indicated that sudden slippage along the San Andreas Fault was the primary cause of the earthquake (Fig. 5-3). It was also shown unequivocally that the displacement was lateral, not vertical. Before 1906 the importance of that type of fault movement was only dimly appreciated.

All of the evidence, including a nearly continuous trail of furrowed ground, fractured barns, and other features that stretched for hundreds of kilometers, indicated that the western side of the San Andreas Fault had moved horizontally northward with respect to the eastern side, which moved southward. The fault has a total length of about 965 km (600 mi) on land; during the 1906 quake, slippage occurred along about one-half its length—from Point Arena, north of San Francisco, to San Juan Bautista to the southeast.

The maximum offset of 6.4 m (21 ft) was near Tomales Bay, north of the city; elsewhere offsets of about 4.6 m (15 ft) were common. Most observers noted that the maximum displacement was in rather soft ground, which appeared to have lurched, thereby amplifying the offset. The true maximum offset appears to have been about 5 m (16 ft).

On the basis of the offset data, the geologist Harry Fielding Reid (1859–1944) formulated a theory to explain the generation of earthquakes. Blocks of the earth's crust are imagined to slide past one another along faults in response to large-scale forces. As long as the blocks on either side of the fault are free to slide, slippage is continuous and smooth. If, however, the walls on opposite

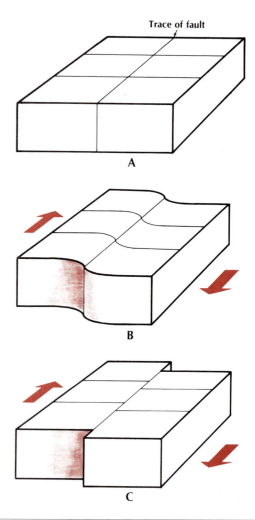

**Fig. 5-4.** Three stages in the movement of two blocks of the earth's crust past one another, as postulated in Reid's theory of elastic rebound. **A.** Undeformed blocks with imaginary straight lines that extend across the trace of the fault. **B.** The two blocks move as indicated by the arrows, but the fault is locked, resulting in bent lines near the fault trace. **C.** The blocks continue to move, leading to sudden slippage along the fault and an earthquake. The lines are again straight, but offset along the fault trace.

sides of the fault become locked together because of irregularities along the fault surface, slippage occurs in fits and starts.

During the times when the walls are locked, so the theory goes, the rocks adjacent to the fault are progressively bent and accumulate **elastic strain energy.** When the crust fails and rapid movement occurs, the strained rocks snap back (rebound)— much like a diving board—and an earthquake results. The larger the amount of strain energy released, the larger the quake. Reid's explanation, diagrammed in Figure 5-4, is called the **elastic-rebound theory;** today it is still the accepted model for the generation of most earthquakes.

**New Madrid, Missouri, 1811–12** The San Francisco earthquake was not the most severe quake to jolt the U.S. mainland in the last 200 years. The one with that distinction occurred in the unlikely location of New Madrid, Missouri.

The New Madrid quake was felt from the Canadian border to the Gulf of Mexico and from the Rocky Mountains to the Atlantic Ocean (Fig. 5-5). The quake stopped clocks as far away from its point of origin as Boston, Massachusetts. It originated in the central part of the Mississippi Valley, near Cairo, Illinois, where the Mississippi and Ohio rivers join—a vast lowland plain, remote from any actively growing mountain range.

The main effects resulted from three major shocks: the first early in the morning of December 16, 1811, and then on January 23 and on February 7, 1812. Shocks of a similar magnitude today, would flatten most of the cities of the central Mississippi Valley. Aftershocks continued for a year following the major quake.

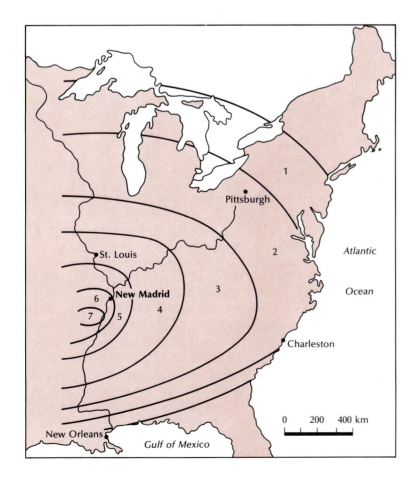

**Fig. 5-5.** The New Madrid earthquake: generalized zones of equal intensity east of the Mississippi. Intensity increases from moderate in Zone 1 to total damage in Zone 7. The shock would have been felt over an area of about 2,500,000 km² (965,250 mi²). (After O. Nuttli, 1973, "The New Madrid Earthquake of 1811 and 1812, Intensities, Ground Motion, and Magnitudes," *Bulletin of the Seismological Society of America,* vol. 63, pp. 227–48, Fig. 1)

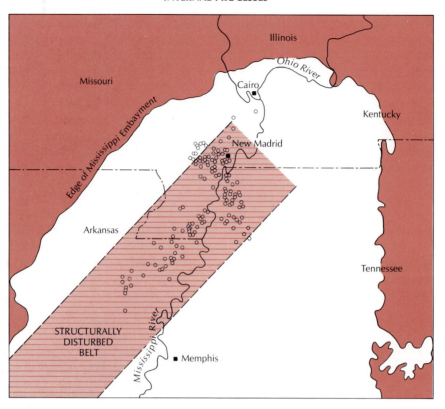

**Fig. 5-6.** Structurally disturbed belt near the apex of the Mississippi Embayment. Recent small earthquakes, shown by circles, fall mostly within the limits of the belt. (After R.M. Hamilton, 1980, "Quakes Along the Mississippi," *Natural History*, vol. 89, pp. 70–74)

The history of that earthquake was better documented than might have been expected because there were several capable observers in the region at the time, including the naturalist John James Audubon (1785–1851). Charles Lyell also visited the region in 1846 when the evidence was still fresh, and left a wealth of observations.

There were three noteworthy geological effects of the earthquake: (1) the appearance of low cliffs cutting across country, very possibly fault scarps, some of which produced waterfalls—around 2 m (6.5 ft) in height—where they intersected the Mississippi River; (2) the elevation of low archlike ridges, up to 24 km (15 mi) in length, along which former swamp areas were uplifted about 3 to 6 m (10 to 20 ft); and (3) the sudden appearance of the so-called sunken ground, a broadly depressed part of the Mississippi flood plain extending along

the river 240 km (150 mi), which is the site of two very large lakes that came into being at the time of the quake—St. Francis and Reelfoot. The latter, now a bird sanctuary, is an imposing sight: Gray trunks of cypress trees, drowned more than a century ago, stand in water up to 6-m (20-ft) deep.

The New Madrid quake was most unusual in two ways: it consisted of three main shocks of extraordinary magnitude occurring over a period of nearly two months and it occurred in an area normally not thought of as earthquake prone. Because of the threat of future large-scale earthquakes, this area has been studied intensively in recent years. The data indicate a wide structurally disturbed belt, containing a number of faults, that trends northeast-southwest (Fig. 5-6). It is thought that the belt is relatively old, having formed at least 500 million years ago, and deep, extending to depths of 40 km (25 mi). Still in question is the source of stress in the region. Some think that it is associated with movements of the earth's tectonic plates; others think that it is related to loading along the southern margin of North America as a result of sedimentation by the Mississippi River during the last few million years.

**Mexico City**   In September 1985 an intense earthquake struck the largest city (population 18 million) in the Western Hemisphere during the morning rush hour. Although an exact death count is difficult to establish, about 8000 people died. To add to the devastation, a second large tremor struck about 1½ days after the first. The quakes actually occurred nearly 500 km (310 mi) west-southwest of Mexico City, just offshore in the Pacific Ocean along the convergence zone that borders Central America and much of Mexico along their Pacific margins. The earthquake was unusual in that it consisted of two nearly equal-sized tremors, each lasting about 16 seconds, with 25 seconds between each. The earthquake waves, propagating outward from the source caused only minor to moderate damage near the origin and in the area between the coast and Mexico City. When the waves encountered the soft lake sediments upon which much of the city is built, however, they were amplified. Many parts of the city shook, and only the stronger, better-reinforced buildings withstood the shaking. Taller buildings, in particular, were affected; in all nearly three hundred multistoried buildings failed. Some of the buildings, weakened by the initial quake, tumbled down during another strong tremor that occurred 36 hours later.

## Japan

The islands of Japan are particularly prone to earthquakes. Since 1700, approximately 200,000 people have died from quakes in Japan; since 1900 there have been more than 25 earthquakes equal in intensity to the San Francisco quake of 1906.

About 100 million people live on the islands, which together make up a land area about equal to that of California. Of necessity, people are crowded into restricted urban areas, a factor that

surely has increased the probability of earthquake damage. Many of the tremors occur just offshore, commonly along the landward side of the deep trench that skirts the islands. Some of the earthquakes are associated with rapid changes in submarine topography, either as a result of offset along faults or slumping of loosely consolidated sediments that have accumulated in bays or other near-shore areas. Not uncommonly, after such seafloor changes, seismic sea waves are generated that devastate coastal population centers. In 1869 an earthquake in the central Pacific sent a wall of water 34-m (111-ft) high against the Japanese coast north of Tokyo, resulting in the death of 27,000 people. The seismic sea wave is now called a *tsunami,* a Japanese word meaning harbor or bay mouth wave, an allusion to the fact that such a wave commonly reaches maximum height and destructiveness in the confines of coastal bays.

## China

China, too, has been beset by earthquakes. As recently as 1976 a large quake struck Tangshan, in northeastern China, producing extensive damage and death to over 300,000 people.

Undoubtedly, the largest loss of life from an earthquake ever recorded occurred during the 1556 tremor in China's Shaanxi region. At least 1 million people perished as a result, most from the collapse of buildings, but many from the famine and pestilence that followed when the countryside lay in ruins.

## Europe

Earthquakes are common in southern Europe, especially around the Mediterranean Sea. Since 1780, Italy, for example, has been struck by six large quakes that caused the death of 150,000 people. As recently as November 1980 a ruinous quake struck southern Italy, devastating many small villages east and south of Naples and killing about 4000 people.

One of the best known of the earth tremors in southern Europe occurred in Lisbon, Portugal, on the banks of the Tagus River, on All Saints' Day, November 1, 1755. There were three major shocks, at 9:40 A.M., at 10:00 A.M., and at noon. Many people were at Mass and perished in the collapse of the many medieval churches, whose spires were so notable a feature of the city. It is no wonder that buildings made as they were of loosely bonded masonry and huddled along narrow, irregular streets collapsed in one thunderous crash.

A startling phenomenon observed at Lisbon and other places along the Portuguese coast was the sudden appearance of a tsunami set in motion by a submarine disturbance off the coast. In Lisbon the wave reached a height of about 6 m (20 ft) as it swept up the Tagus River. Ships and boats were smashed together and sank, and much of the wave's fury was concentrated on a newly constructed marble pier, the *Cais de Pedra,* which was jammed with people fleeing the burning city. Their drowning gave rise to the story that all of them were engulfed and swallowed up by a great fissure— and neither pier nor people were ever seen again. Even Charles Lyell, on the basis of a visit to Lisbon, believed that such an event had occurred. But it is more likely that the quay broke up under the combined effect of the earthquake shock and the attack by seismic sea waves.

## The Middle East and North Africa

In 1978 a very strong quake jolted a remote region in the salt flats of eastern Iran, flattening the city of Tabas, destroying 49 nearby villages, and killing about 25,000 people.

In 1980 Algeria was struck by a devastating quake, which wiped out about 20,000 inhabitants in and around the city of Al-Asnan. This city had been rebuilt after an earthquake in 1954; unfortunately, the new multistoried buildings had not been adequately reinforced. After the 1980 quake, the Algerian cabinet discussed the possibility of moving Al-Asnan to a new location, one less prone to earthquakes, rather than rebuilding the broken city.

## South America

The west coast of South America is also subject to violent earth tremors. Charles Darwin, during his voyage on the HMS *Beagel,* was in Chile in 1835 at about the time that the region near Concepción was struck by an earthquake of moderate severity. On May 21, 1960, a particularly crippling series of earth tremors shook Chile. The first occurred on Saturday and caused widespread damage in and around Concepción; they continued all the next day (Fig. 5-7). At 2:45 P.M., Sunday, an unusually strong shock was felt, and many persons throughout southern Chile left their homes and stood about in the streets. They were still standing there at 3:15 P.M. when the main shock rocked the entire region.

Pierre Saint Amand, a well-known seismologist, provided a lucid account of the first large shock:

The motion of the ground during the main shock was as if one were at sea in a small boat in a heavy swell. The ground rose and fell slowly with a smooth, rolling motion, smaller oscillations being superimposed on larger ones. In Concepción, cars and trucks parked by the side of the road rolled to and fro over a distance of half a meter when they bobbed up and down in response to the movement of the ground. The tops of the trees waved and tossed as in a tempest. Some already damaged buildings fell. The earthquake itself was silent; not a sound came from the earth. The period of vibration was of the order to ten to twenty seconds or more. The shaking lasted fully three and a half minutes and was followed for the next hour by other shocks, all having a slow, rolling motion. . . . In the Region of the Lakes . . . the movement began smoothly and continued for some two minutes, just as in other localities, when suddenly, a loud subterranean noise was heard followed by a sharp jarring motion and a more rapid, less regular vibration of the earth. Similar reports were obtained at other points to the east of the Lakes, and it seems from these that another earthquake took place here . . . while the ground was still shaking from the first shock.*

Aftershocks continued to be felt for months. Saint Amand recorded 119 shocks from the time of the main quake in 1960 until June 1961. Although none was so severe as the main quake, 32 of the aftershocks were very strong. Because the active area was so large, 160-km (100-mi) wide by almost 1600-km (1000-mi) long, and included a wide variety of land features and surface conditions, most of the possible earthquake effects were demonstrated in the Chilean earthquake, with one notable exception. Even though a thorough search was made, both on the ground and from the air, in no place were large offsets of the surface along a fault trace found.

*P. Saint Amand, 1961, "Los Terremotos de Mayo—Chile, 1960," Technical Article 14, U.S. Naval Ordinance Test Station, China Lake, Calif.

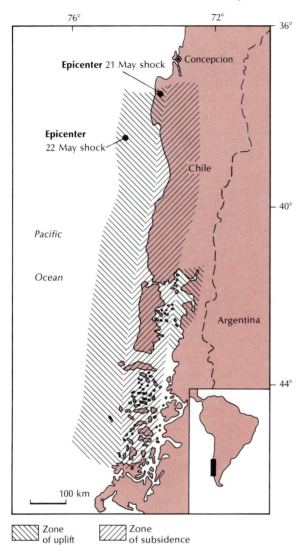

Fig. 5-7. Epicenters of the Chilean earthquake (May 21 and 22, 1960) and the area affected by uplift and subsidence accompanying the earthquake. Incredibly large forces changed the elevation of a considerable area of the earth's crust. (After G. Plafker and J.C. Savage, 1970, "Mechanism of the Chilean Earthquakes of May 21 and 22, 1960," *Geological Society of American Bulletin*, vol. 81, pp. 1001–30)

In the area of the most intense shaking, trees were uprooted and limbs snapped off. In some places the broken branches formed a circular pile of debris on the ground around the trunk.

Flooding by seawater resulted from land subsidence; some areas, however, were elevated 1.5 to 2 m (5 to 6.5 ft) above sea level (Fig. 5-7). The rivers, too, flooded. In places the severe shaking compacted the poorly consolidated or unconsolidated material of the riverbanks, and they were lowered considerably. The earthquake actually shook water out of the ground, so that the rivers were unusually full. Almost continuous rain added to the floodwaters.

Two days after the main shock, the volcano Puyehue in the Andes east of Concepción erupted and continued to be active for several weeks. Steam and ash issued from a fissure about 300-m (985-ft) long and from several smaller openings. The last stage of the eruption was characterized by flows of viscous lava.

The Region of the Lakes lies inland, but it is still well within the area affected by the 1960 earthquake. The lakes exhibited a feature known as a **seiche.** The motion of the quake set up a wave that oscillated back and forth in each lake basin. The water sloshed, much like the familiar wave in a bathtub. Seiches in the Chilean lakes were small compared to those generated in the artificially dammed Hebgen Lake in Montana during the 1959 earthquake. Seismologist John Hodgson wrote:

An eyewitness standing in the moonlight on Hebgen Dam and looking down its sloping face could not see the surface of the water, so far had it receded. Then with a roar it returned, climbing up the face of the dam until it overflowed the top, and poured over it for a matter of minutes. Then the water receded again, to become invisible in the moonlit night. The fluctuation was repeated over and over, with a period of about seventeen minutes; only the first four oscillations poured water over the top of the dam, but appreciable motion was still noted after eleven hours.*

In the Chilean quake a tsunami also developed. Shortly after the main shock, the coastal waters receded well below the lowest low-tide line. The inhabitants along the coast knew from experience what to expect and fled from their homes to nearby

Other surface effects, however, were numerous. Landslides were common; the earthquake in many places triggered the sudden movement of unstable material. Ground cracks were caused by the settling of fill and the subsidence of areas underlain by soil that had liquified. Liquefaction also caused huge earthflows, probably because of the presence of quick clays.

*J.H. Hodgson, 1964, "Earthquakes and Earth Structure," Prentice-Hall, Englewood Cliffs, N.J., p. 34.

**Fig. 5-8.** Adobe buildings in Huaraz, Peru, destroyed by the 1970 earthquake. The epicenter of the main shock was about 150 km (93 mi) west of the town. (USGS)

hills to safety. Within 15 to 30 minutes the sea returned in a mighty wave that reached 6 m (20 ft) at some places and traveled as much as 3 km (2 mi) inland. Large waves continued for the rest of the afternoon, generally diminishing in height, although the highest wave is reported to have been the third or fourth rather than the first one.

Leveling surveys made after the 1960 quake indicated that two strips parallel to the coast, each about 1500-km (933 mi) long by 150-km (93-mi) wide, had changed in elevation; one went up, the other went down (Fig. 5-7). The amount varied locally, but it appears that, on the average, uplift amounted to about 1 m (3 ft) and subsidence to about 1.5 m (5 ft).

Although the 1960 Chilean quake was of greater magnitude, the one that struck parts of Peru on May 31, 1970, caused ten times the destruction and loss of life. That quake was centered about 25 km (15.5 mi) off the coast and caused extensive damage on land. About 30,000 people died, largely as a result of the collapse of buildings (Fig. 5-8). The quake also set in motion a chain of events that led to a catastrophic rock avalanche and the death of thousands more in the towns of Yungay and Ranrahirca (see chap. 11).

## OCCURRENCE AND NATURE OF EARTHQUAKES

Earthquakes occur all over the world. Their distribution, however, as discussed in chapter 4, is not haphazard; they tend to occur in linear belts or zones of seismic activity—zones that mark the boundaries of lithospheric plates. Earth tremors are commonly generated at the plate boundaries where the plates converge, diverge, or slide past one another (Fig. 5-9). Away from the plate edges, earthquake activity is generally absent or minimal.

It is noteworthy that a belt of seismic activity almost completely girdles the Pacific Ocean basin and is generally coincident with a zone of active volcanism (see chap. 8). Both of these phenomena are associated with a discontinuous zone of convergence that also circumscribes the Pacific Ocean. It was the release of strain energy along various portions of this zone that produced the earthquakes in Chile, Peru, Mexico City, and Japan just described.

Many large quakes occur in an east-west swath across southern Europe and Asia. The quakes also apparently mark a zone of convergence of two main plates and perhaps several smaller ones. This

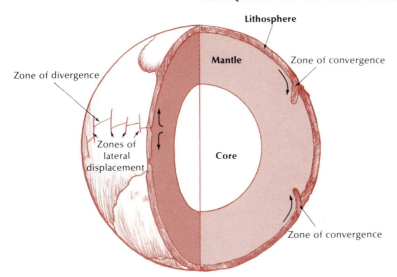

**Fig. 5-9.** The earth, showing the zones where plates converge, diverge, and slide laterally. Zones of convergence are characterized by shallow-, intermediate-, and deep-focus quakes, whereas the other two plate-boundary types are marked by shallow-focus quakes only. (After R. Siever, 1983, "The Earth," *Scientific American*, vol. 233, no. 3, pp. 82–91)

broad belt of activity appears to be associated with convergence of two continental lithospheric plates, one moving from the south, the other from the north.

Several earthquake belts occur within the ocean basins. Those that are coincident with the oceanic ridges and rises, such as the Mid-Atlantic Ridge or the East Pacific Rise, mark zones of divergence, that is, zones where the plates are moving apart. Volcanic activity occurs along such zones as well. One landlocked zone of divergence with both earthquake and volcanic activity runs along the east side of Africa. It is marked topographically by nearly continuous elongate basins—the African Rift valleys.

Earthquakes also occur in zones where two plates slide laterally, one past the other. Such boundaries are relatively abundant in the ocean basins and are represented by the transform faults that cross the oceanic ridges and rises. One of the largest plate-boundary faults along which lateral movement occurs is the San Andreas Fault on the western edge of central and southern California—a structural feature that will be discussed in chapter 6.

Earthquakes rarely originate at the earth's surface; usually, the fault movement starts at some depth below it. The point of initiation of rupture or slippage is called the ***focus.*** The point on the earth's surface vertically above the focus is called the ***epicenter*** and the depth from the epicenter to the focus is called the ***hypofocus*** (Fig. 5-10). Once a rupture starts, it propagates along the fault at several kilometers per second. The total length of the rupture associated with the 1906 San Francisco earthquake was about 430 km (268 mi). Most fault ruptures are much smaller. The earthquake begins when rupture begins and the event continues dur-

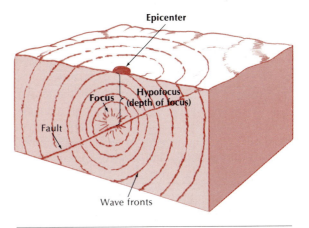

**Fig. 5-10.** Relation of the focus to the epicenter and hypofocus of an earthquake. Also shown are wave fronts of seismic waves radiating out from the focus. The focus represents the point where fault rupture first begins. (After Gilluly et al., 1968, *Principles of Geology, 3rd ed.,* W.H. Freeman)

ing the entire ensuing phase of rupture propagation. In 1983 a quake in southern Idaho was triggered by a rupture that raced for 40 km (25 mi) at a speed of 8000 km/hr (5000 mi/hr).

Arbitrarily, earthquakes that originate at depths less than 70 km (44 mi) are called *shallow focus;* between 70 and 300 km (43 and 187 mi), *intermediate focus;* and between 300 and 700 km, (187 and 435 mi) *deep focus.* Most earthquakes fall into the category of shallow focus. None has ever been recorded from depths in excess of 700 km.

Along the plane of rupture, strain energy is converted and released in two different ways, as heat and as seismic wave motion. About half of the released energy goes into heating the rocks directly adjacent to the fault zone. The seismic vibrations, on the other hand, travel away from the focus in all directions, much as ripples radiate after a pebble is dropped into a quiet pond. As the waves travel outward, the energy they carry is distributed over a progressively larger perimeter, and the intensity of wave motion diminishes accordingly.

Some seismic waves travel through the interior of the earth and are called *body waves;* others travel at, and close to, the surface of the earth and accordingly are called *surface waves.* It is the motion of surface waves that we feel during an earthquake. Such waves can be set in motion either directly, by fault movement at, or close to, the surface or, indirectly, by body waves that emanate from the focus and strike the ground near the epicenter directly above. The violence and complexity of the shaking at the surface is related to the magnitude of the released strain energy, the proximity of the focus, the duration of the quake, and the nature of the surface materials. Unconsolidated mud will develop much larger waves than will crystalline igneous or metamorphic rocks. We will discuss body and surface waves in more detail later in this chapter.

## PEOPLE AND EARTHQUAKES

What can be done to modify the destructive effect of earthquakes? It is inconceivable that urban areas will be moved to regions that are seismically less active. Managua, Nicaragua, has been devastated at least four times by earthquakes, the last in 1972; yet its residents continue to rebuild. California is hit by hundreds of quakes each year, still its population continues to grow. And there is also the possibility even in the seismically quiet regions of the world of a rare but devastating quake, such as the one in New Madrid, Missouri. Hawaii, although in the middle of the Pacific plate, is not free from severe tremors.

At present there are two approaches to the problem of possible earthquakes. One is concerned with establishing general awareness of the effects of quakes and with enforcing building codes; the other is concerned with predicting and controlling earthquakes themselves.

It has become apparent that through careful selection of building sites and the establishment of minimum construction requirements, the effects of earthquakes can be lessened. In major earthquakes buildings will be most severely damaged if they stand on filled, unconsolidated, clay-rich or water-soaked ground and least damaged if they stand on solid, well-cemented sedimentary or crystalline rock.

In the 1985 Mexico City quake, most damage was in areas of soft, wet lake sediments that vibrated at just the right frequency to set the buildings above in violent rocking motion. In destructive 1964 quakes at Niigata, Japan, and at Anchorage, Alaska, clay-rich soil was the culprit, liquefying during the quake and causing large land areas to slump, tilt, and wallow. Selection of a building site is a factor over which we can exercise some control; yet it is surprising how little thought has been given to this, even in regions of high earthquake hazard.

The element over which we have the most direct control is type of building construction. A great deal of information on how to make buildings more earthquake proof has been accumulated since the 1906 San Francisco quake. During that tremor most large structures came through with only minor damage (Fig. 5-11). Of 52 large downtown buildings, only 6 were severely affected. The principal reason most survived, however, was that they had been designed to stand up to strong winds— no one even considered the possibility of devastating ground motion from earthquakes. Even after the San Francisco event, interest in minimizing

**Fig. 5-11.** Most major buildings survived the 1906 San Francisco earthquake, including the Fairmont Hotel, shown here. Many, however, were damaged by the fire that subsequently swept the city. (Wells Fargo Bank, History Department)

earthquake destruction through better building practice lagged in the United States for nearly three decades. Extremely severe quakes in Messina, Sicily, in 1908 and Yokohama, Japan, in 1923, however, precipitated tougher building codes.

The problem with this approach was that by 1923 there was not much hard information about how to build earthquake-proof buildings. The popular logic was to erect strongly braced, heavy buildings on solid-rock foundations. In the 1923 Yokohama quake many buildings so built survived; however, some did not. One maverick American architect, Frank Lloyd Wright (1867–1959), designed the Tokyo Imperial Hotel to "float" on short pilings driven into weak shale. The hotel was constructed of 12 strongly reinforced segments, each designed to move independently. The Imperial came through the quake virtually unscathed and Frank Lloyd Wright became famous.

In the United States little progress in building design occurred until after an earthquake in Long Beach, California, in 1933 killed 120 people and caused extensive building failure. Particularly, there was a public outcry because 38 of 42 major masonry buildings in the public school system were in such bad condition after the quake that they had to be torn down. It was clear that had the tremor hit during school hours, the death toll would have soared. A law passed soon after required that new buildings, especially schools, should be able to withstand maximum lateral pushes equal to one-tenth of the building's weight. Among the more memorable sights of the Long Beach quake were the hundreds of brick chimneys snapped off at the roof line, a result that prompted a change in the building code. Steel reinforcing rods were to be used in masonry construciton, and they had to extend the entire length of all brick chimneys.

By the 1930s it became obvious that engineers lacked adequate data not only about maximum expectable ground motions and accelerations during large tremors, but also about the response of buildings to the ground motion. The installation in California of strong-motion seismographs capable of recording the peak shocks near a quake epicenter soon began to demonstrate expected ground motions. Surprisingly, they were much greater than previously suggested. In the 1971 quake in San Fernando Valley, California, horizontal accelerations were about 1.25 the acceleration of gravity ($9.8$ m/sec$^2$; $32$ ft/sec$^2$) and in the Imperial Valley, California, quake of 1979, peak vertical accelerations nearly twice that of gravity were recorded.

During the 1930s and 1940s seismic-simulation experiments began. Earthquake motions were sim-

**Fig. 5-12.** Shortly after dawn on Thursday, September 19, 1985, Mexico City was rocked by a severe earthquake. Over 800 buildings were totally destroyed and 300 more were seriously damaged. (Michael S. Reichle, Department of Conservation, Division of Mines and Geology)

ulated in the laboratory; various models of buildings were then shaken to assess their abilities to withstand the shocks. In 1934 a device was built at Stanford University that could shake actual buildings. The energy imparted by the machine was not enough to cause building failure but it did enable identification of the principal resonating frequencies of the buildings. It was found that if some of the seismic waves corresponded to the resonating frequency, the building would absorb energy, the amplitude of shaking would increase, and eventually the building would collapse. Destruction during the 1985 earthquake in Mexico City is an example. Lake beds beneath the city, in response to the surface waves emanating from the focus nearly 500 km (310 mi) to the west, were sensitive to wave frequencies of about one wave cycle every 2 seconds. This energy, absorbed by the sediments, increased until the ground was moving back and forth about 40 cm (16 in.) every 2 seconds. Buildings with resonating frequencies near that of the shaking ground collapsed (Fig. 5-12). An additional factor was that the quake was of unusually long duration.

After the 1964 earthquake in Anchorage, Alaska, it became apparent that large buildings needed extra reinforcement to withstand peak seismic stress. A new design concept also emerged. It was determined that less total damage would occur if a building were to absorb some seismic energy through controlled partial failure. In Los Angeles, the building code was changed, and in 1966 the city removed the 13-story limit that had been placed on buildings. Soon after, the 64-m (210-ft) Sheraton-Universal Hotel went up and other tall buildings followed, dramatically changing the skyline of downtown Los Angeles.

The 1971 San Fernando Valley earthquake also brought about a change in building codes. Until then, buildings were constructed with a flexible, structurally soft floor near ground level to absorb ground motion, thereby preventing the top floors from excessive swaying. During the 1971 quake, however, several new buildings in the Olive View area—among them hospitals and medical centers—were heavily damaged; nearly all of them had utilized soft floors. Later experiments showed unequivocally that soft floors concentrate stress, thus increasing the chance of failure. Roadways were also affected during this quake, which prompted the California Highway Department to change the standards for new freeway bridges and ramps and to examine 13,000 pre-1971 bridges, one-tenth of which were subsequently strengthened (Fig. 5-13).

Engineers are experimenting with design changes to make buildings even safer. Foundations set on rubber pads or roller bearings to insulate buildings from seismic shock are being tested, as is the use of compressed air delivered in powerful jets to counteract building sway.

In earthquake-prone regions, it is necessary to consider the geological structure when determining the placement of major transportation routes, aqueducts, pipelines, water-storage facilities, power plants, and the like. Geologists now know the major fault traces in most areas and usually can tell which ones are active or are likely to have been recently active. Most municipalities construct their vital industries with that knowledge in mind. For example, in the selection of building sites for nuclear power plants, areas with faults that appear to have been recently active are avoided.

During the last century, then, much has been done in construction practices to lessen potential earthquake damage, and building codes in earthquake-prone cities reflect that progress. Unfortunately, where quakes are rare, such codes are not generally in use. Yet the record over the last 200 years shows that almost any area in the United States could be struck by a devastating tremor.

Regardless of how carefully building sites are selected and how well structures are built, a major earthquake in a populated area will cause loss of property and life (Fig. 5-14). In fact, as urbanization increases, so does the risk from earthquakes. In Mexico City the death toll and property loss from the 1985 quake, which damaged only a relatively small area, amounted to about 8000 people and $5 billion dollars. A 1980 study of the effects that would likely be caused by such an earthquake near Los Angeles indicated about $69 billion in damages and a death toll of about 20,000.

The answer to this dilemma is **earthquake prediction.** If only we could foretell earthquakes, population centers could be evacuated and some preparations could be made to lessen loss of life and property. Earth scientists, particularly in those

**Fig. 5-13.** Earthquake damage to freeway under construction, San Fernando earthquake, 1971. Geological hazards in California have led to public safety legislation. (Newhall Signal, California Division of Mines and Geology)

countries where earthquakes are prevalent and money is available (China, Japan, United States, and the USSR), are trying to reach these goals, and earthquake prediction has been given a high priority. In China earthquakes have been designated the "number-one naturally occurring enemy of the people," and a great deal of effort has been directed toward reliable prediction. One type of long-range seismic forecasting is based on analyzing the distribution and intensity of past earthquakes. From such analysis, seismic-risk maps can be made that indicate the likelihood of earthquakes of a certain size for any geographical area (Fig. 5-14). Such maps, however, do not show the recurrence rates of large earthquakes. Both western California and the area surrounding Charleston, South Carolina, are shown as zones of poten-

tially severe quakes. However, only a single major quake has occurred near Charleston in the last 100 years—that in 1886, which lasted 8 minutes and killed 60 people. On the other hand, many quakes have been recorded over the years in western California. A part of eastern Massachusetts is a high-risk area because of the isolated occurrence in 1755 of an earthquake in the area of Cape Ann, about 100 km (62 mi) from downtown Boston.

Studies of small-scale features in some areas have made it possible to determine general recurrence rates of large quakes. For example, the average length of time between large earthquakes along the San Andreas Fault near Los Angeles is about 200 years. Some occurred at 50-year intervals, others at 250-year intervals. The last great movement along the fault in 1857 was one of the largest seismic events in California. An earthquake of the same intensity today would cause much death and destruction in Los Angeles.

Along the active Wasatch Fault in central Utah, the recurrence rate is about one major quake each 450 years. At this time the experts can only say that the next large-scale rupture is less than 500 years off. They can predict, however, that the fault segments likely to tear will be near Salt Lake City or Provo.

Another long-range prediction technique, generally applied to regions and seismic belts in which big quakes have commonly occurred, is to identify the points of origin of recent large events and their associated aftershocks; these can then be plotted on a map. The technique is based on the premise that strain uniformly accumulates along the length of any seismic belt but is intermittently released only over small segments of the belt. Segments of a belt that have not undergone recent strain release are called **seismic gaps** and are considered prime regions of potential seismic activity. Once such gaps are identified, they can be closely monitored.

In 1979, two U. S. government scientists determined the existence of a seismic gap just off the shore of Peru. They predicted three quakes for the last half of 1981 that would be capable of devastating the entire western coast of the country. Yet this area continues to be free from large quakes. Needless to say, at the time, the prediction caused quite a bit of consternation among the local inhabitants.

**Fig. 5-14.** Seismic risk map of the United States, based on the known distribution and extent of damaging earthquakes, evidence of strain release, and the study of major geological structures and provinces believed to be associated with earthquakes. From the standpoint of seismic risk, Alabama, Florida, Louisiana, and southern Texas are the least hazardous places to live in the United States. (After Uniform Building Code, 1982 ed. Copyright © 1982, with permission of the publisher, the International Conference of Building Officials, Whittier, California)

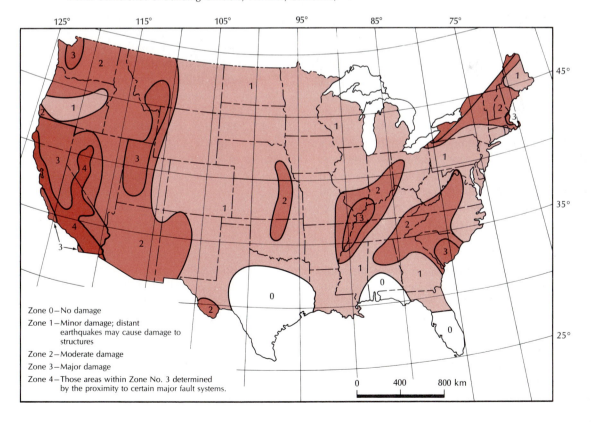

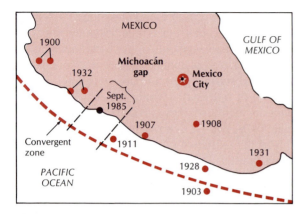

**Fig. 5-15.** Michoacán seismic gap in the convergence zone along the west coast of Mexico. This zone, which had long been inactive, produced a very large, double-event earthquake in September 1985, killing about 8000 people in Mexico City. The circles represent epicenters of quakes of large intensity that have occurred during this century.

The focus of the 1985 Mexico City earthquake was on the convergence zone along the Pacific margin of Mexico. As shown in Figure 5-15, there were relatively few major quakes in this century along that part of the belt, which was designated the Michoacán seismic gap.

Hopes for finding a short-range prediction method have centered on the measurement of the amount of strain that has accumulated along a specified portion of a locked fault plane and the rate at which it continues to accumulate. Earth scientists watch for changes, such as tilt or strain, in the relative and actual positions of points on the earth's surface—either by geodetic or other measurement. They also look along the fault for alterations in the physical properties of the rocks, which are affected by strain buildup. Such changes may be determined by measuring the local magnetic field, the electrical conductivity of the rocks, or the velocities of seismic body waves (Fig. 5-16).

Commonly, the water levels in wells and rates of groundwater flow are anomalously affected during strain buildup. It has also been discovered that in some wells and along some faults the concentration of the radioactive isotope radon (one of the daughter products of uranium decay) greatly increases prior to strain release.

Scientists also set up sensitive seismic instruments along a fault zone to "listen" to the crack-ling of the earth as strain builds prior to fault slippage. As the time of slippage approaches, the number of tiny earthquakes over a given time interval usually goes up. An increase in microactivity may account for the common observation that dogs and horses become nervous and jittery just before a quake. Recently, in fact, different animals have been watched to see if their behavior alters prior to an earthquake. In California, for example, dogs, horses, laboratory rats, and primates are housed near the San Andreas Fault, where they are continually observed. Unusual behavior immediately preceding a shock is closely examined and recorded. No clear-cut results have been obtained so far, but some researchers are convinced that this line of study eventually may prove to be one of the best ways to foretell earthquakes.

Many short-term predictive techniques are being tested, but none has been found to signal the advent of an earthquake. Hindsight, however, suggests that some earthquakes might have been predicted, whereas others could not have been. For example, in 1975 about 90,000 people were evacuated from the Chinese city of Haicheng in response to the prediction of an imminent large earthquake. Shortly thereafter, a very large quake did occur and thousands of lives were saved. In 1976, however, over 300,000 people perished in the city of Tangshan in an earthquake that was preceded by few signals.

In 1979 the prospect of accurate earthquake forecasting was further dampened by a moderate quake in an area well monitored by geophysical

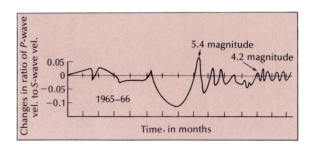

**Fig. 5-16.** Two earthquakes in the Garm district, USSR (1956–66), were preceded by drops in the ratio of P-wave velocity to S-wave velocity. (From C. Kissingler, 1974, *Earthquake Prediction*. By permission of C. Kissingler)

equipment, about 100 km (62 mi) south of San Francisco along the Calaveras Fault—a major branch of the San Andreas Fault. No precursor signals whatever were recorded.

By 1985 hopes of finding a reliable short-range prediction method began to glimmer again. They are in part pinned on the observation of the recurrence rates of moderate-sized earthquakes along designated sections of seismically active zones. By comparing the quake activity over a lengthy time interval, the shorter, unusually quiescent intervals can be identified. The data indicate that these times of lowered activity are commonly followed by earthquakes. Thus far this procedure has been successful in foretelling one large quake in the Aleutian Islands and two moderate-sized ones along the central portion of the San Andreas Fault. The approach is related to the long-range technique of identifying seismic gaps; however, it seems better able to provide predictions over geographically local regions and in time intervals as short as a year.

Another potentially useful technique is based on precise geodetic measurements of changes in baseline distances across specified portions of active major faults. Light beams, (e.g., laser pulses reflected by mirrors or satellites, star and quasar light received from billions of light years in space) are used to make the measurements.

Currently, earthquake predictions must be channeled through the scientists of the National Earthquake Prediction Council. The council then advises the U.S. Geological Survey (USGS). If the prediction is deemed valid, USGS issues an alert and contacts the governors of the affected states. California, because of its high earthquake potential, has its own evaluation council that cooperates with USGS.

Earthquake prediction has a long way to go before it will be a useful tool to humankind. If we do acquire the ability to predict accurately an earthquake of a certain magnitude, another problem arises: How will citizens in the affected area respond? For example, if a potentially disastrous earthquake were predicted for the San Francisco Bay area, what would be the effect on the populace? Many, perhaps most, would panic and flee the danger area. Accidents and deaths would probably occur in the mass exodus. Who would

be held responsible? What if the alarm were false and, on returning, people found their homes ransacked, their pets lost or dead, their city looted? Who would shoulder the blame? What if, after the citizens returned to San Francisco, a severe earthquake occurred, that caused extensive destruction? How would the survivors react? What would be the effect on the economy of a city such as Los Angeles if a large earthquake were predicted within the next 2 years?

These and other questions have led some scientists to conclude that the only way to deal with earthquakes is not to predict them but to control them. However, is there any evidence that we can do so? The answer, to a degree, appears to be "yes," and the approach to controlling them centers around a seemingly unlikely substance—water. About 20 years ago, a deep well was drilled into the Precambrian rocks underlying the Rocky Mountain Arsenal near Denver, Colorado. The purpose of the well was to dispose of radioactive waste fluids. Beginning in 1962 and continuing for 3 years, many thousands of gallons of waste fluids were pumped down the well. Shortly after pumping began, earthquakes began to occur in the Denver area. Officials at the arsenal denied any connection between the pumping and the earthquakes, but a Denver geologist noted that the frequency of quakes was directly related to both the amount of wastewater pumped in during any particular period and the pressures used to pump it in (Fig. 5-17). In 1965 pumping was stopped and earthquake activity slowly ceased.

Most earth scientists have concluded that the Denver earthquakes resulted from the release of tectonic strain in the Precambrian rocks adjacent to a fault. Before pumping, the walls were locked, and no faulting or quakes, resulting from fault movement, were recorded. By forcing wastewater down the hole, water pressure at depth was generated, sufficient to force the walls of the fault apart and thereby effectively "lubricating" it. The fault became unlocked and slippage took place.

Such incidents have been documented elsewhere, leading geologists to speculate that pumped water might be used to control fault movement. Take the San Andreas Fault as an example. In a segment of the fault, three equally spaced wells could be drilled along the fault plane to a depth

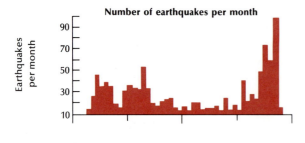

**Number of earthquakes per month**

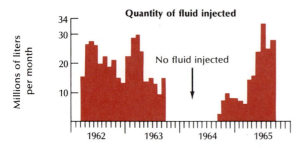

**Quantity of fluid injected**

No fluid injected

1962     1963     1964     1965

**Fig. 5-17.** Correlation between quantity of wastewater pumped into a deep well and the number of earthquakes near Denver, Colorado, 1962–65. (After D.M. Evans, 1966, "Man-made Earthquakes in Denver," *Geotimes,* vol. 10, pp. 11–18)

of 4600 to 6100 m (15,000 to 20,000 ft). Groundwater could be pumped out of the two wells at either end of the segment, thus drying up those parts of the fault and effectively locking them. Then enough water could be pumped down the center well to unlock the fault and permit movement. Supposedly a small-magnitude quake would result; afterward that section of fault would be temporarily free from strain. The process of drying out and lubricating could be repeated up and down the fault until strain energy had been released along its entire length.

Many unknowns surround such an approach to earthquake control. The principal concern is that large, potentially destructive earthquakes could be triggered during the early stages of strain release. In our politically sensitive environment, no government agency is ready to accept the responsibility of risk associated with this method.

The same could be said about earthquakes that is said of the weather: Everyone talks about it, but no one does anything about it. Perhaps in the near future this will change—but don't count on it for some time to come.

## MEASURING THE HEARTBEAT OF THE EARTH—SEISMOLOGY

Earthquakes, simply because tthey are dramatic natural events, have long invited speculation as to their origin. Greek philosophers devised many explanations that today read like mythology. For example, Aristotle, who lived in the fourth century B.C., believed that earthquakes resulted from the escape of air trapped deep within the earth.

By 1700 observations of swinging chandeliers led to the conclusion that earthquakes were associated with wave motion in the rocky crust of the earth. In 1859, Robert Mallet (1810–81), an English engineer, visited the hill towns in the Apennines east of Naples, Italy, to observe the destruction wrought by a quake in 1857. Although he believed, incorrectly, that earthquakes were explosive in origin and were somehow related to volcanic activity, he laid the foundations of observational *seismology,* (from the Greek word *seismos,* "earthquake"). By observing the directions buildings and monuments fell and the nature of cracks in the ground, he worked out a rough method for locating the source of an earthquake. Mallet was among the first to try an experimental approach to earthquake study by setting off explosions and measuring the travel times of the resulting waves. He also started an earthquake catalog, achieving some understanding of the geographical distribution of earthquakes.

### The Seismograph

The person generally given credit as the founder of modern seismology was an English mining engineer, John Milne (1850–1913). Through his efforts the seismograph, the instrument used to record the vibrations set up by an earthquake, was transformed from a scientific curiosity into a precision instrument.

The instrument was perfected over the many years after it was first invented in Italy in 1855 by Luigi Palmieri (1807–96). The initial problem was how to measure the vibrations from an earthquake at the earth's surface when the instrument itself is attached to the surface and is shaking. The solution was relatively simple. A large mass was suspended from a hook on a thin filament (forming a pen-

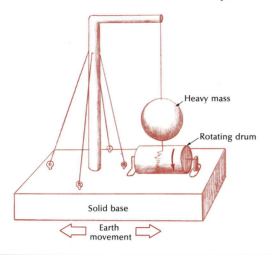

**Fig. 5-18.** A simple model of a seismograph. A heavy mass is suspended from an arm on a thin wire. During an earthquake the base (solidly attached to the earth) moves as does the arm, but the mass because of its inertia is relatively unaffected. (After Arthur N. Strahler, *The Earth Sciences, 2nd ed.,* p. 394, Fig. 23–14A. Copyright © 1963, 1971 by Arthur N. Strahler. Reprinted by permission of Harper & Row)

be determined to the hour, minute, and second. For ease of comparison of records, all stations around the world have agreed to use the same time base—Greenwich civil time.

A more complicated problem occurred with the early seismographs. During a quake the ground moves simultaneously in three dimensions: east–west, north–south, up–down. A single pendulum suspended vertically cannot record this three-dimensional movement. In the late 1800s, however, Milne hit on the idea of using three separate seismographs at each station, one for recording each of the motions (Fig. 5-19). To measure the horizontal movements he mounted two pendulums in horizontal positions at right angles to each

dulum), thereby virtually isolating the mass from effects at the earth's surface. The mass was, in essence, in space (Fig. 5-18). When earthquake waves set the ground into motion at the recording site, the mass, because of its large inertia, tended to remain at rest while the ground beneath it and the hook on which the filament was suspended shimmied and shook. A stylus, or pen point, was attached to the mass so that it could trace the motion of the ground as it moved to and fro.

The first primitive seismographs consisted of a single, vertically suspended pendulum, and the record was traced in smoothed sand or on a stationary piece of smoked paper. Because the stylus traced its path back and forth past the same spot on the ground, each earth tremor produced a bewildering array of lines. The problem was overcome when the recording paper was attached to a plate or a drum that moved beneath the pen at a known rate; thus the pen could not write over its previous trace (Fig. 5-18). In most modern seismographs the pen records on a drum that turns at four revolutions per hour. As the drum revolves it slowly spirals ahead; in essence it is a spiraling clock. The pen also makes a tick mark after each minute, so that the onset of any motion can easily

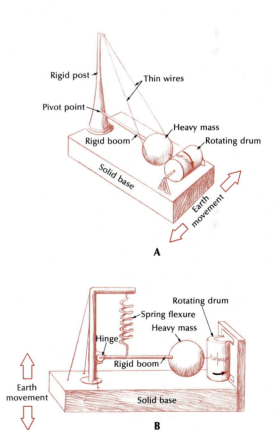

**Fig. 5-19.** Basic design of modern seismographs. **A.** Horizontal-type pendulum seismograph for measurement of north–south or east–west motions. **B.** Hinged pendulum seismograph for measurement of vertical motion. (After Arthur N. Strahler, *The Earth Sciences, 2nd ed.,* p. 401, Figs. 23–15 and 23–14B. Copyright © 1963, 1971 by Arthur N. Strahler. Reprinted by permission of Harper & Row)

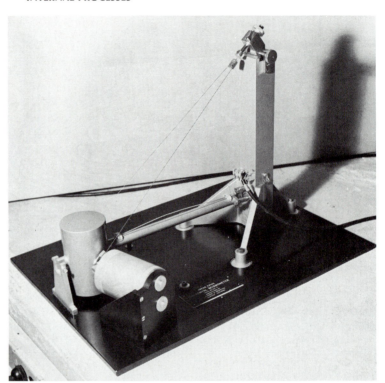

**Fig. 5-20.** Modern seismometer used to record either the east–west or north–south horizontal ground motion associated with seismic waves. (Lamont-Doherty Geological Observatory)

other; to measure vertical movement, he suspended a pendulum vertically. This breakthrough made possible quantitative seismology. By 1882 the new seismographs had been installed at 50 cooperating stations around the world. Scientists were then able to measure and compare records of the most complicated earth movements.

Other problems, such as amplification of signals and damping of vibrations set up in the seismograph by a quake, also had to be overcome as the seismographs of the nineteenth century gave way to more modern instruments. Several recent improvements in design have resulted from research in electronics and experimental physics.

Today's precision seismographs are relatively easy to use (Fig. 5-20). Most are installed at permanent stations, but portable models are also set up to record quakes at certain strategic locations for a few days, weeks, or months and then moved elsewhere.

Some seismographs are built specifically to monitor the intensely strong vibrations close to the epicenter of an earthquake; others are made to measure extremely weak seismic signals. Most seismographs can be tuned to pick up long or short seismic waves. Special instruments record the exceedingly long wavelengths associated with very large quakes; the passage of one wave takes about an hour. These waves, called *free oscillations,* represent the fundamental vibrations of the earth and are similar to the ringing of a bell. They are excited only by the energies of very large quakes.

More than 1000 seismographs are operating around the world today. Since the early 1960s about 120 of the stations have transmitted records to the USGS Earthquake Information Center at Golden, Colorado. Within minutes of an earthquake anywhere in the world, a realistic estimate of the epicenter location, hypofocus, magnitude, and damage can be determined. Most seismologists today use the USGS services and do not operate seismograph stations themselves.

**Seismograph Records** Most of the time the pens on seismographs are quiet, recording only low-

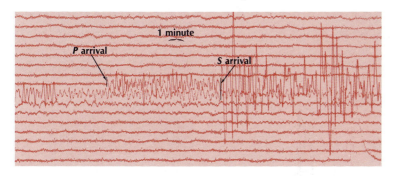

**Fig. 5-21.** Portion of a seismogram from Berkeley, California, recording two seismic events. Succession of dots along each line of the record represents 1-minute intervals. As shown in the first event, the *S* wave arrives about 3 minutes after the *P* wave. Notice that before and after passage of the earthquake waves, the record contains only low-level background noise.

level background vibrations from such sources as traffic, wind gusts, and even the pounding of heavy surf on a distant coast (Fig. 5-21).

From time to time, however, the seismographs will record waves emanating from an earthquake somewhere in the world. If a quake is relatively close by, the instrument generally will record three major wave pulses arriving in succession. The first to arrive, generally a relatively low-amplitude wave, is called the *primary* or *P wave*. The second pulse, the *secondary* or *S wave*, is distinguished by its sudden onset and is normally of greater amplitude than the *P* wave. The last wave to arrive is of higher amplitude and longer wavelength—the *long* or *L wave*. Depending on where the quake is located relative to the recording station, each of the three seismographs (E–W, N–S, and vertical) will show some variation in character in each of the three sets of waves.

The *P* and *S* waves travel through the earth and accordingly are called *body waves,* whereas the *L* waves follow paths in the outer layers of the earth and are accordingly called *surface waves.*

The nature of motion for each type of wave has been studied theoretically and in the laboratory. It has been found that *P* waves travel through a solid as an accordianlike push–pull sort of oscillation (Fig. 5-22). As the wave travels, the rock at any one place is compressed at one moment, expanded at the next. This type of wave motion is characteristic of sound waves as they move through the air and into our ears. Sound, as we hear it, is

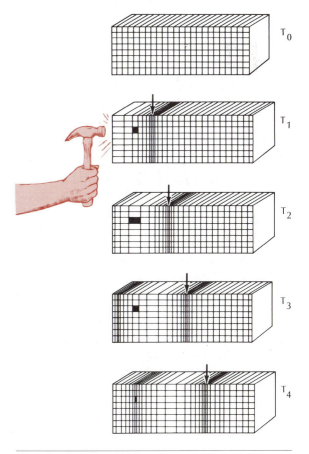

**Fig. 5-22.** Successive stages in the small-scale deformation of a rock by a *P* wave. As the sequence progresses, a wave of maximum compression (arrows) passes through the rock. As wave after wave pass through the block, the darkened square (representing any small volume of the rock) shakes back and forth, undergoing alternate compression and expansion. (After O.M. Phillips, 1968, *The Heart of the Earth,* W.H. Freeman)

simply the recording of the to-and-fro motion of our eardrums in response to the push–pull waves moving through air. Sound travels in a similar way but more quickly in water than in air. Sonar (the beaming and receiving of sound signals underwater), can locate nearby ships, pinpoint obstacles, and gain a knowledge of the topography of the ocean floor.

Compressional sound waves travel fastest in a solid because the atoms are tightly bonded to each other and, thus, are highly elastic. Any increase in degree of elasticity will cause the $P$ waves to be transmitted faster. For instance, when a rock is compressed in a hydraulic press, the atoms in the rock are pushed closer to one another, thereby increasing both the rock's elasticity and the speed with which $P$ waves travel through it. Therefore, $P$ waves travel more quickly deep inside the earth—where pressures are extremely high—than they do near the earth's surface. On the other hand, increasing the temperature of a rock specimen will tend to reduce its elasticity and will correspondingly reduce the speed of $P$ waves. Obviously, at any point inside the earth, the elasticity will reflect rock type, confining pressure, and temperature.

The $S$ wave always travels more slowly than the $P$ wave because of a fundamental difference in the manner in which it travels. In a solid, seismic energy can be transmitted by a shaking motion transverse to the direction of wave travel (Fig. 5-23). In that mode, the rock at any one point moves back and forth sideways, whereas the $S$ wave travels forward in a snakelike fashion. It is much like the motion made by shaking a rope; the waves travel in S-shaped patterns as the rope moves from side to side. Rocks exhibit less elasticity to S-wave motion, therefore the $S$ wave is slower than the $P$ wave. The $S$ waves will also increase in speed with increased pressure. Thus at depth in the earth, the $S$ wave is expected to move more quickly than it does near the surface. An increase in temperature, of course, will decrease a rock's elasticity and slow the $S$ wave. Fluids and gases completely lack the type of elasticity that will permit the passage of $S$ waves; therefore such waves cannot travel through water, air, or any other fluid, such as magma.

The $L$ waves, slowest of the three, travel in complex ways close to the surface of the earth.

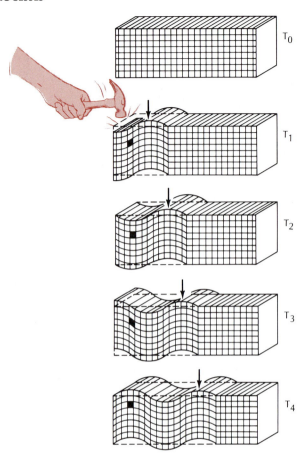

**Fig. 5-23.** Successive stages in the small-scale deformation of a rock by an $S$ wave. As the sequence progresses, the $S$-wave crest (arrows) passes through the block, and the rock shakes sideways. Any particular volume (darkened block) undergoes cyclic changes in shape in response to the passage of the waves. (After O.M. Phillips, 1968, *The Heart of the Earth*, W.H. Freeman)

Because their passage is restricted to the outer levels of the earth, their speed is relatively constant. Surface waves commonly vary in wavelength. Waves of shorter lengths travel close to the surface, whereas longer waves move through the deeper layers of the near-surface zone.

All earthquakes generate all three types of waves. How they appear on a seismograph will depend on (1) the length and nature of the rupture, (2) the distance from the focus to the seismograph station, (3) the depth of the focus, and (4) the kind of material through which the waves travel before

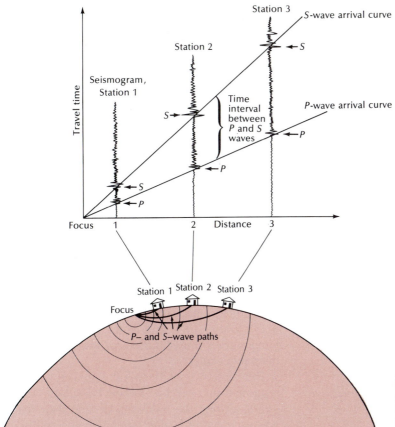

**Fig. 5-24.** Increase in the arrival time of *P* and *S* waves with distance from the focus. Notice that the time interval between the arrivals also increases. As shown, the *P* wave is faster than the *S* wave.

reaching the station. By determining the first three aspects, seismologists can gain a great deal of information about the last one.

Seismologists can gain insight into how seismic waves travel by plotting the arrival time of earthquake waves at stations located at increasing distances from an epicenter (Fig. 5-24). If many stations are used, it is possible to construct a ***travel-time curve*** (Fig. 5-25). As shown, the plot for each wave is decidedly different. The L-wave plot is almost a straight line, whereas the *P* and *S* waves plot along curved paths. Of the body-wave plots, that of the *P* wave bends more sharply. Moreover, from about 10,000 km (6220 mi, equal to about 102° of earth circumference) to about 16,000 km (9950 mi, equal to about 143° of earth circumference) from the epicenter, no *P* and *S* waves are directly recorded. It is as if seismographs in that

portion of the globe were in the shadow of some seismic obstacle. Accordingly, that zone is called the ***shadow zone.*** At arc distances beyond 143°, seismic waves are again received, but their travel time is much longer than expected, as if they had been slowed during their passage through the central part of the earth.

During an earthquake both *P* and *S* waves begin at the same time. However, a *P* wave is about twice as fast as an *S* wave, and the farther the two waves travel, the wider becomes the gap separating the wave fronts (Figs. 5-24, 5-25). The difference in arrival times between *P* and *S* waves provides seismologists with a means of determining the distance the earthquake focus is from a seismograph station. For example, in a seismogram recorded at Berkeley, the *S* wave arrived about 8 minutes after the *P* wave. The travel-time

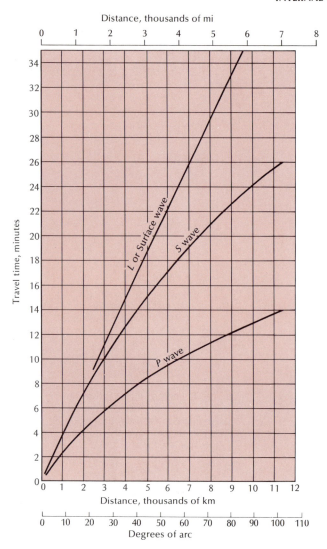

**Fig. 5-25.** Travel-time curves for earthquakes originating at a depth of less than 100 km (62 mi). Notice how the *P*- and *S*-wave curves bend, indicating an increase in speed with distance. (After C.F. Richter, 1958, *Elementary Seismology*, W.H. Freeman)

contain all the possible points of origin (Fig. 5-26). If the same quake is recorded at a second station, however, it is possible to narrow down the location of the quake. From the second record, one can similarly determine the distance from the second station to the epicenter and draw a circle of the appropriate radius around it. This circle will intersect that drawn around the first station at either one or two points. If information is available from a third seismograph station, a single map intersection—the epicentral location—can be determined. In practice, the three circles do not usually intersect at a single point. For one thing, many earthquakes, although originating at a point, involve total movement all along a fault line. It has also been demonstrated that the earth's interior is somewhat heterogeneous, such that wave velocities in different parts are not the same. And last some quakes originate at depths as great as 700 km (435 mi). The travel-time curves in Figure 5-25 are useful for shallow-focus quakes, but different curves must be used for intermediate- and deep-focus quakes.

The existence of deep-focus quakes was first suggested in 1922 by Herbert H. Turner (1861–1930). During the following decade their reality was accepted only after much debate and discussion. Part of the evidence for their existence is the different sort of seismogram they write when compared to that written by shallow-focus earthquakes. The *L* waves are either lacking or ambiguous, and the *P* waves arrive sooner than they would have, had they started near the surface. After all, they have the advantage of a considerable head start in their race toward a seismograph station. As already noted, deep-focus quakes commonly originate in zones of convergence. The nature of rupture associated with intermediate- and deep-focus quakes is not exactly clear, because at depths of several hundred kilometers the rocks are probably so hot that they tend to flow rather than accumulate elastic energy. Yet the quakes do occur, and the amount of energy released during any one event is about the same as that from a near-surface quake.

Nowadays high-speed computers analyze data from many stations. After the focus is approximately determined by the use of travel-time curves, the position of the focus is figuratively moved in small steps laterally and vertically and the travel

curve (Fig. 5-25) indicates that the difference in arrival times is appropriate for a quake originating about 6600 km (4105 mi) from the recording station. Seismologists, however, cannot tell from which direction the quake came nor, on the basis of a single record, pinpoint the epicenter. They can only say that the epicenter was about 6600 km from the station. A circle of that radius drawn around the seismograph station on a map will

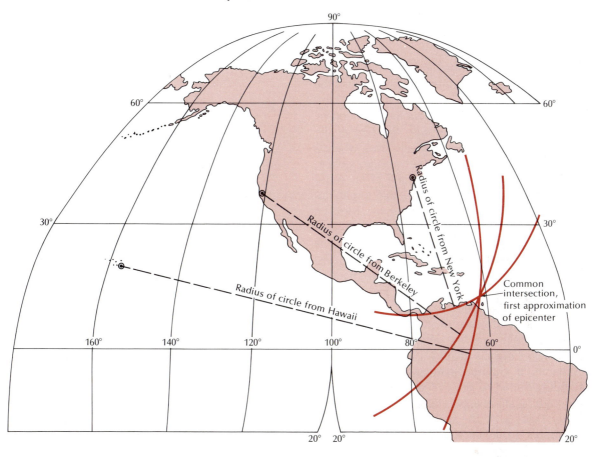

**Fig. 5-26.** From the time lag between *P* and *S* waves recorded at three separate stations, the approximate epicenter of an earthquake can be determined. The distance to the event is calculated, and a circle of that radius is drawn around each of the three stations. The common intersection marks the approximate epicenter.

times are recalculated—based on assumed speeds of the waves—for all stations. The best estimate of the position of the focus is obtained when the calculated travel times to all stations best fit the actual travel times measured directly from the records.

### First Motions

Seismic wave patterns can also aid in determining the type of the relative fault movement—lateral and parallel to the fault, downward or upward along the fault plane, or some combination of these motions—that occurs during the initiation of the seismic waves. This is best understood by referring to Figure 5-22. Note that in response to

a hammer-blow (i.e., a push) a compression wave moves away from the point of initiation. The first part of the wave signal received at any point will be a zone of compression. However, the wave could have been started just as easily by pulling the block sharply to the left. Then the first part of the wave signal would be a zone of rarefaction, or expansion. On the seismographs the pen would initially move up or move down to start its wiggly trace, depending on whether the first part of the *P* wave was a zone of compression or a zone of expansion, respectively. The first motions recieved on the three different seismographs at a station, therefore, provide information on the orientation of the fault and the relative motion of the sliding blocks during earthquake initiation.

## Intensity Versus Magnitude

As soon as the earth starts to rupture, $P$ and $S$ waves begin their outward journey, carrying the vibrational energy in all directions. The size of the waves (their amplitude—measured from the crest of one wave to the trough of another, divided by 2) at any one point varies with the size of the event. Small ruptures produce small waves, whereas large ruptures produce large ones. It is possible to obtain some idea of the magnitude of an earthquake at its source by measuring the amplitudes of the $P$ and $S$ waves as they are recorded on distant seismographs. Of course, the amplitude diminishes as the waves travel away from the disturbance and as the vibrational energy is dispersed. Therefore, a correction must be made to take into account the distance of the recording station from the seismic event.

A scale of earthquake magnitude based on that line of reasoning was first devised by the seismologist Charles Francis Richter (1900–85) and bears his name. It enables seismologists at a distant recording station to get a pretty good idea of the amount of strain energy released during an earthquake (Table 5-1). The scale is roughly exponential, so that a difference of 1 in magnitude corresponds to a difference in source energy equal to an approximate mutiple of 30. During a Magnitude-4 event, for example, the energy released is not just twice that released during a Magnitude-2

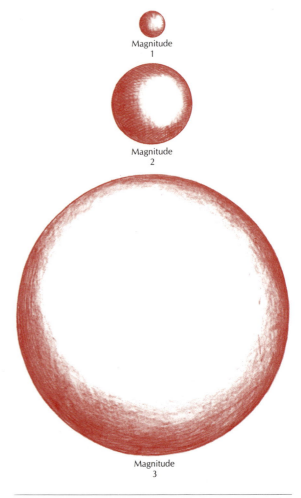

**Fig. 5-27.** Relation of Richter magnitude to energy released—for three magnitudes. The volume of the spheres is roughly proportional to the amount of energy released by earthquakes of the given magnitudes.

**Table 5-1** Earthquake Magnitudes, Statistics, and Energies

| Approximate Richter Magnitude | Number of Earthquakes per Year | Energy (approx. equivalent metric tons of TNT) |
|---|---|---|
| ≥8.0 | 0.1–0.2 | >5,668,750 |
| ≥7.4 | 4 | 907,000 |
| 7.0–7.3 | 15 | 181,400 |
| 6.2–6.9 | 100 | 27,210 |
| 5.5–6.1 | 500 | 907 |
| 4.9–5.4 | 1,400 | 136 |
| 4.3–4.8 | 4,800 | 27 |
| 3.5–4.2 | 30,000 | 1.8 |
| 2.0–3.4 | 800,000 | $6 \times 10^{-3}$–0.8 |

Data from B. Gutenberg, 1945, "Amplitudes of Surface Waves and Magnitudes of Shallow Earthquakes," *Bulletin of the Seismological Society of America,* vol. 35, pp. 3–12.

event, but about 900 ($30 \times 30$) times as great (Fig. 5-27). The largest earthquakes ever recorded have measured about 9.0 on the ***Richter scale:*** they correspond to the energy produced by detonation of about 100 million metric tons of TNT. In contrast, a Magnitude-1 quake would be equivalent to the energy released by the detonation of less than 0.5 kg (1lb) of TNT. You might wonder why no quake larger than 9.0 has ever been recorded. It appears that there is a limit to how much strain energy can be stored before the rocks fail and slippage occurs.

Each year, many hundreds of thousands of magnitude-1 and magnitude-2 earthquakes take place; only one larger than magnitude 8 occurs about every 5 to 10 years (Table 5-1). Any quake larger than magnitude 6 can be considered major and capable of causing great damage. In 1986 a 5.3 quake struck El Salvador. Although moderate in intensity, it killed about 1500 perople and virtually destroyed the capital city of San Salvador. The 1964 earthquake in Alaska released strain energy approximately equal to 12,000 Hiroshima-type atom bombs.

With the continuing study of earthquakes, seismologists have refined the Richter magnitude scale. At present no single type of wave will indicate the source of energy and intensity over the range of rupture lengths. Thus, for most medium-sized quakes, the magnitude $M_b$ is based on $P$ or $S$ (body) waves; for large events, the magnitude $M_s$ is estimated from surface-wave amplitudes; and for the largest quakes, the magnitude $M_w$ is estimated from the size of the rupture zone that produced it. Generally, the magnitude reported by the media will be a body-wave magnitude $M_b$, or sometimes a surface-wave magnitude ($M_s$).

The extent of earthquake damage depends (as discussed earlier) on a variety of factors, such as proximity of the focus, duration, amount of energy released (Richter magnitude), type of rock underlying the site, type of structures, and density of population. The amount of damage for any particular quake will vary as all those factors together vary. An earthquake scale—based on the damage wrought by a quake—is called an **intensity scale,** and it ranges from a minimum number, corresponding to a barely discernible event, to a maximum number, corresponding to total collapse of buildings and great loss of life (Table 5-2). The **modified Mercalli scale** is perhaps the intensity scale most commonly used today. It was formulated in 1902 by the Italian seismologist Giuseppe Mercalli (1850–1913) and modified in 1931 by American seismologists.

Actually, no single intensity value is associated with any particular earthquake. Rather, the damage to surface areas, generally those relatively close to the epicenter, will differ—depending on all of the variable factors—and accordingly will be given different Mercalli intensity ratings. In

**Table 5-2** Modified Mercalli Scale

| Level | Characteristic Effects of Shallow Shocks in Populated Areas |
|---|---|
| XII | Nearly total destruction |
| XI | Few masonry buildings stand; rails are bent |
| X | Some well-built wooden buildings fall; most masonry structures fall; numerous landslides |
| IX | Extensive building damage; buildings shift from foundations |
| VIII | Poorly constructed buildings fall; heavy furniture overturned |
| VII | Most people alarmed and run outside; moderate-to-heavy damage to poorly made buildings |
| VI | Felt by everyone; some heavy furniture moved; plaster falls |
| V | Felt by nearly everyone; some windows and dishes broken; light objects overturned |
| IV | Felt by many people indoors, by some outdoors; dishes and windows rattle |
| III | Felt by some people indoors, autos rock slightly |
| II | Felt by few people; suspended objects may swing |
| I | Generally not felt; detectable mostly by seismographs |

After H.O. Wood, and F. Neuman, 1931, "Modified Mercalli Intensity Scale of 1931," *Bulletin of the Seismological Society of America*, vol. 21, pp. 277–83.

many cases the intensity values will form a bull's-eye effect, progressing inward from a peripheral minimum value to a maximum close to the epicenter (see Fig. 5-5).

Obviously, damage is related to the energy released by an earthquake and those that bring about nearly complete destruction in some areas are generally those with the very high Richter magnitudes.

## Seismic Waves and the Earth's Interior

Our knowledge of the earth's interior is gained from observations of rocks once buried deep below the surface and now exposed through erosion; to cuttings and cores from drill holes thousands of meters deep; and to rock fragments ripped from the walls of volcanic conduits—at most 200- to 250-km (125- to 155-mi) deep—and carried to the surface during episodes of eruption. Direct acquisition of data is restricted to the upper 4 percent of the earth. Fortunately, we can make realistic

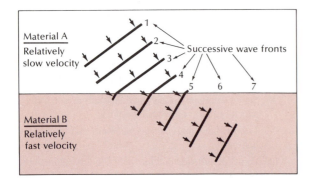

**Fig. 5-28.** Refraction of a light wave as it moves from one material (**A**) to another (**B**). The wave travels more quickly in (**B**) than in (**A**); therefore, when it strikes the interface and is in both materials, it is bent, or refracted.

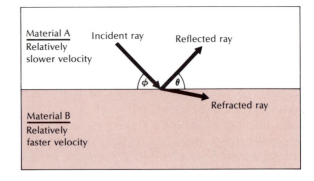

**Fig. 5-29.** Refraction and partial reflection of an incident light wave as it encounters a surface of discontinuity (sharp break) between one material (**A**) and another (**B**). The angle at which the ray is reflected from the interface ($\theta$) equals the angle of impingement ($\phi$) of the incident ray.

inferences about the other 96 percent, especially through seismology. From countless earthquake records, earth scientists have put together a model of the structure and composition of the earth. Of course, we know that the model is only a first approximation and that it will be modified as new data become available.

Seismology has contributed most data because both *P* and *S* waves radiate outward in all directions from an earthquake. Many waves traverse the deeper parts of the earth as they make their way toward distant seismograph stations, and their tracings carry information about the materials through which they pass.

Body waves, like all traveling waves, are subject to refraction (bending) and reflection. In a homogeneous material at constant temperature and pressure, a body wave will travel at a constant velocity, one that depends on the density and elastic properties of the material. If the wave passes into a material of different density or elastic properties, its velocity generally will change and the wave will be refracted (Fig. 5-28). For example, light rays, as they pass into the lens of an eyeglass and back out again, are refracted, or bent, in such a way that the rays are focused upon the retina of the eye.

Reflection of waves will occur when they encounter a drastic velocity change (discontinuity) in the material through which they are passing. Light rays, for example, are reflected when they strike a highly polished surface. Reflection surfaces generally are not totally reflective; some of the wave energy passes through the discontinuity and into the material beyond—undergoing extensive refraction—where it continues (Fig. 5-29).

**Fig. 5-30.** Transmission of three earthquake waves in a hypothetical three-layered earth. Ray 1 travels only in layer **A** and is the slowest. Ray 2 travels both in layers **A** and **B**, moving more quickly through **B**, the intermediate-speed layer. Ray 3 travels through all three layers and moves with the highest average velocity through layer **C**. Wherever a wave encounters a velocity discontinuity, it bends.

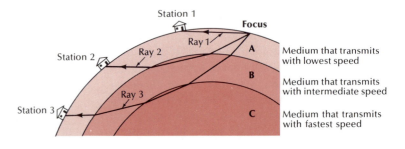

As mentioned earlier, the waves that arrive directly at progressively more distant stations travel at increasingly greater speeds. This can only mean that waves that go progressively deeper into the earth move through materials that permit progressively faster wave propagation. Figure 5-30 shows how seismic waves that increase in velocity downward would behave as they pass through a three-layered body. The implication from the actual *P*- and *S*-wave arrival data is that the earth's physical properties change progressively with depth so that seismic waves move faster and faster.

By studying travel-time and distance data from earthquakes all over the world, seismologists have put together a consistent picture of how body-wave velocities vary with depth (Fig. 5-31).

The results are intriguing, for they show that sharp velocity discontinuities exist within the earth; that *S* waves cannot be propagated between 2900 and 5000 km (1800 and 3110 mi); and that the increase in *P*- and *S*-wave velocities in the outer 700 km (435 mi) of the earth is far from uniform.

These general concentric variations appear to be present around the world.

One of the major velocity discontinuities lies near the outer margin of the earth. It is the **Mohorovičić discontinuity,** named for the Yugoslavian seismologist (1857–1936) who first suggested its presence in 1909. The name has since been shortened to the **Moho** or **M-discontinuity.** The Moho separates rocks with *P*-wave velocities of 6 to 7 km/sec (4 to 4.3 mi/sec) from those with *P*-wave velocities of about 8 km/sec (5 mi/sec). By tradition, rocks above this seismic-velocity break are called the **crust** and those below it, the **mantle.** Under the continents the crust varies from about 30- to 70-km (19- to 44-mi) thick, whereas under the ocean basins it is about 6- to 8-km (4- to 5-mi) thick (Fig. 5-32). The discontinuity is very sharp and is thought to represent a change in rock

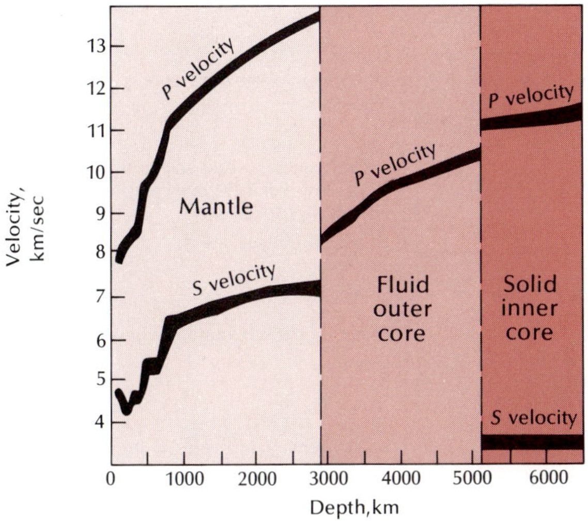

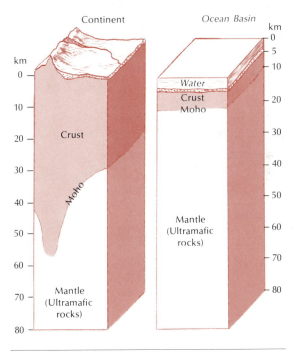

**Fig. 5-31.** How *P*- and *S*-wave velocities change with depth. The variation in speeds is represented by the variable thickness of the lines. Notice the lack of *S* waves and the reduced velocity of *P* waves in the fluid outer core, the several abrupt changes in body-wave velocities in the outer 600 km (375 mi) and the low *S*-wave velocities between about 100 and 250 km (62 and 155) in depth. (After F. Press and R. Siever, 1974, *Earth*, W.H. Freeman)

**Fig. 5-32.** Exaggerated schematic view of the relationships between crust and mantle beneath the continents and the ocean basins. The continental crust, which has a relatively low density (2.7 g/cm³) floats isostatically in a high-density mantle and projects downward as much as 70 km (44 mi) beneath some high mountains. The oceanic crust, with a density of about 3.0 g/cm³ is relatively thin. (After Arthur M. Strahler, *The Earth Sciences*, 2nd ed., p. 401, Fig. 23–27. Copyright © 1963, 1971 by Arthur N. Strahler. Reprinted by permission of Harper & Row)

composition, from more heterogeneous sodium-, calcium-, magnesium-, and iron-bearing silicates and aluminosilicates above to magnesium- and iron-bearing silicates below.

The oceanic crust—similar in composition to a basalt—is richer in calcium, magnesium, and iron than is the continental crust.

The continental crust, however, particularly in its middle and lower parts, is structurally much more enigmatic. Recently information has come from studies of reflections of seismic waves generated by strong surface detonations. This method, called **seismic-reflection profiling,** has shown that large dipping faults are common in the upper part of the continental crust. With depth, many of these flatten out and may even become horizontal. In contrast, the strong reflection layers in the lower crust appear to be close to horizontal. It has been postulated that the lower crust is less brittle than the upper crust above or the uppermost mantle

below. Under stress, possibly, this zone can flow, thereby producing a near-horizontal layering.

Some of the uncertainties about the middle, and possibly even the lower crust, should be answered by results of deep continental drilling. Deep-hole drilling has been going on in the USSR since 1970 and is only now getting underway in the United States. The deepest penetration to date in Russia is nearly 12,200 m (40,000 ft) at a well in the Baltic region. The results already show that the previous model of the crust near the base of the hole, developed from seismic refraction and reflection studies, is incorrect. It is also apparent that fluids associated with metamorphism are much more abundant than had been previously thought. In the United States the first deep hole—to be drilled to a depth of 10 km (6.2 mi)—is slated for the southern Appalachians.

In the outer part of the mantle, down to about 700 km (435 mi), there are several abrupt changes

**Fig. 5-33.** Estimated nature of the outer part of the earth based on variations of the S-wave velocity with depth. The velocity break at 100 km (62 mi) represents the boundary between the lithosphere and asthenosphere. Increases in velocity near 400 km (250 mi) and 650 km (405 mi) apparently represent changes of state.

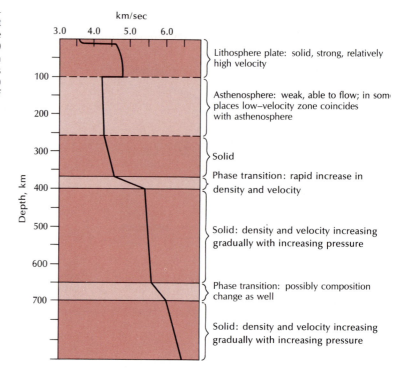

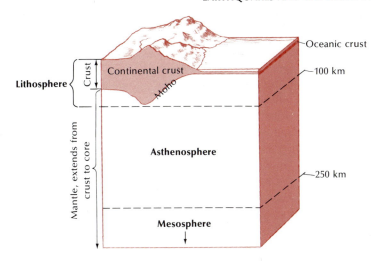

**Fig. 5-34.** The threefold division of the outer part of the earth based on its physical properties. The lithosphere, which includes both the crust and the upper part of the mantle, is strong and relatively brittle; the asthenosphere is relatively weak and capable of flowing; and the mesosphere is strong.

in *P*- and *S*-wave velocities. Beneath the ocean basins, in particular, S-wave velocities are lower than expected—between about 100 and 250 km (62 and 155 mi)—which has led to the designation of this discontinuous layer as the ***low-velocity zone.*** Most geophysicists think that the relatively low S-wave velocities indicate a less brittle, perhaps even partially molten (2 to 5 percent), rock layer.

Two additional abrupt changes in body-wave velocities occur in the outer mantle, one at about 400 km (250 mi) and one at about 650 km (405 mi).

In the outer 700 km (435 mi) of the earth, the physical state of the iron–magnesium silicates in the mantle varies, apparently in response to variations in temperature and pressure (Fig. 5-33). Below 250 km (155 mi), as confining pressure becomes the dominant factor, the material becomes more solid; in the next 450 km (280 mi) it consists of more dense crystalline forms of the iron–magnesium silicates. One such ***change of state*** occurs just above 400 km (250 mi) and another near 650 km (405 mi). Each change is associated with an increase in body-wave velocity. That at 650 km is also characterized by a strong seismic reflection, indicating that the velocity change is also, in part, the result of a change in chemical composition. Below 700 km no further

abrupt changes in velocities are evident; body-wave velocities increase steadily—apparently in response to increasing compaction.

Recently, on the basis of seismic properties and mechanical strength, the outer 5 to 150 km (3 to 93 mi) of the earth have been designated the ***lithosphere,*** the layer extending from the base of the lithosphere to 250 km (155 mi) as the ***asthenosphere*** (Fig. 5-34), and the layer below 250 km as the ***mesosphere.*** The lithosphere (meaning rock sphere) is cooler, more brittle, and stronger than the asthenosphere. The large, mobile tectonic plates of the earth consist of lithospheric material. Because of temperature differences, the lithosphere is thickest under the older, cooler continents, extending to depths of 125 to 150 km (78 to 93 mi), whereas under the oceans it is generally only 50- to 80-km (31- to 50-mi) thick. At the oceanic ridges, where heat flow is very high, the thickness may only be 5 to 10 km (3 to 6.2 mi). It is believed that in the asthenosphere (meaning sphere of weakness), the effects of increasing temperature predominate over those of increasing confining pressure. Thus the rocks are less rigid. This relatively weak zone, thought by many to be necessary to the mobility of the lithospheric plates above, occurs worldwide. Its thickness varies, however, depending on the local temperature and thickness of the overlying lithosphere. The previously discussed low–velocity layer is defined only on the basis of variations in body-wave velocities; it is commonly absent beneath continents and is most conspicuous under the oceans, particularly the ridges. The mesosphere (middle sphere) represents the mantle zone beneath the asthenosphere, where the pressure again becomes high enough to make the rock strong.

There is now evidence that lateral inhomogeneities also exist in the mantle from one part of the earth to another. Seismic tomography, described in chapter 4, suggests the existence of local regions of higher- and lower-velocity seismic waves that correspond to colder and warmer zones, respectively. The apparent thermal pattern has been interpreted as evidence of mantle convection and plume activity to maximum depths of about 650 km (405 mi). Whether convection involves the entire mantle is unclear.

Small–scale mass differences in the mantle are also shown by precise measurements of the sea surface by orbiting satellites—the sea satellite

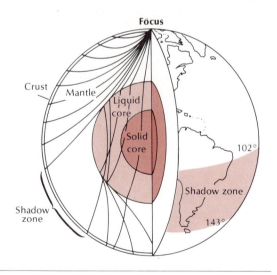

**Fig. 5-35.** The shadow zone, between 102° and 143°, for an earthquake originating at the North Pole. As shown in the cutout, P and S waves travel normally until they intersect the liquid core. Then the P waves slow down abruptly and bend toward the center of earth (accounting for the shadow zone) and the S waves disappear entirely. Beyond 143° of arc from the focus, P waves are again received directly.

(SEASAT) project. The surface of the sea is not smooth, but bumpy. High spots represent concentrations of water over zones of greater gravitational attraction that correspond to greater concentrations of subsurface mass. Again, the pattern is interpreted in terms of a dynamic system in the mantle associated with local convection and plume activity.

At a depth of 2900 km (1800 mi), the P-wave velocity drops drastically to a value close to what it is near the earth's surface; the S wave disappears entirely. Such behavior is exactly what would be expected if at 2900 km the rock changed from solid to fluid. The change in rock properties appears to be very sharp and has been designated the **core-mantle boundary.** The solid mantle is thought to consist of iron magnesium silicate minerals; the liquid core principally of iron. Within the core and at about 5100 km (3170 mi), the P-wave velocity appears to increase abruptly, and there is even a hint of an S wave. Seismologists think this is caused by a change from a fluid **outer core** to a solid **inner core.** Recent studies indicate, in fact, that the top few-hundred kilometers of the inner core is mushy, as if composed of mixed crystals and liquid. This suggests that the top of the inner core is at the melting point.

The existence of a sharp velocity discontinuity at the core-mantle boundary also explains the so-called seismic shadow zone mentioned earlier (Fig. 5-35). Earthquake waves can move only through the mantle, to a depth of 2900 km (1800 mi); beyond that depth, however, they strike the core-mantle boundary and much of the energy is sharply reflected back to the surface. But some energy moves into the core and is greatly refracted as it does so. Because of the extreme refraction at the core-mantle boundary, no direct P waves reach the earth's surface between 102° and 143° arc distance from the epicenter, thereby producing the shadow zone. Only P waves can traverse the fluid part of the core; because they do so at reduced velocity, the time necessary for waves to travel through the central part of the earth is increased.

Seismologists have also been able to estimate density variations inside the earth by combining (1) data from laboratory experiments, (2) theories that relate density and body-wave velocities, and

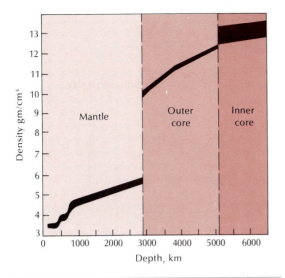

**Fig. 5-36.** Curves depicting the estimated density of the earth. Notice the sharp increase at the mantle-core boundary. The density at the earth's center is thought to be about four to five times that of surface rocks. The thickness of the curves is indicative of the uncertainties of the estimates.

(3) a knowledge of the general distribution of mass in the earth. Figure 5-36 shows that density increases with depth, paralleling the steps in the body-wave velocity distribution. The greatest jump in density occurs at the core-mantle boundary. At the center of the earth, rocks apparently have specific gravities as high as 12 or 13, that is, they are four-to-five times more dense than are most rocks at the surface.

What kinds of rocks could account for the variations in body-wave velocities and density? The elements making up the rocks must be those that are seemingly abundant within our solar system. Such a conclusion is based on our knowledge of the elements that occur in nearby stars (particularly the sun) and in meteorites. The earth's atmosphere continually is hit by meteorites, most so small that they are consumed before they reach the surface. But some are large enough to resist destruction and fall far and wide over the earth. Most of those that are recovered fall into two main groups: the **iron meteorites** and the **stony meteorites.** The former consist primarily of iron–nickel alloys, whereas the latter consist of silicate min-

erals and of lesser amounts of iron–nickel alloys and sulfides. Stony meteorites have a relative abundance of nonvolatile elements—such as magnesium, silicon, aluminum, calcium, and iron—in about the same proportion as in the sun and other stars. Earth scientists have concluded, therefore, that an average stony meteorite is a good representation of the average composition of the planetary bodies in our solar system, including the earth (Table 5-3).

The mantle, accounting for about 67 percent of the earth's mass, appears to be made of silicates—mostly those of iron and magnesium, with only small quantities of aluminum, calcium, and sodium. In the outer part of the mantle, the rock is chiefly olivine, with lesser amounts of magnetite and pyroxene. This conclusion is confirmed by analysis of nodules and clasts found in the basaltic lavas and kimberlites that were ripped from the mantle and carried to the surface during volcanic activity.

The relatively dense core of the earth, which accounts for about 30 percent of its mass, is thought to be composed of iron and, to a lesser extent, nickel. To approximate the density of the core, it is also necessary that small amounts of less dense elements, such as silicon, oxygen, and sulfur, are

**Table 5-3** Average Composition of Stony Meteorites and the Hypothetical Composition of the Earth

| Element | Stony Meteorites, Weight | Earth, Weight |
|---|---|---|
| Oxygen (O) | 33.24% | 29.50% |
| Iron (Fe) | 27.24 | 34.60 |
| Silicon (Si) | 17.10 | 15.20 |
| Magnesium (Mg) | 14.29 | 12.70 |
| Sulfur (S) | 1.93 | 1.93 |
| Nickel (Ni) | 1.64 | 2.39 |
| Calcium (Ca) | 1.27 | 1.13 |
| Aluminum (Al) | 1.22 | 1.09 |
| Sodium (Na) | 0.64 | 0.57 |
| Chromium (Cr) | 0.29 | 0.26 |
| Manganese (Mn) | 0.26 | 0.22 |
| Phosphorus (P) | 0.11 | 0.10 |
| Cobalt (Co) | 0.09 | 0.13 |
| Potassium (K) | 0.08 | 0.07 |
| Titanium (Ti) | 0.06 | 0.05 |

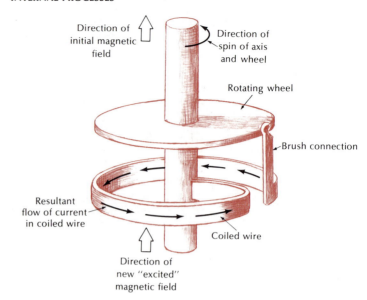

**Fig. 5-37.** Model of a simple, self-exciting dynamo. If the dynamo is to function, the wheel must be turning and an initial magnetic field must be present. The resultant electric currents produce a new "excited" magnetic field, which is sufficient to keep the dynamo functioning even if the initial magnetic field is removed. (After H. Takeuchi, S. Uyeda, and M. Kanamori, 1970, *Debate About the Earth,* Freeman, Cooper)

present. In some models the core is considered to be composed largely of either iron sulfide or iron oxide.

However, most geophysicists favor the idea of a core composed mostly of metallic iron because of the long-lived presence of the dipolar geomagnetic field. It is well established that the earth's field is internally generated, most likely in the core. If the outer core were metallic iron, it could conduct electric currents, and being fluid, it is

able to flow. Geophysicists now believe that in response to local heating or local density gradients, the liquid core flows convectively while the earth, including the core, spins on its axis. If a weak magnetic field is initially present, electric currents will be created during the flow, which, in turn, will create new magnetic fields in a complex interaction of electromagnetic coupling.

Once formed, the new magnetic field will be amplified and constrained to lie more or less along

**Fig. 5-38.** Estimated variation in temperature and pressure with depth inside the earth. Notice that initially the temperature increases rapidly, but it quickly tapers off. One atmosphere of pressure equals about 1 kg/cm$^2$ (14.7 lb/in.$^2$), the air pressure at sea level. (After P.J. Wyllie, 1975, "The Earth's Mantle," *Scientific American*, vol. 232, no. 3, pp. 50–63)

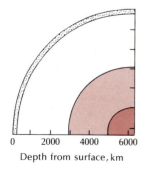

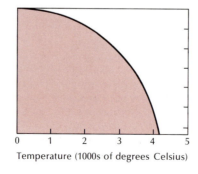

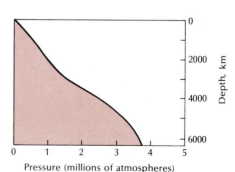

Depth from surface, km

Temperature (1000s of degrees Celsius)

Pressure (millions of atmospheres)

the axis of rotation of the earth; it is therefore dipolar. The mechanism just outlined can be crudely modeled in the laboratory and has been named a *self-exciting dynamo* (Fig. 5-37). Because the dynamo action is controlled, in part, by the rate and scale of movement in the liquid outer core, the field will vary in intensity and to some degree in direction as changes in the fluid motions occur. Thus, although the field appears relatively stable from day to day, over hundreds of years it will vary. In fact, the field can vary greatly in intensity, even dropping nearly to zero. When it regenerates the magnetic poles may even be reversed.

Paleomagnetic studies of ancient continental rocks indicate that a dipolar field has existed for at least 3 billion years. This implies that the core has also existed at least this long. Probably most of the onionlike density layering of the earth reflects an early phase of the earth's development when it was partially to completely molten, thereby enabling radial redistribution of materials on the basis of density.

Temperatures and pressures within the earth also vary with depth. Reasonable estimates of the variation in these two parameters are given in Figure 5-38.

## SUMMARY

1. The San Francisco earthquake of 1906 resulted from slippage along the San Andreas Fault, which amounted to about 5 m (16 ft). Fire after the quake caused most of the destruction. In trying to explain the earthquake, the *elastic-rebound theory* was postulated whereby elastic strain is slowly accumulated over a relatively long time and suddenly released.

   One of the largest quakes ever felt in the United States centered in New Madrid, Missouri, in 1811–12. The death toll and property damage were low because of a low population density at that time. In 1985, Mexico City was struck by a devastating quake, emanating from the west coast subduction zone, that killed about 8000 people and caused property damage of $5 billion.
2. Japan, China, southern Europe, and South America also have experienced severe and destructive quakes.
3. *Shallow-focus* quakes are defined as those down to 70 km (44 mi); *intermediate-focus* quakes, between 70 and 300 km (44 and 187 mi); and *deep-focus* quakes between 300 and 700 km (187 and 435 mi). Earthquakes tend to occur in linear belts that delineate the edges of tectonic plates that are converging, diverging, and sliding laterally. Shallow-, intermediate-, and deep-focus earthquakes are associated with convergent zones, whereas shallow-focus quakes are associated only with zones of divergence and lateral translation. Some quakes occur relatively far from plate margins. This currently is inexplicable within plate-tectonic theory.
4. The control or mitigation of earthquake damage is being approached in two ways: through *earthquake prediction* and *earthquake control*. At present prediction appears more feasible than control.
5. The first waves to arrive on a seismograph are *P waves*, then *S waves*, and finally *L waves*. *P* waves are *compressional* waves and can travel through a solid or liquid. *S* waves are *transverse* waves that can only travel through a solid. *L* waves are the slowest and travel at, or close to, the earth's surface. Because *P* waves travel faster than *S* waves, it is possible to estimate the distance to a seismic disturbance by the time lag in the arrival of these two waves.
6. The energy released during an earthquake is measured on the *Richter magnitude scale*. The largest magnitude ever recorded is about 9.0. Nearly a million quakes occur yearly; of them only about 10 are above magnitude 7. A quake of magnitude 8 or larger occurs every 5 to 10 years.
7. Earthquake damage depends on several factors—magnitude, proximity, duration of quake, type of ground, and type of building.
8. Seismic data indicate that the earth is a layered body. The *crust* extends down to the *Moho discontinuity*, which occurs as deep as 70 km (44 mi) beneath the continents and 8 km (5 mi) beneath the oceans. Below the crust, the *mantle* is composed of ferromagnesian silicates and extends down to 2900 km (1800 mi). Below that is the *core*, which is liquid down to 5100 km (3170 mi) and solid down to the center of the earth.
9. On the basis of rock strength, the outer, brittle layer of the earth has been designated the *lithosphere*.

Under the continents it may be as much as 150-km (93-mi) thick, whereas under the oceans its thickness is less than 100 km (62 mi) and under the ocean ridges can be as little as 5 km (3 mi). The layer between the base of the lithosphere and 250 km (155 mi) is the *asthenosphere,* which is relatively weak and is thought to be necessary to the mobility of the lithosphereic plates above. The *mesophere* is the relatively strong mantle zone beneath the asthenosphere.

10. Seismic speeds and laboratory data indicate that two *changes of state* occur in the outer part of the mantle in response to increased pressure. (Each is associated with an increase in body-wave speed.) One occurs at about 400 km (250 mi) and the other at about 650 km (405 mi). The second one may also represent a change in composition.

11. In some areas of the earth, particularly the ocean basins, *S* waves in the depth range of 100 to 250 km (62 to 155 mi) exhibit a relative slowing; accordingly this zone has been designated the *low-velocity zone.* Temperatures in this zone are considered to be great enough to cause weakening of the mantle rocks and possibly 2 to 5 percent melting.

# QUESTIONS

1. Discuss the global distribution of earthquakes.
2. Why do some earthquake belts have shallow-, intermediate-, and deep-focus quakes, whereas others only have shallow-focus ones?
3. Why is the E–W earthquake belt across Europe and Asia so wide?
4. How do scientists explain earthquakes, such as the New Madrid quake of 1811–12, that occur far from plate margins?
5. What factors make some earthquakes more devastating than others?
6. What is the elastic-rebound theory of earthquake generation?
7. What two approaches are being taken to reduce earthquake threat?
8. Discuss some of the code requirements and considerations presently being made in quake-prone cities.
9. How far along are we in being able to accurately predict earthquakes?
10. Describe a simple seismograph and how it developed through time.
11. What are the principal waves generated by an earthquake? How do they differ?
12. How is the intensity of an earthquake measured? The magnitude?
13. Describe how the velocities of the *P* and *S* waves vary with depth in the earth.
14. Compare the lithosphere and the asthenosphere with the crust and the outer mantle.
15. Describe the variation with depth within the earth of temperature, pressure, and density.

# SELECTED REFERENCES

Baranzangi, M., and Dorman, J., 1969, World seismicity maps compiled from ESSA, epicenter data, 1961–67, Seismological Society of America Bulletin 59, pp. 309–80.

Bath, M., 1973, Introduction to seismology, Wiley, New York.

Bolt, B.A., 1978, Earthquakes—a primer, W.H. Freeman, San Francisco.

Bronson, W., 1959, The earth shook, the sky burned, Doubleday, Garden City, N.Y.

Burchfield, B.C., 1983, The continental crust, Scientific American, vol. 249, no. 3, pp. 130–45.

CDMG Note, 1979, How earthquakes are measured, California Division of Mines and Geology, Note 23, pp. 35–37.

Chiles, J.R., 1986, Standing up to earthquakes, Invention and Technology, vol. 2, pp. 56–63.

Davison, C., 1936, Great earthquakes, Thomas Murby, London.

Evans, D.M., 1966, Man-made earthquakes in Denver, Geotimes, May–June, pp. 11–18.

Fuller, M.L., 1914, The New Madrid earthquake, U.S. Geological Survey Bulletin 494.

Hamilton, R.M., 1980, Quakes along the Mississippi, Natural History, vol. 89, pp. 70–74.

Iocopi, R., 1964, Earthquake country, Lane Book Company, Sunset Books, Menlo Park, Calif.

Jeanloz, R., 1983, The earth's core, Scientific American, vol. 249, no. 3, pp. 56–65.

Kendrick, T.D., 1956, The Lisbon earthquake, Methuen, London.

Lawson, A.C., et al., 1908, The California earthquake of April 18, 1906, Report of the State Earthquake Commission, Carnegie Institute of Washington, D.C.

McCann, W.R., et al., 1980, Hakataga gap, Alaska: seismic history and earthquake potential, Science, vol. 207, pp. 1309–15.

McKenzie, D.P., 1983, The earth's mantle, Scientific American, vol. 249, no. 3, pp. 66–113.

Plafker, G., Ericksen, G.E., and Concha, J.F., 1971, Geological aspects of the May 31, 1970, Peru earthquake, Bulletin of the Seismological Society of America, vol. 61, pp. 543–78.

Plafker, G., and Savage, J.C., 1970, Mechanism of the Chilean earthquakes of May 21 and May 22, 1960, Geological Society of America Bulletin, vol. 81, pp. 1001–30.

Reichle, M.S., 1986, Mexico earthquake damage, California Geology, vol. 39, pp. 75–79.

Reid, H.F., 1914, The Lisbon earthquake of November 1, 1755, Seismological Society of America Bulletin, vol. 4, pp. 53–80.

Richter, C.F., 1958, Elementary seismology, W.H. Freeman, San Francisco.

Robertson, E.C., ed., 1972, The nature of the solid earth, McGraw-Hill, New York.

Saint Amand, P., 1961, Los Terremotos de Mayos—Chile, 1960, Technical Article, 14, U.S. Naval Ordinance Test Station, China Lake, Calif.

Shepard, F.P., 1933, Depth changes in Sagami Bay during the great Japanese earthquake, Journal of Geology, vol. 41, pp. 527–36.

Silver, P.G., et al., 1985, Mantle structure and dynamics, EOS, vol. 66, pp. 1193–98.

Stein, R.S. and Bucknam, R.C., 1986, Quake replay in the Great Basin, Natural History, vol. 95, pp. 28–35.

Sutherland, M., 1959, The damndest finest ruins, Coward-McCann, New York.

Verhoogen, J., 1956, Temperatures within the earth *in* Physics and chemistry of the earth, vol. 1, Pergamon, New York.

Witkind, J.S., 1962, The night the earth shook: a guide to the Madison River Canyon earthquake area. Department of Agriculture, U.S. Forest Service Miscellaneous Publication 907.

Wyllie, P.J., 1975, The earth's mantle, Scientific American, vol. 232, no. 3, pp. 50–63.

**Fig. 6-1.** Sedimentary strata deformed on a large scale: Murdafil, Iran. (Aerofilms, Ltd.)

# 6

# DEFORMATION OF ROCKS AND MOUNTAIN BUILDING

When the crust of the earth is subjected to stress, it buckles and breaks and the rocks become permanently deformed. In some places, originally horizontal strata are tilted and folded (Fig. 6-1); in others the rocks crack or become offset along faults (Fig. 6-2). The field of geology that deals with the deformation of the earth is called *structural geology*.

Rocks in the outer part of the earth are even now being bent, broken, uplifted, and depressed at numerous places in response to forces acting deep beneath the surface. The fact that the earth is in dynamic readjustment is brought home to us forcibly every time there is a large earthquake—a tangible indication that stresses in the earth have built to the point at which rocks fracture and suddenly shift.

Rates of deformation are slow. Geologists, who have been making observations for only a short span of the earth's history, have witnessed only minor structural changes in the rocks at the earth's surface. Yet, given long periods of geological time, those nearly imperceptible small-scale changes can accumulate, to bring about deformation on a grand scale. There is evidence of surface rocks having been broken and separated along faults by

as much as tens or even hundreds of kilometers; of sedimentary rocks having been crenulated into great and small folds resembling those of a gigantic rumpled tablecloth; of spectacular mountain chains, like the Himalayas, having been pushed up where none existed before, and of continents having been ripped apart.

The realization that forces working inside the earth can bring about such drastic changes came about slowly. Leonardo da Vinci (1452–1519), the Renaissance genius, observed fossil shells preserved in the rocks of the Tuscan hills of Italy and concluded that the Apennine Mountains must have once been under the sea. The same kind of observation was made later in Florence by Nicolaus Steno in the seventeenth century, who made the additional observation that sedimentary rocks were deposited in horizontal layers, and therefore those that deviated from the horizontal must have been tilted after the sediments were laid down (see Fig. 6-1).

The idea that deformation of rocks is accomplished by small deformational increments over long periods of geological time is a uniformitarian view. Charles Lyell was the champion of that idea in the early nineteenth century and, in fact, pro-

**Fig. 6-2.** Surface rupture resulting from an earthquake on the Motagua fault, Guatemala, in 1976. (R.C. Bucknam, USGS)

**Fig. 6-3.** Temple of Jupiter Serapis near Naples, Italy, showing the zone of marine shellfish borings made in the columns during submergence. The top of the zone, at present 7 m (23 ft) above sea level, represents the depth to which the region was most recently submerged below sea level.

vided one of the classic examples. Buildings had been constructed along the shores of the nearly tideless Mediterranean Sea for millennia; the more durable ones were made of stone and had withstood the ravages of time and weather in the relatively dry climate. Those built close to the shores became unusually sensitive recorders of changes in the sea level.

Lyell noted that all three of the surviving columns of the Temple of Jupiter Serapis, not far from Naples (Fig. 6-3), had lines encircling them about 7 m (23 ft) above sea level. Below the line, each column was riddled with small holes bored by shallow-water marine shellfish in a band almost 3-m (10-ft) wide. He concluded that the uppermost line on each column represented a former stand of sea level. And the holes bored by the marine shellfish further supported his belief that the temple was once submerged in ocean waters.

The historical record bears out Lyell's conclusion. Apparently the temple was built around the second century B.C., and it must have started to subside shortly thereafter. Continued subsidence of the land allowed the sea to invade the entire

**Fig. 6-4.** Map showing land area in the Netherlands that has sunk below sea level.

structure, eventually reaching the 3-m (10-ft) band on the columns. Later, the land again began to rise, and by A.D. 1503 the uplift was well under way. Most of the rest appears to have occurred at the time of the destructive eruption of Monte Nuovo in A.D. 1638.

Subsidence at an alarming rate is now affecting much of the Netherlands; the coastal part of Holland is sinking at a rate of about 21 cm (8 in.) per century. At present, the coast is protected by dikes, but much of the country, including the cities of Rotterdam and Amsterdam, is below sea level (Fig. 6-4).

The hills along the coast near Los Angeles, California, provide another illustration of a segment of land that has been uplifted from the sea. A person approaching the Palos Verdes Hills, the bold headland that partially shields the harbor of San Pedro in Southern California, is impressed by the promontory's seaward slope, which rises above the sea like a Cyclopean stairway. Wave-cut terraces separated from one another by steep cliffs rise in 13 steps to 396 m (1300 ft) above the sea (Fig. 6.5). They are evidence that the coast in that

**Fig. 6-5.** Wave-cut terraces at Palos Verdes Hills, California. (Photo John S. Shelton and R.C. Frampton. From John S. Shelton, *Geology Illustrated,* W.H. Freeman. Copyright © 1966)

area has been sporadically elevated so recently in geological history that fossil seashells preserved on the flat terraces are identical to the shells of marine organisms living today in the ocean.

Another even more impressive set of uplifted wave-cut platforms exists along the coast of Peru. Some are 16- to 24-km (10- to 15-mi) wide and are littered with marine shells that appear as if they had lived in the sea only yesterday.

And—as a final example—at the world's highest elevation, on the flanks of lofty Mount Everest, occur water-deposited rocks containing marine fossils that lived in the sea 60 million years ago.

## RATES OF MOVEMENT

The Coast and Geodetic Survey ran leveling lines in 1906, 1924, and 1944 across Cajon Pass between Victorville and San Fernando in southern California. The measurements show that during the 38-year span of the surveys the area around

the pass where it crosses the San Gabriel–San Bernadino mountains rose as a gentle arch by about 20 cm (8 in.)—that is, at the rate of about 53 cm (21 in.) per century. Although this rate may appear modest, it yields an uplift of 133 m (440 ft) in 25,000 years for the San Gabriel–San Bernadino moutains. Mount Everest, rising at a comparable rate, could have reached its present height in about 2 million years.

Deformation of the earth need not be a uniform, continuous process; it can, and usually does, occur in abrupt jumps. For example, in 1899, during a severe earthquake in the area around Yakutat Bay, Alaska, a large portion of the earth's surface was uplifted as much as 14.4 m (47 ft). And during the 1964 Alaskan earthquake, a maximum uplift of 11.9 m (39 ft) was recorded in the region of Patton Bay.

So far, all of these historic examples have demonstrated structural deformation in a vertical sense—that is, sinking or rising of the land. Horizontal

**Fig. 6-6.** Local offset of fence line of 2.6 m (8.5 ft) along the San Andreas Fault, resulting from displacement during the 1906 San Francisco earthquake. Location is near Woodville, Marin County, north of San Francisco. (G.K. Gilbert, USGS)

shifts have been recorded, too. In the San Francisco earthquake of 1906, roads and fences were offset along the San Andreas Fault (Fig. 6-6) by as much as 6.4 m (21 ft). Repeated surveys since that time are indicative of continued creep of the land areas on either side of the fault moving past each other at about 2 cm (0.8 in.) per year.

## STRUCTURAL RESPONSE OF ROCKS

Rocks in the outer portion of the earth, in response to large- and small-scale processes associated with the movements of lithospheric plates, are sub-

jected to compression, extension, shearing (lateral slippage), and torsion (twisting). Some rocks fracture, others break and slide past one another, still others fold to varying degrees. How any particular body of rock responds structurally to stress depends on a number of factors—confining pressure (the all-sided pressure that increases with depth), temperature, the time span of stress application, and the amount of reactive pore-space fluids. In general, at the low confining pressures and temperatures within a few kilometers of the earth's surface, rocks are brittle; they respond by *jointing* (fracturing with no movement of blocks) and *fault-*

**Fig.6-7.** Plastic deformation in limestone along the Rio Extorax, Querétaro, Mexico. Some beds have been thickened through tight chevron folding or thinned through stretching. (K. Segerstrom, USGS)

Fault

**Fig. 6-8.** Steeply tilted strata adjacent to a fault in the Dinosaur National Monument, Utah. The Green River is in the foreground. (Philip Hyde)

*ing* (fracturing with slippage between adjacent blocks). At higher confining pressures and temperatures at greater depths, rocks tend to soften, becoming ductile and, ultimately, folding and flowing like toothpaste (Fig. 6-7). At depths where temperatures are relatively high (particularly if fluids are present), the rocks undergoing folding and flowage also commonly recrystallize (metamorphose).

Both observation and experimentation demonstrate that jointing, faulting, and some types of folding are characteristic of rocks a few kilometers in depth, whereas ductile folding and flowage are characteristic of rocks at depths of more than 3 km (2 mi).

## EVIDENCE OF ANCIENT STRUCTURAL DEFORMATION

The first step in the study of ancient structural deformation is to describe the geometry and complexity of local structural features, such as folds, fractures, and faults (Fig. 6-8). When structures are small, they can be investigated by direct obser-

**Fig. 6-9.** Strike-and-dip of a tilted bed. The strike is represented by the line of intersection of the water surface with the dipping bed. The dip is perpendicular to the strike and represents the path a marble would take as it rolls down the bed. (William Estavillo)

vation; some are so large, however, that they can only be studied piecemeal through the making of a geological map.

Of the three rock types, sediments best lend themselves to structural study. Most strata were deposited horizontally; deviations from that orientation or offsets in the continuity of beds indicate deformation. Moreover, strata that possess distinctive rock properties can be traced easily for long distances, so that deformation can be studied over wide areas.

If a sedimentary layer is tilted at an angle to the horizontal, it is said to be a **dipping bed:** The amount of **dip** is the angle between the bed and the horizontal, as determined by a level. The imaginary line of intersection made by the dipping stratum and a horizontal surface is called the **strike** of the bed. The dip is measured in a vertical plane that is perpendicular to the strike.

One of the best ways to visualize dip and strike is to imagine a dipping bed of rock that projects above the surface of a sea or lake (Fig. 6-9). The strike is the intersection of the surface (a horizontal plane) with the bed; the dip is the amount of tilt of the bed in a plane perpendicular to the strike line, measured downward from the horizontal. A marble placed on the inclined bed would roll down the dip direction, showing that the line of dip is directional (as shown by the arrowhead in Fig. 6-9).

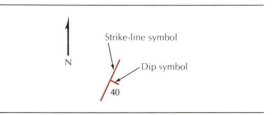

**Fig. 6-10.** Map symbol for a bed striking N 30° E and dipping 40° toward the southeast.

The two ends of the strike line point in opposite compass directions. That is, if one end of the line points northeast, the other end points southwest. By convention the strike is chosen as the direction that makes an acute angle with true north. A strike line given as N 30° E would be a line in which one end points to a position 30° east of north and the other 30° west of south. On a map the geologist would draw a short segment with that orientation, see Figure 6-10. The dip can be represented on the map by a short line attached to the strike line drawn in the direction of dip. The appropriate dip angle is written close to the dip-and-strike symbol, as shown in Figure 6-10.

Vertically dipping beds are denoted on the map by ⨯. Horizontal beds, of course, lack both a dip and a definable strike; in such cases geologists use the symbol ⊕. If a bed is structurally rotated past 90°, in other words is an **overturned bed,** the geologist measures the dip and strike as before but denotes it on the map with the symbol ⌐⊢.

By measuring the dips and strikes of sedimentary strata and plotting them on a map, it is generally possible to determine the large-scale fold geome-

**Fig. 6-11.** Large-scale anticlinal and synclinal folds of the Grande Chartreuse north of Grenoble in the French Alps. (Swissair-Photo)

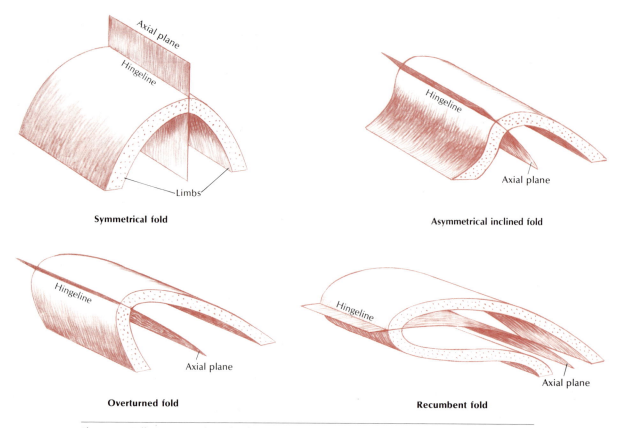

Symmetrical fold

Asymmetrical inclined fold

Overturned fold

Recumbent fold

**Fig. 6-12.** Different types of anticlinal folds, showing in each case the relation of the axial plane and the axis to the fold geometry.

**Fig. 6-13.** Recumbent folds in slate folded about 350 million years ago. (Geological Survey of Canada)

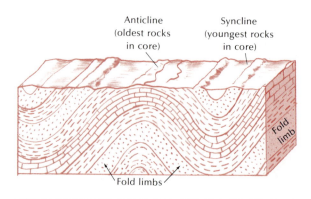

**Fig. 6-14.** An open symmetrical anticline and syncline. As a result of erosion the oldest rocks are present in the core of an anticline and the youngest in the core of a syncline.

164

try. Offsets in the geometry or abrupt changes in the continuity of key beds will indicate a pattern of faulting.

## Folds

Some structural processes result in **folds** which are small- to large-scale wavelike features in the strata (Fig. 6-11). Folds exist in a multitude of shapes and forms, so we now define the terminology used in their geometrical description.

Folds resembling wave crests are called **anticlines**—from the Greek, meaning ''to be inclined against itself'', those resembling wave troughs are called **synclines**—also from the Greek, meaning ''to lean together'' (Fig. 6-11). The sides of such folds are called **limbs,** or **flanks,** and a line drawn along the points of maximum curvature of a particular bed is called the **hingeline.** A plane connecting the hingelines in successive beds in a fold is called the **axial plane.** It can be flat or curved, and vertical to horizontal (Fig. 6-12), depending on the straightness of the axis and the difference

in dip angle between the two flanks. Folds in which the axial plane is close to vertical are termed **upright folds;** those in which the axial plane is moderately to greatly tipped from upright are termed **inclined folds;** and those whose axial planes are within 10° of being horizontal are termed **recumbent folds.** Small-scale recumbent folds are shown in Figure 6-13. Large-scale recumbent anticlinal folds, largely containing deformed metamorphic and igneous rocks in their centers, are called **nappes;** they are conspicuously developed in the European Alps.

If the two limbs of a fold make nearly equal angles with the axial plane, the fold is **symmetrical;** if the angles are significantly dissimilar, however, the fold is **asymmetrical** (Fig. 6-11). In all symmetrical and many asymmetrical anticlines, the limbs dip away from the hingeline, whereas in all symmetrical and some asymmetrical synclines they dip toward it (Fig. 6-14). The hingelines of most anticlines correspond to the ridge lines of a roof, for example, just as the hingelines of most synclines correspond to the keel of a ship.

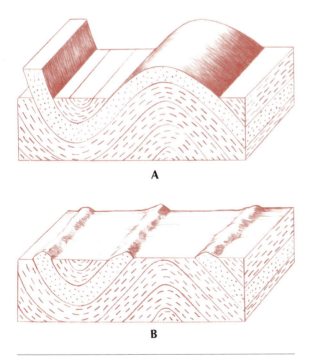

**Fig. 6-15.** Symmetrical nonplunging folds **(A)** before erosion and **(B)** after erosion. Notice how the resistant layers form parallel ridges after erosion.

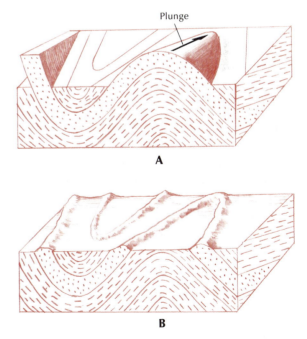

**Fig. 6-16.** Symmetrical plunging folds **(A)** before erosion and **(B)** after erosion. Notice how the more resistant layers form series of arcuate, V-shaped ridges after erosion.

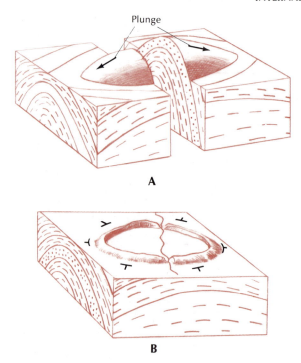

**Fig. 6-17.** Doubly plunging fold **(A)** before erosion and **(B)** after erosion. Dip-and-strike symbols show attitudes of the layers.

**Fig. 6-18.** (*Opposite*) Little Maverick Dome, northwest of Riverton, Wyoming. Aerial photo of a doubly plunging anticline, or elongate dome, on which are shown the trace of the doubly plunging axis and appropriate dip-and-strike symbols. (John S. Shelton)

Folds in which the hingelines are horizontal are said to be ***nonplunging*** (Fig. 6-15); they could be represented by a piece of roofing tile placed on a table. In many cases, the hinglines dip at an angle to the horizontal and the folds appear to dive, or dip, into the ground (Fig. 6-16). Such types are called ***plunging folds.*** Some folds plunge downward at both ends and others upward at both ends. These are called ***doubly plunging folds*** (Figs. 6-17, 6-18).

In fold structures that have been truncated by surface erosion, the following relationships are evident (as shown in many of the preceding figures):

1. Anticlines plunge in the direction in which their sides converge (come together); correspondingly, synclines plunge in the directions in which their sides diverge (spread apart).
2. In anticlines the rocks become progressively older toward the axis; in synclines, they are younger.
3. The dip measured on the hingeline is a measurement of the fold plunge.

Anticlines that lack a well-defined elongation and that plunge from a point in all directions are called ***domes*** (Fig. 6-19); the corresponding synclinal structures are called ***basins.***

One other fold type is the ***monocline*** (from the Greek, meaning "one inclination"), a one-limbed structure with horizontal strata on either side of steeper-dipping strata (Fig. 6-20). This structural type is common in cases in which vertical fault displacement occurs at depth beneath a sequence of subhorizontal strata, thereby producing a step-like upward or downward draping of the strata overlying the fault. In some cases a monocline will laterally merge with a fault. Monoclinal structures are typical of the Colorado Plateau.

## Joints

Nearly all rocks at, or close to, the earth's surface exhibit cracks and fractures called ***joints*** (Fig. 6-21). The word is thought to come from British coal miners who imagined that rocks were joined along these fractures, like bricks in a wall.

A small proportion of joints results from the polygonal cracking of thin sheets of igneous rock during cooling. Such fractures, termed *columnar joints,* are discussed in chapter 7. Most joints result from deformation of the rocks in which they occur; the details of how they originate are not always clear. Tensional stress perpendicular to the joint surface is thought to be a major factor, however.

**Fig. 6-19.** (*Opposite*) Upheaval dome, Canyonlands National Park, Utah, taken vertically downward; an almost perfect example of a circular dome. (Eros Data Center, USGS)

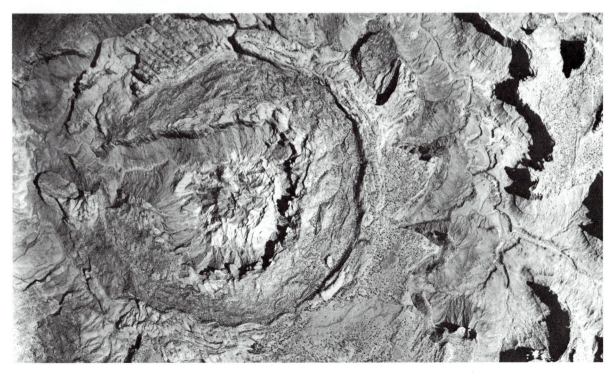

**Fig. 6-20.** Looking south along the monocline that forms the west flank of Raplee anticline, Colorado Plateau, southern Utah. In the foreground is the San Juan River. (D.L. Baars)

Many joints occur in a closely spaced subparallel alignment—called a *joint set*—not too uncommonly, three such sets are mutually perpendicular (Fig. 6-22). Jointing is less likely in laminated, or foliated, rocks than it is in texturally massive varieties.

One common type of jointing, particularly well developed in the more massive rock types, occurs in proximity to the earth's surface. This joint pattern consists of a series of parallel, broadly curved surfaces—called **exfoliation joints**—that roughly parallel the surface topography and become more numerous the closer they are to the surface. Exfoliation joints and the gently rounded domical topography forms associated with them are typical features in regions underlain by granitic rocks. Classic examples, including Half Dome, abound in the upper reaches of Yosemite National Park (Fig. 6-23).

Joints are of more than academic interest. From a practical standpoint, they are surfaces of incipient weakness and must be given consideration by quarry, mining, and civil engineers. Subsurface joints also affect the movement of groundwater; consequently, they assume importance in the siting of dams and in estimating groundwater reserves. In some cases joints have also played a key role in the formation of ore deposits, providing the avenues along which the ore-bearing solutions moved.

**Fig. 6-21.** Closely spaced joints in nearly horizontal sandstone strata in eastern Utah. Differential erosion has emphasized the jointing pattern. (J.R. Balsley, USGS)

**Fig. 6-22.** Three sets of joints nearly at right angles (two are perpendicular to the rock face, the third is parallel to it) in granitic rocks on the east face of Mount McAdie in the Sierra Nevada, California. (Tom Ross)

**Fig. 6-23.** Exfoliation joints in granite, central Sierra Nevada, California. Notice man for scale. (N.K. Huber, USGS)

## FAULTS

*Faults* are fractures along which movement has occurred. Sometimes the offset amounts to a centimeter or less; in others, it may amount to hundreds of kilometers.

Because fault movement involves the sliding of one block past another, many fault surfaces are smoothly polished and grooved (Fig. 6-24). These features, called *slickensides,* provide one clue to the direction of fault slippage, but they record only the most recent event. Earlier events may have involved slippage in totally different directions,

but the slickensides associated with them will have been obliterated by the most recent slippage.

Often, too, the rocks adjacent to a fault will have been pulverized, forming a claylike soft material called *fault gouge.* In some instances the rocks in the fault zone may be broken and sheared, creating a coarse *fault-zone breccia.*

If fault movement occurs close to the earth's surface, the surface itself may be broken and offset. The resulting low linear cliff is called a *fault scarp* (Fig. 6-25). Such low, fault-induced cliffs are common throughout the world wherever earthquakes are prevalent.

**Fig. 6-24.** Slickensided fault surface. The grooves indicate that the last fault motion was parallel to the dip of the fault. Sumter County, Alabama. (W.H. Monroe, USGS)

Because fault planes have all the geometrical attributes of a dipping stratum, we can describe their orientation in space by means of strike-and-dip notation. For example, a fault plane might strike N 45° W and dip 52° toward the southwest.

Many faults, because they provide avenues for the circulation of underground fluids, have been mineralized by ore-bearing solutions. Accordingly, numerous tunnels and shafts have been dug to reach the ores trapped in the fault zones. Miners working a mineralized zone along a moderately dipping fault actually stand on the block on one side of the fault, with the face of the other block hanging above their heads. In miners' jargon, the block beneath their feet is called the **footwall;** that above their heads, the **hanging wall** (Fig. 6-26). These aptly descriptive terms have carried over into the general terminology used in fault nomenclature.

### Apparent Relative Movement

It is almost always impossible to tell the actual direction of movement of the two blocks on either side of a fault just by observing the offset. In Figure 6-27, for example, the final pattern (Fig. 6-27D)—

**Fig. 6-25.** Nearly 6-m (20-ft) high, this prominent fault scarp developed in the foothills along the east base of the Sierra Nevada Range, near Lone Pine, California, during the Owen's Valley earthquake on March 26, 1872. Many consider this the largest earthquake in California history. (David B. Slemmons)

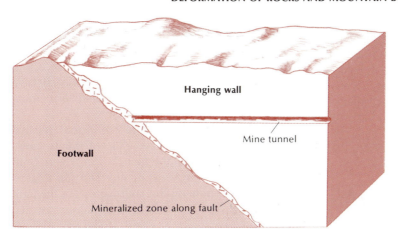

**Fig. 6-26.** Relation of a hanging wall and a footwall along a fault.

**Fig. 6-27.** Two possible displacements on a fault leading to identical offset of a dipping stratum. **A.** Unfaulted dipping bed, with incipient fault plane shown. **B** and **C.** Two types of fault motion that could produce offset. **D.** Faulted dipping bed as observed in field. The relatively uplifted block in **B** has been eroded.

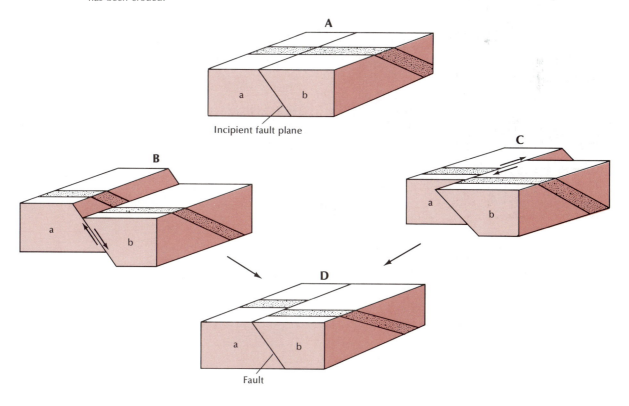

which is all we see—could have resulted from (1) slippage of block *a* upward, in relation to block *b*—as seen in Figure 6-27B; (2) the horizontal slippage of block *a* past block *b*—as seen in Figure 6-27C; or (3) slippage at an oblique angle, involving some vertical and some horizontal displacement. In most cases, then, we can determine only an apparent relative movement.

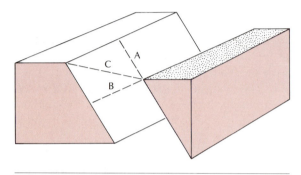

**Fig. 6-28.** Types of slip on a fault plane: **(A)** dip slip, **(B)** strike slip, and **(C)** oblique slip.

## Known Directions of Relative Displacement

In those cases where a line—for example, a fold axis—or a volume of small dimensions is cut by a fault, the exact nature of the relative displacement can be determined.

The term used to denote the actual relative displacement of the once adjacent points, measured in the plane of the fault, is *slip*. Figure 6-28 illustrates the three kinds of slip. **Dip slip** (Fig. 6-28A) is movement on the fault parallel to the fault dip; **strike slip** (Fig. 6-28B) is a measure of displacement parallel to the fault strike; and **oblique slip** (Fig. 6-28C) is movement at an angle to the dip and strike of the fault. A fault in which the displacement is primarily dip slip and in which the footwall has moved up relative to the hanging wall is called a **normal fault** (Fig. 6-29B). If the displacement is primarily dip slip, but the footwall has moved down relative to the hanging wall, the fault is called a **reverse** or **thrust fault** (Fig. 6-

**Fig. 6-29.** Types of faults: **(A)** Block of unfaulted stratified rock, **(B)** normal fault, **(C)** reverse fault, and **(D)** strike-slip fault.

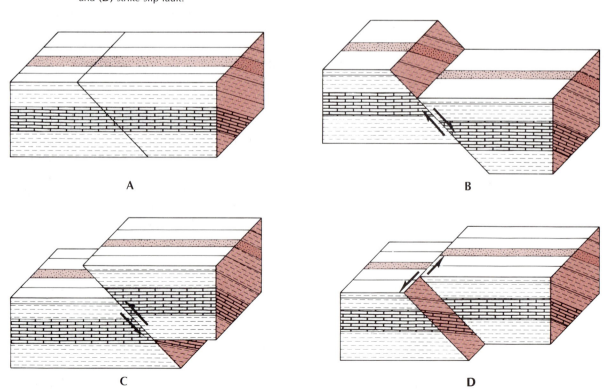

A

B

C

D

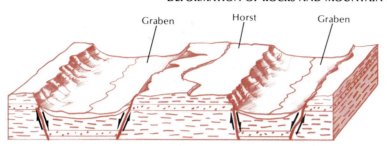

**Fig. 6-30.** Structural and physiographic relations of a horst and adjacent graben.

29C). Depending on the dip angle of the fault, it can be described as a low- or high-angle fault: if less than 45°, **low angle;** if more than 45°, **high angle.** If the dip angle on the thrust fault plane is less than 10°, it is called an **overthrust fault.** If the slip is primarily parallel to the strike of the fault, it is called a **strike-slip** or **lateral fault** (Fig. 6-29D).

**High-Angle Normal and Reverse Faults** Many faults show displacements that are primarily of the dip-slip variety. Along some of the larger faults of this type, displacements of more than 2 km (1 mi) have occurred.

If a block rises above the blocks on either side of it, the fault-bounded upland block is called a **horst** (from German, meaning—among other things—"a ridge"). The lower-lying linear blocks between horsts are called **graben** (from German, meaning "trough" or "trench") (Fig. 6-30). The Rhine graben, which is followed by the Rhine River from Basel to Mainz, represents a well-known example. The graben lowland is linear and trenchlike, with marginal fault-scarp walls that rise to highlands—the Schwarzwald (Black Forest) to the east in Germany and the Vosges Range to the west in France.

There exists an impressive set of graben represented by a nearly continuous string of fault-bordered troughs, that extends the length of Africa and part of the Middle East, from Mozambique on the south, northward to the Dead Sea. These great down-dropped segments of the earth's crust are commonly termed **rift valleys.** One of the better-known rift valleys holds Lake Tanganyika, with a length of 672 km (418 mi) and a width from 32 to 64 km (20 to 40 mi). The water surface is 771 m (2530 ft) above sea level, but the bottom is 506 m (1660 ft) below sea level.

**Strike-Slip Faults** Most of the world's faults that exhibit the greatest amounts of relative movement—in some cases up to hundreds of kilometers—are of the strike-slip type; the displacement is primarily parallel to the strike of the fault. In most of the large strike-slip faults, the fault plane is nearly vertical.

Very large strike-slip faults have been recognized in California, Canada, Japan, New Zealand, the Philippines, Scotland, and Switzerland. Given their extensive geographical range, it is not surprising that a widely varied terminology has been used to describe these faults. British geologists are likely to call them *transcurrent faults,* or *wrench faults.* In the United States they are commonly called *lateral faults.* Movement along lateral faults is relative. If you are facing a fault and the ground on the other side of the fault moves to your left, it is said to be a *left-lateral fault;* if the ground moves to your right, it is a *right-lateral fault* (Fig. 6-31).

The San Andreas Fault is one of the world's more renowned examples of a strike-slip fault. The map in Figure 6-32 shows the trace of the fault and reveals its great length—roughly 960 km (600 mi) from where it emerges from the ocean along the coast north of San Francisco, until its several branches disappear beneath the waters of the Gulf of California. The map also shows that it is not a single fracture at its southern end but rather a splayed system of faults.

The San Andreas is of particular structural interest because it represents the demarcation plane between the North American lithospheric plate to the east and the Pacific plate to the west (Fig. 6-32). At present movement along the fault has been right lateral, that is, the North American plate has been shifting relatively southward at a rate of about 2 cm (1 in.) per year. At this rate, San Francisco and Los Angeles, which are on the North

SAN ANDREAS FAULT

Large right lateral offset
in drainage pattern

Small-scale drainage pattern
also shows right lateral offset

**Fig. 6-31.** The easily recognized trace of the right-lateral San Andreas Fault, Carrizo Plain, California. (A.M. Bassett)

American and Pacific plates, respectively, will be nearby suburbs in about 20 million years. Mason L. Hill and Thomas W. Dibblee, Jr., well-known California geologists, have estimated that during the last 60 million years, on the basis of observed separations of conspicuous rock formations along the fault, lateral movement has amounted to about 193 km (120 mi).

Another well-documented example of a strike-slip fault along which considerable lateral movement has occurred is the Great Glen Fault in Scotland that extends from the North Sea to the Atlantic Ocean. The valley is also the site of Loch Ness, home of the legendary sea serpent. Fortunately, the rocks and structures on either side of the Great Glen Fault can be matched up with some confidence and, according to the British geologist W. Q. Kennedy, they indicate a strike-slip movement of about 104 km (65 mi), with the northern part of the Scottish Highlands being displaced relatively southwestward.

**Overthrust Faults** Early in the nineteenth century geologists struggling to interpret the perplexing geological relationships of the Northwest Highlands of Scotland were puzzled to find sandstone, shale, and limestone—sedimentary layers—interbedded with metamorphic rocks like gneiss and schist. Even that astute observer Charles Lyell accepted this unlikely combination as a comformable succession. By the 1860s geologists began to accept the view that the aberrant geological relationships in the northwest Highlands resulted from the gliding of great sheets of rock over one another

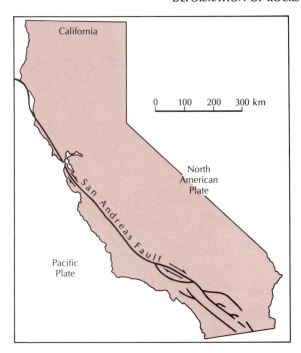

**Fig. 6-32.** The trace of the San Andreas Fault in California. Notice how the fault splays at its southern end. This fault is a fundamental lithospheric suture that separates the North American and Pacific plates.

in a fashion similar to that recognized in 1849 in the European Alps. At least three overthrust faults were recognized in the Scottish Highlands, along which regionally metamorphosed rocks were carried over younger unmetamorphosed sedimentary rocks.

In the United States the pioneer work on overthrust faults was done around 1900, in large part through the efforts of the geologist Bailey Willis, who had been doing field work in Glacier National Park, Montana. He came to realize that the eastern edge of the Rocky Mountains in that area consisted of older (Precambrian) sedimentary rocks that rested on younger, (Cretaceous) sedimentary rocks. The older rocks are much more resistant to erosion and form the castellated ridges and steep cliffs that make the park landscape so renowned. The less-resistant Cretaceous shales and sandy shales are characteristic of the gently rolling, undulating landscape of the high plains of Montana to the east. Willis concluded that the older sediments

slid over the younger sequence along a large overthrust fault (Fig. 6-33), to which he gave the name Lewis Overthrust in commemoration of the pioneering efforts in 1803–6 of Meriwether Lewis (1774–1809). Chief Mountain, an isolated peak in the eastern part of the park, is an erosional remnant of the hanging wall of the fault—literally a mountain without roots (Fig. 6-34). Such an erosionally isolated fragment of a thrust sheet is called a **klippe** [from the German word for "cliff" (pl. klippen)]. Conversely, erosion may cut through the overthrust plate, exposing the rocks beneath, thereby producing a **fenster** (from the German word for "window") (Fig. 6-35).

Many thrust faults are not regular geometrical planes but are complexly curved—in places, the dip may be 10° or less; in others it may steepen to 45° or more (e.g., Fig. 6-34). Some even have been folded, or bent, after initial formation. Many overthrust faults appear to have very gentle dips near the leading edge of what may once have resembled tonguelike lobes and steeper dips in the root zone to the rear.

There is a major question about how large overthrust sheets are emplaced. Are they shoved from behind or are they the result of gravitational gliding of great masses of rock down gentle slopes? The observational data are somewhat equivocal. Recent investigations, however, have shown that if the pore spaces within rocks near the base of a potential thrust sheet are filled with water under abnormally high pressure—so that a buoyant effect results—less force than had formerly been thought necessary is required to overcome friction and set such a mass in motion. Large overthrust sheets, then, may be able to slide downhill on relatively gentle slopes.

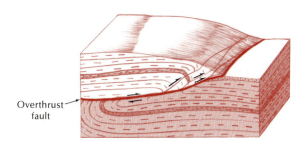

**Fig. 6-33.** Cross section of an overthrust fault.

**Fig. 6-34.** Looking northwest toward Chief Mountain with Cable Mountain directly to its left, Glacier National Park, Montana. Chief Mountain, composed of sediments over 500 million years old, sits on top of rocks ranging from 60 to 130 million years old (forming the lower, gentler slopes); it is a remnant of the upper fault block that moved for many kilometers over the Lewis Overthrust Fault. (Doug Erskine, Glacier National Park)

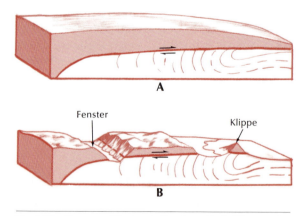

**Fig. 6-35.** Diagram of the Lewis Overthrust Fault **(A)** soon after movement along it ceased and **(B)** after erosion brought about partial dissection. The erosional remnant, or klippe, is isolated from the thrust sheet and the window, or fenster, is eroded through the sheet.

Many overthrust sheets occur along the margins of some of the world's larger complex mountain chains. Possibly, as mountain building progressed, the central part of the range was uplifted, thereby providing the slope necessary for overthrust slippage—and the blocks of sediment slid, like large single landslide blocks, downslope for many kilometers or even tens of kilometers. Quite commonly, the rocks in the upper plate, especially if they are relatively weak sediments, will be folded complexly and cut by numerous smaller thrust faults. Such an overthrust sheet, composed primarily of rumpled sediments, is called a **décollement.** Many occur in the Jura Mountains along the edge of the Alps (Fig. 6-36) and in parts of the Appalachian Mountains in the eastern United States.

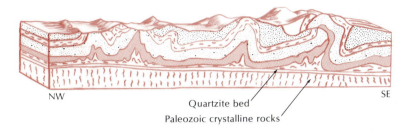

**Fig. 6-36.** Décollement in the Jura Mountains, western Europe. The lowest, nearly vertical structure consists of Paleozoic crystalline rocks (mostly metamorphics). Directly above is a thin bed of flat-lying Triassic quartzite; and above that is a structurally weak unit consisting largely of layers of shale and salt along which most of the sliding occurred.

NW

SE

Quartzite bed

Paleozoic crystalline rocks

## MOUNTAINS AND MOUNTAIN BUILDING

The description of such structural features as folds, joints, and faults is only a prelude to a discussion of mountains and mountain-building processes.

Of all the landforms and structural features on the earth's surface, surely none is closer to the hearts of geologists than mountains. The greatest variety of rocks is visible in their valleys and on their ridges and peaks. In some long, linear mountain belts, metamorphic and plutonic rocks are present in close association; in the outer portion of these ranges, sedimentary rocks are broken, overthrust, and complexly folded. Dramatic erosional landforms are common in mountains: Landslides and other kinds of mass movement have their maximum development here; streams are more powerful because their gradients are steeper; and, of course, in the high mountains, glaciers make the montane scenery a source of joy and inspiration.

The scientific study of mountains began as late as the mid-eighteenth century. Now, with the rise of plate tectonic theory, we are coming to understand better how they are formed.

Every mountain and mountain range is unique, however it is judged. Therefore, like many other natural phenomena, mountains are difficult to classify. We are prone to assign to rigid pigeonhole categories features that may have had more than one kind of origin or that tend to merge with another feature. The short classification that follows serves to differentiate the more distinctive types.

Volcanic mountains are built up of an accumulation of volcanic products, such as ash and lava flows.

Fault-block mountains owe their elevation to differential movement along faults, so that some parts of the crust are raised and others are lowered relative to one another.

Oceanic ridges and rises form a nearly continuous elongate mountain system throughout the world's ocean basins.

Folded and complex mountains generally consist of igneous, strongly deformed sedimentary, and metamorphic rocks. They usually occur in great elongate belts and require (on the average) several hundred million years to form.

### Volcanic Mountains

Volcanoes are among the world's best-loved and most scenic peaks. Among the more familiar are Fujiyama, Honshū, Japan; Mount Rainier, Washington State; Mount Etna, Sicily; Mauna Loa, Molokai, Hawaii; and lofty Andean summits, such as El Misti in Peru and Cotopaxi in Ecuador. Mauna Loa—counting the submerged as well as the visible part—rises about 9150 m (30,012 ft) from a base 145 km (90 mi) in diameter on the seafloor; it is the world's largest single mountain mass. Active, dormant, and recently extinct volcanoes constitute a large number of the world's mountains (Fig. 6-37). If the waters of the sea were removed, we would be doubly impressed because the peaks of many of the volcanic islands would loom far above the surrounding abyssal lowlands; we would also become aware of numerous submarine volcanoes.

Volcanic mountains differ fundamentally from the others in our classification in that they are accumulations of material piled up on the earth's surface. Characteristically, therefore, volcanoes rise as conical or domelike mountains above their surroundings; they may be grouped in clusters or

**Fig. 6-37.** Volcanic peaks of the central Cascade Range, Pacific Northwest, looking southward from the glaciated peak of Mount Rainier, Washington (foreground) to Mount Adams (far left), Mount Hood, Oregon, just left of Mount Rainier, and Mount St. Helens, Washington (right). (Special Collections Division, University of Washington Libraries, photo by H. Miller Cowling, UW Neg. 4832)

even in chains, as in the Cascades of the Pacific Northwest and the Andes of Peru and Chile.

Most of the world's active or recently active volcanoes occur in convergent zones. They are ephemeral features, the surface expression of igneous activity that develops episodically as part of the total mountain-making activity along these zones.

### Fault-Block Mountains

We have already discussed some mountains of this type in connection with the description of normal and reverse faulting. Probably the best examples of fault-block mountains anywhere in the world are those of the basin and range province of the United States, which extends from the Wasatch Range, Utah, westward to the Sierra Nevada Range in California, and from Idaho southward to Arizona. In this area hundreds of elongate north-south to north-northwest trending ranges alternate with broad, gravel-floored basins. The first person to solve the riddle of the origin of this unusual proliferation of ranges and basins was Grove Karl Gilbert (1843–1918), a largely self-taught geologist who accompanied an exploring party of the U.S. Army Corps of Engineers in 1872.

Gilbert realized that the unusual topographic form of the generally straight-margined mountains was most plausibly explained by the presence of one or more faults along their margins, faults along which they had been uplifted. Bounded at one or both lateral margins by large, high-angle normal or reverse faults, such features are called **fault-block mountains** (Figs. 6-38, 6-39). His interpretation, based on limited observational data, has stood the test of time and is the accepted theory

of origin for fault-block mountains throughout the world.

Most geologists agree on the nature of the boundary faults in the basin and range province—they are relatively straight along the strike; the dip is steep, perhaps 60° to 70°; and they resemble normal faults more than other types. There is evidence that the surfaces of many such faults decrease in dip downward. Several kilometers beneath the surface, they may be inclined at angles of only 40° to 50° or less (Fig. 6-39). Over the past

**Fig. 6-38.** Looking northwestward across the Paradise Range, Nevada, a mountain block that has been relatively uplifted along high-angle faults that more or less parallel the margin of the range. Approximate trace of the fault is shown by the lines. (J.R. Balsley, USGS)

Ancient landslide

Fault

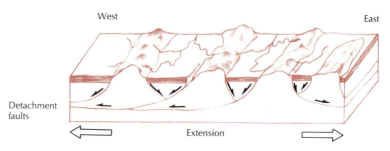

**Fig. 6-39.** Typical fault-block mountains of the basin and range province, western United States. Many of the range-border faults decrease in dip downward. Some faults coalesce at depth to become widespread, near-horizontal slip surfaces called detachment faults. Large arrows at the bottom indicate the directions of extension. Smaller arrows indicate direction of relative movement along the faults.

5 years seismic reflection profiling by a consortium of scientific institutions has demonstrated that many of the larger basin and range faults are almost horizontal at depth. This type of fault has been named a ***detachment fault*** (Fig. 6-39).

Basin-and-range topography was not formed by compression or shortening of the earth's crust. Instead, there is evidence of extension and crustal thinning across the entire width of the region—about 700 km (435 mi). If one were to construct a model of the numerous horsts and graben and to slide the blocks along their low-angle, inclined fault planes, and to move them back to their original unbroken condition, the east-west dimension would be reduced by at least 64 km (40 mi). This is the amount of extension that has occurred in the basin and range province during the last 25 million years.

In addition to east-west stretching and thinning, abundant geological evidence points to regional uplifting over the same time span, particularly during the last 10 million years. No one really knows what could bring about both the uplift and stretching of a large segment of continental crust. Some geologists have suggested that the East Pacific Rise does not die out where it butts into the North American Plate near the southern tip of Baja California, but that it continues northward beneath the continent (Fig. 6-40). The extension and uplift of the basins and ranges, then, are the cracks and rifts of the crustal blocks above the landlocked zone of divergence. But one drawback to this explanation is the extremely wide zone of extension. Most geologists think that the zone of stretching associated with a diverging plate boundary should be concentrated in a narrow band along its crest—as in the rift systems in the Red Sea area and eastern Africa. Moreover, many earth scien-

tists do not believe that the East Pacific Rise extends beneath the continent. They believe either that the rise ceases to exist where it abuts Baja California or that it has been offset northward along the San Andreas Fault system, which would represent a transform fault. They point to the small segment of a divergence zone off the Oregon/Washington shore as the other end of the severed rise system.

There is one piece of geophysical data that probably has some bearing on the origin of the basin and range province. The entire western part of the United States, including all the areas that have been structurally and volcanically active during the last 30 million years, is underlain by an anomalous upper mantle that extends 100 km (62 mi) or so downward, below the base of the crust. The mantle is anomalous because it is less dense than it should be. Some geophysicists have speculated that in some way the upper mantle has changed chemically to produce a less dense (and, accordingly, more voluminous) material. In response to the increase in the mantle's volume, the crust above has been uplifted and stretched. The cause of such a chemical change is unknown.

Fault-block mountains larger than those in the basin and range province [commonly rising to elevations over 3800 m (12,465 ft)] are found in Colorado, New Mexico, and Wyoming. In fact, most of the large, generally north-south to northwest-southwest trending ranges in these states, which include the Sangre de Cristo, Front, Gore, Sawatch, Owl Creek, Wind River, and Big Horn ranges, partly owe their existence to block faulting along one or both sides of the mountain blocks. In many places the mountain block has been raised so high that erosion has stripped off most of the overlying Paleozoic and Mesozoic strata to

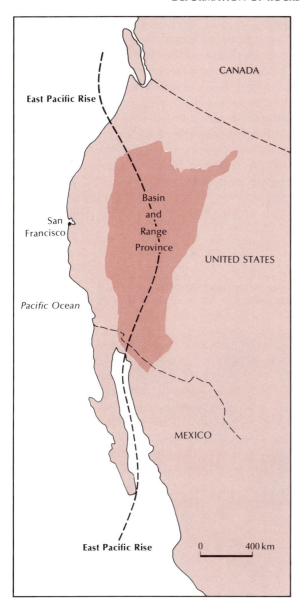

**Fig. 6-40.** Western North America, showing the general area of the basin and range province and the possible location, at depth, of the East Pacific Rise divergence zone.

indicates that the boundary faults along the eastern side of the Wind River Range in Wyoming are thrust faults that flatten downward becoming almost horizontal at depths of 25 to 30 km (16 to 19 mi). Most geologists, therefore, have interpreted these faults, and the other range-margin faults in the vicinity, to be the result of east-west compression related to plate-margin convergence that occurred along the western edge of the North American Plate about 70 million years ago. The biggest problem with this interpretation is that such compressional deformation would have occurred about 1600 km (1000 mi) inland from the plate margin.

Undoubtedly, in the next several years, we will learn a great deal more about the deep structure of the Rocky Mountains and how the range-margin faults came to be.

## Oceanic Ridges and Rises

The world's longest mountain chain, consisting of the oceanic ridge-and-rise system, lies mostly hidden beneath the ocean's surface. It stretches, almost unbroken—nearly 64,000 km (39,800 mi)—throughout the world's ocean basins. The range, unlike any on the continents, appears to be caused by a broad, linear rise of the ocean floor. It is cut by abundant normal faults that parallel the ridge axis and is commonly offset for distances up to 1000 km (620 mi) and more along transform faults that cross it at nearly right angles.

It is now well known that oceanic ridges and rises are located astride the zone where two plates of lithosphere are being pulled apart—the zones of divergence discussed in chapter 4. Such oceanic mountain chains, then, are closely associated with the origin and movement of lithospheric plates.

The ridge system is the site of many shallow-focus earthquakes that occur directly beneath the ridge axis and along parts of the cross-cutting, transform faults. Volcanic activity is common along the ridge axis. Many earth scientists feel that the elevation of the ridge results, in part, from the accumulation of basaltic lavas and from the injection of numerous dikes at shallow depths. Some of the elevation of the linear welt may also be caused by a volume increase either as a result of alterations of olivine-rich rocks by hot water percolating upward along the faulted and fractured

expose the Precambrian crystalline rocks beneath—rocks that were at least 7620 m (25,000 ft) lower in elevation only 70 million years ago. Oil-well drilling and field studies have shown that most of the faults vary from high- to low-angle thrusts and, in some places, become overthrusts. Deep seismic-reflection profiling by (COCORP)

ridge axis or because the newly formed lithosphere along, and near, the divergent zone is relatively hot and, therefore, thermally expanded. Proponents of the latter view argue that the lithosphere cools and contracts in volume as it moves away from the ridge axes—and accordingly, the ocean floor occurs at greater water depths.

## Folded and Complex Mountains

When geologists think of mountain-building processes, they usually have in mind those that produce **alpine chains,** the complex linear mountain belts common to all continents (Fig. 6-41). Because of their complexity, no two alpine chains are exactly alike, either in their features or in their history of development.

Alpine chains generally are elongate, extending for several hundred up to 1600 km (1000 mi) or more in length. Most have a core of structurally complex, regionally metamorphosed rocks (generally sedimentary or sedimentary and volcanic initially) that are locally intruded by, and intimately associated with, granitic plutonic rocks. Also characteristic are elongate belts of thick sections of relatively unmetamorphosed sedimentary

**Fig. 6-41.** Alpine mountain terrain near Banff, Alberta, Canada. The high peak is Mount Assiniboine. (This aerial photograph (c) 20-10-24 Her Majesty the Queen in Right of Canada, reproduced from the collection of the National Air Photo Library with permission of Energy, Mines and Resources Canada. Photograph number CA114-24)

**Fig. 6-42.** Main structural features of the Appalachian Mountains.

rocks—usually on one side but in places on both sides of the core—that have been complexly folded, overturned, and thrust faulted. Décollement structures are common. The degree of folding and faulting decreases with distance from the mountain core until, at a sufficiently great distance, the rocks are essentially flat lying. The thickness of the sedimentary section decreases progressively from the core region outward.

Most alpine mountains, just after they are formed, apparently rise high above sea level to form linear topographic prominences, such as the European Alps and Himalayas. Even after erosion has essentially reduced them to low-relief features once again, if the characteristic elements are recognizable, they still can be described as alpine mountain chains. The Appalachians in eastern North America certainly do not have the surface grandeur of many modern mountain ranges. Yet they possess all the features just enumerated, in a nearly classical manner, and can be classified as an alpine chain even though the last mountain-building activity there occurred about 225 million years ago (Figs. 6-42, 6-43).

**Fig. 6-43.** Cross section A–A′ across the Appalachian Mountain belt. In the axial portion are high-grade metamorphic and plutonic rocks; outward from the axis, the degree of metamorphism and the intensity of deformation progressively decrease.

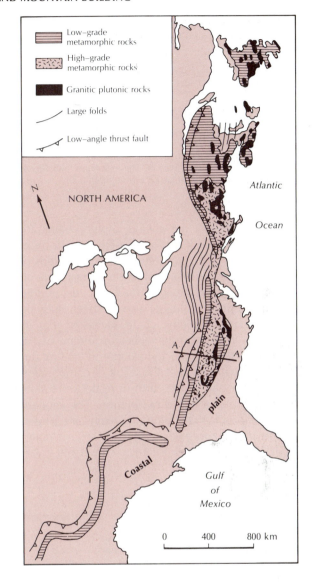

Low-grade metamorphic rocks

High-grade metamorphic rocks

Granitic plutonic rocks

Large folds

Low-angle thrust fault

NORTH AMERICA

Atlantic

Ocean

A

A′

Coastal

plain

Gulf
of
Mexico

0    400    800 km

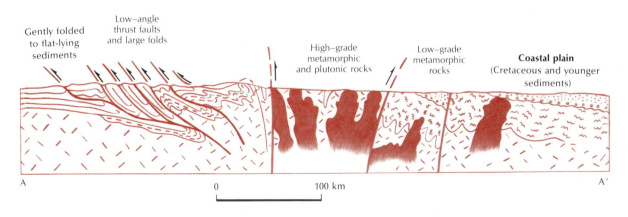

Gently folded
to flat-lying
sediments

Low-angle
thrust faults
and large folds

High-grade
metamorphic
and plutonic rocks

Low-grade
metamorphic
rocks

**Coastal plain**
(Cretaceous and younger
sediments)

A

0    100 km

A′

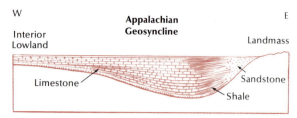

**Fig. 6-44.** Hypothetical cross section through the Appalachian geosyncline as it filled with sediments, showing the thickening of strata toward the axial portion. The maximum thickness of sediments amounted to about 12,200 m (40,000 ft).

Commonly, but not always, alpine chains are formed where thick sections of sedimentary rocks have accumulated, over tens or even hundreds of millions of years, in an immense elongate trough called a **geosyncline** (Fig. 6-44). In the middle 1800s American geologists who had been studying the Appalachian alpine chain suggested that its formation was the result either of folding of the geosynclinal sediments in response to loading under their own weight or the result of compression in the outer part of the earth in response to slow contraction of the earth's interior. The modern interpretation is that alpine mountains are formed in convergent zones where two lithospheric plates push together. Many present-day convergent zones

(Fig. 6-45) are characterized by active volcanism, seismic activity down to 700 km (435 mi), and linear geosynclinal troughs (deep-sea trenches) deep enough to enable the accumulation of great thicknesses of sediment derived from the erosion of nearby continents and volcanic islands. Such areas correspond to the island archipelagoes that border much of the Pacific Ocean and the mountain chains that run intermittently along the west coast of South, Central, and North America. Sediments that fill the geosynclines can be carried down to great depths in the active zone of convergence (subducted), where they can be heated, metamorphosed, melted to produce granitic magmas, and folded and faulted. Figure 6-46 depicts how the process might unfold with time.

One island-arc segment, that of Japan, has had a long and complex geological history. The rocks exposed in the islands are complexly deformed, show signs of regional metamorphism, contain numerous granitic igneous rocks interspersed with the metamorphic rocks, and include considerable thicknesses of sediments derived both from the Asian mainland and from the eruption and erosion of island-arc volcanoes. Thus many geologists point to Japan as a site of active mountain building through the process of plate convergence. Its development, however, has not yet reached the max-

**Fig. 6-45.** Deep ocean trenches, shown in black, demarcate zones of convergence. Notice that most border the Pacific Ocean basin.

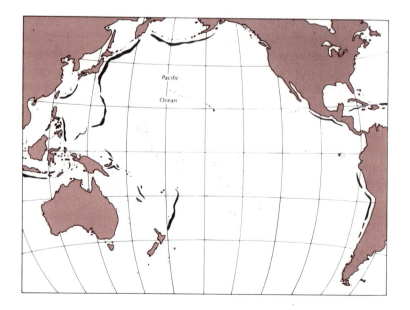

# DEFORMATION OF ROCKS AND MOUNTAIN BUILDING

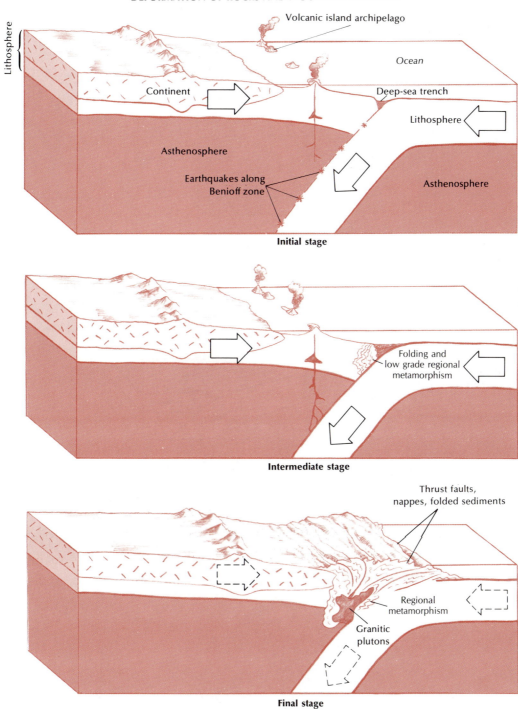

**Fig. 6-46.** Postulated sequential development of an alpine mountain system along a zone of convergence according to the plate-tectonics theory.

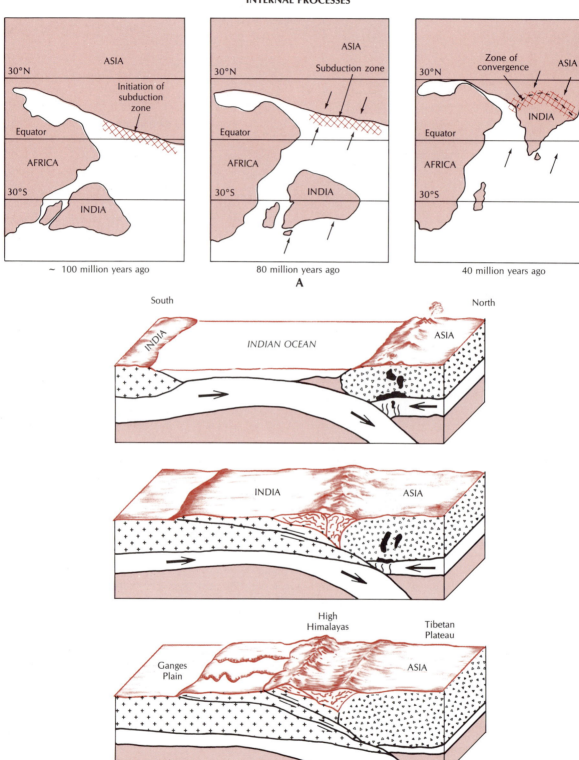

~ 100 million years ago

80 million years ago

40 million years ago

**A**

South

North

**B**

imum stage of alpine-mountain splendor. For that we must turn to the most magnificent range on earth, the Himalayas.

Geological evidence indicates that about 100 million years ago, the south coast of Asia was the site of an east-west convergence zone that separated a continental-type lithospheric plate to the north from an oceanic-type plate to the south (Fig. 6-47). At that time, as determined by paleomagnetic directional data, paleontology, and the character of the rocks, the subcontinent of India was attached to, or in close proximity to, the eastern shore of Africa at about the present location of Madagascar. At that time, the convergence zone would have been similar to the modern western coast of South or Central America—with a deep-sea trench, an accretionary wedge of sediments continuing to build and to be subducted along the southern edge of the Asian lithospheric plate, and a line of volcanoes (like the Andean volcanoes today) being fed by magmas generated at depth along the subduction zone.

As plate motion and subduction of the southern, oceanic lithospheric plate continued, the Indian subcontinent, which was embedded in the lithospheric plate, broke free about 80 million years ago and began to be carried northward toward the zone of convergence. As determined from paleomagnetic data, the northward rate of translation between 65 and 50 million years ago amounted to about 15 to 20 cm (6 to 8 in.) per year. At about 45 million years ago, the rate slowed abruptly to approximately 5 cm (2 in.) per year, the rate at which the southern plate has continued to move northward until the present.

The time of abrupt slowing is thought to have coincided with the arrival of the front edge (northern) of the Indian subcontinent at the convergence zone. Apparently, because of its more buoyant

nature, the Indian continental crust could not be easily subducted beneath the Asian plate. Initially, the sedimentary wedges that had existed along the northern edge of India and the southern margin of Asia were compressed, severely folded, thrust faulted, and at depth metamorphosed. Following this, as subduction of the southern plate continued, the front edge of India was carried downward—but only with great difficulty. The lower portion of the continental mass was carried beneath the southern edge of the Asian plate; the upper portion, however, resisting subduction, sheared off along a large, low-angle thrust fault, as shown in Figure 6-47B. As northward translation and subduction continued at a rate of about 5 cm (2 in.) per year, so did underplating of the Asian plate by the lower part of the Indian subcontinent. In addition, other large thrust faults developed. The Himalayas as we see them today are composed primarily of the stacked, sheared-off upper levels of the northern edge of the old Indian subcontinent.

Parts of the lower Indian crust have been carried northward and are plastered beneath the Tibetan Plateau. Moreover, during the difficult continent-continent convergence of the last 45 million years, the southern edge of the Asian continent has been pushed northward, perhaps as much as 1500 km (938 mi).

As far as geologists can tell, the processes that have built the Himalayas are continuing. Shallow-focus earthquakes abound in the region and are typically of intermediate-to-major magnitude. In association with these quakes, the mountains continue to lurch upward, as India moves north, pushing beneath and into the southern margin of Asia.

One question still remains concerning the Himalaya Range: Why is it so high? The probable answer relates to the thicknesses of the lower-density continental crust and the higher-density lithospheric plate underlying the range. The crust can stand tall if it is thick in comparison to the lithospheric mantle beneath (consequently "floating" on the mantle like an iceberg in water), or if it rests on a relatively thick, strong mantle that holds it up.

The former explanation was the result of a land survey on the plain just south of the Himalayas under the direction of Sir George Everest (1790–

Fig. 6-47. Development of the Himalaya Range during the last 100 million years. A. Plan view of eastern Africa and southern Asia, showing translation of Indian subcontinent from 30° south latitude 100 million years ago to 20° north latitude by 40 million years ago when it was brought into juxtaposition with southern Asia. B. Block diagram, showing the progressive convergence of the Indian and Asian lithospheric plates and the development of the complex alpine Himalayas. (After Molnar, 1986)

1866), for whom the highest peak in the range is named. The distance between two stations—one close to the mountains (Kaliana), the other about 600 km (375 mi) out on the plain (Kalianpur)—was precisely measured by two different standard methods. But it was found to differ by 153 m (502 ft). John H. Pratt (1809–71), a British mathematician, concluded that the discrepancy was caused by a slight difference in the vertical direction at the two localities as determined by hanging a weight on a cord. He reasoned that the mass of

the Himalayas must exert a substantial lateral gravitational force, that would be greater nearer the mountain, thereby causing the vertical directions between the two sites to differ (Fig. 6-48).

Pratt first estimated the mass of the Himalayan mountain block and calculated the lateral gravitational effect it would have at Kaliana and Kalianpur. However, his results were puzzling—the difference in angular deflection of the two plumb bobs at the two sites was about three times greater than the surveyed distance between the two cities

**Fig. 6-48. A.** Deflection of the vertical direction (as determined by a plumb bob) resulting from the lateral gravitational attraction of the mass of the Himalayas above the level of the plain. **B.** Airy's model, showing the root of the less dense material that must exist beneath the Himalayas to account for the actual discrepancy of the vertical as measured at Kaliana and Kalianpur.

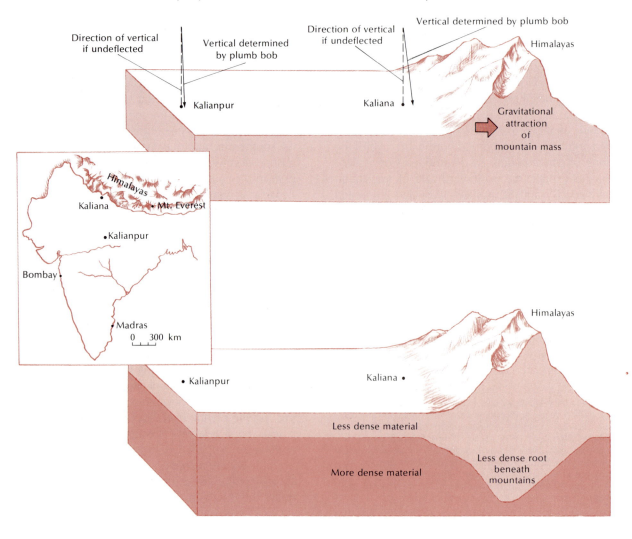

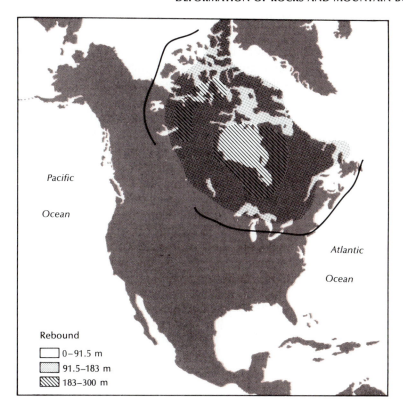

Rebound

☐ 0–91.5 m
▦ 91.5–183 m
▨ 183–300 m

**Fig. 6-49.** North America, showing rebounding of the northeast region after the rapid melting of the ice following the last glaciation. The regions of maximum uplift correspond to areas of maximum ice thickness.

would suggest. Somehow, he realized, there was less mass in the Himalayan block than he had estimated from its topographic dimensions.

Sir George Airy (1801–92), a noted English astronomer, read about Pratt's quandary of the "missing mass" and after some thought proposed the idea that a "root" of relatively low-density crustal material extended below the Himalayas into the denser mantle beneath. In essence, he suggested, the Himalayas "float" on the denser rocks below and, like an iceberg in water, the blocks that project upward the farthest extend the farthest downward (Fig. 6-48). Pratt had considered only the part of the Himalayas above the Indian plain when he considered its mass and was unaware that a lower-than-normal density root stuck far down into the mantle. At about the time that Airy formulated his hypothesis, Pratt came to almost the same conclusion. Since that time the idea that mountains and continents have roots has been tested over and over again—and it has been shown to be valid. Conclusive evidence has come from studies of seismic waves reflected back from

the base of the crust and from measurement of the variation of the gravity field across a mountain range. In some places, for example, the Sierra Nevada Range in California, the crustal root extends nearly 65 km (40 mi) below the surface.

The term applied to floating gravitational equilibrium of the crust is ***isostasy*** (from the Greek *isos,* "equal," and *stasis,* "standing still"). That the less dense surface rocks of the earth are in at least partial isostatic equilibrium is now well established. If the equilibrium is disturbed through some sort of redistribution of the mass at the earth's surface, the lower density, crustal blocks will move up or down and the denser mantle material at depth will slowly deform and flow in order to adjust to the surface movement. If material is loaded on a portion of the earth's surface, that portion will sink to some degree; if material is removed, that portion will rise until a new isostatic equilibrium is reached.

During the ice ages of the recent past, portions of the continents were loaded with great masses of ice, up to about 3000-m (9840-ft) thick (see chap. 14), and the earth's surface was pushed downward. Today most of that ice has melted away, and the former ice-laden areas have risen and will continue to rise until isostatic equilibrium is reestablished. As shown in Figure 6-49, the upward rebound in northeastern North America has already amounted to as much as 275 m (900 ft). The rebound is still progressing, but its rate is decreasing. The last time the area was ice covered was about 10,000 years ago. The large amount of rebound that has already taken place shows how quickly isostatic readjustment can occur.

High mountains are subject to high rates of erosion, and as material is carved from a mountain top, the block undergoes continual isostatic uplift. Because of isostatic recovery, it takes a considerable length of time, perhaps several tens of millions of years, to wear down a high mountain range after structural growth of the mountain has ceased.

The crust and mountain ranges can also rise to great heights because the lithospheric plate is stronger than normal and does not isostatically bow under the weight of the surface protuberance. This idea gained acceptance only in the last few years as geologists found out more about crustal thicknesses around the world. The strength of a lithospheric plate varies directly with its thickness, which itself is a function of temperature. In areas

of the world where the outer portion of a plate is relatively cooler, the lithosphere is thicker and stronger. It is now known that the lithosphere beneath the Himalayas is relatively thick and strong; it is capable, therefore, of supporting some of the load of the high Himalayas. The crust beneath the range is about 55-km (34 mi), thick, which is greater than that under much of India. However, in order to be fully supported isostatically, the range would have to be nearly 80-km (50-mi) thick. Obviously, then, the great height of the Himalayas is partly the result of isostasy and partly the result of support by a strong, thick lithosphere.

In the European Alps, the crust is about as thick as it is in the Himalayas, but the lithosphere is only about one-half as thick. Accordingly, the Alps stand only about one-half as high.

In the Andes of Chile and Peru, which reach heights above 6700 m (22,000 ft), the crust is abnormally thick (resulting from subcrustal underplating and crustal shortening), but the lithosphere is relatively thin. There, the range stands high mostly because of isostatic floatation.

Dynamic forces can also act during convergence to raise or depress blocks within a zone. When the plate motion finally subsides, the height of the resultant alpine range depends solely on the thickness of the continental crust and lithosphere.

## SUMMARY

1. Rocks in the outer part of the earth are currently being bent, broken, uplifted, and down-dropped in response to the action of internal forces.

2. Evidence that deformation occurred in the geological past is seen in *jointed, folded,* and *faulted* rocks. Folding produces a variety of *synclines and anticlines. Faults* are classified on the basis of known directions of relative displacement as *normal faults, reverse* or *thrust faults, strike-slip faults,* and *overthrust faults.*

3. There are many kinds of mountains: volcanic mountains, fault-block mountains, oceanic ridges and rises, and folded and complex mountains. The last, called *alpine chains,* include such ranges as the Appalachian Mountains; they are complex in structure, and require up to hundreds of millions of years to form. Most contain a core of metamorphic and associated igneous rocks flanked on one or both sides by folded

and faulted sedimentary rocks. The degree of deformation in the strata progressively lessens from the core outward.

4. Alpine mountain ranges are thought to have been formed in convergence zones. The Japanese Islands appear, for example, to be an alpine range in the process of forming.

5. The highest range on earth, the Himalayas, apparently was formed during the "collision" of the Indian subcontinent with the Asian continent along a convergence zone bordering the southern Asian coast. The mountain-making process began about 50 million years ago and is still going on. In the process portions of the front edge of India have been sheared off and pushed relatively southward over the front edge of the descending lithospheric plate. The southern margin of Asia has been pushed northward.

6. The height of a mountain depends on the thickness

of the low density crustal root, the thickness and strength of the underlying lithosphere, and whether dynamic processes related to subduction are in operation. The Himalayas stand high because the root is thick and the lithosphere is thick and strong. In contrast the Andes overlie a thin lithosphere but possess a thick root.

7. Low density crustal material isostatically floats on the underlying mantle. If any process disturbs the isostatic equilibrium, such as loading by ice or sediments or unloading by erosion, the crust moves down or up, accordingly, to establish a new equilibrium position. Because of isostatic adjustments, high mountains take a long time to erode away.

## QUESTIONS

1. Give some tangible evidence that the earth's outer portion has been structurally deformed.
2. Discuss the factors that influence how a body of rock will respond to deforming forces.
3. Describe the various kinds of folds. How do domes, anticlines, and monoclines differ?
4. Compare joints with faults.
5. How does the nomenclature of faults differ when only the apparent relative movement is known as compared to when the relative displacement directions are known?
6. Describe lateral faults in general and the San Andreas fault in particular.
7. In what type of plate-tectonic environment do you think overthrust sheets would be most likely? Why?
8. Discuss the various kinds of mountains, giving examples and outlining the principal features that differentiate one kind from another.
9. Outline a reasonable explanation for the Himalaya Range.
10. What are the factors that determine how high a mountain range stands?
11. What is isostasy and what affects isostatic readjustment?
12. After the erosion of the superstructure of an alpine range, what features would indicate its former existence?

## SELECTED REFERENCES

Billings, M.P., 1960, Diastrophism and mountain building, Geological Society of America Bulletin, vol 71, pp. 363–98.

Burchfield, B.C., and Davis, G.A., 1975, Nature and controls of cordilleran orogenesis, western U.S., American Journal of Science, vol. 275A, pp. 363–96.

Clark, S.P., Jr., 1971, Structure of the earth, Prentice-Hall, Englewood Cliffs, N.J.

Collett, L.W., 1927, The structure of the Alps, R.E. Krieger, Huntington, N.Y. (Repr. 1974.)

Curtis, B.F., ed., 1975, Cenozoic history of the southern Rocky Mountains, Geological Society of America, Memoir 144.

Davis, G.H., 1984, Structural geology of rocks and regions, Wiley, New York.

Dewey, J.F., and Bird, J.M., 1970, Mountain belts and the new global tectonics, Journal of Geophysical Research, vol. 75, pp. 262–65.

Dott, R.H., Jr., and Batten, R.L., 1981, Evolution of the Earth, 3rd ed., McGraw-Hill, New York.

Eardley, A.J., 1951, Structural geology of North America, Harper & Bros., New York.

Gilluly, J., 1949, The distribution of mountain building in geologic time, Geological Society of America Bulletin, vol. 60, pp. 561–90.

———, 1970, Crustal deformation of the western United States, in The Megatectonics of continents and oceans, H. Johnson and B.C. Smith, eds., Rutgers University Press, New Brunswick, N.J., pp. 47–73.

Hill, M.L., 1947, Classification of faults, American Association of Petroleum Geologists Bulletin, vol. 31, pp. 1669–73.

Hill, M.L., and Dibblee, T.W., Jr., 1953, San Andreas, Garlock, and Big Pine faults, California, Geological Society of America Bulletin, vol. 64, pp. 443–58.

Kennedy, G.C., 1959, The origin of continents, mountain ranges, and ocean basins, American Scientist, vol. 47, pp. 491–504.

King, P.B., 1959, The evolution of North America, Princeton University Press, Princeton, N.J.

Molnar, P., 1986, The geologic history and structure of the Himalaya, American Scientist, vol. 74, pp. 144–54.

———, 1986, The structure of mountain ranges, Scientific American, vol. 255, pp. 70–79.

Oakeshott, G.B., 1966, San Andreas fault: geologic and earthquake history, 1966, Mineral Information Service, vol. 19, no. 10, pp. 159–65, California Division of Mines and Geology, Sacramento.

Ramsay, J.G., 1967, Folding and fracturing in rocks, McGraw-Hill, New York.

Rubey, W.W., and Hubbert, M.K., 1959, Role of fluid pressure in mechanics of overthrust faulting, Geological Society of America Bulletin, vol. 70, pp. 167–206.

Suppe, J., 1985, Principles of structural geology, Prentice-Hall, Englewood Cliffs, N.J.

# 7

# IGNEOUS PROCESSES AND ROCKS AND PLUTONIC ROCK BODIES

Igneous rocks usually form when minerals crystallize during the cooling of a fiery hot molten material called **magma,** or **melt.** Magma originates deep below the earth's surface and accumulates in molten bodies, magma chambers, which are surrounded mostly by solid rocks. In some cases magma makes its way to the earth's surface where it feeds the eruption of a volcano (Fig. 7-1). **Volcanic rocks,** which are formed during such an event, represent one of the two principal types of igneous rocks. The subsequent cooling and crystallization of magma in subsurface chambers or in conduits that once fed volcanoes occur slowly, generally leading to the formation of coarser-grained **plutonic rocks,** the second major igneous rock type.

In this chapter, we discuss not only the origin of magma, the processes that affect its movement and subsequent chemical modification, and the

nature of the main igneous rock types, but also the characteristics of the principal types of plutonic rock bodies. Chapter 8 is concerned with the nature and origin of volcanic rocks.

## THE NATURE OF MAGMAS

Melting of preexisting rocks—resulting in the formation of magmas—is largely restricted to the outermost mantle and crust. Accordingly, magmas are rich in elements common to that zone, particularly oxygen and silicon, with lesser amounts of the metal ions, including those of aluminum, iron, magnesium, calcium, sodium, potassium, and titanium. All contain a small percentage of volatile constituents, especially water and carbon dioxide. For the most part the chemical composition of parent rock determines the initial composition of the various magmas.

Igneous rocks have a somewhat limited range of composition (Table 7-1). If rich in magnesium, iron, and calcium, they are called **mafic.** This term is derived from the word *ma*gnesium and the Latin word for iron, "ferrum" (hence, *fer*rous and "fer-ric"); the classification refers to an abundance of these elements in the magma. Those magmas rich

**Fig. 7-1.** Frozen torrent of lava, south coast of Kealakoma, Hawaii. Molten material (magma) at a temperature near 1200° C (2192° F) formed initially at 50 km (31 mi) beneath the surface and eventually welled up beneath Kilauea volcano to feed a volcanic eruption, during which the lava flow was formed. (Donald Swanson, USGS)

**Table 7-1** Typical Compositions of Magmas (wt %)

| Oxide Component | Felsic[a] | Intermediate | Mafic[a] | Ultramafic[a] |
|---|---|---|---|---|
| $SiO_2$ | 73.86 | 62.27 | 50.83 | 43.54 |
| $TiO_2$ | 0.20 | 0.57 | 2.03 | 0.81 |
| $Al_2O_3$ | 13.75 | 17.17 | 14.07 | 3.99 |
| $Fe_2O_3$ | 0.78 | 1.62 | 2.88 | 2.51 |
| FeO | 1.13 | 3.58 | 9.06 | 9.84 |
| MnO | 0.05 | 0.10 | 0.18 | 0.21 |
| MgO | 0.26 | 2.38 | 6.34 | 34.02 |
| CaO | 0.72 | 5.46 | 10.42 | 3.46 |
| $Na_2O$ | 3.51 | 4.11 | 2.23 | 0.028 |
| $K_2O$ | 5.13 | 1.71 | 0.82 | 0.005 |

[a]D.M. Hyndman, 1972, Petrology of Igneous and Metamorphic Rocks, 2nd ed., McGraw-Hill, p. 12.
From M.G. Best, 1983, Fundamentals of Igneous and Metamorphic Petrology, W.H. Freeman, San Francisco, p. 612–615.

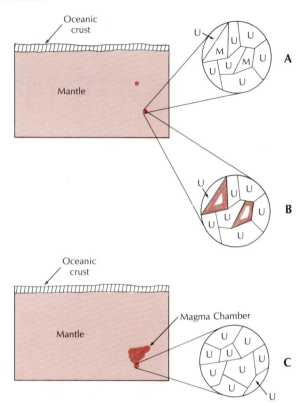

**Fig. 7-2. A.** The outer mantle primarily consists of ultramafic (U) and mafic (M) silicate minerals in the ratio of 3:1. **B.** During the early melting stage, the mafic minerals, which generally have lower melting points, selectively melt. **C.** Melt of mafic composition collects into a magma chamber, leaving behind a residuum greatly enriched in ultramafic minerals.

in sodium, potassium, and silicon are called **felsic** (from *fel*dspar and *sil*ica) and contain larger amounts of feldspar-producing elements. Magmas that are transitional in composition between mafic and felsic are called **intermediate,** and those containing extremely large amounts of magnesium and iron are called **ultramafic.** Each of these terms can be applied to igneous rocks as well as magma.

Most felsic magmas are generated within the continental crust where felsic parent rocks are abundant, whereas mafic and ultramafic magmas are likely to be formed in the outer mantle from parent materials rich in magnesium, iron, and calcium. Intermediate magmas commonly form at depth near the lower portions, or at the margins, of continents.

The initial composition of a magma depends, in part, on the amount of melted parent rock. Most parent rock contains several different minerals, each with a different **melting point,** or melting point range. As the temperature rises locally, those minerals with lower melting-point ranges will begin to melt first. If the amount of heat available is limited, only partial melting will occur and the resulting magma will be primarily derived from those minerals that melted at the lower temperatures. For example, in the mantle beneath the continental and oceanic crusts, parent rocks appear to be made up of a mixture of ultramafic and mafic materials in the ratio of 3:1. Because the melting-point ranges of mafic minerals are gener-

ally lower, partial melting in this zone produces mafic magma (Fig. 7-2).

Magma, which is less dense than the surrounding solid parent material, is buoyant. Once enough magma accumulates to form a pool or blob, it rises toward the earth's surface.

## RECENT VOLCANIC ACTIVITY

The active and recently active volcanoes of the world occur within fairly well-defined zones or linear belts. Perhaps the most renowned is the **Ring of Fire** that girdles much of the Pacific Ocean (Fig. 7-3). In that ring are found about four-fifths of the world's active volcanic centers. Most of the other active volcanoes occur in the Caribbean and

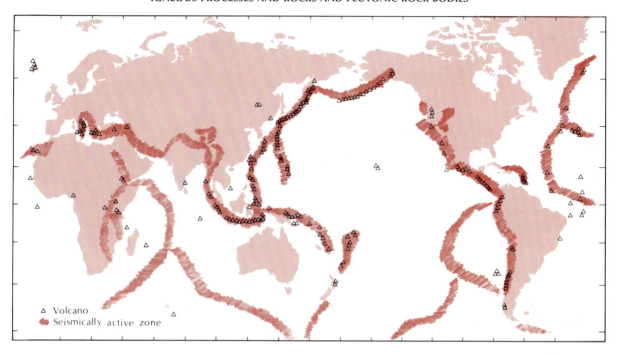

**Fig. 7-3.** Distribution of the active volcanoes of the world in relation to the active seismic zones. Each triangle represents one or more active or recently active volcanoes, of which there are about 516. Most volcanoes occur in a belt around the Pacific Ocean, often called the *Ring of Fire.*

Mediterranean regions, in Asia Minor and the vicinity of the Red Sea, and in central Africa.

Also shown on the map of the world's volcanoes (Fig. 7-3) are the belts of modern earthquake activity. The similarity in areal pattern between the zones of volcanism and seismicity indicate a close relationship between the two phenomena. This is not to say, however, that major earthquakes cause volcanic eruptions or vice versa. Instead, it seems that volcanism and earthquakes are two different ways the earth responds to large-scale forces within it. We are now certain that the linear belts of volcanoes more or less coincide (as do the earthquake belts) with the margins of the lithospheric plates (see chap. 5). Away from the plates' edges, volcanism and earthquake activity are minimal.

Where the plates converge, blocks push and slide past one another creating enough heat to melt rocks and produce magma. In these zones of convergence, earthquakes originate as far down as 700 km (435 mi), whereas magma sources

appear to extend no deeper than about 125 km (78 mi)—many are much shallower.

The Ring of Fire owes its existence primarily to the convergence of plates. The volcanoes tend to be mafic to intermediate in composition and steep-sided. Explosively violent eruptions are common. The volcanoes of the Mediterranean region also occur at a zone of convergence; they pour forth eruptive products that largely are intermediate to felsic in composition.

Volcanic activity is also notable along the divergence zones of the world—areas where plates move apart. Except for isolated localities, such as Iceland and the Azores, most of the activity occurs in the oceans and is not easily observed. Only in the last 10 years or so, with the advent of more sophisticated deep-water equipment, have geologists begun to appreciate the extent of volcanism along the oceanic ridges and rises. Volcanoes in Africa's Rift Valley (Fig. 7-3) also appear to be the result of a diverging plate boundary. Unlike the

Ring of Fire, lavas associated with these divergence zones are mostly mafic. They seem to spring from a magma source at depths of 30 to 70 km (19 to 44 mi) within the outer mantle.

Many of the large basalt fields of the earth, where erosion has cut deeply, show evidence of having been fed by magma that moved upward along fractures possibly related to rifts that developed along ancient zones of divergence.

Volcanism has also been detected along transform-plate boundaries—areas where the plates slide laterally past each other. Most of the activity occurs below the ocean's surface but is revealed by strings of small volcanic cones along the traces of the transform faults.

Not all active volcanoes or volcanic areas are classified on the basis of convergence, divergence, or transform faulting. In some cases, such as the volcanoes of the Hawaiian Islands and the geothermally active region of Yellowstone, Wyoming, magmatism occurs far out in the middle of the plates. Such localized regions of volcanic activity are called *hot spots;* most geologists agree that these spots occur above a plume of relatively hot mantle that is rising from great depths.

## DISTRIBUTION OF PLUTONIC ROCKS

The tectonic settings in which plutonic rocks form are less easily determined because of the time lag between emplacement of the magma and the exposure of the solidified rock through erosion of overlying rock material.

A study of how coarse-grained felsic plutonic rocks are distributed on the continents, however, can provide some insights. Many of these rocks occur (in association with metamorphic rocks) in chains or belts, and many of the geologically more recent ones are found in the cores of mountain ranges—leading to the conclusion that they are related to alpine mountain building—and, therefore, to convergence of lithospheric plates. Long chains of old batholiths stretching across a nonmountainous, highly eroded terrain, as in large parts of Canada, probably are the deeply eroded roots of fossil mountains. Geologists hypothesize that British Columbia's Coast Range and the string of batholiths that runs the length of the Sierra Nevada Range and through Baja California (Fig.

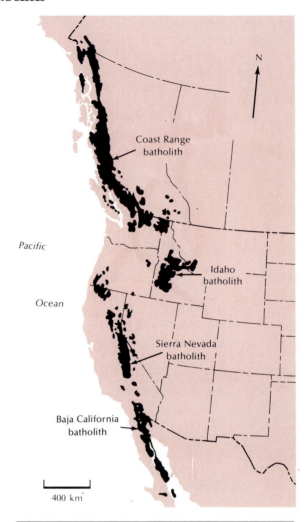

**Fig. 7-4.** Granitic batholiths (black) occur in large elongate bodies along the western margin of North America. Most of the plutonic rocks in these bodies were intruded during the Jurassic and Cretaceous periods.

7-4), are plutons that formed along a zone of convergence over a period of 90 million years. The peak of activity occurred between 100 and 80 million years ago; at that time the surface above these zones was most likely dotted with steep-sided volcanoes—like those so conspicuous in convergence zones today. In essence the plutonic rock bodies are the frozen remains of the subsurface magma bodies that once fed the volcanic fields above.

Most convergence zones are found along the

margins of continents. If this were also true in the past and if granitic rocks formed periodically along the margin, it would seem that the continents would grow outward as new batholithic material is added. In fact, some continents do show a decrease in age of igneous and metamorphic rock belts toward their margins (see Fig. 4-17). This has led many geologists to suggest that early in the earth's history the continents were small or even nonexistent, and that they have been undergoing accretion at their margins ever since.

## MELTING OF ROCK

Melting most commonly occurs when temperatures are locally increased in the earth's mantle and crust. The source of heat may be (1) from convection of the deeper, hotter mantle; (2) from the upward movement of hot magma bodies; or (3) from energy produced as rocks bend and shear along subduction zones. Heating related to convective rise would be significant in hot-spot areas, divergence zones, and regions of back-arc upwelling.

Because of its buoyancy, magma tends to rise, carrying heat upward into cooler levels above. If the density of the upper layers—particularly in continents—is less than that of the magma, upward movement ceases and the magma stagnates and pools. The heat carried with it from deeper levels then causes partial melting of the surrounding crustal rocks. Many of the felsic-to-intermediate magmas in continental regions may have formed in this way.

The melting of parent rock is also affected by changes in **confining pressure** and by volatile constituents, particularly water. Confining pressure is the all-sided pressure at a particular depth that results from the load of overlying rocks. As such it increases with depth. Higher confining pressures tend to favor the solid state. Thus it follows that with a decrease in confining pressure, a rock's melting range will be lowered and some melting may take place if other conditions are appropriate. Therefore, when mantle material rises through convection—in plumes and beneath divergence zones—some melting is to be expected.

Volatile materials, especially water, aid melting

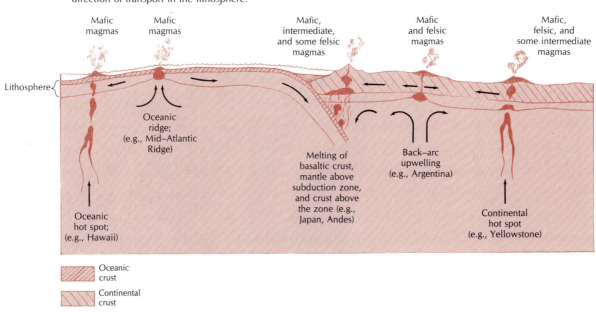

**Fig. 7-5.** A summary showing principal regions of melting in outer portion of earth. Arrows indicate direction of transport in the lithosphere.

and counteract increases in confining pressure. The greater the confining pressure, the more water a rock or magma can contain. Even at high temperatures in the outer 300 km (187 mi) of the earth, the confining pressure is great enough that up to 6 percent water can be contained.

Water acts as a catalyst, reducing the stability of silicate bonds and enabling melting to occur at much lower temperatures. The presence of even small amounts of water (approximately 1 percent) will reduce the melting ranges of most silicate minerals as much as 200° to 300° C (392° to 572° F). All igneous rocks contain some water, suggesting that it was present at the time of magma generation. Measurement of volcanic gases shows that the principal volatile constituent accompanying eruptions is water. Some of the water appears to be from the mantle, part of the primordial volatile fraction still escaping from the earth; some is derived from water-bearing minerals, such as mica and amphibole; and some from sediments subducted along convergence zones.

Figure 7-5 shows the tectonic environment in which most of the magma-producing mechanisms mentioned here might occur.

## CRYSTALLIZATION OF A MAGMA

At some time after a magma is emplaced beneath the earth's surface or erupts at the surface, it will begin to crystallize into a mineral aggregate—an igneous rock. Crystals start to form as temperature decreases and heat no longer disrupts the bonds between the various ions. The minerals usually crystallize at different temperatures—the first to crystallize are those with the highest temperature stability range (Fig. 7-6); those that follow have successively lower ranges. The process thus proceeds until all of the magma is transformed into a solid.

Studies of the order of occurrence of common silicate minerals during crystallization were begun in the early 1900s by Norman L. Bowen (1887–1956) and his colleagues at the Geophysical Lab-

**Fig. 7-6.** Crystals of calcic plagioclase (elongate, white) and olivine (equant, gray) are the first to form during crystallization of a mafic magma. The black material between the crystals is uncrystallized glass. Microscope photograph. (Edwin E. Larson)

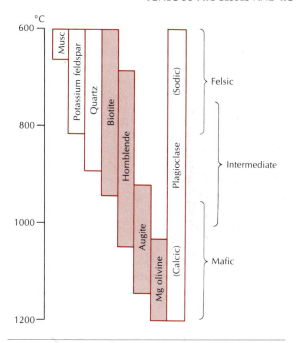

**Fig. 7-7.** Generalized formational ranges and succession of common silicate minerals that crystallize from mafic, intermediate, and felsic magmas—based on Bowen's reaction series and field observations. Ferromagnesian minerals are shaded; Musc = muscovite; Mg olivine = magnesium-rich olivine.

oratories in Washington, D.C. Their work led to the formulation of an idealized mineral succession, **Bowen's reaction series,** which is basic to the understanding of the wide variety of igneous rocks.

The Bowen series represents the entire range of common rock-forming silicates in igneous rocks—(Fig. 7-7). During cooling, however, a particular type of magma—mafic, intermediate, or felsic—will follow only a small part of the series. For example, during crystallization of a mafic lava, probably originating within the outer mantle, well-formed crystals of olivine and calcic plagioclase are commonly the first to develop (Fig. 7-6), followed by augite, which arises at lower temperatures in the spaces between the earlier-formed crystals. By the time augite has crystallized, less than 5 percent of the magma will remain. On the other hand, a felsic magma will contain potassium and sodium ions as well as silica in abundance at a temperature of about 700° to 800° C (1292 to

1472° F); first-formed crystals might be hornblende, biotite, and sodic plagioclase followed, if any magma remains, by quartz and potassium feldspar. As shown in Figure 7-7, the different rock types overlap with regard to temperatures of formation and mineralogy. This is because (1) the amount of cations in actual melts vary (i.e., magnesium may be abundant in one mafic magma, whereas iron may be the dominant metallic cation in another and (2) the melts crystallize under different confining pressures (in relation to depth) and their water content varies (thereby affecting freezing-temperature ranges). The last is a most important factor. Magmas under moderate confining pressures that contain large amounts of water can exist in the molten state at temperatures of 200° to 300° lower than can water-deficient magmas. Water is also critical in the crystallization of minerals—such as hornblende and the micas—that contain water in their crystal structures.

One complication of this rather simple succession of crystallizing minerals is that some of the ferromagnesian minerals—particularly olivine, augite, and hornblende—that initially crystallized at higher temperatures from more-mafic melts may react chemically with any remaining melt and change to one of the lower-temperature ferromagnesian minerals. This reaction seems to depend both on the rapidity of crystallization and the initial composition of the magma. It is less likely to occur in mafic magmas that are rich in magnesium and deficient in silica. In some cases mineral grains may be caught in the act of transformation, so that a core of high-temperature ferromagnesian material will be rimmed by a later-formed, low-temperature material.

In Figure 7-7 plagioclase is shown as a stable, crystallizing material produced from all three magma types. Calcic plagioclase, however, forms at high temperatures, intermediate plagioclase at intermediate temperatures, and sodic plagioclase at low magmatic temperature. As magma cools, therefore, the stability range of the precipitating plagioclase changes progressively, so that the earlier-formed, more-calcic crystals are no longer in equilibrium with the melt at lower temperatures. Equilibrium is regained, however, when these crystals react chemically with the melt, converting into plagioclase crystals richer in sodium. This

**Fig. 7-8.** Microscope photograph of zoned plagioclase crystal in a mafic lava flow. This crystal has a core relatively richer in calcium and aluminum and a margin richer in sodium and silicon. (Edwin E. Larson)

temperature minerals that crystallize from any magma will tend to be richer in calcium, magnesium, and iron than the remaining melt, which will be enriched in potassium, sodium, and silicon (Fig. 7-7). At any intermediate stage in the crystallization, bulk composition of the crystals will differ from that of the remaining magma. If some process occurs that physically separates the remaining magma from the previously formed crystals, the starting magma will be subdivided into two chemically different parts. The processes that lead to this subdivision are lumped under the term *magmatic differentiation.* The process responsible for most magmatic differentiation is related to an intermediate stage of crystallization. The process is called *crystal fractionation.*

Crystals can be separated from the residual magma in a number of ways. Some are heavy enough to sink through the melt and accumulate on the chamber floor. The results of this process are well displayed in the Palisades Sill, a tabular plutonic body exposed in a bold cliff along the Hudson River in New York and New Jersey (Fig. 7-9). In addition, crystals in the lower layers of accumulations commonly are unable to enter in subsequent melt-crystal reactions (such as between the melt and calcic plagioclase or some of the ferromagnesian minerals). In some cases earlier formed crystals may float, again producing a gravitational separation between crystals and melt.

If the residual magma in some way separates from the crystal mush (by draining, compression of the liquid-crystal mixture, or inability of the crystal portion to flow readily), the initial magma will have been completely separated into two different compositional parts. And it is entirely possible that the magma portion resulting from one stage of crystal fractionation will, in turn, undergo subsequent stages during continued cooling and crystallization.

Over an extended period of time, crystal fractionation can lead to substantial modification of the composition of the melt. On Hawaii, for example, the most commonly erupted lava is mafic, crystallizing to form a dark rock composed mainly of olivine, calcic plagioclase, and augite. In the melt remaining in the underground chamber, crystals of olivine and calcic plagioclase normally form with time; these then settle out and fall to

process is slow, requiring sold-state diffusion of calcium and aluminum cations out of, and sodium and silicon cations into the crystal lattice. It has a better chance of occurring if solidification is slow, as it generally is in the formation of plutonic rocks. If solidification is too rapid to permit completion of this process, as is common during volcanism, the resulting plagioclase crystal will be zoned, with a calcic core formed at high temperature, armored by layers that are progressively more sodic (Fig. 7-8).

## MAGMATIC DIFFERENTIATION

Because of the variation in mineral stabilities with temperature, the earlier-formed higher-

In a cooling magma, nuclei can begin to arise spontaneously once a critical temperature has been reached. The rate is slow, however, and in some cases nucleation is delayed until further cooling takes place. As cooling continues, the rate increases up to a maximum. Thus, the rate of formation of nuclei on which crystals can grow is greatest at some temperature considerably below that at which crystallization could theoretically begin.

On the other hand, the rate of crystal growth, once a nucleus is present, depends on the efficiency of diffusion—that is, the ease with which the ions needed in crystal growth move through the magma to the surfaces of the growing crystals.

The maximum rate generally occurs only slightly below the theoretical crystallization temperature, and it steadily decreases with falling temperature.

If cooling is slow, much of the crystallization takes place just below the critical temperature. Few nuclei form, but diffusion rates are high, so that the crystals grow to large size. With more rapid cooling, more nuclei develop, but the diffusion rate is lower, resulting in smaller crystals. If the cooling rate is very rapid, as it commonly is for lavas, both the rate of nucleation and diffusional crystal growth eventually drop to near zero. Then, few crystals can develop and the remaining magma quenches and becomes a volcanic glass.

Diffusion rates depend on the fluidity of the

**Fig. 7-12.** Porphyritic lava photographed under the microscope. Large white tabular crystals of calcic plagioclase are set in a fine-grained groundmass composed of plagioclase (well-formed elongate to blocky, white to light gray), augite (nonelongate, medium to dark gray), and magnetite (black). A texture such as this is usually the result of two distinct phases of cooling, a slow one followed by a rapid one. The smaller plagioclase crystals in many areas are in subparallel arrangement as a result of alignment during flowage. (Edwin E. Larson)

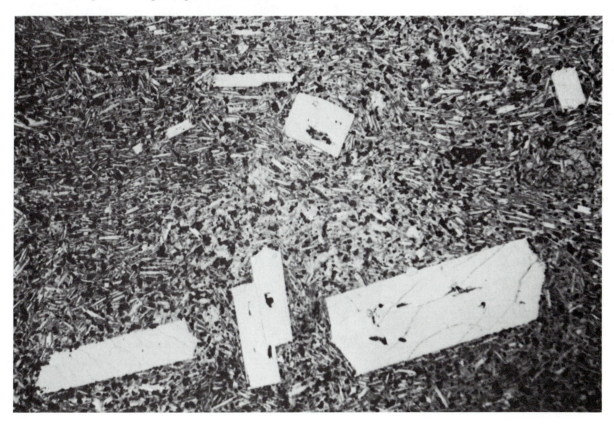

magma; the more fluid it is, the faster the rate of diffusion. The mafic magmas are the most fluid, in part because they are much hotter than intermediate and felsic magmas. Felsic magmas, on the other hand, contain larger amounts of silica, which decreases their fluidity. Diffusion rates drop accordingly.

Some igneous rocks are **porphyritic;** the term derives from **porphyra,** the Greek word for imperial purple—a highly prized dye extracted from an eastern Mediterranean shellfish. By extension, the name was applied to a specific kind of rock— **porphyry,** a dark, igneous rock from Egypt that contains small white felspar crystals embedded in a purplish, fine-grained matrix.

Today the word is applied to any igneous rock with crystals of two markedly different sizes (Fig. 7-12). A porphyritic texture commonly is the result of two stages of cooling—an early, slow one during which some crystals with well-formed faces grew to a fairly large size, followed by a later, more rapid one when smaller grains crystallized. In many cases the early phase of crystal formation took place in deep magma chambers, whereas the later, more rapid phase occurred after the magma's injection into near-surface chambers or even after its eruption. In porphyritic rock the larger crystals

are called **phenocrysts** (from the Greek *phainein,* meaning "to show," plus *cryst* for "crystal"), and the material in which they are embedded is called the **groundmass.** Usually, the groundmass is finely crystalline or glassy (when the last stages of cooling are very fast).

Because the larger crystals in a porphyry arise while the magma is still largely fluid, they grow without interference and often may achieve a nearly perfect external form. The minerals crystallizing later come out of solution more rapidly; because the growth of each crystal can be hampered by neighboring crystals, few are well formed.

Clastic rocks can also be formed as a direct result of volcanism. The term applied to most of these rocks is **pyroclastic** (Greek words for "fire" and "broken"), which differentiates them from normal sedimentary rocks and emphasizes their mode of origin. Most pyroclastic rocks develop from volcanic material blasted into the atmosphere during the explosive discharge of a vent and consist principally of fragments of crystalline rocks, glass, and phenocrysts (Fig. 7-13).

Some volcanic clastic rocks are deposits of debris flows that accumulated on the flanks of a volcano. These deposits—composed mainly of subrounded, subangular, and angular fragments—

**Fig. 7-13.** Volcanic ash, viewed microscopically, composed predominantly of larger glass shards (white to light-gray subequant, elongate, and bifurated) and biotite and magnetite grains (elongate dark gray to equant black) set in a matrix of finer glass shards. Much of the dark- to light-gray interstitial material, which appears hazy, is very fine-grained volcanic dust. (Edwin E. Larson)

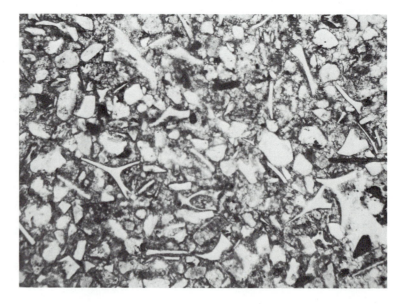

can range from boulder-size to clay-size grains. Because they are associated with volcanism, they are considered to be volcanic with a clastic texture.

## CLASSIFICATION OF CRYSTALLINE-TO-GLASSY IGNEOUS ROCKS

All igneous rocks can be classified on the basis of mineralogical makeup and texture. Scores of names have been applied to igneous rocks, but we will discuss only three plutonic and three volcanic divisions. We will consider six major rock groups (Fig. 7-14): granite and rhyolite, which form at low temperatures from felsic melts; diorite and andesite, which form at intermediate temperatures from magmas of intermediate composition; and gabbro and basalt, which form at high temperatures from mafic melts.

**Fig. 7-15.** Granite from San Bernardino County, California. The large crystal (center) is potassium feldspar, the banded elongate crystals are mostly plagioclase, and the light- to dark-gray to black equally spaced, subequant grains are mostly quartz. Microscope photograph. (J.C. Olson, USGS)

| Crystalline igneous rocks | | | |
|---|---|---|---|
| | Felsic | Intermediate | Mafic |
| Extrusive fine grained (volcanic) | BASALT | ANDESITE | RHYOLITE |
| Intrusive coarse grained (plutonic) | GRANITE | DIORITE | GABBRO |

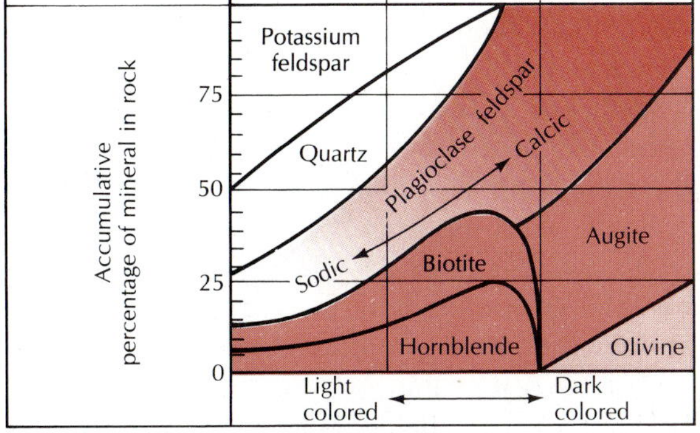

**Fig. 7-14.** The mineral composition and names of six common crystalline igneous rocks. The degree of shading generally corresponds to the darkness of the mineral. Felsic rocks contain only a small percentage of dark minerals, mafic rocks a large percentage. The size of lettering of rock names indicates the volumetric importance in the extrusive and intrusive rock groups.

### Felsic Rocks

**Granite**  The name for granite derives from the Latin word for "grained." It is a relatively light-colored plutonic rock, coarse- to fine-grained. Most of the rock is composed of potassium feldspar and sodic plagioclase (Fig. 7-15). Quartz, normally present in amounts up to 25 percent, is the last mineral to crystallize and occurs as gray rounded-to-irregular masses that fill the spaces between earlier-formed minerals. Muscovite is also present in some granites. Black minerals, chiefly amphibole and biotite, occur in small amounts only.

Granite has a characteristic speckled appearance, with a white-to-gray background of feldspar and quartz flecked with dark spangles of mica and needles of hornblende. In the coarser varieties,

**Fig. 7-16.** A microscope photograph of rhyolite from Rio Grande County, Colorado. Phenocrysts are mostly quartz and potassium feldspar with a few plagioclase, biotite, and magnetite (black) crystals. The groundmass, originally glassy, has recrystallized into a network of extremely tiny grains of quartz and potassium feldspar. Compare the grain size of this rock with that of its plutonic, compositional equivalent in Figure 7-15; both photos are at the same scale. (P.W. Lipman, USGS)

some minerals (chiefly the feldspars) may be 1 cm (0.5 in.) or more across.

**Rhyolite** A volcanic rock, rhyolite's name derives from the Greek words for *"lava torrent"* and *"stone."* It has the same compositional range as granite, but is porphyritic, fine-grained, or glassy. The minerals in the nonglassy types can only be seen under a microscope, if at all. Phenocrysts in the porphyritic varieties are sodic and potassic feldspars, quartz, biotite, and (rarely) other ferromagnesian minerals (Fig. 7-16). Quartz can crystallize early in the formation of a rhyolite, and when it does, the crystals appear completely formed, often with triangular faces at both ends. Rhyolite

ordinarily is light colored, but it may be white, light gray, or various shades of red. A characteristic textural feature is a streaked pattern known as *flow banding.* As the name implies, the banding results from the concentration of colored material or glass in layers during flowage of the highly viscous lava. In many cases it appears that the lava must have flowed very slowly, much as heavy molasses does.

Rhyolite magma sometimes cools so rapidly that crystallization of minerals is impossible, resulting in a volcanic glass that lacks long-range internal crystal ordering. A common felsic glassy rock is *obsidian* (from the Latin *obsidianus,* after its supposed describer Obsius). It commonly fractures conchoidally or irregularly into fragments with sharp edges (Fig. 7-17)—a property that made it an effective weapon when flaked or chipped into arrowheads, spear points, and knife blades by the people of ancient cultures. Recently, obsidian has been used in making scalpels for delicate surgery. Obsidian also displays spectacular flow-banding patterns (Fig. 7-18).

Over long periods of time, the ions in obsidian undergo solid-state diffusion (usually in association with water), leading to crystallization. As diffusion takes place slowly, only tiny crystals can grow. The process is called *devitrification;* the resultant rock is called *devitrified obsidian* or *devitrified glass.* Because of this process, true obsidians are virtually unknown in the rock record farther back than the Cretaceous Period (135 million years).

A special variety of rhyolitic volcanic glass—rather like petrified froth, resembling the foam on the top of beer—is *pumice* (an ancient name from a Greek word meaning "worm eaten"). Because of an abundance of gas cavities, the rock is extremely porous and light in weight. Some pumice will even float on water; if pumice is blown out of coastal or oceanic volcanoes, it may drift for thousands of kilometers before becoming waterlogged and sinking to the bottom.

## Intermediate Rocks

**Diorite** A coarse- to fine-grained plutonic rock, diorite has a mineral composition that places it between granite and gabbro. Its name, which is

**Fig. 7-17.** Obsidian, the most common type of volcanic glass, showing the characteristic conchoidal fracture. (Ward's Natural Science Establishment, Inc.)

**Fig. 7-18.** Flow banding caused by viscous flow in a felsic lava that largely congealed as obsidian. A volcanic dome south of Mono Lake, California. (John Haddaway)

**Fig. 7-19.** Hand sample of diorite from the Front Range, Colorado. The light-gray patches are composed of intermediate plagioclase or (less commonly) quartz; the black areas are made up of grains of hornblende, biotite, and magnetite. (Janet Robertson)

**Fig. 7-20.** Porphyritic andesite from Conejos Peak, Colorado. Most of the phenocrysts and small crystals are intermediate plagioclase. The groundmass is largely a mixture of glass and lath-shaped fine crystals of pyroxene. Microscope photograph. (P.W. Lipman, USGS)

derived from the Greek, meaning "to divide," indicates its intermediate composition. Intermediate plagioclase felspar is its chief constituent; there is little or no quartz or potassium feldspar present. Hornblende is its predominant dark mineral, although pyroxene and biotite can be abundant. Together the dark minerals can be nearly as plentiful as the feldspar (Fig. 7-19).

Because it contains little quartz and potassium feldspar as well as almost equal amounts of plagioclase and ferromagnesian minerals, diorite tends to be a gray rock.

**Andesite**  Named for its common occurrence in the Andean summit volcanoes of South America, andesite is generally a gray to grayish-black fine-grained volcanic rock that is typically porphyritic.

Visible quartz is usually absent; the chief feldspar is intermediate plagioclase; and the dark minerals are principally augite, hornblende, and biotite (Fig. 7-20). The same minerals also occur as phenocrysts in the porphyritic varieties. Andesite covers the same range of intermediate composition as its plutonic counterpart, diorite. Of the volcanic rocks, andesites are more abundant than the rhyolites, but they are less abundant than basalts.

### Mafic Rocks

**Gabbro**  An old name for the dark rocks used in many Italian Renaissance palaces and churches, gabbro is a plutonic rock consisting of a coarse-grained intergrowth of crystals, typically pyroxene and calcic plagioclase (Fig. 7-21). Many gabbros also contain olivine; some contain small amounts of hornblende. Unlike diorite and granite, gabbro contains larger amounts of ferromagnesian minerals than feldspars. There are exceptions, of course, and one variety, **anorthosite,** consists almost entirely of calcic plagioclase interlocked in a coarse-grained texture.

**Basalt**  One of the most ancient names in geology, basalt apparently dates back to Egyptian or Ethiopian usage; in fact, one of the first references to it by name is in the writings of Pliny the Elder (A.D. 23–79). It is by far the most abundant of all volcanic rocks, forming the upper part of the crust throughout the earth's oceans. Many regions of the world, such as the plateau bordering the Columbia River in the northwestern United States, were once almost completely inundated by vast outpourings of basaltic lavas. The volume of basalt in the Columbia Plateau is estimated to be about 300,000 km$^3$ (72,190 mi$^3$). In addition, many volcanic islands—for example, Samoa, Hawaii, and Tahiti—are composed entirely of basalt.

Basalt is ordinarily coal black to dark gray; its texture ranges from fine-grained to aphanitic. The two principal mineral constituents are pyroxene and calcic plagioclase (Fig 7-22). Olivine is also common. Any of the three minerals may be present as phenocrysts in porphyritic varieties. Basalt is commonly frothy and cellular and filled with

**Fig. 7-21.** Gabbro from Victoria Land, Antarctica. The large, equant-to-oblong dark-gray crystals with conspicuous cleavage cracks are pryroxene, the light-gray elongate- to irregular-shaped crystals are mostly calcic plagioclase, and the small black crystals are iron oxides. Microscope photograph. (W.B. Hamilton, USGS)

**Fig. 7-22.** Moderately coarse-grained basalt from Victoria Land, Antarctica. Lath-shaped, banded calcic-plagioclase crystals are present along with crystals of pyroxene (irregular- to blocky-shaped, dark gray) between. Magnetite (black) is also present. Microscope photograph. (W.B. Hamilton, USGS)

**Fig. 7-23.** Cellular or scoriaceous basalt, a fine-grained to glassy basalt filled with tiny bubble holes. (Hal Roth)

innumerable small holes: vesicles—gas bubbles trapped in the lava as it solidified. These rocks are often called *scoria* when the vesicles are abundant (Fig. 7-23).

## CLASSIFICATION OF PYROCLASTIC AND VOLCANICLASTIC ROCKS

During the eruption of a volcano, many chunks and pieces of lava are hurled into the air (Fig. 7-24). Some fall back to the earth near the erupting vent, some fall farther out on the volcano's flanks, and some of the smaller pieces are carried long distances from the vent by prevailing winds. All such explosively ejected material is *pyroclastic*. Some material consists of large and small incandescent clots of liquid lava, which cool and largely consolidate during their flight. Some of these are spindle-shaped, but most occur in rounded, irregular forms. All such clots are called *volcanic bombs* (Fig. 7-25). Many have a crust a few centimeters thick that looks much like that on a loaf of French bread. Some have a flattened snout, which apparently results from impact with the earth's surface, indicating that the bomb did not solidify com-

pletely during its flight. Vocanic bombs cool rapidly and therefore are usually aphanitic to glassy in texture; they can be of any chemical composition, felsic to mafic.

Equant, subangular to subrounded fragments that are highly vesicular are common in pyroclastic material. Felsic clasts are called pumice, whereas mafic clasts are called scoria. Fragments of scoria that are exposed to water-rich vapors at high temperature tend to be bright red; such clasts are called *cinders*.

Tiny pyroclastic fragments of chilled lava [about 2 cm (1 in.) in diameter] are called *lapilli* (an Italian word meaning "little stones"); rocks composed dominantly of this material are called *lapilli tuffs*. Smaller pieces, down to the size of dust, are called *ash*.

Generally, bombs and cinder-size fragments are so heavy that they cannot be blown far from the vent; they fall mostly on the volcano's flanks. Deposits of angular bombs, cinders, lapilli, and ash, in an unsorted mixture but possessing a crude layering, are called *volcanic agglomerates* or sometimes *volcanic breccias.* If the ashy material dominates, they are called *tuff breccias.* Lapilli

**Fig. 7-24.** Recent eruption of Cerro Negro, Nicaragua, that produced the lava flow in the middle background and, subsequently, the steep-sided cinder cone. Large blocks, blasted from the vent, fall near the crater and tumble downslope trailing white plumes of vapors (mostly water vapor). (USGS)

**Fig. 7-25.** A large elongate volcanic bomb, amid other smaller-sized pyroclastic material, on the northeast flank of North Sister volcano in the Cascade Range. (Edwin E. Larson)

and ash can be blown high over an erupting cone; subsequently, some will fall back on the cone flanks and some—particularly the ash—can be carried by prevailing winds and later deposited in measurable amounts thousands of kilometers away. Ash that is carried into the air and that later falls back to earth to form layered deposits is called *air-fall ash,* and the consolidated rock composed of that material is called an *ash-fall tuff.* This tuff, fine-grained in texture, is composed of pieces of pumice, jagged fragments of volcanic glass and phenocrysts (usually potassium feldspar, quartz, amphibole, and biotite), and small angular clasts of volcanic rock ripped from the throat of the volcano during the explosion. Highly explosive volcanoes that pour forth great quantities of ash are as a rule felsic to intermediate in composition.

One variety of pyroclastic rock composed largely of glass-rich ash is so hot when it is deposited that the angular fragments of glass are able to weld together after they come to rest. Under their own weight, the deposits—up to 300-m (985-ft) thick and laterally extensive—settle and become compact, even to the point that softened pumice fragments lose their porosity and completely collapse into dark-colored flattened blebs of volcanic glass. The resulting rock, called a *welded ash-flow tuff,* is dense and strong, with an easily visible banding (Fig. 7-26). Impressive layers of welded tuff occur

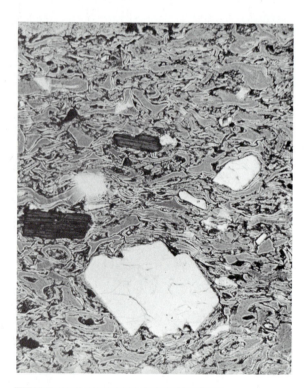

**Fig. 7-26.** Microscope photograph of welded ash-flow tuff from near Gunnison, Colorado. Partly broken phenocrysts of sodic plagioclase (white) and biotite (very dark gray to black) are set in a matrix of fused, angular fragments of glassy ash (medium gray rimmed by light gray to white). Notice that the crude layering in the rock is parallel to the long dimension of the biotite grains. (J.C. Olson, USGS)

**Fig. 7-27.** Laharic material on the slopes of volcanoes in Yellowstone National Park accumulated about 40 million years ago and buried a stand of trees—only the petrified trunks remain. The fragmental character of the material can be seen in the outcrop above the geologist's head. Notice the angularity of the clasts, the lack of good stratification, and the mixture of large and small particles. (J.P. Iddings, USGS)

**Fig. 7-28.** Volcanic breccia—angular fragments of andesite set in a matrix of fine-grained andesite fragments, ash, and mud. From Gunnison County, Colorado. (Janet Robertson)

in the San Juan Mountains, Colorado; southern New Mexico; Nevada; and in Yellowstone National Park, Wyoming. In these regions, single beds, resulting from catastrophic emissions of immense volumes of hot ash, are more than 160-km (100-mi) long.

Some bedded volcanic deposits are composed of angular, subangular, and subrounded fragments, ranging in size from boulders down to silt and clay. Such deposits are the result of sedimentary processes acting on the slopes of a volcanic upland. The upper slopes of a volcano commonly are piled high with pyroclastic debris and interlayered lava flows that are particularly susceptible to slumping and stream erosion. When suddenly saturated with water (e.g., from rainfall induced either by a volcanic eruption or occurring through normal precipitation, drainage of a summit-crater lake, or melting of a summit ice cap, etc.), the slopes often become very unstable. The water-soaked material moves downslope, within the confines of stream channels, in torrents of muddy ash and coarser clastic and pyroclastic debris, and out onto the gentler slopes at the perimeter of the volcano. Such slides were once called *volcanic mud flows*,

a term recently replaced by **lahar**—a word from Indonesia (Fig. 7-27).

Moreover, through the continued activity of streams on the sides of the volcano, material can be eroded from the steeper slopes of the volcanic vent and deposited in aprons surrounding its base (Fig. 7-27). Such deposits, composed of stream-worn clastics, are truly sedimentary rocks and are so named. Often stream-deposited material will be interlayered with strata composed of lahars and lapilli- and ash-rich pyroclastic materials.

The term *volcanic breccia*, introduced earlier with respect to certain pyroclastic deposits, can also be applied to any coarse-grained volcanic rock layers in which most of the fragments are angular (Fig. 7-28).

## PLUTONISM

### Intrusive Rock Bodies

Bodies of plutonic rock occur in a great variety of sizes and shapes. Many of the largest are elongate bodies over 1600 km (1000 mi) in length, whereas the smallest can be less than 1-cm (0.5 in.) across.

Some are characteristically tabular or cylindrical; others are irregularly ellipsoidal, circular, or even shaped like a hotdog. All such bodies, regardless of size or shape, are known as **plutons**—most did not result from the melting and later solidification of rocks in one place. Usually, their chemical composition is completely different from that of the host rocks (referred to as **country rocks**) that surround them. Instead, magma at depth, once formed, is relatively mobile. It tends to rise, pushing aside preexisting rocks and forcing its way into cracks, some of which it may actually produce—a process called **forceful injection.** In some crack systems, magma has moved distances of 50–100 km (31–62 mi). It can also move upward by melting away the surrounding country rocks and by wedging or prying out solid chunks of rock, which

sink into the magma. More will be said about how magma moves upward later in this chapter.

When magma makes its way into preexisting rock, it is said to have **intruded** the host rock. Plutonic rocks, then, can be called **intrusive rocks** and the surface between the pluton and country rock, the **intrusive contact.** Depending on conditions that prevailed during emplacement of the magma, the contact can be very sharp or hazily gradational.

Magma can move and intrude both laterally and vertically. It is through vertical rise that it comes close enough to the surface to be uncovered by erosion. If magma can rise high enough, it will eventually break through the surface in a volcanic eruption.

The reasons that magma moves and intrudes

**Fig. 7-29.** A mafic dike, about 3 m (10 ft) thick, cuts across shaly sediments, Hance Rapids, Grand Canyon, Arizona. (P.E. Patterson)

where it does are varied and not always completely clear. One good reason for its tendency to rise—its buoyancy—is that liquid magma is normally less dense than the solid country rocks around it.

Once magma has stopped moving, it becomes stagnant and begins (or continues) to cool. At great depths it cools slowly and produces a coarsely crystalline texture; material crystallizing near the surface, where cooling is more rapid, forms rocks of finer-grained texture. Magma that has risen close to the surface may cool so quickly that the grain sizes of the resulting rocks are about the same as those of volcanic rocks.

Porphyritic plutons—those with textures indicating different rates of crystallization—are typical of intrusive rocks from shallow to intermediate depths where there may have been a considerable

movement of magma from one environment to another in a complex conduit system.

**Small Plutons** *Dikes* and *sills* are tabular intrusions in which two dimensions are large compared to the third—they have about the same geometry as that of a sheet of notepaper. The two types of intrusions differ in their relationship to the layering of country rocks (usually sedimentary rocks). Dikes are ***discordant intrusions***—they cut across the layers (Fig. 7-29); sills are ***concordant intrusions***—they are more or less parallel to the layering (Fig. 7-30). Concordance does not mean that sills follow a single stratum: It is not at all uncommon for sills to angle up or down from one stratum to another and to follow it, for perhaps hundreds or thousands of meters, before making another step-like change in level (Fig. 7-30).

**Fig. 7-30.** Diabase sill on Banks Island, Northwest Territories, Canada, showing an abrupt change in level of intrusion along a fracture. All the rock layers now above the sill and those that have since been eroded away were uplifted during the emplacement of the sill. (Geological Survey of Canada)

**Fig. 7-31.** Swarm of subparallel dikes of felsic rock cutting a dark-colored gneiss in the face of Painted Wall, Black Canyon of the Gunnison National Monument, Colorado. The Gunnison River (left) flows about 685 m (2250 ft) below the top of the cliff. (W.R. Hansen, USGS)

Dikes come in all sizes; they are seldom more than 30-m (100-ft) wide, but some are hundreds of kilometers long. An exceptionally long one, probably the longest in the world, is the Great Dike of Rhodesia in southeast Africa. More than 480-km (300 mi-) long, it has an average width of only about 8 km (5 mi).

Dikes can occur singly, but most often they occur in groups called *dike swarms* (Fig. 7-31). Sometimes they are aligned on roughly parallel courses; sometimes they radiate from centers, such as the host of basaltic dikes that lace the northern part of Great Britain or those that produce the eye-catching ridges at Spanish Peaks, Colorado. The focal point for a radial swarm of dikes usually is the throat, or conduit, of an extinct volcano or a cylindrical pipelike intrusion.

Dikes that are more resistant to erosion than country rocks may form semicontinuous topographic walls that extend far across the countryside (Fig. 7-32).

Dikes most commonly form when magma intrudes a fracture, the walls being pushed apart as the magma moves ahead. In some cases, the crack is actually formed and propagates just ahead of the intruding magma by the concentration of stress at the front, wedgelike edge of the developing dike. Crack formation and dike intrusion are most common in terrain undergoing tectonic extension rather than in regions of compression. Injection of mafic magma along swarms of parallel dikes, as you might expect, occurs to great extent in divergence zones. The total thickness of the many individual dikes is a rough indication of the amount of lateral extension in the region.

Tabular igneous bodies, including dikes, sills, and lava flows, are prone to a distinctive pattern of fracturing called *columnar jointing. Columnar joints,* which tend to be more or less uniformly spaced, outline sharply angular polygonal—generally four to eight faces—prisms, or columns, the long axes of which are essentially perpendicular to the areally extensive surfaces of the dike, sill, or flow (Fig. 7-33). In flows and horizontal sills, the columns are vertical, but in dikes they are horizontal and look much like an immense stack of cordwood. Tabular bodies of any composition can be affected by columnar jointing, but the effect is most common in mafic rocks.

Columnar joints result primarily from contraction of the tabular bodies during cooling. In the cooling of a basalt dike from 1000° C (1832° F) to surface temperature [about 25° C (77° F)] a 100-m (328-ft) length (parallel to the walls) contracts about

**Fig. 7-32.** Four wall-like dikes of different sizes cutting nonresistant sedimentary rocks on the northwest side of Spanish Peaks, Colorado. (G.W. Stose, USGS)

25 cm (10 in.). Because the walls are firmly attached to the country rock, however, the dike (or sill or flow) cannot accommodate the contraction by slippage along the walls. Therefore, equally spaced, slightly open cracks develop over the 100-m distance that total 25 cm in open space. Moreover, cooling of the tabular body begins at the surface and progresses inward. Soon after cooling commences, the areally broad surfaces may be several hundred degrees cooler than the inner portion; accordingly, the igneous rock near those surfaces will have undergone much more contraction than the inner hotter region. The outer rock shells adjacent to the broad surfaces, therefore, contract, stretching over the more-voluminous in-

ner part and leading to the development of cracks that propagate inward as cooling progresses.

Sills are comparable to dikes in size. An excellent sample is the Great Whin Sill, Northumberland, in northeastern England—a dark, north-facing ledge about 30-m (100-ft) thick that dominates the countryside. The Romans, with their experienced eye for the military potential of the terrain, seized upon this natural barrier as a foundation for Hadrian's Wall, which was built to keep the Picts from ravaging northern Britain. Probably the most familiar of all sills in the United States is exposed in the Palisades Sill, the abrupt cliff that follows the New Jersey shore of the Hudson River.

Sills usually are restricted to the outermost part

of the earth, particularly in thin sedimentary rock sequences that cover crystalline rocks. The American geologist M.R. Mudge has studied nearly one hundred concordant structures (mostly sills) in the western United States and found that none has been emplaced below a depth of about 2300 m (7545 ft). Sills are confined to the outer layers of the earth because, during intrusion, all the rocks above the sill must be lifted up by a distance equal to its thickness. The force necessary to lift such immense weights comes from fluid pressure in the magma body. In a way the sill and associated magma system act like a huge hydraulic jack. Sill formation, therefore, can occur only when the

fluid pressure is equal to, or greater than, the overburden weight, a condition that can only normally be met in the outer part of the earth where the overburden rock layer is thin.

Most sills are offshoots from a dike. Magma intrudes upward until its fluid pressure exceeds the overburden weight; then sills may branch off if the layered-rock section is conducive to sill formation.

Small volcanic vents are vulnerable to erosion; they consist of loosely consolidated ash and pyroclastic material, their slopes are easily gullied, and, if they are high enough, they intercept more snow and rain than the surrounding countryside. Many

**Fig. 7-33.** Giants Causeway in northern Ireland consists of the eroded remains of countless well-formed basalt columns, most of which are six-sided in horizontal cross section. (By permission of the Director, British Geological Survey, British Crown copyright reserved)

**Fig. 7-34.** Devils Tower, northeastern Wyoming. The remnant of a volcanic neck or small stock that, through erosion, stands high above the relatively soft sedimentary strata (foreground). Each of the columns making up the tower is 1- to 1.5-m (3- to 5-ft) across. (N.H. Darton, USGS)

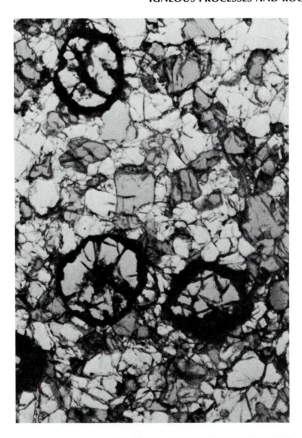

**Fig. 7-35.** Microscope photograph of ultramafic rock composed of garnet (thick black rims), olivine (white to light gray), and pryoxene (medium to dark gray)—derived from the mantle and carried to the surface during the intrusion of a kimberlite pipe, near the border between Chile and Argentina. (C. Stern)

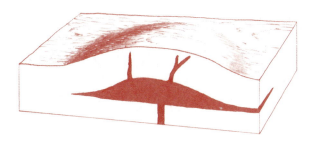

**Fig. 7-36.** Laccolith fed from a central pipe or dike.

small volcanoes, therefore, are quickly eroded after they lapse into dormancy. Their internal skeleton of radial dikes often stand up as partitions, and the solidified magma of the conduit forms a central tower known as a ***volcanic neck.*** A well-known example is Devils Tower in Wyoming (Fig. 7-34).

Volcanic necks are usually found in clusters. Prime examples in the United States are the buttes of the Navajo–Hopi country near the so-called Four Corners area in the southeast, where more than one hundred volcanic necks interrupt the surface of the plateau.

Another group of volcanic conduits that have achieved a measure of fame are the pipes of *kimberlite,* a rock consisting predominantly of fer-romagnesian minerals and one of the sources of diamonds. In South Africa kimberlites are deeply weathered near the surface into what is called blue ground, a sticky clay from which the diamonds were at one time separated by washing. These weathered, pipelike columns of dark rock were mined in the Kimberley pit to depths of more than 1000 m (3280 ft) before those workings were abandoned. Diamond-bearing kimberlites also occur in the United States in Arkansas and near the Colorado–Wyoming border.

Diamonds require high pressures and moderate temperatures for their formation—conditions that naturally exist only at depths hundreds of kilometers below the earth's surface. The diamonds in the kimberlite pipes are therefore "foreign" phenocrysts that have risen along with the magma from deep below the surface. Many kimberlites also contain rock fragments that apparently were ripped from the pipe's walls as the magma moved toward the surface (Fig. 7-35). From the presence of the diamonds and the nature of the foreign fragments, geologists have estimated that kimber-litic magma must have originated at depths near 250 km (155 mi). The way in which this magma rises to the surface is not entirely clear. Most geologists, however, contend that it rises as buoy-ant, tadpole-shaped blebs for much of the journey and only moves upward along cracks and fractures where it approaches the more brittle outer part of the lithosphere.

***Laccoliths,*** concordant igneous bodies, are more or less circular in outline with a flat base and a dome-shaped upper surface (Fig 7-36). Generally

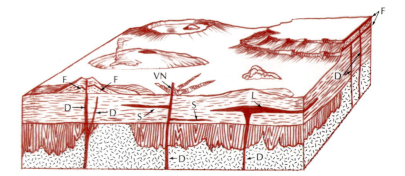

**Fig. 7-37.** Relationships among dikes (D), sills (S), a volcanic neck (VN), a laccolith (L), and flows (F).

up to 3-km (2-mi) across at the base, with maximum thicknesses of several thousand meters, laccoliths begin development much like sills. However, the magma injected between the rock layers is so viscous that the lateral flow cannot keep up with magma input. More magma moves into the developing body than is transported marginally; thus the body swells (or domes) upward, carrying the sediments above it into a domical arch. Because felsic magmas are generally more viscous than mafic ones, most laccoliths are felsic in composition. Commonly, several laccoliths will occur in a group, all of which were fed from a single magma body.

In the United States laccoliths are conspicuously developed in southwestern Colorado and eastern Utah. Some of them—for example, the LaSal and the Henry mountains in Utah—stand in scenic beauty high above the surrounding countryside owing to initial doming and subsequent erosion.

Figure 7-37 is a composite diagram depicting the characteristic formation of dikes, sills, a volcanic neck, laccoliths, and flows.

**Large Elongate or Irregular Plutons** The largest pluton bodies are composed of granite and diorite. They are found on all continents and range in age from Tertiary to Precambrian. In Labrador and northeastern Canada, where erosion has laid bare huge expanses of the earth's surface, plutonic rocks extend over hundreds and even thousands of square kilometers. Large bodies composed mostly of granite with lesser amounts of diorite—we will lump them together and call them granitic rocks—are found also in the cores of many mountain ranges. For example, the granitic plutons in British Colum-

bia, California, and Baja California (Fig. 7-4), which extend north-south for more than 3200 km (2000 mi), are located in close association with a zone of regional metamorphism along the axis of a Paleozoic–Mesozoic alpine-range complex. As noted earlier in this chapter, most geologists think that regions of granitic batholiths in association with alpine-mountain belts represent ancient convergence zones between lithospheric plates.

The huge bodies of igneous rock just described are called **batholiths** (a word derived from the Greek words for "depth" and "stone"). In modern usage a batholith is defined as a pluton with a surface exposure of more than 100 km$^2$ (40 mi$^2$). Irregularly shaped plutons of less areal extent are called **stocks.** Many stocks may be partially eroded incomplete batholiths. If erosion were to proceed further, a batholith might well be unroofed and exposed.

Much has been learned about the physical configuration of a batholith's upper parts because erosion has cut down to different levels at different places within the same body. Deep mining operations have also provided information. We lack definitive data, however, on the lower regions of batholiths. The walls of many of them are very steep sided, which has led many geologists to hypothesize that such bodies extend downward to great depths.

Studies of earthquake waves and measurement of the gravity fields over these plutons have given us some insight into their structure. For instance, it is suspected from geophysical studies that a batholith, perhaps still molten in part, lies just beneath the surface in Yellowstone National Park. This magma body has spewed large volumes of felsic magma at three different times: 2 million, 1.2 million, and 0.6 million years ago.

It is thought that an ancient frozen batholith also lies under the pile of volcanic rocks that form most of the San Juan Mountains in southwestern Colorado (Fig. 7-38). Gravity measurements indicate that it is 240-km (150-mi) long by 160-km (100-mi) wide. The top of the pluton is thought to be 2 to 7 km (1 to 4 mi) below the earth's surface and to extend downward for another 19 km (12 mi).

Generally, batholiths are not simple bodies; most large ones consist of smaller bodies of different kinds of plutonic rocks. One small body may be a white granite almost entirely lacking in ferromagnesian minerals; another may be a standard granite; another may consist of diorite or rocks intermediate between granite and diorite. The nature of the boundaries between the rock bodies as well as radiometric dating indicate that the smaller bodies were formed at different times and that the formation of the batholithic complex took many millions of years. Each smaller pluton represents an individually generated pulse of magma.

An indication of the heat and volatile content of a batholithic magma is provided by the halo, or **aureole,** of metamorphosed rocks that surround many batholiths. As they are intruded, there is, in some cases, a release of a great deal of hot, water-rich fluids that travel long distances through the host rocks; these fluids carry ions in solution that have been derived from the magma body. These liquids are called **hydrothermal solutions.** They are chemically very active and react readily with many of the minerals in the country rock. The hot fluids not only act as catalysts in the recrystallization of the country rock, but also bring in new chemicals that are incorporated in the recrystallizing minerals or are deposited in veins in marginal fracture systems as ore deposits.

Hydrothermal activity is generally most active during the last stages of solidification of a felsic-to-intermediate pluton, after most of the silicate minerals have been crystallized. At this stage the residual magma contains large concentrations of silica, water (and other volatiles) and those ions that could not readily fit into the crystal structures of the rock-forming silicate minerals because of size, charge, or other reasons. Most hydrothermal ore deposits, therefore, consist largely of "misfit" elements and ions. In some instances, the hydrothermal solutions can react with the host rock to remove chemical constituents, which are then carried away as the fluids pass through and beyond the rock.

Not all solutions in the recrystallization process come from the magma itself. Some fluid is groundwater that is heated and driven in convective circulation (like water boiling in a coffeepot) by the hotplatelike action of a nearby, underlying hot magma body.

Batholithic magmas, like other magma bodies, also tend to move upward. In some instances the magmas appear to shoulder their way bodily into the country rock, displacing it. The boundary of the intrusion is knife sharp and the granitic rock is uncontaminated and homogeneous. Study of sedimentary rocks pierced by salt domes has provided a good example of how such upward displacement could occur.

In some parts of the world, salt (usually sodium chloride or calcium sulfate), which precipitated from a narrow arm of the sea, lies in thick beds thousands of meters below clastic sedimentary

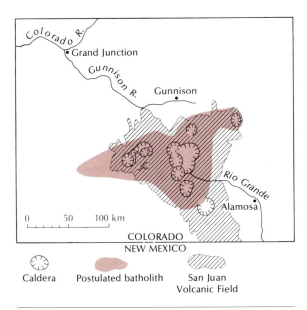

**Fig. 7-38.** Areal extent of a postulated batholith relative to the distribution calderas in the San Juan volcanic field, southwestern Colorado. Repeated explosive volcanism that produced the calderas was the result of near-surface activity of the batholith. (After T.A. Steven, 1975, "Middle Tertiary Volcanic Field in the Southern Rocky Mountains," *Geological Society of America Memorandum* 144, pp. 75–94)

## DEVELOPMENT OF SALT DOME

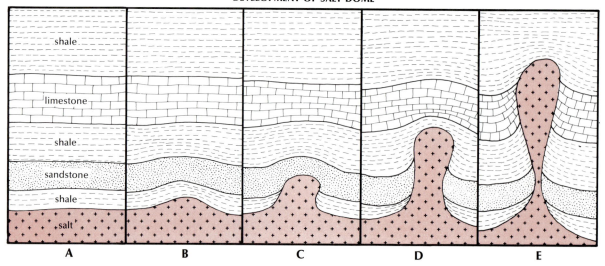

shale

limestone

shale

sandstone

shale

salt

A        B        C        D        E

**Fig. 7-39.** Sequential development of a salt dome, beginning with undeformed sediments overlying the salt layer **(A)** and progressing to a teardrop-shaped salt intrusion **(E)**.

strata. Salt can deform easily and has a low density; therefore, it tends to rise (Fig. 7-39). In this respect it is similar to a magma body. After a while bumps of salt begin to protrude into the sediments above. With more time the bumps rise higher and higher until an elongate cylindrical mass of salt projects above the original beds. Sometimes the cylinder pinches off, forming a long, thin tail section—looking somewhat like a tadpole. Eventually the tadpole separates from the salt layer below and rises right to the surface. The surrounding sedimentary rock remains solid, but it slowly deforms to accommodate the rising plug of salt. It is a slow process. Because granitic magma is less dense than the surrounding rocks and is easily deformable, many geologists feel, given enough

time and the right circumstances, that magma can also displace the solid rock above, just as salt domes do in sections of strata. In the lower portions of the continental crust, where temperatures are higher and the country rocks more ductile, this process should be most effective. In fact, many of the individual plutons making up the Baja California batholith appear to have solidified fairly deeply in the crust and most of them exhibit the tadpole shape.

In the upper part of the crust, where the rocks are strong and brittle, such upward movement of the magma is less likely. Geologists have suggested that magma, under its internal hydraulic pressure, could pry loose blocks from the roof and walls of the magma chamber (Fig. 7-40). This

**Fig. 7-40.** Cross section that shows stoping, subsidence of blocks, and hydraulic lifting of surface blocks by batholithic magma in the near-surface realm.

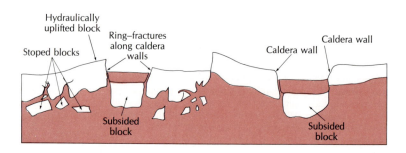

Hydraulically
uplifted block

Stoped blocks

Ring–fractures
along caldera
walls

Caldera wall

Caldera wall

Subsided
block

Subsided
block

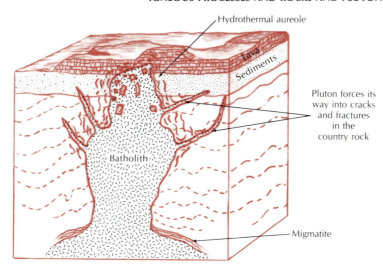

Hydrothermal aureole

Lava

Sediments

Pluton forces its
way into cracks
and fractures
in the
country rock

Batholith

Migmatite

**Fig. 7-41.** Cross section of a hypothetical batholith depicting most of the ways a batholith intrudes the country rock. At the bottom, on either side, the granite is shown without boundaries, indicating that either it reacts with the other rocks or it is produced when they melt. This would also be the zone of migmatite development. At the top the batholith has penetrated sedimentary rocks and lavas and stoped blocks sink into the magma. The batholith can melt the country rock at any level. (After H. Cloos, 1931, ''Neuer Jahrbuch für Mineralogie, *Geologie und Paläontologie,* vol. 66, B)

process is termed **stoping.** If the stoped blocks are more dense than the melt, they will slowly subside into the magma that will, in turn, move into the space vacated by the country rock. Close to the surface the internal pressure of the magma can hydraulically jack up entire blocks of the crust, enabling the magma to move upward (Fig. 7-40). Stoping and hydraulic lifting of large near-surface blocks have been established as the principal mechanisms enabling emplacement of individual plutons of the complex Peruvian batholith.

Is it also possible for batholiths to move by melting the rocks above? First, one should keep in mind that when a material changes from a solid to a liquid, heat is absorbed, and when it changes from a liquid to a solid, heat is given off. It is generally thought that most melts contain little more heat than that needed to keep the magma molten. Most earth scientists, therefore, consider melting to be of only minor significance. A diagrammatic summary of the various ways a batholith may make room for itself as it intrudes into the country rock is provided in Figure 7-41.

The basal portions of some batholiths that formed and solidified deep in the crust have been exposed through continued erosion over a long span of geological time. In these cases the transition between the invading and the invaded rocks is not sharp at all. The contact between the granite and its encasing shell may be blurred by a zone of **migmatite** or mixed rocks—part igneous and part

**Fig. 7-42.** Migmatite from Summit County, Colorado, showing lit-par-lit structure. Light-colored bands, composed of quartz, potassium feldspar, and muscovite, are plutonic; the dark layers, composed of intermediate plagioclase and biotite, are foliated metamorphic rocks. (M.H. Bergendahl, USGS)

metamorphic. Migmatite typically exhibits a **lit-par-lit** (French for ''bed-by-bed'') structure—one layer of a rock may be granite, the next a high-grade metamorphic rock, the next granite, and so on (Fig. 7-42). Or crystal clusters of potassium feldspar or other minerals typical of granite may appear in the enclosing metamorphic rocks some distance out from the main body of granitic rock.

Such phenomena have been interpreted by some geologists to mean that the pregranitic rocks, whatever they may initially have been, were converted in place by solid-state metamorphic processes into granite. Advocates of this process, called **granitization,** believe that most granitic rocks are formed in place through the alteration of large volumes of sedimentary rock by the chemical activity of solutions ascending from greater depths.

Some batholiths, in fact, grade laterally from what appears to be plutonic rock in the center to plutonic/metamorphic rock, then to metamorphic rock, and finally to virtually unaltered rock that in some cases is sedimentary. However, because such occurrences are so rare most earth scientists do not endorse granitization, except on a minor scale.

## SUMMARY

1. The kinds of minerals and their proportions in an igneous rock depend largely on the initial composition of the magma and on differentiation that occurs as the magma cools and solidifies. The initial composition of magma depends on where in the outer part of the earth it originates. *Felsic magmas* are generally restricted to continental regions and are generated by melting of continental crust above hot-spot plumes; most *intermediate-to-felsic magmas* originate in a number of ways along convergent margins; and *mafic magmas* most commonly originate from the outer mantle in divergent zones, back-arc settings, or above hot-spot plumes.

2. *Differentiation* of a magma mostly occurs by *crystal fractionation,* which is the result of the separation of earlier-formed crystals that are generally more mafic and calcic from the remaining more-felsic magma. Some differentiation can also occur by *compositional zonation,* which is primarily restricted to large felsic-magma bodies; it is thought to be the result of density, thermal, and convectional stratification, with the lighter, less-hot felsic constituents on top and the more-dense, hotter mafic constituents below.

3. Most igneous rocks are *crystalline* in texture, that is, the minerals have an interlocking structure. Grain size depends primarily on the cooling rate: plutonic rocks tend to be coarser grained than volcanic rocks. If cooling occurs so quickly that crystallization is minimal, the resulting rock is a *volcanic glass.* A *porphyritic* texture implies two cycles of cooling, one in which it was relatively slow followed by one in which it was rapid.

4. Some volcanic rocks possess clastic textures as a result of deposition by lahars, reworking of volcanic rocks by streams, or accumulation of volcanic material *(pyroclastics)* blown into the air during volcanic eruptions.

5. The three most common plutonic rocks are *granite, diorite,* and *gabbro;* the three most common volcanic rocks are *rhyolite, andesite,* and *basalt.*

6. Magma originates in small-to-large pools inside the earth and, once formed, moves laterally and toward the surface primarily because it is more buoyant than the surrounding rocks.

7. Relatively small *plutons* can be subdivided into the following classes: discordant tabular bodies *(dikes),* concordant tabular bodies *(sills),* volcanic necks, and domed concordant bodies *(laccoliths).*

8. The largest plutons, called *batholiths* and *stocks,* consist almost exclusively of granitic rocks. They are found on all continents in bodies that may cover thousands of square kilometers. Commonly, they are present in the cores of alpine-mountain ranges in association with metamorphic rocks.

9. Batholiths intrude into the surrounding country rock in a variety of ways: *forceful injection, stoping,* and *melting.* In the lower crust, where the rocks are more ductile, they may rise through the surrounding country rock much as salt domes do through overlying sedimentary strata. Although there is some evidence to suggest that some batholiths originate through *granitization,* most appear to result from melting.

10. Batholiths and associated metamorphic rocks in many cases appear to be the result of mountain-making processes marginal to a continent in convergence zones.

## QUESTIONS

1. What is the basic system for naming igneous rocks? How well do you think the system works?
2. Which is more important in determining the type of igneous rock to be formed—initial magma composition or magmatic differentiation?
3. Looking at a world map, outline the regions where magmas are most likely to form and the types of magma—mafic, intermediate, or felsic—that are generated.
4. Is there any relation between type of magma and the tectonic environment where it originates?
5. What are several prime factors or conditions that must be considered in any discussion of the melting of rocks to produce magmas?
6. What is Bowen's reaction series and why is it significant to fractional crystallization?
7. What evidence is there that fractional crystallization actually occurs?
8. Name as many rock textures as you can and give the important factors determining their occurrence.
9. What factors determine the fluidity of a volcanic rock?
10. How does rhyolite differ from basalt? From andesite? From granite?
11. Compare and contrast a sill, a dike, and a laccolith.
12. Describe the physical and chemical characteristics, and tectonic setting of alpine-mountain batholiths.
13. Describe the ways that granitic batholiths move upward in the earth's crust.
14. What is granitization? What are the arguments for and against its occurrence?

## SELECTED REFERENCES

Barker, D.S., 1983, Igneous rocks, Prentice-Hall, Englewood Cliffs, N.J.

Best, M.G., 1982, Fundamentals of igneous and metamorphic petrology, W.H. Freeman, San Francisco.

Billings, M.P., 1972, Structural geology, Prentice-Hall, Englewood Cliffs, N.J.

Bowen, N.L., 1928, The evolution of igneous rocks, Princeton University Press, Princeton, N.J.

Daly, R.A., 1933, Igneous rocks and the depths of the earth, McGraw-Hill, New York.

Ehlers, E.G., and Blatt, H., 1982, Petrology: igneous, sedimentary, and metamorphic, W.H. Freeman, San Francisco.

Hamilton, W.B., and Myers, W.B., 1967, The nature of batholiths, U.S. Geological Survey Professional Paper 554–C.

Hyndman, D.W., 1985, Petrology of igneous and metamorphic rocks, 2nd ed., McGraw-Hill, New York.

McBirney, A.R., 1984, Igneous petrology, W.H. Freeman, San Francisco.

Mudge, M.R., 1968, Depth control of some concordant intrusions, Geological Society of America Bulletin, vol. 79, pp. 315–22.

Read, H.H., 1957, The granite controversy, Interscience, New York.

Tuttle, O.F., and Bowen, N.L., 1958, Origin of granite in the light of experimental studies, Geological Society of America Memoir 74.

Walton, M., 1960, Granite problems, Science, vol. 131, pp. 635–45.

Wyllie, P.J., 1971, The dynamic earth, Wiley, New York.

**Fig. 8-1.** Steep-sided stratovolcano, Kliuchevskoi, Kamchatka, U.S.S.R., roars into eruption on February 7, 1987. (Anatoly P. Khrenov, Institute of Volcanology, Petropavlovsk-Kamchatsky, U.S.S.R.)

# 8

# VOLCANISM

## TYPES OF VOLCANOES

At present more than five hundred active and recently active volcanoes dot the earth's surface (Fig. 8-1). The word *volcano* comes from Vulcan, the ancient Roman god of fire and metalworking who manufactured the thunderbolts of Jupiter and the armor of the gods. One active volcano in the Mediterranean region carries the name Vulcano, and it was considered by the ancient Romans to be the forge at which Vulcan practiced his blacksmithing.

Most volcanoes mark a local center of eruption, leading to the development of a **volcanic cone.** The cone is essentially a piling up of the products of eruption around a volcanic vent. Some volcanoes have a variety of eruptive styles, pouring forth flows of lava at one time and volumes of pyroclastic material at another resulting in **composite cones,** or **stratovolcanoes** (Fig. 8-2). Inasmuch as coarse- to intermediate-grained pyroclastic material can stand at steep angles of repose, most of the composite cones are rather steep-sided (Fig. 8-3). They account for most of the scenically magnificent volcanoes in the world, such as Fujiama in Japan, Mount Hood in Oregon, and Mount St.

Helens and Mount Rainier in Washington. The rocks erupted from stratovolcanoes are mostly intermediate to mafic; felsic types are present but only in minimal quantities. Composite cones are the characteristic type at convergent plate boundaries.

In contrast some cones are built almost entirely of fluid lava flows, one upon the other. Cones of this type generally are basaltic. With gently rounded profiles and a nearly circular outline, they resemble Viking shields. As a consequence, they are called **shield volcanoes** (Fig. 8-4). The Hawaiian Islands are some of the finest examples of shield volcanoes on earth.

At times eruption is not restricted to a single localized vent, but occurs simultaneously along a linear trend, forming an elongate welt rather than a symmetric cone.

Almost all individual composite and shield volcanoes are built over a considerable period of time—generally a million years or less. Some vents—commonly on the flanks of larger cones—go through a complete eruptive cycle in a few weeks, months, or years and attain a height of only 100 m (328 ft) or so. A particular type of small cone, steep-sided and composed almost en-

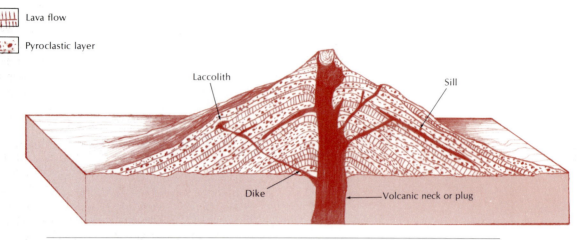

Lava flow

Pyroclastic layer

Laccolith

Sill

Dike

Volcanic neck or plug

**Fig. 8-2.** Diagram of a composite cone (stratovolcano), showing volcanic neck, dike, sill, and a laccolith. (After G.A. MacDonald, 1972, *Volcanoes*, Prentice-Hall)

tirely of reddened scoriaceous blocks, bombs, cinders, and lapilli, is called a ***cinder cone*** (Fig. 8-5). It occurs as a solitary feature, but quite often many cinder cones occur together, each marking a single short-lived eruption. The growth of some cinder cones is associated with the eruption of small lava flows.

The birth and growth of a larger-than-normal cinder cone was recorded in Mexico beginning in 1943. The eruption lasted about 9 years and built a cinder cone, called Parícutin, almost 370-m (1215-ft) high. During its later stages, torrents of lava broke through the flanks of the cone and flowed outward, eventually reaching the nearby town of San Juan Parangaricutiro where it covered all but the church steeple (Fig. 8-6).

The apexes of nearly all volcanic cones are characterized by a relatively small funnel-shaped depression called a ***crater*** that marks the top of the conduit through which the products of eruption are channeled (Fig. 8-7).

There is considerable variety in the way volcanoes erupt. Even a single volcano may go through several phases of eruption, each one somewhat different than the one before. Nonetheless, volcanoes, in general, or any single eruption can be categorized into broad types based on degree of explosivity. Some volcanoes erupt violently, with the explosive release of great clouds of pyroclastic

material, whereas others quietly erupt fluid lava. Even the quietest eruptions are spectacular to behold. Of course, there are eruptions that fall between these two extremes.

In general, volcanoes that erupt andesitic material tend to be more explosive than those that erupt basalt. Historically, the most explosive volcanoes in the world occur in Indonesia and Central America; moreover, the degree of explosivity is much greater in the Pacific Ocean's Ring of Fire than in other parts of the world. Felsic eruptions tend to vary in explosivity with the volume of the eruptive products. Eruptions of small volume are generally not very explosive whereas eruptions of large volume are among the most violent ones known.

Explosive eruptions are the result of the sudden release of gases contained (dissolved) in the magma. All magmas contain gases, mostly in the range of 1 to 4 weight percent. By far the largest proportion of the gases (frequently 50 to 80 percent) is water vapor, accounting for at least some of the clouds that swirl above an erupting volcano (Fig. 8-8). Water has a low molecular weight, so that weight percentages of 1 to 4 represent a large water content. Carbon dioxide, nitrogen, and sulfur dioxide also contribute measurably to the gas content, but in varying proportions. Small amounts of carbon monoxide, hydrogen, sulfur, chlorine, fluo-

**Fig. 8-3.** Plume of ash and vapor escaping from the summit of Shishaldin Volcano, Aleutian Islands, Alaska. Note the steep flanks of this stratovolcano. (USGS)

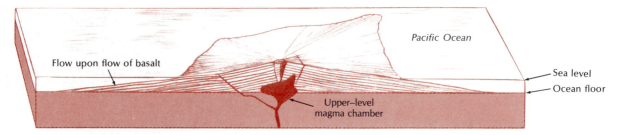

**Fig. 8-4.** Diagram of a Hawaiian shield volcano composed almost entirely of thin basalt flows. Because the magma is fluid, explosive activity occurs rarely and pyroclastic material forms only a small part of the cone. As a result the slopes are gentle and the volcano resembles a shield.

**Fig. 8-5.** Classic form of a recent cinder cone, Sunset Crater, northeast of Flagstaff, Arizona. The cone, which is an accumulation of cinders, lapilli, ash, and bombs, formed in a few weeks' time in A.D. 1064. (Wupatki National Monument, National Park Service)

**Fig. 8-6.** Parícutin Volcano, Michoacán, Mexico. This photograph, taken in 1950, shows the still-active vent of the cinder cone. (F.O. Jones, USGS)

rine, and hydrogen chloride are also detectable. In 1984 El Chinchón, a felsic volcano near Mexico City that had been dormant for 250 years, erupted, spewing huge quantities of sulfur dioxide into the atmosphere. A cooling of the global temperatures by about 1° C (2° F) soon followed, and some volcanologists and meteorologists have suggested that a catastrophic change in the circulation of ocean waters—called El Niño—was triggered by the sudden drop in temperatures (see chap 13).

Some of the dissolved gases are undoubtedly derived from the country rocks that the magma contacts or engulfs as it moves upward in the crust; some, however, are indigenous to the magma and represent ancient volatile material trapped inside the earth during the early stages of its formation. In 1955 the geochemist William W. Rubey (1898–1974) reached the rather startling conclusion that virtually all the water on the face of the earth and in the atmosphere could be accounted for by volcanic activity throughout the history of the planet.

More gas is soluble in magma at higher confining pressures than at lower ones. Therefore, magma deep within the earth can contain more gas than can the same magma at a shallower depth. As the magma moves upward and nears the surface of the earth and as crystallization of those silicate minerals lacking volatiles occurs, the gas pressure may increase to the maximum possible for that confining pressure. Subsequently, the gases begin to "boil" away. If there is an open channel to the surface and if the magma is fluid, the gases easily escape into the atmosphere. If the magma is less fluid or if a plug of solidified material blocks the vent, however, escape of the gases is impeded; with time pressure builds up. If the vent is suddenly cleared, either by release of the gases under high pressure or by eruption of magma, the pent-up gases escape violently and explosively (Fig. 8-9). It is like popping the cork on a bottle of champagne. Expanding rapidly, the gases propel blobs of magma and blocks of solid material upward to great heights. The nearly instantaneous release of the gases can cause the top of a magma chamber to froth, forming pumice or scoria, depending on the composition of the magma. In some cases continued expansion of the gases within the vesicles in the chilled glassy froth can cause its fragmentation into fine ash and lapilli. When large quantities of gas are suddenly discharged, the magma chamber can be emptied nearly instantaneously.

As we discussed in chapter 7, the viscosity of a magma with a relatively uniform water content is determined by the magma's temperature and chemical composition. Mafic magmas are hotter and form fewer complex silicate bonds; therefore, they are usually more fluid. In contrast the more felsic, silica-rich magmas are cooler and form larger proportions of sheet and three-dimensional silicate bonds. Accordingly, these magmas are less fluid and the volatiles, at lowered pressure and high temperature, do not dissociate easily from the magma. Intermediate to felsic magmas, therefore, are more prone to explosive eruptions.

From the human standpoint most eruptions are spectacular and awesome. However, a few are particularly noteworthy because of their significance to geological understanding or, possibly,

**Fig. 8-7.** Asama Volcano, Honshu, one of Japan's most active and dangerous volcanoes. Notice the sharp central crater marking the top of the active volcanic vent. Recent ash covers the flanks of the volcano. Andesite lava issued from the vent in 1783 to produce the molasses-like flow to the right. (USGS)

**Fig. 8-8.** Eruption of Hekla Volcano, Iceland. The billowing plume, carried windward, is composed of volcanic ash and volatile constituents, mostly water. (Thorsteinn Josepsson)

their destructiveness or intensity. Now we will single out a few of the more significant eruptions.

### Explosive Eruptions

**Krakatoa**  A century ago crews of sailing ships beating their way through the Sunda Strait that separates the islands of Djawa (formerly Java) and Sumatra in the East Indies knew the island of Krakatoa well. Its green-clad slopes rose uninterrupted about 793 m (2600 ft) to its summit. Although there had been volcanic activity on the island since May 1883, it seemed innocuous enough to the crew of the British ship *Charles Bal*—until they reached a point about 16 km (10 mi) south of the island on the afternoon of August 26, 1883. Minutes later the mountain exploded, disappearing in clouds of black smoke. The air was charged

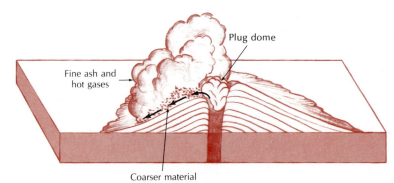

Fine ash and
hot gases

Plug dome

Coarser material

**Fig. 8-9.** Pyroclastic eruption of a volcano. Viscous lava congeals in the throat, plugging the vent. Pressure builds up and eventually the pent-up gases are released explosively, completely or partially clearing the vent.

with electricity—lightning flashed continuously over the volcano—and the masts of the ship glowed in St. Elmo's fire. Immense quantities of hot ash fell on the deck, and as the ship plowed through broken seas and squalls of mud-laden rain, the explosions continued, accompanied by a crackling sound that resembled the tearing of gigantic sheets of paper. The last effect was interpreted as the rubbing together of large rocks hurled skyward by the explosions. With dawn the *Charles Bal* set sail, driven by a rapidly rising gale, and was able to leave the smoking mountain far astern.

Paroxysms of volcanic fury continued to shake the mountain until the final culmination of four prodigious explosions on the morning of Monday, August 27. The greatest of them, the third at 10:02 A.M., was one of the most titanic explosions recorded in modern times—greater in intensity than some nuclear blasts. The sound traveled great distances and was heard at Alice Springs in the heart of Australia, in Manila, in Sri Lanka (formerly Ceylon), and on the remote island of Rodriguez in the southwest Indian Ocean [about 800 km (500 mi) east of Madagascar], where it arrived 4 hours after the explosion had occurred 4800 km (2983 mi) away.

The explosion profoundly disturbed the atmosphere, affecting barometers around the world. The atmospheric shock wave traveled at least seven times around the world before it became too faint to register.

A more impressive visual phenomenon was the huge cloud of pumice, volcanic debris, and ash that blew skyward to a height of 80 km (50 mi) on August 27, blanketing an area of 768,000 km²

(296,525 mi²), including Djakarta (formerly Batavia) 133 km (83 mi) away as well as part of the Indian Ocean.

Volcanic ash hurled into the upper levels of the atmosphere entered the jet stream and completely encircled the earth in the equatorial regions in 13 days. The ash eventually spread across both the Northern and Southern hemispheres to produce a succession of spectacular sunsets over most of the world for the 2 years that it took the finest dust particles to settle. The ash in the atmosphere also effectively screened the sun's rays, producing a worldwide cooling of about 1° C (2° F).

The violent explosion on the morning of August 27 also set in motion destructive sea waves that spread out in ever-widening circles from Krakatoa, much as though a gigantic rock had been hurled into the sea. About half an hour after the eruption, the waves reached the shores of Djawa and Sumatra and surged inland, cresting at a height of about 37 m (210 ft). Nearly 37,000 people died from this cause alone.

After the eruption died down, returning observers were amazed to find that the site of Krakatoa was a depression whose bottom was 275 to 305 m (900 to 1000 ft) below sea level. All that remained of the island were three tiny islets. The loss of material associated with the eruption was estimated to have been about 21 km³ (5 mi³) in volume.

Popular accounts of the eruption commonly give the impression that the volcano blew up and its fragments were strewn far and wide. But if that had been the case, most of the debris that covered the little islands nearby would have been pieces of the fragmented volcano. However, few pieces of the original volcano are to be found; instead, the islands are covered with pumice in layers up to 60-m (200-ft) thick. The pumice—also observed floating in great rafts over the nearby open ocean— is original magmatic material frothed up by gases released from the magma.

The abundance of pumice and the absence of pieces of Krakatoa lead logically to the conclusion that the volcanic cone collapsed on itself after a catastrophic, magma-draining eruption rather than being blown to bits. This explanation advanced in 1929 by Dutch volcanologist Reinout van Bemmelen and refined by the American volcanologist Howell Williams in 1942 is still held today.

The major geological lesson learned from Krakatoa is that a volcanic eruption can be so violent that it may cause a volcanic cone to collapse through loss of magma. The oversize crater resulting from this type of collapse is called a ***caldera.***

**Crater Lake**  Crater Lake, Oregon, at an altitude of 1846 m (6055 ft) is a caldera similar to Krakatoa's (Fig. 8-10) that formed about 6000 years ago.

**Fig. 8-10.** Crater Lake, Oregon, looking southwest. The circular caldera is about 10-km (6.2-mi) across; the lake is nearly 610-m (2000-ft) deep; the highest points on the rim are nearly 610 m above the lake. Wizard Island, a cinder cone that erupted on the floor of the caldera, is visible at the far right of the lake. Before the caldera collapsed, the area was the site of Mount Mazama, a steep-sided strato-volcano. (Ray Atkeson)

**Fig. 8-11.** Stages in the collapse of a volcano to form a caldera.

Stage I    The cycle starts with fairly mild explosions of pumice. The magma chamber is filled and magma stands high in the conduits. As the violence of the explosions increases, magma is drawn off more and more rapidly.

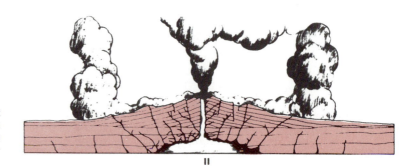

Stage II    The culminating explosions clear out the conduits and rapidly lower the magma level in the chamber. Pumice is blown high above the cone and pumice-laden pyroclastic flows sweep down the flanks.

Stage III    With support removed, the volcanic cone collapses into the magma chamber below, leaving a wide bowl-shaped caldera. (After H. Williams, 1942, *Geology of Crater Lake National Park, Oregon*. By permission of of the Carnegie Institution of Washington)

Howell Williams, after extensive study of the caldera concluded:

When the culminating eruptions were over, the summit of Mount Mazama had disappeared. In its place there was a caldera between 5 and 6 miles wide and 4000 feet deep. How was it formed? Certainly not by the explosive decapitation of the volcano. Of the 17 cubic miles of solid rock that vanished only about a tenth can be found among the ejecta. The remainder of the ejecta came from the magma chamber. The volume of the pumice fall which preceded the pumice flows amounts to approximately 4.5 cubic miles. Only 4 per cent of this consists of old rock fragments. . . . Accordingly 11.75 cubic miles of ejecta were laid down during these short-lived eruptions; in part, it was the rapid evacuation of this material that withdrew support from beneath the summit of the volcano and thus led to profound engulfment. The collapse was probably as cataclysmic as that which produced the caldera of Krakatau in 1883.*

Williams's diagram (Fig. 8-11), shows the sequence of eruptive events that very likely was responsible for the disappearance of ancient Mount Mazama and the formation of a deep caldera in its place.

With the acceptance of Crater Lake as a caldera, came the realization that catastrophic eruptions of this type are relatively common.

*H. Williams, 1942, ''Geology of Crater Lake National Park, Oregon,'' Publication 540, Carnegie Institution of Washington, D.C., p. 162.

**Thíra (formerly Santorini)** Recently, it has been postulated that a violent volcanic eruption similar in intensity and eruptive style to those at Krakatoa and Crater Lake occurred about 3460 years ago in the Aegean Sea, about 200 km (125 mi) north of Crete. Coincidentally, there was a wholesale destruction of the advanced Minoan civilization on the island of Crete. Plato, apparently, wove the story of the fall of the Minoans into a legend about a lost island, Atlantis.

Thíra (ancient Thera) today is a picturesque arcuate island that encloses a beautiful bay (Fig. 8-12). Along the inner wall of the arc, which rises sharply from the sea, are exposed thick layers of light-colored volcanic ash. A study of bottom sediments around the island indicates the presence of a continuous layer of 3460-year-old volcanic ash that is thickest close to Thíra and diminishes outward (Fig. 8-13). This data, and the geological evidence from the island itself, support the idea that Thíra was the site of a volcanic eruption that

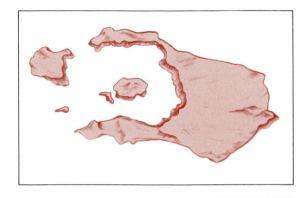

**Fig. 8-12.** Thíra, a volcanic island in the Aegean Sea. Steep-walled arcuate bay marks the site of the caldera formed by an explosive eruption about 3460 years ago.

**Fig. 8-13.** Geographical relations of Thíra Island, showing the zone of significant ash fall (shaded).

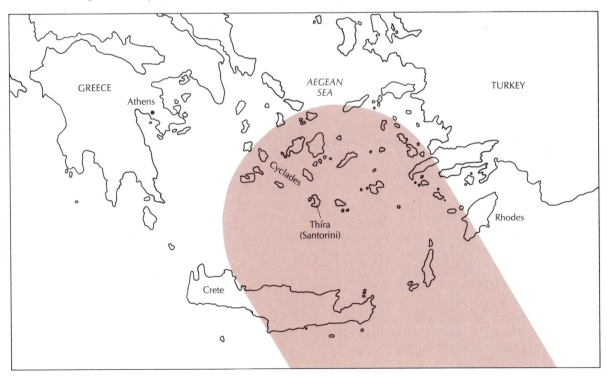

produced a caldera (Fig. 8-12) of 80 km$^2$ (31 mi$^2$). The steep walls of the island correspond to part of the rim of the caldera. It is conjectured that, as at Krakatoa, the eruption spawned large sea waves that swept the coasts of Crete and other islands nearby, bringing great destruction to the Minoans, most of whom, because they were fishermen and merchants, lived close to the shore. Such an eruption also would have deposited a thick layer of ash on the ground, making it impossible to farm.

**Mont Pelée**   Martinique is one of the scenic island-stepping stones across the Caribbean that joins Cuba with the mainland of South America. Perhaps its most notable distinction is that it was the birthplace of Josephine (1763–1814), empress of France and wife of Napoleon.

In the 1900s Saint-Pierre was the leading commercial town of the island. Most of the natives were Carib Indians or descendants of Africans brought to work in the sugar plantations, sugar factories, and rum distilleries. The population of 28,000 also included a substantial number of French and Americans.

The volcano, Mont Pelée, about 8 km (5 mi) north of town, had smoldered quietly for several centuries. On April 23, 1902, there began occasional rumblings, clouds of smoke, and spasmodic outbursts of ashes and cinders. More violent activity began on May 4 when a flow of hot mud, steam, and lava broke through the crater wall, coursed down one of the radial stream canyons, and killed 24 persons.

By that time Saint-Pierre was thoroughly aroused. Frightened country people and villagers poured into town. A few days passed and early in the morning of May 8, 1902, according to the few eyewitnesses, the top of the mountain vanished in a blinding flash. Soon thereafter a rapidly moving, intensely hot cloud engulfed the city, whose population, swollen with refugees, probably numbered more than 30,000 (Fig. 8-14). All but two died in a flash in a cloud hot enough to melt glass [650° to 700° C (1200° to 1292° F)].

Eighteen vessels were in port that day; of them, only the *Roddam,* with more than half her crew dead, was able to weigh anchor and escape. The cable ship *Grappler,* directly in the path of the incandescent cloud, capsized and blew up. The purser of the *Roraima,* then approaching the harbor from the sea, left the most complete narrative of the event. The *Roraima* was enveloped in a wall of flame that had previously incinerated the town. Its masts and stack were sheared off; the captain was blown overboard from the bridge and killed; and the ship itself burst into flames not only from the heat of the glowing cloud, but from the thousands of gallons of blazing rum that had poured from warehouses through the streets of Saint-Pierre and spread out over the waters of the harbor.

Within the town all were dead except for Auguste Cyparis, the occupant of an underground dungeon, who languished in a state of shock for four days until his rescuers arrived, and Léon Compère-Léandre, who was covered with burns.

The violent escape of entrapped high-temperature gases was the chief reason for the hot cloud that overwhelmed Saint-Pierre. This type of cloud bears the French name **nuée ardente** (translated as "glowing cloud").

Although it devastated the city, the nuée ardente was only a minor aspect of the total eruption, which included an avalanche of incandescent rock and pumice that shot out of the mountain top, cascaded down the canyon of the Rivière Blanche, and sped down to the ocean 3 km (2 mi) away. The nuée ardente consisted of hot dust and vapors that rose above the volcanic avalanche. While the avalanche moved down the river's canyon, the fiery cloud jumped the ridge on its south side and continued on to Saint-Pierre.

The 1902 eruption of Mont Pelée was an unparalleled example of the explosive discharge of pyroclastic material. Until then geologists were unsure of how this material, so common in the layers of most composite cones, was deposited. Since then geologists have witnessed the fiery discharge of volcanic avalanches and nuée ardentes from numerous volcanoes around the world. In most eruptions, the avalanche of the larger, heavier blocks and the billowing fiery cloud above it do not become physically separated, so that the total cascade of material is often referred to as the nuée ardente. The less ambiguous term for the glowing avalanche and billowing cloud, together, is a **pyroclastic flow.**

**Fig. 8-14.** Saint Pierre, Martinique (foreground) was gutted by a pyroclastic flow emanating from Mont Pelée in May 1902. Nearly 28,000 inhabitants died. About 6 months later, the volcano again became active, ultimately spewing out a series of billowing pyroclastic flows, one of which is shown here. (By permission of the Director, British Geological Survey: British Crown copyright reserved.)

The nature of deposits left by pyroclastic flows was uncertain before it was demonstrated at Mont Pelée. Stratification in these deposits is typically chaotic; large blocks are mixed with finer particles unlike the uniform layers of air-fall ash deposits that settle more gradually through the atmosphere. During the caldera-forming event at Crater Lake, pyroclastic flows (Fig. 8-15) swept down Rogue River canyon for 55 km (35 mi). They appear to have attained velocities of 160 km (100 mi) per hour and were capable of carrying pumice blocks 2 m (6.5 ft) in diameter a distance of at least 32 km (20 mi).

The origin and movement of pyroclastic flows is still not completely understood. Some flows, such as the Mont Pelée eruption, start with a lateral blast from the crater (Fig. 8-9). Others, probably the most common, begin with a vertical eruption: Large blocks, bombs, lapilli, and ash are blasted into a rapidly ascending, vertically expanding cloud. In time its ascent is slowed by gravity and the cloud collapses with much of its fragmented material falling onto the flanks of the volcano (Fig. 8-16). Driven by the energy acquired in their fall, the fragments move downslope in torrents that attain velocities of more than 160 km (100 mi) per hour. The coarser fragments move close to the ground, whereas the finer material and gases form a billowing cloud that rises just above the coarser material (Fig. 8-16).

**Fig. 8-15.** Ash-flow tuff of consolidated volcanic ash with lapilli and pumice blocks. They are the deposits from pyroclastic flows that erupted prior to the collapse of Mount Mazama to form Crater Lake. Since eruption, the relatively nonresistant deposits have been eroded into spires. View of the Pinnacles along Sand Creek, Crater Lake National Park. (Oregon State Highway Department)

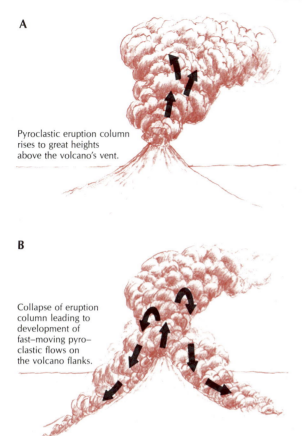

**A**

Pyroclastic eruption column rises to great heights above the volcano's vent.

**B**

Collapse of eruption column leading to development of fast–moving pyro–clastic flows on the volcano flanks.

**Fig. 8-16. A.** Pyroclastic eruption column rises to great heights above the volcano's vent. **B.** Collapse of eruption column that leads to the development of fast-moving pyroclastic flows on the volcano's flanks.

Most volcanologists believe that the high rates of speed of these flows are evidence of a fluid condition in which hot gases buoy up the individual fragments and keep them from pressing against each other. Some of the expanding vapors are emissions of dissolved gas from the hot, quenched pyroclastic material; some is air ingested at the front edge of the flow lobe, which is heated by contact with the hot interior of the flow, thereby undergoing rapid expansion.

**Vesuvius** One of the world's most famous volcanoes, Vesuvius is the only one active on the European mainland today. Its renown probably results from its well-publicized eruption of A.D. 79 and the accompanying destruction of the cities of Pompeii, Herculaneum, and Stabiae (now Castellammaie di Stobia). Although the mountain had been active in prehistoric times, it had been so long dormant that the Romans were unaware of its real character.

Fortunately, a description of the eruption has come down to us through two letters from the 17-year-old Pliny the Younger to his friend Tacitus, the Roman historian:

Gaius Plinius sends to his friend Tacitus greeting.

You ask me to write you an account of my uncle's death, that posterity may possess an accurate version of the event in your history. . . .

He was at Misenum, and was in command of the fleet there. It was at one o'clock in the afternoon of the 24th of August that my mother called attention to a cloud of unusual proportion and size. . . . A cloud was rising from one of the hills which took the likeness of a stone-pine very nearly. It imitated the lofty trunk and the spreading branches. . . . It changed color, sometimes looking white, and sometimes when it carried up earth or ashes, dirty or streaked. The thing seemed of importance, and worthy of nearer investigation to the philosopher. He ordered a light boat to be got ready, and asked me to accompany him if I wished; but I answered that I would rather work over my books. . . .

In the second letter Pliny the Younger comments on his uncle's activities and his attempts to calm his associates:

Ashes began to fall around his ships, thicker and hotter as they approached land. Cinders and pumice, and also black fragments of rock cracked by heat, fell around them. The sea suddenly shoaled, and the shores were obstructed by masses from the mountain. . . .

My uncle, for whom the wind was most favorable, arrived, and did his best to remove their terrors. . . . To keep up their spirits by a show of unconcern, he had a bath; and afterwards dined with real, or what was perhaps heroic, assumed cheerfulness. But meanwhile there began to break out from Vesuvius, in many spots, high and wide-shooting flames, whose brilliancy was heightened by the darkness of approaching night. My uncle reassured them by asserting that these were burning farmhouses which had caught fire after being deserted by the peasants. Then he turned in to sleep. . . .

It was dawn elsewhere; but with them it was a blacker and denser night than they had ever seen, although torches and various lights made it less dreadful. They decided to take to the shore and see if the sea would allow them to embark; but it appeared as wild and appalling as ever. My uncle lay down on a rug. He asked twice for water and drank it. Then as a flame with a forerunning sulphurous vapor drove off the others, the servants roused him up. Leaning on two slaves, he rose to his feet, but immediately fell back, as I understand choked by the thick vapors. . . . When day came (I mean the third after the last he ever saw), they found his body perfect and uninjured, and covered just as he had been overtaken.*

Most of the 20,000 inhabitants of Pompeii and Herculaneum escaped during the early and middle stages of the eruption. About 2000 people perished—most in Pompeii and in the main slaves, soldiers, and those too avaricious to leave their worldly goods. Until recently, the common speculation was that they had been suffocated by hot falling ash, by volcanic gases, and in the case of Herculaneum, by a hot lahar. It is now well established that most of the devastation in Pompeii was the result of ground-hugging pyroclastic flows triggered by collapse of the eruption column.

Several pyroclastic flows and layers of air-fall ash eventually buried Pompeii and its inhabitants to a depth of about 3 m (10 ft). Herculaneum was covered by laharic material. The two cities slept undisturbed for nearly 17 centuries until the discovery of one of the outer walls in 1748 ushered in the period of modern archaeology. During the intervening years the pyroclastic deposits at Pompeii hardened while the entombed bodies slowly disintegrated and sloughed away. Later, when plaster of paris was poured into the cavities once occupied by bodies, allowed to harden, and then excavated from the ash, their shapes were revealed—as well as the shapes of dogs and cats, loaves of bread, and all sorts of objects in similar cavities. Hundreds of papyri were also preserved in the library, as were murals on the walls of houses.

Enough magma was rapidly ejected during the A.D. 79 eruption of Vesuvius that the top foundered, forming a poorly defined caldera of moderate size.

The principal difference between the eruption of Mont Pelée and those of Krakatoa, Thíra (Santorini), Crater Lake (ancient Mount Mazama), and

Vesuvius is one of scale only. The latter group erupted such large quantities of pumice and ash that their weakened cone superstructures collapsed along ring-shaped fractures to produce calderas several kilometers across. Eruptions from Mont Pelée, on the other hand, were restricted to the explosive discharge of small tongues of incandescent material. The magma chamber was never emptied to any great extent and the cone did not collapse.

**Mount Saint Helens** In this Pacific Northwest mountain range, composite cones are strung like beads on a north-south chain that extends more than 1000 km (600 mi)—from Mount Garibaldi in southern British Columbia to Mount Lassen in northern California. Within the range are such well-known cones as Mount Shasta in California, Crater Lake and Mount Hood in Oregon, and Mounts Adams, Rainier, Baker, and Saint Helens in Washington. The chain is locally and intermittently active. Minor eruptions occurred at Mount Lassen from 1914 to 1917 and both Mounts Baker and Rainier have produced steam and ash during the last 60 years. Mount Saint Helens was sporadically active for several decades during the last century.

Studies of volcanoes within the Cascade Range indicate that eruptions of moderate volume are likely to take place here as often as once every 1000 to 2000 years, with large eruptions occurring once every 10,000 years.

In 1980 an eruption occurred at Mount Saint Helens in Washington that provided an unprecedented example of explosive volcanism. In late March of that year numerous small tremors began to emanate from directly beneath the 2950-m (9677-ft) mountain and on March 27 the first cloud of steam and ash was explosively ejected from the small crater atop the cone. During the first days of the eruption, vents close to the summit spewed volcanic ash, steam, and other gases. Ash samples contained no newly formed magmatic material— only fragments of "old" cone material pulverized and blown out by the steam explosions.

On April 1 the first harmonic tremors were recorded on nearby seismographs, signaling underground movement of magma. Whereas a normal earthquake produces a burst of wave energy,

*F.C.T., Bosanquet, ed., 1903, "Pliny's Letters," George Bell & Sons, London, pp. 194–198.

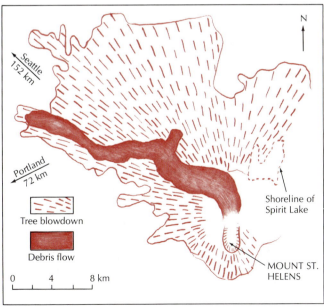

Fig. 8-17. A. *(Above)* Mount Saint Helens on May 18, 1980, from the northwest looking across jagged rim of the broken crater. Mount Hood is in the background. (USGS) B. *(Left)* Damage in the vicinity of the volcano owing to the catastrophic eruption shown in A. (From *Earthquake Information Bulletin,* 1980, vol. 12, July–August, pp. 146–147, Fig. 6)

the flow of magma in cracks and conduits below the surface is associated with a near-continuous release of wave energy, called a **harmonic tremor.** A stronger, more lengthy episode of this type occurred on April 2, alerting scientists that a full-fledged magmatic eruption might soon take place.

Almost from the first days of activity, the volcano's shape had been carefully watched. By the end of April it was confirmed that the upper part of the north flank had pushed outward about 82 m (270 ft) since the first eruption. This area, called the North Bulge, continued to grow laterally at an average displacement of about 1.5 m (5 ft) per day. Scientists interpreted the swelling as the result of magma pushing upward into the cone, and they foresaw a potential danger in this area. An updated hazard warning was issued on April 30, 1980, restricting access to the region around the volcano, particularly the north slope. By May 17 the North Bulge projected outward nearly 120 m (400 ft).

On Sunday morning, May 18, there was a strong earthquake beneath the cone, and the unstable North Bulge began to quiver and slide down the slope. Almost immediately the volcano exploded violently. David Johnston, a geologist for the USGS, watched from a position about 8 km (5 mi) from the peak as the sequence unfolded and radioed his last words: "Vancouver, this is it! . . . She's going!" Dr. Johnston along with about 60 others died in the holocaust on the north face. As pieced together by the observing scientists, the earthquake triggered two landslides, one after the other, on the unstable oversteepened north flank of the volcano. With the release of pressure, the gases dissolved in the magma escaped explosively in a lateral blast—much like the bursting of a tire once the wall has been breached. As the eruption continued, what remained of the volcano's summit collapsed piecemeal into the vent, was blown into bits and pieces, and was carried upward with the rest of the pyroclastic material (Fig. 8-17a).

The lateral blast leveled the forest on the north and northeast flanks outward as far as 24 km (15 mi) away, stripped the soil from the ground, and debarked the timber (Fig. 8-17B). Many of those who died that day were killed by the initial blast. The landslide carried pieces of the north flank of the volcano northward nearly 27 km (17 mi). Pyroclastic flows, which began soon after the land-slide and blast, continued throughout the day and night of May 18.

About 2 km³ (0.5 mi³) of material was erupted, leaving a snaggle-toothed cone 400 m (1310 ft) lower in elevation with a horseshoe-shaped crater 3.2-km (2-mi) long, 2.4-km (1.5-mi) wide, and 1.6-km (1-mi) deep. The lateral blast was unprecedented in the history of the volcano, as determined from the rock record.

Loss of life from this eruption was only moderate. It is conservatively estimated, however, that if the hazard warning had not been issued and enforced, 10,000 lives would have been lost.

Since the eruption, activity has slowed, but by no means abated. Several times, sticky magma has pushed into the cone to produce a lava dome resembling crusty rising bread, which subsequently has been fragmented and destroyed during explosive events.

**Yellowstone, Long Valley, and the San Juan Mountains** As spectacular as the eruptions that produced the Krakatoa, Crater Lake, and Thíra calderas must have been, they pale by comparison with the larger prehistoric ones that have occurred on the face of our planet. For example, careful mapping in the Yellowstone National Park region, Wyoming, at the eastern end of the Snake River Plain, has demonstrated the existence of three large calderas, the smallest of which measures 29-km (18-mi) wide by 37-km (23-mi) long and the largest, 64-km (40-mi) wide by 80-km (50-mi) long. The latter is one of the largest calderas on earth. Yellowstone Lake itself occupies a portion of the youngest of these calderas. The formation of each caldera was associated with the cataclysmic, explosive eruption of hundreds of cubic kilometers of pumice-rich pyroclastic flows. The temperature of the material as it left the vent must have been in the range of 750° to 800° C (1382° to 1472° F). The ashy debris traveled as far as 80 km (50 mi) so quickly that there was still enough heat left in the pyroclastic blanket after it came to rest—to bring about the softening and sticking together of the glassy ash fragments and the collapse of the porous pumice fragments. The result was a dense, resistant rock called a *welded ash-flow tuff* (Fig. 8-18). Rivers, such as the Yellowstone, have incised canyons into the three different ash-flow tuffs, revealing that in most places each deposit is more than 60-m (200-ft) thick.

The finest ashy material, carried high into the atmosphere during each of the three eruptions, was carried eastward by the prevailing winds and forms recognizable deposits as far east as Kansas.

Potassium–argon radiometric dating indicates that the three eruptions occurred at 2.0 million, 1.2 million, and 0.6 million years ago, respectively. Each eruption is separated from the last by about 600,000 to 800,000 years. Can the bubbling hot springs and fountaining geysers in the Yellowstone National Park area, therefore, be the harbingers of the next eruption?

Long Valley, California, is an area well known for its scenery and resorts, including Devil's Postpile National Monument and the Mammoth Mountain Ski area. It is also thought to be the area where volcanic activity in the United States is most likely in the near future. About 730,000 years ago Long Valley was the site of an immense eruption in which about 600-km³ (146 mi³) evacuated the magma chamber, leading to the development of a caldera about 13 km (19 mi) by 18 km (11 mi). It is estimated that the entire eruption lasted only about a week. The pyroclastic flows resulted in welded ash-flow tuffs up to 1500 m (4920 ft) in thickness with one lobe extending 65 km (40 mi) down Owens Valley. Air-fall ash reached thicknesses of more than 1 m (40 in.) up to 120 km (75 mi) from the eruption center (Fig. 8-19). Even as far east as Kansas and Nebraska, the ash formed a layer 1-cm (0.5-in.) thick.

The extraordinarily large explosive eruptions just described appear to be near-surface manifestations of granitic magma chambers of batholithic proportions. Granitic batholiths, once formed, tend to rise buoyantly. Many solidify far below the surface, but some remain molten while rising toward the surface. Many of these, under reduced confining pressures and after partial crystallization of anhydrous silicate minerals, reach saturation in water content. If the water pressure becomes great enough or if the brittle rock roof overlying the

**Fig. 8-18.** Welded ash-flow tuff from an area east of the Sierra Nevada, California. Shows characteristic streaky layering resulting from vertical compaction of the ash flow after it came to rest. The elongate black blebs, now obsidian, were initially porous fragments of light-colored pumice. (Karl Birkeland)

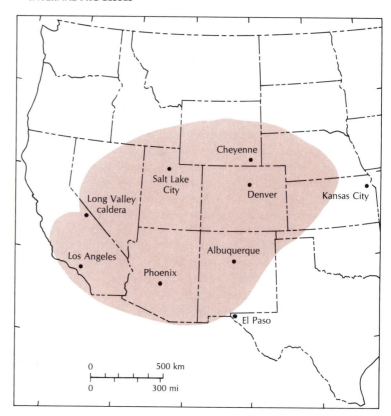

**Fig. 8-19.** Distribution of ash deposits from the eruption at Long Valley, California, about 730,000 years ago. Eastward to 120 km (75 mi) from the vent, the air fall thickness was 1 m (40 in.) or more. At the eastern edge of the ash fall, the thickness would have been about 1 cm (0.5 in.). (After C.D. Miller et al., 1982, "Potential Hazards from Future Volcanic Eruptions in the Long Valley–Mono Lake Area, East–central California—A Preliminary Assessment," USGS Circular no. 877, p. 9)

batholith cracks for any reason, an eruption can begin during which the upper part of the magma chamber is rapidly drained as the volatiles explosively expand (Fig. 8-20).

In the San Juan Mountains of southwestern Colorado, about 15 large calderas, each with one or more associated sheets of welded ash-flow tuff, formed about 30 million years ago (Fig. 8-21). Like those at Yellowstone and Long Valley, they appear to have been closely associated with a large, granitic batholith that lay close to the surface.

## Quiet Eruptions

### Mafic Type

***Shield Volcanoes*** The Hawaiian Islands, one of the most idyllic archipelagos in the world, owe their entire existence to volcanism. They are part of a chain of extinct, dormant, and active volcanoes built up from the sea and trending southeastward across the Pacific for 2560 km (1600 mi)—from Midway on the north to the largest island, Hawaii, on the south. The eight largest islands are at the southeastern end and their relative age, based on erosional appearance of the landscape, decreases southward. Hawaii, the only one of these islands with active volcanoes, appeared above the sea more recently than Oahu, where Honolulu is located. The best evidence of the progressive decrease in age of the islands is provided by potassium–argon dating of the lava flows on the largest islands. Kauai, the northwesternmost of the eight, is about 5.3 million years old, whereas Hawaii, at the southeast end, is about 750,000 years old.

There are five major volcanic centers on Hawaii. Two of them include immense volcanoes: Mauna Kea, 4205 m (13,792 ft) above sea level, and Mauna Loa, 4170 m (13,678 ft) above sea level. The bases of these mountains rest on the ocean floor, about 4600 m (15,088 ft) below the water's surface, so that they are as tall as Everest. Yet they are far greater in bulk; the circumference of Mauna Loa is about 320 km (200 mi) at its

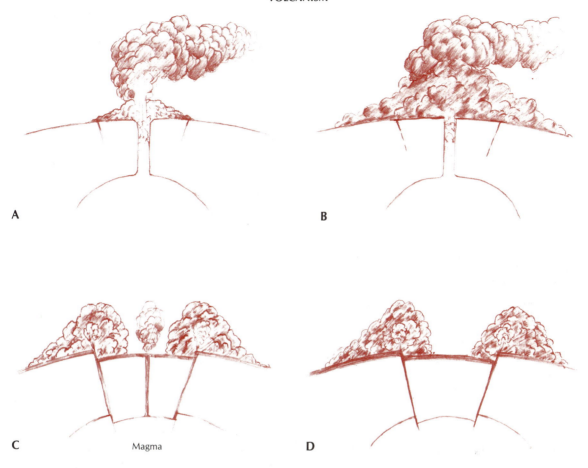

A

B

C   Magma

D

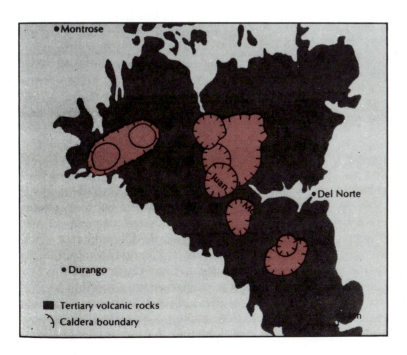

Montrose

Del Norte

Durango

■ Tertiary volcanic rocks
⌒ Caldera boundary

**Fig. 8-20.** Eruption and formation of a large caldera above a batholith-sized felsic magma chamber. Patterned after the interpreted sequence of events in the Jemez Mountains, north-central New Mexico, about 1.1 million years ago. (After S. Self et al., "Explosive Rhyolitic Volcanism in the Jemez Mountains," *Journal of Geophysical Research,* vol. 91, pp. 1779–98)

**Fig. 8-21.** Distribution of Tertiary volcanic rocks and calderas in the San Juan region, Colorado. (After P.W. Lipman et al., "Volcanic History of the San Juan Mountains, Colorado, as Indicated by Potassium–Argon Dating," *Geological Society of America Bulletin,* vol. 81, pp. 2329–52)

**Fig. 8-22.** Fire fountain at Kilauea, November 18, 1959. The lighter-colored parts of the fountain are glowing, fluid basaltic lava. The lava cools in the air and turns dark at the top of the fountain. (G.A. MacDonald)

base. They are by far the largest volcanoes in the world.

Although some pyroclastic material is included in the mass of these huge volcanic piles, for the most part, they consist of thousands of superimposed, relatively thin flows of basalt that at the time of their eruption were extremely fluid. The result is that the slopes of these classic shield volcanoes are gentle because they are made up of thousands of overlapping, tonguelike sheets of material rather than loose piles of heaped-up volcanic fragments.

Mauna Kea is dormant for the time being, but Mauna Loa is highly active. Most of the historic flows on Mauna Loa have occurred on its flanks rather than from the summit.

Lava flows on Mauna Loa seldom issue from a single vent, but almost always break out from great cracks, or **fissures.** The first phase of such an outbreak may be a line of *fire-fountains*—geyserlike columns of lava—along the fissure that spurt as high as several hundred meters into the air (Fig. 8-22). The basalt lavas that stream from the fissures are of a high temperature, close to 1200° C (2192° F)

**Fig. 8-23.** Kilauea caldera looking westward to the summit of Mauna Kea, Hawaii. The depression is about 5-km (3-mi) long by about 3-km (2-mi) wide. The smaller, sharply defined countersunk oval within the main caldera is the fire pit of Halemaumau. Mauna Kea has the classic rounded form of a shield volcano. (USGS)

and are usually very fluid. They may flow down river beds with velocities approaching those of the rivers; where there are sharp irregularities, the lavas plunge over them in lavafalls. If the lava streams reach the sea, immense clouds of steam boil upward and the waters seethe like a gigantic cauldron. Sometimes the lavas are quenched so abruptly that they form a tawny, cellular sort of volcanic glass.

Kilauea, which stands at about 1220 m (4000 ft), is a satellite vent on the flank of the larger Mauna Loa. At the top of Kilauea is an elliptical caldera (Fig. 8-23). Today only a part of the caldera is active.

Not all the Hawaiian basalts flow in torrents; blocky flows, that move forward much like a tank, are common, too. The advancing crust breaks up at the leading edge of the flow, and the blocks cascade over the front to make a track over which the still-molten flow interior can advance. The top and bottom of such a flow will become volcanic breccia when the whole mass has solidified and, for the most part, the interior will be uniformly textured basalt.

Two Polynesian terms are commonly used in geology to describe the surface of some lava flows. Basalt with a rough, blocky appearance, much like furnace slag (Fig. 8-24), is called *aa* (pronounced

**Fig. 8-24.** Blocky (aa) lava flow, slowly advancing over a field during the 1955 eruption on Hawaii Island. The flames and smoke are primarily from burning vegetation. Inside the blocky exterior, the lava is still liquid. (G.A. MacDonald)

**Fig. 8-25.** Contorted surface of ropy (pahoehoe) lava on Footprint Trail, Kilauea Volcano, Hawaii. (USGS)

ah-ah); more fluid varieties with smooth, satiny, or even glassy surfaces that are commonly contorted and wrinkled (Fig. 8-25) are given the more euphonious name of **pahoehoe** (pronounced pa-hoi-hoi). It is not uncommon for a pahoehoe flow—as it loses gases and cools while moving downslope—to turn into an aa flow.

When lava enters the sea or when eruptions occur beneath the surface of the sea, **pillow lavas** may be formed. As the name suggests, pillow lavas are lobes of lava stacked one upon the other that resemble a pile of bed pillows (Fig. 8-26). They almost always form from pahoehoe-type basalt flows. The exact mechanism of their formation is not clear, but it appears that small extrusion lobes, pushing out from cracks in the advancing pahoehoe flow, chill on contact with water and form an elastic, glassy rind. The rind continues to grow into a pillow form, which either remains attached to the flow front or breaks away and bounces down to the base of the slope.

The island of Hawaii has been the most inten-

**Fig. 8-26.** Field of whale-backed pillows, Gulf of Alaska region; notice photographer for scale. The pillows formed during extrusion of basalt into oceanic water. In some cases erosion has exposed the radial internal structure and concentric glassy rinds (F.H. Moffitt, USGS)

sively studied volcano in the world. The Hawaiian Volcano Observatory of the USGS, which sits at the edge of the caldera atop Kilauea, is staffed by scientists who not only study the flows and gases of each eruption, but also monitor almost undetectable changes in the surface of the volcano as well as earthquakes from deep beneath the island. Principally from earthquake data, scientists have concluded that the magma starts about 50 km (31 mi) or more beneath the surface and moves upward periodically to collect in shallow magma chambers within the volcanic cones themselves. As the upper chamber is filled, the cone just above it inflates like a balloon, causing measurable tilting and stretching of the volcano's surface. The stretching continues until eruption occurs and the tumescence declines in response to the draining of the upper chamber. Scientists are now able to predict when an eruption is likely to begin, how

long it will last, and if the eruptive phase has definitely ended or has merely become quiescent.

### Fissure Eruptions

In the geological past, several regions on earth have been inundated by vast floods of lava that obviously did not come from shield volcanoes. Rather, the widespread flows appear to have been the result of eruption from innumerable cracks or fissures, commonly subparallel in their orientation, that extend over a large area. Prominent examples of **flood lavas,** (also called fissure flows), which are entirely basaltic, are found in western India inland from Bombay, in South America near the Paraná River, in Britain, in South Africa, around Lake Michigan, and in the Pacific Northwest (Fig. 8-27).

In the Pacific Northwest an extensive lava flood

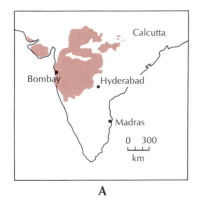

**A**

**B**

**Fig. 8-27.** The extent of **(A)** the Deccan basalts in India (After A.K. Baksi and N.D. Watkins, 1973, "Volcanic Production Rates: Comparison of Ocean Ridges, Islands, and Columbia Plateau Basalts," *Science,* vol. 180, pp. 493–96) **(B)** The Columbia River basalt and its equivalents, northwestern United States. (After A.C. Waters, 1955, *Volcanic Rocks and the Tectonic Cycle,* Geological Society of America Special Paper 62, pp. 703–22)

**Fig. 8-28.** Flow upon flow of the Columbus River Basalt in eastern Washington. Some flows exhibit columnar jointing. (Ray Atkeson)

underlying the Columbia Plateau is designated the Columbia River Basalt. It covers an area of 200,000 km² (77,375 mi²), extending from western Idaho on the east to the Cascade Range and the Pacific on the west. In places it is about 2-km (1-mi) thick and is composed of flow upon flow of basalt, each layer less than 120-m (400-ft) thick (Fig. 8-28). The composition of the flows, areally and vertically, is remarkably uniform, especially in view of the fact that the enormous volume of basalt erupted over a time span of about 10 million years. The volume of lava erupted was more than 300,000 km³ (72,200 mi³).

The lava welled up through thousands of fissures that are present today as subparallel basaltic dikes (dike swarms) that trend north-south to northwest-southeast and are primarily in southeastern Washington, northeastern Oregon, and western Idaho. The land buried by the lava floods was of moderate relief. The walls of canyons that cut across the basaltic plateau show that the individual flows filled valleys and depressons, overtopped ridges, and ultimately coalesced to form a nearly uniform plain.

There is little doubt that the initial magmas originated in the outer mantle of the earth, most likely at depths of 100 to 50 km (62 to 31 mi). But geologists are baffled by the relatively rapid generation of such remarkably homogenous magma. The tectonic setting in which the magma was generated is also unclear. The dike swarms could represent either an aborted divergence zone or a zone of back-arc upwelling inland from a convergence zone near coastal Washington and Oregon.

The only modern example of a fissure eruption is the Laki eruption on Iceland, at volcano Skaptar Jökull that began on June 8, 1783. For two years a stream of basalt about 3 km³ (0.7 mi³) in volume poured out from a fissure about 24-km (15-mi) long, covering some 90 km³ (35 mi³).

Iceland, which sits astride the Mid-Atlantic Ridge divergence zone and apparently above a hot spot as well, contains more than 100 volcanic centers, of which at least 20 are active. In the long cultural history of the island, extending back to A.D. 874, there have been many recorded eruptions, but few have been as disastrous as that of 1783. Lava issued from the length of the huge fissure, poured down the slope, filled the deep canyon of the Skaptar to overflowing, and completely displaced a lake that lay in its path. The eruption produced two major lava flows, 64- and 80-km (40- and 50-mi) long. Their average depth was 30 m (100 ft), but where canyons were filled to overflowing, they were as much as 185-m (600-ft) thick. Where the lava overtopped a stream valley and spread out across the plain it advanced along a front 19- to 24-km (12- to 15-mi) wide.

The lava, blocking and diverting rivers and melting snow and ice, caused huge floods that destroyed much of the island's limited agricultural land. Twenty villages were overrun by the lava, and many others were swept away in the floods. About 10,000 people—20 percent of the population—died; 80 percent of the sheep (190,000), 75 percent of the horses (28,000), and more than 50 percent of the cattle (11,500) perished.

**Felsic Type**   Quiet eruptions of felsic magma can also occur if the volume is small and if the magma either lacks appreciable volatile constituents or the temperature is so low and the viscosity so high that the volatiles cannot escape explosively.

After the explosive phase had ended at Mont Pelée in 1902, a viscous, stiff felsic lava was extruded into the summit crater, which was followed by the protrusion of a dome of blocky lava encrusted with lesser spires and pinnacles. By September 1903 the main spire had reached a height of 305 m (1000 ft) and a diameter about twice that. Its volume was estimated at 100 million m³ (3550 million ft³).

At Lassen Peak in northern California, a protrusion of blocky lava, now standing about 770 m (2525 ft) above the crater rim and with a volume of approximately 2.5 km³ (0.6 mi³), was formed when the peak was last active in 1914–17. Steep-fronted, mushroom-shaped bodies of small volume are called *volcanic domes;* they are largely composed of obsidian. Fine examples of volcanic domes are the Mono Craters in east-central California (Fig. 8-29), some of the vents of the Valley of Ten Thousand Smokes in Alaska, and the Puys of the Auvergne region of France.

At Mount Saint Helens dome building has been the principal volcanic activity in the vent area since the eruption in May 1980.

**Fig. 8-29.** Mono Craters, California, looking southwest toward the Sierra Nevada. Mono Lake is in the foreground. At the lower right is Panum, a volcanic dome surrounded by a cone of pumice fragments. Above and to the left is a larger dome of obsidian surrounded by a thick lava flow of obsidian bordered by a steep slope. Above and to the left of that is a lava flow of obsidian that has almost buried the dome that was its source. Above that are three more domes and to the right a large flow of obsidian. The chain of eruption centers continues to the base of Sierra Nevada, forming a line of volcanoes nearly 16-km (10-mi) long. (Roland von Huene)

## VOLCANOES AND SOCIETY

Throughout recorded history, volcanism and man's activities have been intertwined and the association has not generally been one-sided.

In some parts of the world—notably Iceland and Hawaii—volcanism has produced the very land on which the people live. In many places the fields on which the farmers raise their crops are the slopes of volcanoes. In fact, some of this planet's most fertile soils are weathered volcanic materials. In Iceland, Italy, Japan, New Zealand, and the United States (to name a few), the thermal energy associated with recent volcanism has been tapped to provide heating and electrical power. Moreover, geysers and hot springs, which result from geothermal emanations, provide natural recreational and scenic areas.

Many of the world's most scenic mountains are volcanic cones. The image of Japan is incomplete without Fujiama. One of the greatest natural assets of Arequipa, Peru, is its backdrop, the almost perfectly formed snow-capped cone of El Misti and the attractiveness of Portland, Oregon, is no

**Fig. 8-30.** Volcanic-hazard assessment of the area around Long Valley and Mono Lake, California. (From C.D. Miller et al., 1982, "Potential Hazards from Future Volcanic Eruptions in the Long Valley– Mono Lake Area, East-central California—A Preliminary Assessment," USGS Circular no. 877)

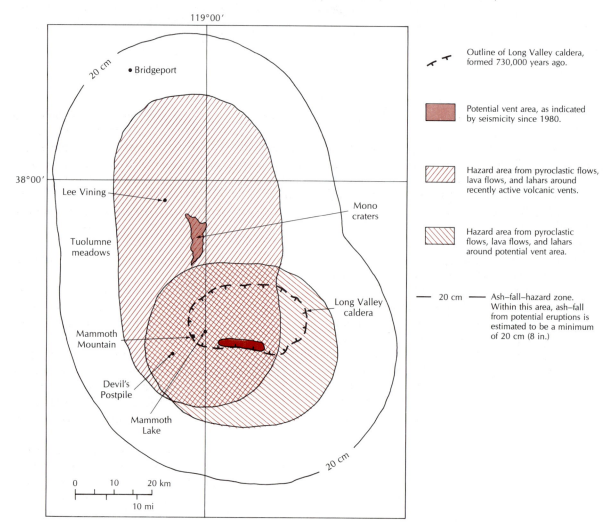

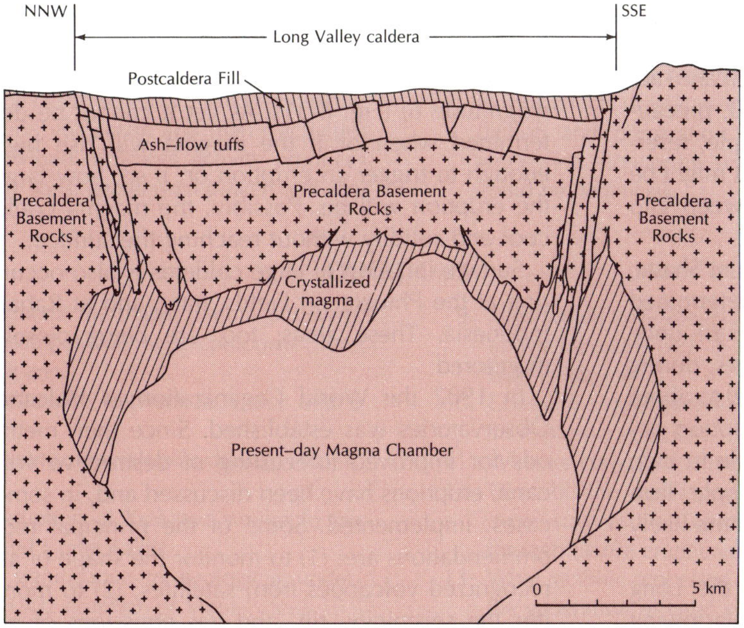

**Fig. 8-31.** Approximate relations (determined from geophysical techniques) of crustal rock and magma underlying Long Valley Caldera, the site of a 600 km³ (145 mi³) ash-flow eruption. (After D.P. Hill and R.A. Bailey, 1986, "Active Tectonic and Magmatic Processes Beneath Long Valley Caldera, Eastern California: An Overview," *Journal of Geophysical Research,* Vol. 90, No. B13, p. 11, 114, Fig. 2)

doubt enhanced by Mount Hood in the Cascade Range to the east.

On the downside, volcanoes are potentially dangerous, and threaten the safety of people nearby. In A.D. 79, the eruption of the long-dormant Mount Vesuvius, Italy, brought about the complete destruction of Pompeii and Herculaneum. Tens of thousands of people died during the eruptions of Krakatoa in 1883, and of Mont Pelée in 1902. In November 1985, 20,000 people, most of them from the village of Armero, Columbia, were killed by lahars that swept down the Lagunilla River Valley for 55 km (34 mi). The lahars apparently were triggered by the rapid melting of summit snows following two small explosive eruptions at Nevado del Ruiz, a 5432-m (17,822-ft) high Andean composite cone.

Geologists are becoming aware that the potential threat is greater than once anticipated. To date civilization has witnessed eruptions of only moderate intensity. The geological record is clear, however: During the last 3 million years, at least six eruptions have occurred in the western United States alone that were about one hundred times larger than the largest historic ones. There is little doubt that if a maximum-intensity eruption occurred at Long Valley, the western one-third of the nation would be disastrously affected by the ash cloud that would spread downwind. The ash would foul the atmosphere and cover the soil, bring with it harmful vapors and chemicals that would be absorbed by plants, adversely affect crops, and bring transportation to a standstill. There also is a strong possibility that dust and aerosols injected into the atmosphere would drastically lower the earth's temperature and adversely affect the weather for several years.

There seems little possibility of controlling volcanoes; the only reasonable solution lies in predicting eruptions and preparing for them. This tactic is being followed in several countries in which volcanic activity constitutes a hazard. Making appraisals of volcanic hazards involves determining the most-likely sites for imminent activity, establishing recurrence rates between eruptions, and estimating the type and intensity of activity that might occur.

In the United States volcanic-hazards assessment is conducted by the USGS. Until now they have concentrated their efforts on Hawaii, Alaska, and the Cascade Range in the Pacific Northwest. Just a few years before Mount Saint Helens erupted,

it w
mos
haz
tual
tion
voir
spa
the

In
Cal
that
eve
the
onc
curr
the
flow
loca

A
Vall
to i
and
30).
of a
sive
prel

1. To
ac
p
C
ar
vo
tiv
2. M
co
po
tic
m
co
m
p
3. In
of
In
up
pu
4. In
Se

**Fig. 9-1.** Banded, metamorphosed sedimentary rocks near Convict Lake on the east side of the Sierra Nevada, California. (John Haddaway)

# 9

# METAMORPHISM

A rock is formed within specific temperature and pressure ranges. If, after formation, the conditions alter drastically, the minerals in the rock may become unstable and, given sufficient time, change to minerals that are more stable under the new conditions. How the rock alters depends on the new conditions and the chemistry (mineralogical assemblage) of the material prior to alteration.

The changes that occur at, or near, the earth's surface—at relatively low temperatures, near 1 atm of pressure, and with abundant water, carbon dioxide, and oxygen—are included under the general term *weathering*. The changes that occur at slightly greater confining pressures and at temperatures up to about 200° C—conditions commonly reached at relatively shallow depths, particularly within sequences of strata—fall within the category of **diagenesis.** Beyond these limits, at even greater temperatures and pressures, but below the conditions necessary for complete melting, is the realm of **metamorphism** (from the Greek, meaning "to change in form") within which preexisting rocks may undergo changes in grain size, mineralogy, or both, and become metamorphic rocks.

The conditions necessary for metamorphism, the tectonic environments in which they are at-tained, and the changes undergone by rocks during the process are the topics of this chapter.

When metamorphism occurs, the changes described can come about in a number of ways. For example, the temperature of a rock can rise through the close approach of a hot pluton. Or rocks can be subjected to unusually high levels of shear stress within fault zones. Or rocks, once they enter a convergence zone, can be carried downward into an environment of extremely high confining pressure, directed stress, and elevated temperatures (Fig. 9-1).

In most cases the change in mineralogy is **isochemical,** that is, the overall chemical composition of the rock before and after the metamorphic event is the same. In some cases, however, chemicals from circulating fluids are added during mineral transformation. This occurs most often adjacent to a cooling stock or batholith.

The chemistry of the original rocks sets limits to what can be formed during metamorphism. For example, if limestone, composed solely of grains of calcite ($CaCO_3$), is metamorphosed, the resultant rock will be composed only of recrystallized grains of calcite. On the other hand, if a shale, composed largely of clay (hydrous aluminum silicate) and quartz ($SiO_2$) particles, is metamorphosed, silicate minerals form, but calcite will be absent.

## FACTORS IN METAMORPHISM

Any kind of rock—igneous, sedimentary, or metamorphic—can serve as the starting material in a metamorphic event, and the combinations of temperature, pressure, and circulating volatile and chemical constituents that can bring about this transformation are many.

### Temperature

Thermal energy is probably the single most important factor involved in the metamorphism of rocks. By increasing a rock's temperature, energy is added that can activate recrystallization of the minerals. As the temperature rises, ions diffuse more quickly, increasing the speed and efficiency of the process. Moreover, as a rock becomes hotter, minerals containing structurally bound, volatile components, particularly water, become less stable and eventually break down, releasing those components. Recrystallization is facilitated by the presence of free volatile constituents. The recrystallized minerals, of course, become proportionately less rich in volatile components.

As the temperature of a rock is increased, the rock tends to decrease in strength, that is, it is more likely to deform or even to flow in response to directed stresses. Many metamorphic rocks are highly contorted—thus providing evidence that they had been subjected to high temperatures.

If temperatures become high enough, rocks can melt. At first, only those minerals with lower thermal stabilities will melt, but eventually, if enough heat is available, all minerals in the rock body will do so. With the generation of magma, the realm of metamorphism gives way to that of plutonism.

### Pressure

Much of the pressure to which a rock is subjected comes from the load of the rocks above it. The thicker the layer of rock above, the greater the pressure. The situation in a swimming pool is similar: The deeper you dive, the more pressure you feel in response to the increasing load of the water over you. At any point in the pool, the pressure is equal in all directions—otherwise the water would flow. This all-sided pressure is called hydrostatic pressure. The analogous all-sided pressure created at depth in the earth—from the rock load above—is called *confining pressure.*

Some of the pressure exerted on rocks results from the presence of volatile components (mostly water). Some of the volatiles may have been initially present as interstitial constituents in sedimentary rocks; some may be formed by decomposition of water-bearing minerals in response to thermal activation (discussed earlier). Such pressure is commonly called **pore-fluid pressure,** and it, too, is an all-sided pressure. Under increased confining and pore-fluid pressures, the mineral grains in a rock are subjected to squeezing that can lead to recrystallization and the formation of new minerals with more tightly packed atomic structures and greater densities.

In many instances **directed pressure,** in response to a differential stress, also plays a part in forming metamorphic rocks. This pressure, operating in a particular direction, is commonly generated in response to alpine-mountain building. No one seems to know quite how intense directed pressure may be, but all agree that it is important to the development of the foliated and lineated rock textures that are so characteristic of metamorphic rocks. In a rock recrystallizing under directed pressure, many of the newly forming mineral grains—especially those of platy species, like mica and chlorite—are constrained to grow with their cleavage planes in parallel alignment, perpendicular to the direction of the pressure. Bands of elongated minerals, such as hornblende, also commonly form under directed pressure, with their long axes in parallel alignment and in planes perpendicular to the directed stress.

### Chemical Activity

The chemical environment is also of great importance to the recrystallization of rocks, but its exact role in the process is hard to evaluate. For the most part, chemical activity is related to the presence of small amounts of volatile components (mostly water and carbon dioxide) that permeate

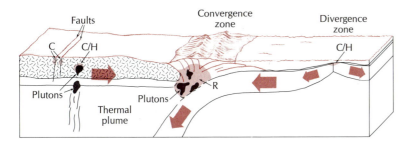

**Fig. 9-2.** Common environments for development of specific types of metamorphism: C = cataclastic, along fault zones; C/H = contact/hydrothermal, marginal to plutons; R = regional, at depth, axially along convergence zones.

and fill pore spaces within and between grains in the metamorphosing rock. Some of these are derived (as previously discussed) from the breakdown of minerals as metamorphism progresses. Others, however, undoubtedly come from fluids already present in the cracks and pore spaces in the unmetamorphosed rock. Sediments, which are formed mostly in aqueous environments, generally contain an abundance of initial pore fluids.

The permeating fluids increase the efficiency of the recrystallization process, facilitating the breakdown of unstable grain and aiding the migration of ions to the growing margins of newly forming grains. As the environment becomes progressively hotter, the fluids become more active. All earth scientists agree that without the presence of the pervasive, hot pore-space fluids, metamorphism would proceed at very slow, perhaps insignificant, rates. Laboratory studies support this view.

As metamorphism proceeds, the fluids progressively diminish in abundance; they either become structurally bound in some of the crystallizing minerals, as in micas and hornblende, or they are squeezed out and driven off. At the end of the metamorphic process, therefore, free volatiles are scarce. Because of this decrease in volatile constituents, chemical alteration and recrystallization during the ensuing end-stage phases of metamorphism—cooling and depressurization—are minimal. The mineral associations and textures exhibited by a metamorphic rock, therefore, generally represent the highest level of metamorphism experienced by the rock.

## METAMORPHIC ROCKS

Metamorphic rocks occur both on the continents and in the ocean basins. Some, such as the *con-tact/hydrothermal* and the *cataclastic metamorphic* types are limited in extent. Most, however, are exposed over thousands of square kilometers; these are called **regionally metamorphic** and **mixed metamorphic-plutonic rocks.** The formation of each of these kinds of metamorphic rock requires an unusual set of circumstances as shown in Figure 9-2.

### Realm of Metamorphism

In contact/hydrothermal metamorphism, heat plays the dominant role in association with chemically active solutions. In most cases the anomalous heat is supplied by the near approach of a pluton. The hot solutions arise either as emanations from the magma body or by convectional circulation of water in the rocks surrounding the pluton—driven by the heat from the pluton—or by both. Hot water solutions capable of bringing about hydrothermal alteration are also prevalent in the outer parts of the oceanic crust in the axial portions of oceanic ridges. For cataclastic metamorphic rocks to form, localized shearing-type strain must occur at depth and the confining pressures and temperatures must be sufficiently great to facilitate recrystallization processes. This type of metamorphism occurs along faults, particularly the larger ones that experience continued movement over long periods of time.

Regional metamorphism requires a variable combination of factors for its occurrence—heat, all-sided and directed pressures, and chemically active fluids. The most probable environment for the simultaneous development of all these factors is in zones of plate convergence. During continued subduction of one plate beneath another, rocks are carried downward into regions of progressively increasing temperature and confining pressure. Thermal energy is liberated both as a by-product of mechanical deformation near the active zone of subduction and by the rise of hot, volatile-rich magmas from depth. Volatiles are supplied either as interstitial water in subducted sediments or by dehydration of water-bearing minerals as they are carried deeper, and directed stresses are generated perpendicularly to the axis of convergence. As evidence of their formation in convergence zones, most regional metamorphic rocks occur in the axial portions of elongate alpine-mountain chains in close association with batholithic felsic plutons. The intensity of metamorphism declines from the mountain-chain axis, where deformation is greatest (in response to the maximum confining and directed pressures and temperatures), outward toward the range margins where deformation is least.

### Metamorphic Rocks of Local Extent

***Contact/Hydrothermal Metamorphic Rocks*** At the margins of smaller plutonic bodies, the surrounding country rocks are heated to such an extent that some degree of metamorphism occurs in a thin sheath around the pluton. The recrystallized rocks are called contact metamorphic rocks, and their formation is primarily the result of the thermal energy derived from the hot magma within the plutonic body. Directed pressure plays an insignificant role in the recrystallization process and, therefore, these rocks generally lack foliation or banding.

In most cases the thinnest contact-metamorphic zones are found at the edges of dikes and sills. Because of their small size and the rapidity with which these igneous bodies cool, the contact zones are usually only a few centimeters wide at most. Larger plutons, such as batholiths and stocks, which take many thousands to a million years or more to cool, make their influence felt over a much wider zone. Characteristically, contact metamorphism results in a dense, hard, nonfoliated rock called **hornfels.** Most hornfels are fine-grained and composed of nearly equidimensional recrystallized minerals. Some of the newly formed mineral grains result from increased growth and recrystallization of grains of one mineral species already in the rock; other grains are created through

chemical reactions between two or more mineral species during recrystallization. In the latter case, for example, if a rock containing dolomite ($CaMg(CO_3)_2$) and quartz ($SiO_2$) is heated, the reaction between the two mineral species will result in the production of magnesium-rich olivine ($Mg_2SiO_4$) and calcite ($CaCO_3$), with the liberation of carbon dioxide ($CO_2$).

Rocks that are particularly susceptible to being transformed into hornfels are the finer-grained, clay-rich sedimentary types, including shale, mudstone, and siltstone. In the transformation to hornfels, the clay completely recrystallizes to form muscovite, whereas quartz or feldspar particles recrystallize and grow only at their margins.

At the margins of some felsic stock and batholithic intrusions, large volumes of hot, water-rich fluids from the magma are released, particularly as the pluton undergoes solidification. These hot fluids (liquids and gases), called *hydrothermal solutions,* can travel long distances through enclosing host rocks and can carry ions in solution derived from the magma body. These fluids are chemically active and react readily with many of

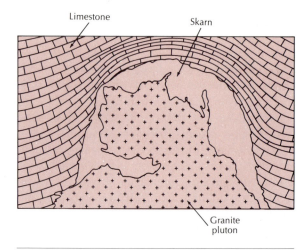

**Fig. 9-3.** Contact/hydrothermal halo (skarn) developed at the expense of limestone country rock marginal to a granite intrusion. The skarn is composed of calciums, iron-, and aluminum-rich silicates. The calcium was derived from the limestone; however, the iron, aluminum, and silicon ions were derived from hydrothermal solutions emanating from the pluton. (After C.W. Burnham, 1959, *Geological Society of America Bulletin,* vol. 70, Fig. 3)

**Fig. 9-4.** Mylonite resulting from cataclastic metamorphism from the Front Range, Colorado. The large broken grains (augen) are quartz and feldspar; most of the streaked-out matrix is composed of very small crystals of biotite and broken grains of quartz and feldspar. Microscope photograph. (W.A. Braddock, USGS)

the minerals in the country rock with which they come in contact. The hot fluids not only aid in the recrystallization of country rock, but also may bring in new chemicals that are also incorporated in the recrystallizing minerals. In some instances the hydrothermal solutions react with the host rock to remove chemical constituents, which are then carried away as the fluids pass through the rock. Not all solutions that enter into the recrystallization process come from the magma itself. Some fluid in the country rocks is groundwater that is heated and driven in convective circulation (like water boiling in a coffeepot) by the hot-platelike action of the nearby hot magma body.

Rocks formed through the combined heat from intrusions and hydrothermal solutions are called **contact/hydrothermal metamorphic rocks,** or more simply, **hydrothermal metamorphic rocks.** If the composition of the rocks has been appreciably altered by the addition of ions from the circulating hydrothermal solutions, the rocks are called **skarns.** A classic example of skarn development around the lateral and upper margins of a felsic pluton is well displayed in a cement quarry at Crestmore, Riverside County, California (Fig. 9-3).

Some minerals are particularly susceptible to alteration by hydrothermal fluids. For example,

rocks rich in olivine ($(Mg,Fe)_2SiO_4$) commonly undergo serpentinization (see chap. 4), a reaction that is principally one of water enrichment by hydrothermal solutions. Many geologists now feel that serpentinization is particularly prevalent near the oceanic-ridge systems, where chemically active solutions, heated by shallowly emplaced dikes, rise along the faults and fractures associated with the ridge.

***Cataclastic Metamorphic Rocks*** In localized zones near the surface of the earth, stresses may build until the rocks shear along faults. Close to the fault planes, the rocks are crushed and granulated. Near the earth's surface, where the rocks are brittle and the confining pressure is minimal, the granulated material lacks cohesion; the name given to this material is *fault-breccia,* or if very fine-grained, *fault gouge.* At somewhat greater depths along the fault zones, where temperatures and confining pressures are higher, some of the more finely ground material remains cohesive and, with time, undergoes recrystallization.

The process that produces partially recrystallized granulated rocks is called **cataclastic metamorphism** (Greek for "broken down").

Not all the minerals in rocks are equally resistant

to shearing stress. Nonresistant, platy minerals, such as mica, become streaked in parallel bands, whereas the more resistant minerals, such as feldspar and quartz, become pulverized. Some of the larger, less-pulverized mineral grains may stand out as rounded or even lenticular, eyelike clots (Fig. 9-4).

A common example of cataclastic metamorphism is the rock **mylonite.** The name comes from the Greek, meaning "mill"—in a figurative sense it is rock that is ground in the geological mill. After the minerals in the original rock are crushed and pulverized, the tiny fragments are partially-to-completely recrystallized into an interlocking network that makes the rock as hard as flint.

**Metamorphic Rocks of Regional Extent**  Regionally metamorphosed rocks are characteristically exposed at the earth's surface over broad areas, amounting in some cases to many thousands of square kilometers. Commonly, they are found in close association with large felsic plutons (batholiths and stocks) in the axes of alpine-mountain ranges.

On essentially all continents, leveling by erosion has produced large, flattened expanses of land floored almost exclusively by plutonic and regional metamorphic rocks of Precambrian Age. Such wide expanses of crystalline rock—both igneous and metamorphic—have been given the name **shields.** Two well-known examples are the Canadian Shield—a broad expanse of igneous and metamorphic rocks on the margins of Hudson Bay that extends southward into Minnesota and Wisconsin and eastward across Labrador—and the Fenno-Scandian Shield—including most of Finland, Sweden, and Norway. Such large expanses of regionally metamorphosed rocks in association with felsic plutons are thought to represent the deeply eroded remnants of several alpine ranges that formed during the early (Precambrian) history of the continents.

In most large bodies of regionally metamorphosed rocks, it is apparent from mineral variations from place to place that the temperatures and pressures were not uniform during the metamorphic event. If, as is common, the starting materials were essentially the same throughout the

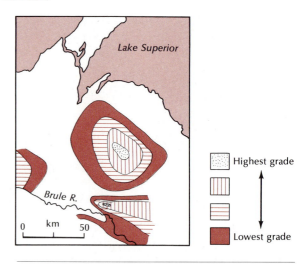

**Fig. 9-5.** Variation in intensity of metamorphism in northern Michigan. Relatively low-grade metamorphic rocks are found toward the outside and high-grade metamorphic rocks toward the inside of each center. (After H.L. James, 1955, "Zones of Regional Metamorphism in the Precambrian of Northern Michigan," *Geological Society of America Bulletin,* vol. 66, pp. 1455–87)

Legend: Highest grade / Lowest grade

region in which metamorphism occurred, it is possible to associate progressive changes in texture and mineral association with a progressive increase in degree of regional metamorphism, from low grade through intermediate grade to high grade (Fig. 9-5). Overall, an increase in metamorphic degree brings about a progressive coarsening in grain size of individual minerals—from submicroscopic at low levels to coarsely crystalline at high levels. Individual grains in some high-grade rocks may be 1-cm (0.5 in) or more across. Parallel alignment of platy minerals, perpendicular to the orientation of directed stress, is also evident as well as an increase in the segregation of mineral grains into distinctive bands. Notably, the nonferromagnesian silicates tend to separate from the ferromagnesian silicates. The fine-to-coarse layering that results is called *foliation.* In the coarser-grained, higher-grade rocks, the bands appear as distinctive light and dark layers (Fig. 9-6).

Changes may also take place in the mineral associations, progressing from a low-to-high degree of metamorphism. These changes reflect differences in the stability of minerals to the increases

**Fig. 9-6.** Biotite gneiss from Uxbridge, Massachusetts, showing light and dark banding, or foliation. Such regionally metamorphosed rocks characteristically show rock cleavage or foliation. (Ward's Natural Science Establishment, Inc.)

in temperature and pressure. Low-grade regional metamorphic rocks generally are similar in mineral makeup to the initial starting materials. At a higher degree of metamorphism, this mineral assemblage becomes unstable and recrystallizes isochemically to produce a different set of minerals appropriate to the new temperature, pressure, and pore-fluid conditions. At even higher levels of metamorphism, recrystallization to another, more appropriate mineral assemblage will occur. And so on. Depiction of these progressive changes on a map shows how the physical and chemical conditions during the metamorphic event varied regionally (Fig. 9-5).

Finer-grained sedimentary rocks, particularly clay-rich types—such as shale and mudstone—are particularly susceptible to regional metamorphism and display a wide range of mineralogic and textural variation in progression from low-to-high metamorphic grade (Table 9-1).

***Lower-Grade Foliated Rocks*** Probably the most familiar foliated metamorphic rock is **slate** (from the Old French, *"esclate,"* or *slat*). Characteristi-

cally, it is dense and of uniformly fine texture; it can be cleaved along smooth, closely spaced parallel surfaces. The latter property is called **rock cleavage** to distinguish it from mineral cleavage (Fig. 9-7). Because of these properties, slate was used for centuries for roofing material and blackboards.

Rock cleavage in slate is the result of an extremely fine foliation that develops during low-grade regional metamorphism. Viewed under the microscope, the cleavage planes of tiny recrystallized flakes of muscovite mica, which are everywhere present in the slate, all appear parallel (Fig. 9-8). Splitting of a slate along one planar rock-cleavage surface is the result of the cleaving of numerous tiny mica flakes in that plane. Although slate is usually derived from fine-grained, clay-rich sedimentary rocks, it can also be formed from other finely textured rocks, such as volcanic tuffs.

The original stratification planes of the original sedimentary or pyroclastic rock are usually still recognizable in slate, but in most cases do not coincide with the rock's cleavage planes (Figs. 9-7, 9-8). During metamorphism, the small planar

**Table 9-1** Progressive Sequence from Low- to High-grade Foliated Rocks Developed from a Shale Parent

| | Rock Name | Metamorphic Grade | Texture | Grain Size | Mineral Assemblage | Rock Cleavage | Foliation |
|---|---|---|---|---|---|---|---|
| | Shale | Parent | Clastic | Very fine | Clay, quartz | — | — |
| | Slate | Lower | Micro-crystalline | Very fine | Mica, quartz, and in some cases, chlorite | Excellent | Very closely spaced |
| | Phyllite | Lower | Micro-crystalline | Fine | Mica, quartz, and chlorite | Uneven, but very good; wavy | Closely spaced |
| | Schist | Higher | Visibly crystalline | Fine to medium; medium to coarse | Mica, chlorite, hornblende, garnet | Very erratic | Medium to coarse; segregation of light and dark minerals |
| | Gneiss | Higher | Visibly crystalline | Medium to coarse | Potassium or sodium feldspar, quartz, biotite and muscovite, hornblende, and garnet | Little to none | Coarse; segretation of light and dark minerals |

Increasing metamorphic grade ↓

**Fig. 9-7.** Relation of rock cleavage to the original bedding. Rock cleavage, which parallels the hammer handle, cuts across bedding in folded Cretaceous slate. Hidalgo, Mexico. (F.S. Simons, USGS)

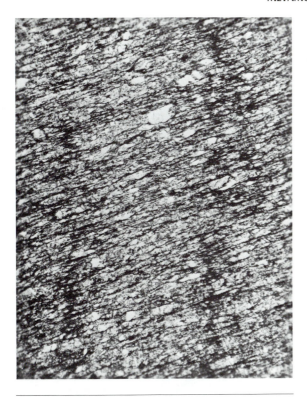

**Fig. 9-8.** The fine structure of rock cleavage shown under a microscope. The bedding, demarcated by dark, poorly delineated and nearly horizontal bands, is cut at a high angle by the closely spaced cleavage planes. Recrystallized flakes of mica parallel the cleavage planes. (W.B. Hamilton, USGS)

clay grains in the shale (which were aligned parallel to the bedding planes) are recrystallized to mica platelets, all with their cleavage planes parallel, but at an angle to the bedding. How the foliation comes about is not clear. Some geologists believe that as the mica plates form, they are constrained to grow with their cleavages at right angles to the directed stress. Other geologists are of the opinion that once formed, the plates are rotated mechanically (i.e., deformed) in response to directed stress. And there are some earth scientists who believe that the original clay particles first are rotated during deformation and then recrystallize to mica.

Three aspects of slate—retention of recognizable original bedding, fineness of foliation, and smallness of the individual mica grains—together indicate that it is formed by gentle regional metamorphism at relatively low temperatures. The met-

amorphic rock looks very much like the original fine-grained rock from which it was derived.

**Phyllite** (from the Greek word *phyllon,* meaning "a leaf"), also a common lower-grade metamorphic rock, is generally more crystalline than slate and intermediate in grain size between slate and schist (described later). The amount of aligned fine-grained platy minerals, mostly muscovite, may reach 50 percent or more. As a result rock cleavage is well developed, and the rock characteristically possesses a silky sheen (Fig. 9-9).

Again, fine-grained clay-rich sedimentary (Table 9-1) and tuffaceous rocks usually constitute the original minerals from which phyllites are derived. In rocks that contain fewer clay minerals than shales and mudstones—and accordingly are less susceptible to alteration—lower-grade metamorphism produces little outward evidence of having occurred except for the development of poorly defined rock cleavage and the spotty growth of grains of mica in subparallel alignment. As is evident from the coarser grain size of the muscovite grains, phyllites are the result of slightly more intense regional metamorphism than are slates.

**Higher-Grade Foliated Rocks**   Probably the most common metamorphic rock is **schist** (from the Greek word *schistos,* meaning "cleft," or *schizein,* meaning "to split"). Phyllites grade into schists with increasing grain size. Whereas the platy minerals in phyllite are tiny, those in schists generally are easily visible to the unaided eye. All schists include tabular, flaky, or fibrous minerals, and the extent to which they develop in parallel orientation determines to a considerable degree the amount of **schistosity,** that is, the characteristically wavy or undulatory rock cleavage. Names given to schists generally reflect the principal, recognizable platy or fibrous minerals present—for example, mica schist, chlorite schist, amphibole schist. Because many schists split readily into tabular blocks, they are widely used throughout Europe in courtyards and castle walls and in North America in fireplaces and patios.

The individual folia laminations of schist are, for the most part, regularly spaced at distances up to about 0.5 cm (0.2 in.). Spacing is the characteristic attribute that separates schists from the more coarsely layered gneisses (to be described

**Fig. 9-9.** Well-developed rock cleavage in a phyllite, glistening in the sun; eastern Nevada. (Karl Birkeland)

next). Sometimes adjacent folia bands have different mineralogies: one may contain platy minerals such as biotite, muscovite, or chlorite—and the adjacent one, mostly quartz and feldspar. In most cases the flaky minerals (those with one well-developed cleavage) make up more than 50 percent of the rock (Fig. 9-10).

Original bedding or other sedimentary structures are almost never identifiable in schists—indicating that their metamorphic grade is much higher than that of slates and phyllites. From the mineral associations common to schists and from the relatively large size and alignment of the grains, most geologists agree that schists were formed over a range of intermediate temperatures and confining pressures under the influence of directed pressure (and to a lesser degree, chemically active fluids). Almost any rock can be transformed into a schist, but likely candidates are shale (Table 9-1), siltstone, and muddy sandstone.

The most-coarse crystalline foliated rock is **gneiss** (pronounced "nice"), an old Saxon miner's term for a rotted, decomposed rock. The sizes of the quartz and feldspar crystals contained in gneiss are about the same as those in granite. During recrystallization of the rock under directed stress at high temperatures and confining pressures, the minerals become rearranged with most of the light-

**Fig. 9-10.** Hand sample of garnet-bearing mica schist, showing the well-developed foliation so characteristic of this rock type. (Department of Library Services, American Museum of Natural History. Photo by J. Kirschner, no. 109299)

**Fig. 9-11.** Gneiss from the north rim of the Black Canyon of the Gunnison River, Colorado. The light-colored bands are composed mostly of quartz and feldspar; the dark-colored layers mostly of quartz and biotite. Notice the crenulations and folds, indicative of ductile folding and flowage at high temperature. (W.R. Hansen, USGS)

colored ones in one layer and the dark ferromagnesian ones in another; the light- and dark-colored bands alternate throughout the rock (Fig. 9-11). Unlike the cleavage plates of slate, the bands do not make uniformly parallel planes that continue for long distances. Gneisses do not have the highly developed rock cleavage of slate or schist. These rocks, despite the roughly uniform spacing of their bands, break in random directions.

**Nonfoliated Rocks** Nonfoliated rocks form the same way as the foliated types, but they are composed predominantly of one nonplaty mineral. Two examples are **marble** and **quartzite**.

Marble, consisting primarily of fine- to coarse-grained calcite ($CaCO_3$), is the metamorphic equivalent of the sedimentary rock limestone. In the transformation of limestone to marble at relatively high temperatures and pressures, the bedding and any visible organic shell fragments are mostly obliterated; the result is a fine- to coarse-grained aggregate of nearly equidimensional calcite grains (Fig. 9-12). In a freshly broken surface, the grains exhibit a texture that looks like sugar. Because marble consists of one mineral only and the grains are essentially all the same size and equidimensional, there is little possibility for the development of foliation or banding.

**Fig. 9-12.** Marble, consisting of medium-sized equant grains of calcite. Microscope photograph. (Edwin E. Larson)

Not all marble is the result of regional metamorphism. Some results from recrystallization during contact metamorphism when, for example, a limestone bed is intruded by a plutonic body of moderate-to-large size. One of the finest white marbles comes from a quarry at the appropriately named town of Marble, in west-central Colorado. It owes its existence to the contact metamorphism that took place when a Tertiary laccolith intruded a Mississippian limestone.

Pure marble is snow white, and one of the most highly prized varieties from ancient times down to our day comes from the quarries of Carrara on the west coast of Italy. Carrara marble is ideal for sculpture because it is uniformly textured and relatively soft (only 3 on the Mohs scale).

**Fig. 9-13.** Deformed pebbles in the Purgatory Conglomerate, Rhode Island. These pebbles, once nearly spherical, became stretched during regional metamorphism. Notice the coin for scale. (Sharon Mosher)

**Fig. 9-14.** Highly contorted layers of a migmatite in the Coast Mountains, British Columbia. This type of rock is commonly associated with the lower parts of batholiths and appears to form deep inside the earth where the temperature falls just short of that necessary for complete melting. (Geological Survey of Canada)

Not all marble, however, is white. Limestone commonly contains clayey or sandy material. When it is metamorphosed, reactions between the calcite and the impurities can produce new minerals, usually calcium-rich silicates and aluminosilicates, which are dispersed throughout the marble.

Quartzite is a common metamorphic rock derived from a preexisting quartzose sandstone. In quartzite the pore spaces that initially separated the individual grains in the sandstone are filled with newly crystallized quartz. In many cases the ghostlike boundaries that separate the original grains from the pore-space silica are barely discernible. Moreover, the interstitial silica often is stronger than the sand grains themselves and, when struck, the rock may break through the grains rather than around them—not the way quartzose sandstones do. Most quartzites are the result of recrystallization at intermediate-to-high temperatures. Because they are predominantly quartz, which occurs in equidimensional grains, they generally are not foliated regardless of the presence of directed pressure during their development. Like marble, quartzite exhibits a sugary-looking texture on a freshly broken surface.

Quartzites are nearly always light colored; white, gray, light pink, and light red are quite typical. With increasing amounts of impurities the colors darken. Very often quartzites are interbedded with marble and other rocks derived from sedimentary sources. Relic sedimentary structures, such as cross-bedding, are sometimes preserved; they are emphasized by slight color differences that superficially may resemble the banding in a gneiss.

A striking metamorphic rock, in many cases interbedded with quartzite, is **metaconglomerate.** As the name implies, this rock was derived by metamorphism of a conglomerate or conglomeratic sandstone. The finer-grained sandy material between the gravel clasts alters to a quartzite. The gravel clasts generally are still recognizable, but in most cases have been distorted, or "stretched," during the metamorphic event (Fig. 9-13).

### Mixed High-grade Metamorphic and Plutonic Rocks
At great depths temperatures can be so high—between 600° and 800° C (1112° and 1472° F)—that in the presence of free water, some minerals can melt.

If the temperature continues to rise after the formation of gneiss, melting may begin. Recall that gneiss consists of light-colored bands of quartz and potassium- and sodium-rich feldspars that alternate with dark-colored bands of ferromagnesian minerals. Melting will occur preferentially in the light-colored bands because they contain minerals with relatively low temperature-stability ranges (see chap. 7). Layers rich in ferromagnesian minerals with higher temperature-stability ranges will become ductile and start to flow and fold, but still remain solid. If the temperature stabilizes and then falls, the resulting rock will be made up of highly contorted bands of dark metamorphic rock (composed largely of ferromagnesian minerals) alternating with light-colored granitic plutonic rock (Fig. 9-14). These mixed plutonic and high-grade metamorphic rocks are called migmatites (Greek, meaning "to mix) and are common throughout the world wherever high-grade metamorphic rocks exist. Rocks that superficially resemble migmatites have, in some cases, been formed by the intrusion of sill-like granitic bodies into preexisting dark-colored, high-grade metamorphic rocks.

Not all geologists agree that migmatites are the result of combined high-temperature melting and metamorphism. Some contend that the granitic layers themselves are metamorphic rocks formed by a process in which solid-state recrystallization is accompanied by large-scale diffusion of ionic material in and out of the layers, that finally become "granitelike." Accordingly, this process has been called granitization, a process previously discussed in chapter 7.

If temperature does not stabilize, but continues to rise, the entire rock unit eventually melts to form a body of magma with a composition representative of the entire rock body—perhaps equivalent to a diorite. The frequent association of granitic plutons (apparently formed from the solidification of a molten magma) with high-grade metamorphic rocks and migmatites lends a great deal of support to the idea that migmatites originated with the partial melting of gneisses.

## SUMMARY

1. Through such processes as plate convergence, faulting, and intrusion, rocks in the earth's outer portion can be subjected to physicochemical environments different from those in which they formed. Minerals recrystallize, grow, or are transformed into new, more stable ones. The process, which occurs while the rocks remain in the solid state, is called *metamorphism*.

2. Metamorphism is brought about by *heat, lithostatic* and *directed pressures,* and the *chemical activity* of fluids.

3. Metamorphic rocks of limited extent include *contact/ hydrothermal* rocks and *cataclastic* rocks. Contact metamorphism is primarily a thermally induced process that occurs adjacent to plutonic bodies. The resultant rocks, which lack foliation and are equigranular, are called *hornfels*. Hydrothermal metamorphism results from the interaction between hot, chemically active, water-rich fluids and the country rocks surrounding large plutons. Country rocks that are chemically altered are called *skarns*. One type of hydrothermal metamorphism, *serpentinization,* takes place along the oceanic ridges. *Cataclastic metamorphism* entails recrystallization of the finer-grained portion of sheared and crushed rock material along fault zones. A common cataclastic rock is *mylonite*.

4. Metamorphism over wide areas results in *regional metamorphic rocks.* Recrystallization in such rocks is caused by heat, pressure, and chemical solutions. Because of directed pressure, most dynamothermal rocks are *foliated*.

5. *Slate* is one of the best-known *lower-grade metamorphic* foliated rocks. Its rock cleavage results from the alignment of tiny recrystallized mica grains. *Phyllite* resembles slate, but its mica grains are more numerous and slightly larger. The higher-grade foliated rocks include the *schists* and *gneisses*. Their *schistosity*, or foliation, is determined by the parallel orientation of tabular or fibrous minerals.

   Schists are more fine-grained than the gneisses. The foliation in both rocks, however, consists of a dark-and-light banding with ferromagnesian grains concentrated in the dark layers; quartz and feldspar grains, in the light ones.

6. *Marble* (recrystallized limestone) and *quartzite* (recrystallized quartz sandstone) are two common nonfoliated rocks.

7. *Migmatites,* which display bands of light-colored plutonic rock that alternate with dark-colored bands of metamorphic rock, represent the highest grade of metamorphism. They are commonly associated with other high-grade metamorphic rocks and granite plutons in the cores of alpine mountain ranges.

## QUESTIONS

1. What are the various realms in which metamorphism can occur?
2. What is meant by isochemical metamorphism? Are there any cases in which recrystallization is not isochemical?
3. Discuss the main factors involved in the metamorphism of a preexisting rock. Would you consider any one of them to be the most important?
4. Name and briefly describe the principal types of metamorphism. Under what conditions do each of these types occur?
5. Describe the differences in rock mineralogy and texture corresponding to the different types and degrees of metamorphism.
6. Which general kinds of metamorphic rocks are the most abundant worldwide?
7. How does a slate differ from a schist? From a gneiss?
8. Does rock cleavage have anything to do with mineral cleavage?
9. What are migmatites, and how are they related to regional metamorphism?

## SELECTED REFERENCES

Barth, T.F.W., 1962, Theoretical petrology, Wiley, New York.

Best, M.G., 1982, Fundamentals of igneous and metamorphic petrology, W.H. Freeman, San Francisco.

Ehlers, E.G., and Blatt, H., 1982, Petrology: igneous, sedimentary, and metamorphic, W.H. Freeman, San Francisco.

Harker, A., 1930, Metamorphism, Methuen, London. (Repr. 1976, Halsted, New York.)

Hyndman, D.W., 1985, Petrology of igneous and metamorphic rocks, 2nd ed., McGraw-Hill, New York.

Mason, D., 1978, Petrology of the metamorphic rocks, Allen & Unwin, London.

Miyashiro, A., 1978, Metamorphism and metamorphic belts, Wiley, New York.

Ramberg, H., 1952, The origin of metamorphic and metasomatic rocks, University of Chicago Press, Chicago.

Spry, A., 1969, Metamorphic textures, Pergamon, Oxford.

Turner, F.J., and Verhoogen, J., 1968, Metamorphic petrology, 2nd ed., McGraw-Hill, New York.

Williams, H., Turner, F.J., and Gilbert, C.M., 1982, Petrography, 2nd ed., W.H. Freeman, San Francisco.

Winkler, H.G.F., 1979, Petrogenesis of metamorphic rocks, 5th ed., Springer-Verlag, New York.

# PART III

# Surficial Processes

**Fig. 10-1. A.** *(Left)* The surface of the obelisk in this photograph—still standing in the desert at Karnak, Egypt—is scarcely marred by weathering over four millennia. (Jean B. Thorpe) **B.** *(Right)* The obelisk of Thutmose III (reigned 1504–1450 B.C.), from the temple of Heliopolis, Egypt, now in Central Park, New York. The lower part of the granitic column shows a loss of detail as a result of weathering. The monument was brought to New York in 1879, and most of the weathering took place in the next few years. (Metropolitan Museum of Art)

# 10

# WEATHERING AND SOILS

For the pharaohs the obelisk along the rich Nile Valley was a cherished status symbol (Fig. 10-1A). These stone columns with their hieroglyphic legends would also become collector's items for later monarchs and conquerors over the centuries—from the Caesars to Napoleon I. By the late nineteenth century even the treasure seekers of America were seeking an obelisk. It was in 1879 that one of Cleopatra's Needles finally landed in New York's Central Park (Fig. 10-1B). And New York was not alone—Paris, London, and Rome also harbored obelisks.

But New York's climate is considerably more humid than that of Egypt. It is no wonder that in about 100 years many of the hieroglyphs flaked off and the surface of the obelisk started to disintegrate. Yet its counterparts still standing in Egypt have survived nearly unscathed beneath the desert sun for almost 4000 years.

This brief aside makes a clear point: Climate is one of the leading factors determining the rate and manner in which rocks disintegrate or decompose, that is to say, *weather*—used in about the same way as when we speak of a weather-beaten face.

Time is also an important factor in weathering, as shown by the toll of more than 600 years of weathering on the Great Wall of China (Fig. 10-2). Aside from time, another crucial factor in weathering is the kind of rock involved. In New England, for example, slate gravestones that carry the salty epitaphs beloved by some of our forebears still survive from the 1700s, whereas the words carved on limestone or marble markers of much more recent vintage may be partly or wholly obliterated (Fig. 10-3). We can use data derived from tombstone studies to help us select the most durable rock for building construction in various areas. Yet the effect of air pollution, especially in industrial areas, is greatly accelerating the natural rate of rock decay. In parts of Europe, for example, centuries of weathering only slightly modified some fine stone sculptures, whereas in the past century, with the rise of intensive industrialization, many of these works of art have been almost destroyed (Fig. 10-4). Acid rain, a by-product of industrialization, has probably accelerated the rate of weathering in recent decades.

Soil is partly a product of weathering and is perhaps the most valuable mineral resource on earth. Without soil, life as we know it would be impossible; needless to say, it is a resource rapidly gaining in importance as the world population

**Fig. 10-2.** U.S. Army trooper at the Great Wall of China in 1928. (Vance T. Holliday)

**Fig. 10-3.** Weathering has rounded off the corners and blurred the inscriptions of these nineteenth-century gravestones—the weathering rate is a function of both climate and rock type. **A.** *(Above)* Weathering has nearly obliterated the inscriptions on this marble gravestone in Damariscotta, Maine. (John Manger) **B.** *(Above right)* In contrast, weathering has barely affected this limestone gravestone in semiarid Saguache, Colorado. (Peter W. Birkeland) **C.** *(Below right)* A granite gravestone displays little evidence of weathering in the relatively wet climate of Christchurch, New Zealand. (Peter W. Birkeland)

expands. Yet in many areas of rapid growth, housing tracts and industrial communities are built on our most productive agricultural lands. Common sense would dictate that such lands should remain agricultural and that any construction should take place on adjacent land with less productive soils. Although some landuse planning is directed toward that goal, it is too slow in coming in the United States and all too often it is simply ignored. Countries with population pressures, however, preserve the best soils for crops and many increase the farmland acreage by extensive terracing (Fig. 10-5). Observant travelers often notice how different the soils are in various parts of the world. In humid areas the soil is thick and may be dark colored at the surface, grading to reddish hues with depth. But in dry regions soils are quite thin and may

**Fig. 10-4.** Decay of a seventeenth-century sandstone sculpture from the Rhine-Ruhr region, Germany. **A.** *(Above)* The sculpture as it appeared in 1908. **B.** *(Above right)* The same sculpture as it appeared in 1969. **C.** *(Below right)* A plot of the amount of decay with time. (Landesdenkmalamt Westfalen-Lippe, Munich) **D.** *(Opposite)* Marble sculptures at the Terrace of the Naxian Lions, Delos, Greece. These date from the late seventh century B.C., and their relatively good state of preservation probably is due to both the climate and the lack of a polluted atmosphere. (Deutsches Archäologisches Institut, Rome)

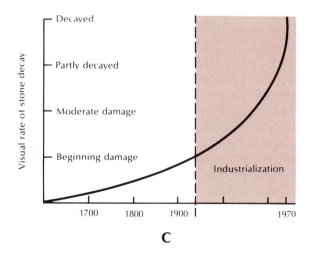

**C**

have a prominent white layer at a shallow depth. Climate and vegetation have long-been known to be responsible for regional variations in soils, but rock type and length of time of formation are major factors as well.

In this chapter we will examine the various processes that cause rock to weather and to form soils. We will also look at the overall effects of the environment on the kind of soil that forms and review some of the practical reasons for studying soils.

## WEATHERING

Most rocks found in the top several meters of the earth's crust are exposed to physical, chemical, and biological processes much different from those prevailing at the time the rocks were formed. Because of the interaction of these processes, the rock gradually changes. Collectively the changes are called *weathering,* and two main types are recognized: chemical and mechanical. Where

*chemical weathering* is dominant, rocks tend to decompose or to decay. When rocks decompose, they are changed into substances with quite different chemical compositions and physical properties then those of the original rock. But where *mechanical weathering* is dominant, rocks break up into smaller fragments, much as if they had been struck a hammer blow. There are few areas where only chemical weathering or only mechanical weathering act alone, but there are many areas where one or the other predominates.

### Mechanical Weathering

Some aspects of mechanical weathering are irritatingly familiar to all of us, such as the wedging apart of sidewalks, foundations, and walls by the roots of grass, trees, and shrubs. The same process goes on in the mountains, where a common sight high on the slopes is an isolated pine clinging to a sheer granite ledge. With no soil in which to take hold, the tree's roots force their way into

**Fig. 10-5.** Terraced landscape along a tributary of the Ganges River, Himalaya Range, India. (Peter W. Birkeland)

crevices; root-volume expansion with growth tends to push the rocks still farther apart. The process is much like the one known millennia ago to the Egyptian slaves who pried out granite blocks for obelisks by using water-soaked wooden wedges.

Almost all rocks are cut by cracks, large and small—sometimes as closely spaced as a fraction of a centimeter, at other times many meters apart, as they are in the stupendous cliff of El Capitan in California's Yosemite Valley. Such cracks, which are called *joints,* allow roots and water to penetrate the rocks and the weathering process starts (Fig. 10-6).

**Frost Wedging** When water freezes, it expands—to about 9 percent by volume. Should water freeze in a confined space, it delivers an enormous pressure against its containing walls—as anyone knows who has contemplated a cracked engine block or a ruptured water pipe. If water freezes in an open crevice in a rock, it can act as a wedge to pry it apart. This process, a major weathering factor in cold climates, is called ***frost wedging.***

Some geologists think that another process is responsible for frost wedging. It has been suggested that the migration of water in microcracks at temperatures of about $-5°$ to $-15°$ C ($23°$ to $5°$ F) creates enough stress to crack rocks.

Common products of long-term frost-wedging are the large areas of broken rocks called ***block fields*** (Fig. 10-7). These are common on gently rolling landscapes at high altitude or in high latitudes. However, not all block fields have been formed under present-day environmental conditions. Indeed, many fields seem to be inactive, as shown by the weathered character of the blocks, the lichens that grow on them, and the well-developed cover of vegetation between them. Such fields were formed some time in the past and probably testify to more rigorous mountain or polar climates during former ice ages.

**Salt Crystal Growth** A mechanical weathering process similar to frost wedging is the result of the growth of salt crystals in rock. In arid regions ground and soil water as well as water within rock pores and cracks commonly contain dissolved salts in ionic form. As the water evaporates, the salt ions are left in the remaining water and their concentration increases. In time the concentration reaches the point at which salt minerals crystallize

**Fig. 10-6.** Vertical and horizontal joints in granite in the Sierra Nevada, California. The joints provide both entry for moisture, which probably aids in mechanical disintegration when it turns to ice, and enables plants to send down their roots, which wedge blocks apart. (Cedric Wright Collection, Sierra Club)

**Fig. 10-7.** Angular blocks produced by frost wedging of well-jointed granitic rock in the Sierra Nevada. (Cedric Wright Collection, Sierra Club)

from the solution. Pressures accompanying the crystallization can be quite high, certainly high enough to dislodge individual minerals or rock fragments (Fig. 10-8).

Another salt-related mechanical weathering process is *hydration,* or the expansion of salt minerals when water is added. If the salts are part of the rock, the pressures generated can break up the rock. An excellent example is the weathering of the obelisk in New York (see Fig. 10-1B). Before it was brought to the United States the monolith rested on its side on Egyptian soil for about 500 years. During that time salt-laden groundwater penetrated the column, and salt minerals crystallized out as the groundwater evaporated into the hot desert air. But in the humid climate of New York, the same minerals absorbed the water from the atmosphere and expanded. The expansion from within caused considerable mechanical weathering in the form of surface flaking over a short period of time. Frost wedging may have contributed to the weathering of the obelisk.

**Surface Unloading** In many parts of the world, the rock close to the surface is cut by joints that more or less parallel the surface, giving it the appearance of an onion skin (Fig. 10-9). Several processes, collectively called **exfoliation** explain the onion-layered appearance of these rocks. A common process is the upward expansion of rock as an overlying or confining rock burden is removed. At depth the rock is under high confining pressures equivalent to the weight of the overlying mass. As erosion removes the overlying rock, the remaining rock can expand—usually only upward or toward the valley walls. This expansion produces joints oriented at right angles to the direc-

tion of the release; thence, they are usually parallel to the land surface. Workers in mines and quarries know the expansive properties of rock all too well. Not uncommonly, as new rock faces are exposed, the pressure release is instantaneous, and the resulting **rock bursts** send dangerous missiles flying through the air.

**Temperature Changes** Rock disintegration may be the result of alternating expansion and contraction of rock induced by severe temperature changes. In deserts, for example, exfoliated rocks exist in great numbers. The theory is that rocks expand drastically under the noonday sun and contract

**Fig. 10-8. A.** Telephone pole damaged by salt crystallization, Bonneville Salt Flats, Utah. Saline groundwater is drawn upward in the pole and, as the water evaporates, the salts crystallize and shatter the wood. (W.C. Bradley) **B.** Salt-shattered bedrock south of the Dead Sea, Israel. (W.B. Bull)

0    5    10    15
Centimeters

**Fig. 10-9. A.** Exfoliation sheets in Independence Rock, Wyoming, a landmark for early travelers in the area. (W.H. Jackson, USGS) **B.** *(Below)* Here the relatively quick-forming exfoliation sheets conform to the lower walls of a small canyon tributary to the Colorado River. (Edwin E. Larson)

sharply with the falling temperature at night, causing the heated and expanded surface layers to peel away in concentric rings. The rock-volume changes would be greatest on the surface of the rock and least in its interior because rocks are such poor conductors of heat. Crude heating and cooling experiments have not duplicated what is seen in the field, and geologists are searching for other causes for exfoliation in rocks.

Very high temperatures associated with fires, however, bring about exfoliation, as many rocks around campfire sites indicate. Surface rocks in forested areas also commonly show signs of exfoliation, probably the result of occasional forest fires; rocks in nonforested areas do not.

### Chemical Weathering

Chemical weathering occurs in all environments but is dominant in lands where temperatures are high, large amounts of water are available, and vegetation flourishes. Acids aid chemical weathering, and a common one is carbonic acid, which results from the union of water and carbon dioxide:

$$H_2O + CO_2 \rightarrow H_2CO_3$$

Generally water is readily available from atmospheric sources, and the carbon dioxide is derived partly from atmospheric sources and partly from root respiration and the decay of organic matter. Although weak as acids go, carbonic acid is common enough in most natural environments to be an effective weathering agent. Another important acid source are organic acids formed in the soil during the decay of vegetation.

Of the manifold processes involved in chemical weathering, three of the most important are *solution, oxidation,* and *hydrolysis.*

**Solution**   Solution is perhaps the easiest process of chemical weathering to visualize because a rock may literally dissolve away, much like a sugar cube in coffee, but at a slower rate (Fig. 10-10). Limestone is especially susceptible to solution, as shown by the following reaction:

$$CaCO_3 + H_2CO_3 \rightarrow Ca^{2+} + 2HCO_3^-$$

Here a solution containing carbonic acid reacts with calcite, the chief mineral in limestone, to form one calcium and two bicarbonate ions. Those ions are removed from the site of weathering by percolating water. Thus, where there was a layer of limestone, there may now be nothing as the calcite has been completely dissolved. This process accounts for the profusion of caverns, underground channels, and disappearing rivers in limestone regions. Indeed, the expansion of such underground voids sometimes leads to the sudden collapse of the overlying rock into a cavern (see chap. 18).

**Oxidation**   Rusting is a process familiar to most of us. In anything but the most severe climates, such as the central Antarctic ice sheet, all unprotected objects made of iron will eventually rust away. In a rainy tropical climate, the struggle to maintain steel bridges, ships, rails, and automobiles is a relentless one.

Most rocks contain some iron-bearing minerals and are originally gray. When they are exposed to atmospheric attack, the rocks are stained a wide variety of colors, such as red, yellow, orange, or red-brown if weathering occurs in an environment

Granite    Limestone    Gypsum

**Fig. 10-10.** In an experiment to demonstrate chemical weathering, acid water was dripped on these rocks for 6 months. Gypsum weathers the most rapidly and limestone shows substantial weathering. In contrast, granite is so resistant to weathering that the chemical effects are barely discernible. (Janet Robertson)

with ample oxygen. A simplified reaction describing the process is

$$4FeO + 3O_2 \rightarrow 2Fe_2O_3$$

| ferrous iron oxide (gray-green) | oxygen | ferric iron oxide (rust-colored) |

Indeed, the discoloration of rock to yellowish brown and red is one of the first visible signs of chemical weathering. It is thought that the exact color is determined by the kind of iron oxide mineral that forms.

**Hydrolysis** The common rock-forming minerals weather by a process called *hydrolysis*. The products of hydrolysis are much different from the minerals being weathered. The weathering of feldspar (an aluminum-silicate mineral) is a common example of hydrolysis and is shown by the following reaction:

$$2KAlSi_3O_8 + 2H^+ + 9H_2O) \rightarrow$$

| feldspar | hydrogen ion from carbonic acid | water |

$$H_4Al_2Si_2O_9 + 4H_4SiO_4 + 2K^+$$

| clay mineral (kaolinite) | silicic acid | potassium ion |

Feldspar eventually breaks down in the presence of carbonic acid, and the aluminum and some of the silicon combine with the hydrogen ions to form a clay mineral. The silicic acid and potassium ions remain in solution and may be carried away, leaving only the clay mineral behind. Some of the potassium, however, might be used by plants or become part of other clay minerals. If the weathering mineral is an aluminum-silicate mineral containing iron, the solid materials resulting will be clay minerals and iron oxide minerals.

Clay minerals—of which there are a great variety—are so small that they can be identified only by X-ray. Most are in the form of platelets (Fig. 10-11). The kind of clay that forms is determined by the environment. Kaolinite, the clay mineral mentioned above commonly forms in areas with humid, warm climates where large amounts of water move through the soil, leaching from it the ions released by weathering. In dry regions where little leaching occurs, the most common clay min-

eral is montmorillonite. Clays vary in their physical and chemical properties. Some are used to make pottery and ceramics or adobe buildings, some are especially useful as muds in drilling wells, and others enhance the fertility of soils.

### Granite Weathering—An Example of the Hazy Boundary Between Mechanical and Chemical Weathering

The boundary between mechanical and chemical weathering is not always clear. The weathering of granite illustrates the point. In many deep road-cuts, small fragments of loose and broken granite (mainly the size of the individual mineral grains) can be seen. Because evidence of chemical weathering—notably the presence of clay minerals and iron oxide colors—may be missing, such weathering has been commonly ascribed to a mechanical process. Recently, however, careful X-ray studies of the minerals in granite indicate that the most common alteration is a subtle change of the biotites to a similar mineral of greater volume. Apparently, the little water that does penetrate granite first reacts with biotite. The accompanying expansion of all the biotites produces enough internal pressure to break up the rock. Thus, volume expansion as a result of a subtle form of chemical weathering mechanically shatters the rock.

Research in the western United States suggests that about 100,000 years of weathering are needed to bring about such a change—an example of the snail's pace at which weathering takes place.

### Relative Weathering Rates of Minerals and Rocks

Minerals and rocks weather at different rates, a phenomena we refer to as ***differential weathering.*** Numerous examples of differential weathering can be found in nature (Fig. 10-12). Rocks that weather more rapidly erode more rapidly, so it is not uncommon to find that the less-resistant rocks form slopes and low areas, whereas the more-resistant rocks form prominent ledges or cliffs (Fig. 10-13). The common rock-forming minerals vary in their resistance to chemical weathering (Fig. 10-14). Calcite, for example, weathers rapidly because it dissolves so readily in water. The atoms

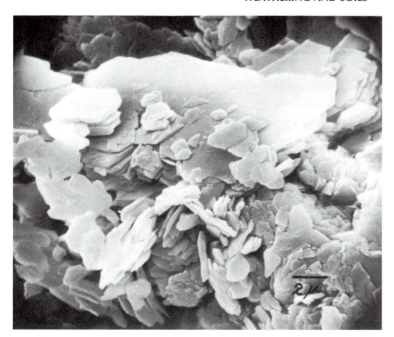

**Fig. 10-11.** Kaolinite crystals, photographed under the scanning electron microscope, were formed from the alteration of feldspars at depth in ancient alluvial fan deposits, Colorado. (T.R. Walker)

**Fig. 10-12.** Betatakin ruins (A.D. 1267–1300) are situated in this large sandstone alcove, Navajo National Monument, Arizona. Differential weathering formed the alcove; it probably resulted from a combination of weathering along springs at the base of the alcove and the failure of exfoliation sheets parallel to the canyon walls to produce arches. (Peter W. Birkeland)

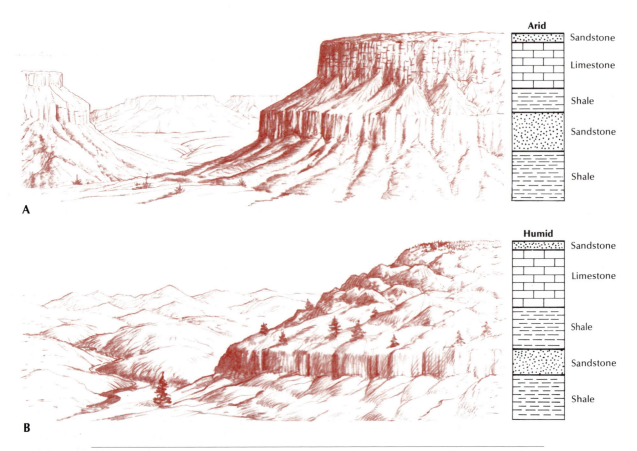

**Fig. 10-13.** Rocks vary in their resistance to weathering and subsequent erosion. Whether a rock forms a steep cliff or a gentle slope depends partly on climate. In an arid climate **(A)**, limestone and sandstone are cliff formers and shale is a slope former, often covered by talus. In a humid climate **(B)**, sandstone also is a cliff former, but limestone weathers by solution to form irregular slopes. Again, shale is a slope former, often covered by a thick soil.

**Fig. 10-14.** Extreme alteration of pyroxene grain, originally rectangular in shape, has produced these delicate needlelike structures. The alteration took place long after sediments were deposited in basins in southern New Mexico. (T.R. Walker)

| Relative rate of mineral weathering | | Relative rate of rock weathering | |
|---|---|---|---|
| | | Coarse Grained | Fine Grained |
| Quartz | | Granite | Rhyolite |
| Feldspar | Na/plagioclase | | |
| Biotite | | Diorite | Andesite |
| Hornblende | | | |
| Pyroxene | Ca/plagioclase | Gabbro | Basalt |
| Olivine | | | |

Most resistant

Least resistant

**Fig. 10-15.** Relative rates of weathering of various common minerals and rocks.

of the silicate and aluminosilicate minerals are relatively tightly bounded, however; thus these minerals are better able to resist weathering. Yet within the group, individual minerals weather at different rates. Quartz and feldspar appear to be the most resistant to weathering, olivine and calcium-plagioclase the least resistant (Fig. 10-15). The resistance of quartz helps to explain the predominance of quartz grains in many sandstones. With repeated cycles of weathering and transportation, the durable quartz grains persist, whereas the less-durable grains weather away, often forming clay minerals and ions.

Rocks weather chemically according to the rate at which their constituent minerals weather. Hence granite, because it contains an assortment of resistant minerals, weathers chemically much more slowly than gabbro (Fig. 10-15).

The size of grains also has an important influence on the rate at which minerals weather. Small particles have a much larger surface area per unit volume than do large particles (Fig. 10-16). Weathering solutions, therefore, have access to much larger surface area if the particles are small; in these cases weathering is more rapid.

## SOIL

The products of weathering can be rearranged into layered materials that are quite different physically, chemically, and biologically from the parent rock. These materials make up soil. Because soil is so fundamental to life, it has been studied intensively for more than a century. In the United States the emphasis of study is on soil genesis, classification, and the relationship between soils and environment as well as crop production. Most recently, there has been a trend to use soil data to help solve environmental problems.

Most soils consist of three basic layers, called **horizons,** the sum total of these layers constitutes what we refer to as the **soil profile.** At the surface is the **A horizon,** usually a dark-colored layer rich in humus—the decomposed organic matter derived from surface vegetative debris and roots. Beneath the A horizon is the **B horizon,** a layer formed by the accumulation of various soil and weathering products. Several common accumulation products are clay (designated Bt horizon), iron and aluminum (Bs horizon), and $CaCO_3$ (Bk horizon). Because of the materials present, most Bt and Bs horizons are brown to red, whereas Bk horizons are white. Beneath the B horizon is the **C horizon,** which is slightly altered parent material. Finally, at greater depth is relatively unaltered

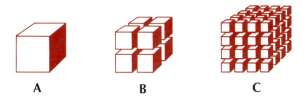

A        B        C

**Fig. 10-16.** An example of the increase in surface area with particle size. If cube **A** is broken into smaller cubes, **B** and **C,** the total weight remains the same, but the surface area is significantly increased. In general, the rate of chemical weathering is proportional to the amount of surface area exposed to weathering solutions.

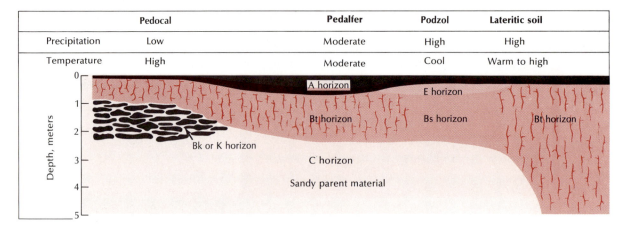

| | Pedocal | Pedalfer | Podzol | Lateritic soil |
|---|---|---|---|---|
| Precipitation | Low | Moderate | High | High |
| Temperature | High | Moderate | Cool | Warm to high |

**Fig. 10-17.** Transect from a dry to a humid environment, showing soil profiles characteristic of each environment on fairly old landscapes. The desert profile could be found in the valleys of the western United States, the humid region profiles in the Great Lakes region and the Northeast, and the warm, humid region profile in the Southeast.

**A horizon** Dark brown to black due to high amount of organic matter, and well aggregated so that rainfall readily enters soil with little surface runoff.

**B horizon** Brown to reddish brown due to iron oxides, and high clay content as shown by vertical shrinkage cracks that develop on drying. Grades downward to slightly altered C horizon material.

A

**Fig. 10-18.** Two different kinds of pedalfers, each about 1-m (3-ft) deep. **A.** Grassland soil formed from loess in Iowa. (Roy W. Simonson) **B.** Podzol formed from a sand deposit in northeastern Minnesota. (Soil Science Department, University of Minnesota)

material, called parent material, thought to have been present in the position of many soil profiles before soil formation.

From early work on soil distribution, it became apparent that soil differed markedly from place to place because environmental conditions vary from place to place. Five factors are generally considered in describing the soil environment: (1) climate, (2) vegetation, (3) age, (4) parent material, and (5) topographic position. (Because the influence of topography is local, we will not discuss its effect on soil formation.)

## Climate and Vegetation

Soil properties are most affected by climate. Because the effects of vegetation are hard to distinguish from those of climate, both factors will be covered here.

*Pedalfers* are the common soils in humid temperate regions that exist under either grass or forest vegetation (Fig. 10-17). The word is derived from the Greek *pedon,* meaning "ground," and the symbols for aluminum (Al) and Iron (Fe) such soils contain. Pedalfers consist of relatively thick A, B, and C horizons; there is a fairly high content of organic matter in the A horizon (Fig. 10-18A). In the cooler climates toward the northern limit of trees, but also in some warm, humid climates, a special kind of pedalfer, known as a ***podzol,*** forms (Fig. 10-17). Podzols are characterized by a whitish layer, called the ***E horizon,*** that lies between the A and B horizons; most of the iron oxides have been removed by downward-percolating

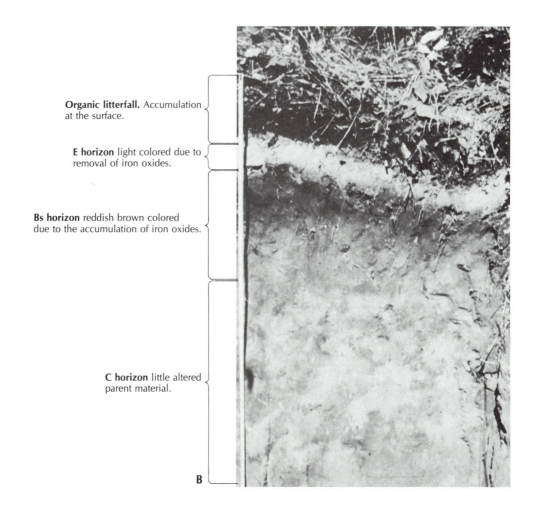

**Organic litterfall.** Accumulation at the surface.

**E horizon** light colored due to removal of iron oxides.

**Bs horizon** reddish brown colored due to the accumulation of iron oxides.

**C horizon** little altered parent material.

B

**Fig. 10-19.** Road-cut about 7-m (23-ft) deep near Rio de Janeiro, Brazil, exposes highly weathered lateritic soil formed from gneiss. The rock is so thoroughly decomposed that it can be readily cut with a knife or a shovel, yet quartz veins (diagonal white lines) are virtually unaltered. (Roy W. Simonson)

waters in this horizon, thus accounting for its white appearance (Fig. 10-18B). The B horizon of podzols is characterized by iron enrichment rather than clay enrichment, and it is termed *Bs*.

On old landscapes in humid, warm climates—especially tropical climates—we see the end product of extreme weathering—the ***lateritic soils*** (Figs. 10-17, 10-19). For the term *laterite* (from the Latin "brick") we are indebted to Buchanan Hamilton, an observant Scotsman who, while traveling in India in 1807, was greatly impressed by the ease with which the red-brown tropical clay could be transformed into a building material. Hindu laborers simply excavated the clay and shaped it into bricks that needed only case-hardening in the sun before they could be used. As construction mate-

rial, laterite has been used in some enduring monuments; indeed, much of the long-forgotten temple at Angkor Wat in Kampuchea is built of laterite (Fig. 10-20). Despite a humid tropical climate, the buildings of that complex not damaged by warfare in recent years are well preserved; this alone is an indication that laterite is virtually insoluable.

Few soils harden into actual laterites, but lateritic soils—deep, highly weathered, and reddish—are widespread in tropical climates. In such soils calcium, sodium, and potassium have been leached out; even the silica of most silicate minerals has been removed. What remains is mostly the iron oxides ($Fe_2O_3$) derived from the weathering of iron-bearing minerals and quartz if the parent rock contained any. The crystallized iron oxides form the bricks, or true laterite. Should the rocks from which lateritic soils are derived have a high content of aluminum, then bauxite ($Al_2O_3 \cdot 2H_2O$), a chief ore of aluminum, can form.

In some parts of the tropics, soils may be mined as ore. Cuba's iron mines and those on Mindanao in the Philippines are examples. The bauxite ores of Little Rock, Arkansas, were formed under climatic conditions that probably were much like those of the tropical savanna today. However, most of the aluminum ore now processed in North America comes from the high-alumina clays of Suriname, Jamaica, and other south American and Caribbean lands.

A much different soil forms in semiarid-to-arid regions. It is named a **pedocal** because an accu-

**Fig. 10-20.** The temple at Angkor Wat, Kampuchea, built from blocks of laterite. They are highly resistant to weathering because they are composed chiefly of residual iron oxide; all the other original materials have been removed by chemical weathering. The sandstone columns and statues, in contrast, show considerable weathering. (Leonard Palmer)

**A horizon**  Thin and light colored due to small amount of organic matter. Most of the A horizon has been eroded.

**B horizon**  Reddish brown due to the kind of iron oxides that form at high temperatures. The clay content is high, as shown by shrinkage cracks.

**K horizon** (caliche)  Contains 50 percent or more $CaCO_3$, and can reach the hardness of concrete, especially between 0.30 and 1 m. Grades downward to the parent material at about 2.2 m.

**Fig. 10-21.** Desert soil (pedocal) formed from alluvium in New Mexico. (L.H. Gile)

mulation of calcium carbonate occurs at some depth (Figs. 10-17, 10-21). As rainfall is slight and vegetation scanty, the A horizon is thin and not too rich in organic matter, and the common clay mineral in the B horizon is montmorillonite. Beneath the B horizon, at about the depth to which water from the annual rainfall penetrates, is an accumulation of white calcium carbonate ($CaCO_3$) known as a **Bk horizon.** On some old landscapes there is so much calcium carbonate that it forms a highly indurated layer known as the **K horizon,** or **caliche.** The $CO_3^{-2}$ in the calcium carbonate is derived from the reaction of carbon dioxide ($CO_2$) and water, whereas the calcium ions are derived from the weathering of calcium-bearing minerals or from atmospheric or groundwater sources.

The three main types of soil—pedalfers, lateritic soils, and pedocals—are all found in the United States. The 100th Meridian is the approximate boundary between pedalfers to the east and pe-

docals to the west. In the West, however, pedocals occur only in dry basins; the adjacent mountains, in contrast, receive enough rainfall for pedalfers to occur. In the East, podzols occur in the Northeast and the Great Lakes area and lateritic soils are widespread in the Southeast, south of the glacial boundary.

A new soil classification scheme, called **Soil Taxonomy,** has been introduced by the U.S. Soil Conservation Service and is used by professionals in the field. Soil Taxonomy, a highly quantitative and rather complex scheme, recognizes 10 major subdivisions, called orders, which are described in Table 10.1. A simplified version of the older classification is used in this book, however.

### Age of Soil

All the key properties of soil take considerable time to form (Fig. 10-22). The A horizon forms most rapidly—several centuries in humid environ-

**Table 10-1** New U.S. Soil Classification Scheme Compared to Older, Less-detailed Scheme

| Order | Generalized Properties | Old Classification |
|---|---|---|
| Entisol | Minimal development, an A horizon may be present | |
| Inceptisol | Weak development, with A horizon, a B horizon that lacks clay enrichment, with or without a Bk horizon | Pedalfer and Pedocal |
| Mollisol | Thick, dark A horizon, high in organic matter, a B horizon that may or may not be clay enriched, with or without a Bk horizon | Pedalfer and Pedocal |
| Alfisol | Relatively thin A horizon overlying a clay-enriched B horizon, with an E horizon separating the A and B layers in places | Pedalfer |
| Spodosol | Highly organic surface horizon above an E horizon that in turn, rests on an iron-enriched B horizon | Podzol |
| Ultisol | A horizon over highly weathered B horizon | Pedalfer and Lateritic |
| Oxisol | A horizon over an extremely weathered B horizon | Lateritic and Laterite |
| Aridisol | Thin A horizon above a relatively thin B horizon, some are clay enriched, with a Bk or K horizon at depth | Pedocal |
| Histosol | Peaty soil | |
| Vertisol | Very high content of clays; shrinks and swells with seasonal moisture variation | |

**Fig. 10-22.** Approximate time necessary in the western United States to develop diagnostic soil horizons from a sandy parent material.

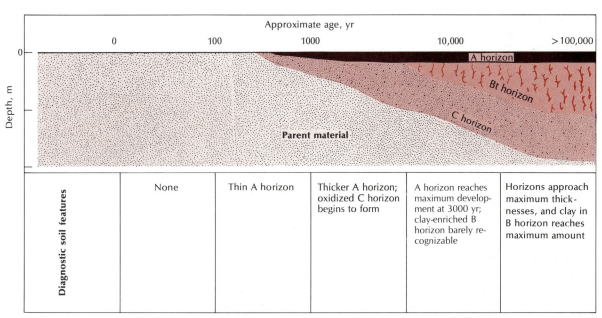

ments to several thousand years in arctic and alpine environments. B horizons take much longer to form because minerals weather so slowly. Discoloration owing to the weathering of iron-bearing minerals shows up in about 1000 years, and the strongest red colors may require 100,000 or more years to develop. Clay-rich B horizons, if formed from sandy parent materials, require at least 10,000 years for initial detection and more than 100,000 years for maximum development. In arid regions, the strongly cemented K horizons may take 100,000 to 500,000 years to form. Deep lateritic soils take the longest to form: Our best guess of how long is in the order of 1 million years or more. No wonder, then, that we should protect our soils from erosion! Once they are lost they are probably lost forever.

## Parent Material

Parent material influences both the rate of weathering and the rate at which clay is produced. Granite, for example, produces a fairly deep sandy soil because, although biotite weathering breaks the rock down to sand-size particles, the minerals undergo chemical alteration very slowly to form clay minerals. In contrast soil on basalt in the same area would be fairly thin because fine-grained rocks weather fairly slowly. What soil there is, however, could be richer in clays because the minerals in basalt are less resistant to chemical weathering.

## BENEFITS OF SOIL RESEARCH

We can benefit greatly from the study and mapping of soils. Some of the benefits may seem academic to us, but others are of immediate concern.

The study of soil development on various deposits has aided earth scientists in estimating the age of those deposits. In areas of glacial deposition, for example, soil studies not only helped to develop the idea of multiple glaciations in the United States, but also helped to determine how often ice ages occur.

We can also estimate the frequency of geological hazards, such as landslides, rockfalls, and major flooding, from the study of soils. For ex-

ample, areas in which such events took place in the recent past would have poorly developed soils, whereas better-developed soils would be associated with areas in which hazards occurred long ago. Soil dating can also provide key data for building sites: Sites for nuclear power plants are probably the most crucial. These installations must be located in areas free of the ground breakage (faulting) associated with earthquakes, and soil studies can aid in dating recently active faults.

Soils are also valuable in determining the nature of past climates (Fig. 10-23). Buried pedocals in a humid region tell us that at one time the climate must have been drier there. In the more arid parts of Australia, pedocals have formed from young deposits, but on ancient landscapes lateritic soils are found. The change seems remarkable—from tropical landscape to parched desert in a few million years, a short time in the earth's history.

Soils vary in productivity and manageability for crop growth. Maps produced by the U.S. Department of Agriculture help us to locate the best—and worst—soils for agricultural use. People involved inland-use planning should consult such maps when laying out subdivisions, taking care to keep the most productive soils for agriculture. Once a good soil is paved over or covered by a house or factory, it is a resource lost. And when crops are planted on less-productive land, it only serves to drive the cost of foodstuffs higher and higher. Citizens and politicians should take heed.

Although lateritic soils are no longer considered a major problem in tropical agriculture (they make up 7 percent of the tropical land area), we still need a better understanding of how they behave under cropping. Such soils are extensively weathered and thus are low in the nutrients necessary to plant growth. Cropping only reduces the nutrients further, and expensive fertilizers are required to restore them. Some lateritic soils harden if dried out; once hardened, they can no longer be tilled. Intensive research, to fit crops to soils rather than soils to crops, is needed to help solve the problems of food production in tropical areas.

Throughout the world we have found that natural soil fertility is related to recent geological activity and to the weathering of minerals in the soil. With time key elements are leached from the ground and soil fertility declines: little-weathered,

**Fig. 10-23.** The study of buried soils (S) can help determine the nature of past landscapes and environments. (Lubbock Lake Project, Museum of Texas Tech University)

young soils are the most fertile, highly weathered, deep old soils the least. The most-fertile soils are in young mountain belts with associated recent volcanism (e.g., the west coasts of North, Central, and South America); relatively young glaciated areas where ground-up fresh rock is exposed or areas of loess associated with the glaciation (e.g., central United States and central Europe); and areas of extensive young river deposits (e.g., the flood plains of the Nile and Ganges rivers). Be-

cause of these findings, some geologists have suggested that one way of increasing the fertility of old, weathered soils in tropical areas would be to add ground-up fresh rock materials.

Some soils (e.g., vertisols) swell and shrink as moisture varies seasonally. The reason for such behavior is the high content of clay in the soil; the clay minerals are mostly montmorillonite, which is prone to swelling on wetting and shrinking on drying. Recognition of soils that shrink and swell

**Fig. 10-24. A.** Dust storm in Prowers County, Colorado, photographed in the 1930s. **B.** *(Below)* Abandoned farmland near Stillwater, Oklahoma; notice the extent of the erosion. (Both, Soil Conservation Service)

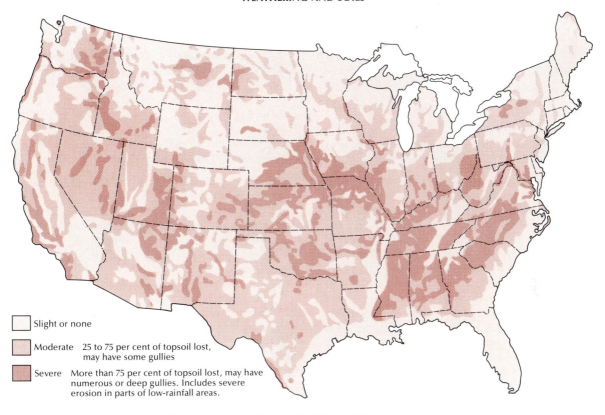

Slight or none

Moderate   25 to 75 per cent of topsoil lost, may have some gullies

Severe   More than 75 per cent of topsoil lost, may have numerous or deep gullies. Includes severe erosion in parts of low-rainfall areas.

**Fig. 10-25.** Various classes of soil erosion in the United States.

is of vital importance to the construction industry. Shrinkage and expansion accounts for more than $2 billion of damage in the United States each year.

## SOIL EROSION

An adequate soil is necessary to protect land from accelerated erosion, as we learned so dramatically in the Dust Bowl in the 1930s (Fig. 10-24). The organic matter, clay particles, and ions in soil have the ability to bind, that is to cause the soil particles to aggregate and form small clumps or blocks of various sizes (see Fig. 10-18). Water will penetrate such spongelike materials more readily than it will run off the surface. And the more water that sinks

into the ground, the less runoff—hence the less the surface erosion. The properties of the A horizon are especially critical to the ability of the soil to take up water. If that horizon is removed or compacted, runoff and erosion may be accelerated. There has been considerable erosion of soil throughout the United States, yet most people do not seem to be aware that an important natural resource is being lost (Fig. 10-25).

Soil erosion is becoming a major global problem. Figures are difficult to obtain because few countries have measured their losses. Worldwatch Institute has estimated that the world is losing 24.5 billion tons of topsoil each year from croplands in excess of what is formed each year. The loss could be as high as 7 percent of the total topsoil for each decade. In effect, soils are being mined. As loss continues there is both less crop production on the same land as well as greater costs to produce the equivalent amount of crops. The United States has the capability to launch a successful soil con-servation program, but so far the effort has not been great enough. Of Third World countries, only Kenya has launched a program that appears to be successful.

## SUMMARY

1. No rocks or sediments are stable at the earth's sur-face. With time they will alter or weather to form products more stable in the surface environment.

2. Two main kinds of weathering are recognized. *Me-chanical weathering* is the disintegration of particles to smaller size without an appreciable change in either the chemistry or mineralogy of the original material. In contrast, through *chemical weathering,* the chemistry and mineralogy of the original material are changed. Clay minerals found in soils and sedi-ments are the result of chemical weathering.

3. Soils are layered materials consisting of an *A horizon,* which is rich in organic matter, overlying a *B horizon* that rests on the only slightly altered *C horizon.*

4. Soils vary from site to site. The main factors control-ling the variation are climate and vegetation, age of soil, parent material, and topographic position. Most well-developed soil profiles have taken tens or hundreds of thousands of years to form, and the very deep ones of the warm humid areas may have re-quired 1 million years.

5. Soils can provide much information on the geologi-cal history of the earth, and their properties and distribution should be taken into account in all land-use decision.

## QUESTIONS

1. What properties of a rock are important in determin-ing its resistance (1) to mechanical weathering and (2) to chemical weathering?

2. How would you go about investigating the relative rate of physical weathering between frost action and a combination of salt crystallizations and hydration?

3. Parts of the arid portions of Australia have old later-itic soils. What are several working hypotheses that might account for this?

4. How would you use soil information to date a sur-ficial deposit, such as a beach, glacial, or river deposit?

5. How would you determine which minerals in the soil weathered to form the clay minerals?

## SELECTED REFERENCES

Bartelli, L.J., et al., eds., 1966, Soil surveys and land use planning, Soil Science Society of America and Amer-ican Society of Agronomy, Madison, Wis.

Birkeland, P.W., 1984, Soils and geomorphology, Oxford University Press, New York.

Brown, L.R., and Wolf, E.C., 1984, Soil erosion: quiet crisis in the world economy, Worldwatch Paper 60, Worldwatch Institute, Washington, D.C.

Buol, S.W., Hole, F.D., and McCracken, R.J., 1981, Soil genesis and classification, Iowa State University Press, Ames.

Carter, V.G., and Dale T., 1974, Topsoil and civilization, University of Oklahoma Press, Norman.

Chesworth, W., 1982, Late Cenozoic geology and the second oldest profession, Geoscience Canada, vol. 9, pp. 54–61.

Hunt, C.B., 1972, Geology of soils, W.H. Freeman, San Francisco.

Jenny, H., 1980, The soil resource, Springer-Verlag, New York.

Sanchez, P.A., and Buol, S.W., 1975, Soils of the tropics and the world food crisis, Science, vol. 188, pp. 598–603.

Singer, M.J., and Munns, D.N., 1986, Soils, Macmillan, New York.

Soil Survey Staff, 1975, Soil taxonomy, U.S. Department of Agriculture, Handbook No. 436, Washington, D.C.

Winkler, E.M., 1973, Stone: properties, durability in man's environment, Springer-Verlag, New York.

# 11

# MASS MOVEMENTS AND RELATED GEOLOGICAL HAZARDS

The movement of large masses of rock and debris downslope is an event known to many of us, especially if life and property are destroyed.

It will be long before the people of Utah can forget the events of spring 1983. Rainstorms and the rapid melting of a thick snowpack combined to produce numerous landslides throughout the mountainous parts of the state. The most newsworthy event was the reactivation of a landslide near the town of Thistle. Masses of debris blocked a canyon and formed a lake, thus cutting off railway and highway access to the central part of the state (Fig. 11-1). Damages were estimated at $250 million, making the Thistle slide the most expensive in the U.S. history. North of Salt Lake City, debris flows moved rapidly down steep mountain fronts and crashed through towns, burying and crushing homes. No fatalities occurred,

thanks to the organized response of state and local governments to the threat.

Unfortunately, most landslides mean loss of life. One estimate puts the human loss worldwide at about 5500 for the years 1947 to 1980. In the United States between 25 and 50 people each year lose their lives in landslides.

Some landslides set into motion a chain of events that result in the loss of many lives. A dramatic one occurred in Italy on the night of October 9, 1963, when a torrent of water, mud, and rocks plunged down a narrow gorge, shot out across the wide bed of the Piave River and up the mountain slope on the opposite side, completely demolishing the town of Longarone and killing 2600 inhabitants there and in adjoining towns (Fig. 11-2). It has been called history's greatest dam disaster, but when it was over, the Vaiont Dam in the narrow gorge was still intact. What could have caused the water in that reservoir, which was not even half full, to rise up over the dam and proceed on its destructive course? One clue is that one shoulder of the dam was supported by Monte Toc, nicknamed *la montagna che cammina*—"the mountain that walks"—by the local inhabitants. Despite assurances by engineers regarding the safety

**Fig. 11-1.** This landslide reactivated in April 1983 was the most destructive of the numerous slides in central Utah that spring: 4 million m³ (5.2 million yd³) of earth were on the move. The slide dammed Spanish Fork Canyon, created the lake that drowned the town of Thistle, and cut off railway and highway access to a large area. (U.S. Department of Agriculture, Forest Service, Ogden, Utah)

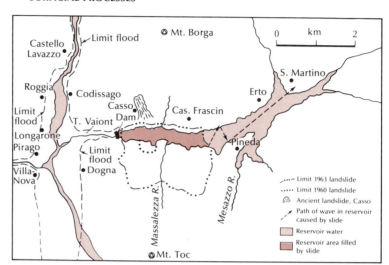

**Fig. 11-2.** Map of Vaiont Reservoir area, Italy, showing limits of the landslide, the area of the reservoir filled by the slide, and the extent of downstream flooding. (After G.A. Kiersch, "The Vaiont Reservoir Disaster," *Mineral Information Service*, vol. 18, no. 7)

of the dam and the expensive work done to stabilize its slopes, Monte Toc did not walk that night in October, it galloped. About 250 million m³ (330 million yd³) of mountainside slid into the lake behind the dam. One wave rode 260 m (850 ft) up the valley wall opposite the slide; another rose 100 m (330 ft) above the dam and dropped into the gorge below. There, constricted by the narrowness of the gorge, the water, carrying tons of mud and rocks, raced on its destructive path. It was all over in 7 minutes. The dam was strong, for it withstood the onslaught of tremendous forces, estimated at 4 million metric tons, and it is still standing today.

A side effect of the Vaiont slide was that towns near the head of the reservoir were also damaged. The destruction there, however, was from slide-generated water waves that ricocheted from one valley wall to the other, wreaking havoc in their path (see wave path in Fig. 11-2).

California has long been known as earthquake country, but it also is renowned for its landslides (Fig. 11-3). A major one occurred in 1956 and for the following 3 years in the Palos Verdes Hills near Los Angeles, where a development had been built directly over an old landslide, the land started to move again. Many houses were totally demolished and the value of property destroyed ran over

$10 million. Eventually, damage suits of several million dollars were collected from the county.

The annual monetary loss from landslides in the United States is estimated at $2 billion. The damage is site specific: One person may lose everything—house and the land it stands on—but a neighbor may be unaffected. A survey of four metropolitan areas between 1969 and 1978 showed per capita annual landslide losses ranged from $1.30 to $5.80. One study in California has shown that with careful geological study and proper engineering practice, it is possible to reduce landslide damages by more than 90 percent.

Material at the earth's surface, be it weathered or fresh rock, can fail under specific circumstances and move downslope under the pull of gravity. Such movements can take place very slowly and with little apparent evidence or they can happen so quickly and involve so much material that they stagger the imagination. In this chapter we will discuss the major kinds of gravity movements, assess their causes, and point out some remedies.

Mass movement is responsible for the downslope transfer of material to rivers, which then act as conveyor belts to carry it away. The manner in which the walls of the Grand Canyon flare outward from the Colorado River is the result largely of the gravity transfer of rock fragments and mineral grains downslope to the river and its intricate network of tributaries, which then carry the material out of the Colorado Plateau.

How much or how little material will be shifted downhill by gravity and how rapidly or how slowly it will move are decided by many factors. These include (1) climate, especially rainfall per storm and per season; (2) amount of water in the material from rainfall and groundwater; (3) kind of bedrock and the presence or absence of bedding planes, joints, or faults; (4) vegetation, for roots can strengthen slopes; (5) amount and kind of weath-

ering, which can weaken rock and produce clay; (6) the steepness of the slope; (7) local relief; and (8) the presence or absence of earthquakes, for quakes can trigger devastating mass movements.

## CLASSIFICATION OF MASS MOVEMENTS

Mass movements are difficult to classify, but it is common practice to subdivide them according to (1) type of movement, (2) type of material involved, and (3) rate of movement (Fig. 11-4). However, the transition from one type of movement to

**Fig. 11-3.** In Laguna Beach, California, the Bluebird Canyon landslide of October 1978 destroyed 23 houses. Its potential for destruction had been predicted nearly 10 years earlier. Movement occurred over an area of 1.5 hectares (3.7 acres) along seams of clay-rich siltstone and sandstone as a result of abnormally high rainfall. The slide, which may have been active in prehistoric times, has now been stabilized. (James P. Krohn, GeoSoils, Inc.)

**FALLS**

Mass travels most of the distance through the air and on impact leaps, bounds, and rolls. Rate of movement is extremely rapid.

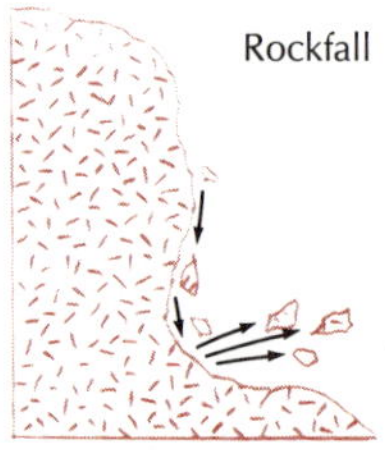

Rockfall

**SLIDES**

May include movement on one or several surfaces or narrow zones. Rate of movement is extremely slow to moderate.

**Slump** movements are along curvilinear surfaces.

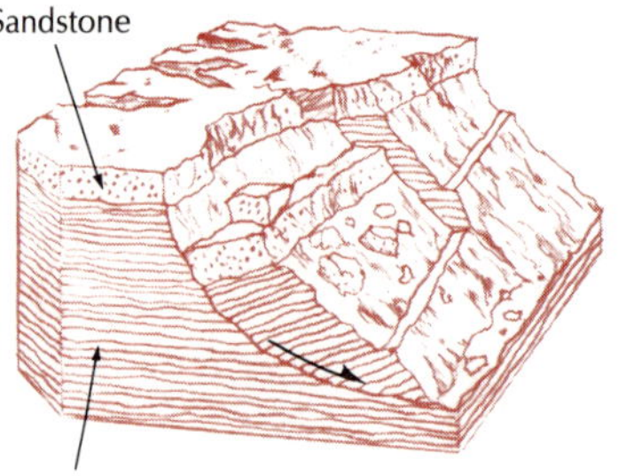

Sandstone

Shale

**Slumps**

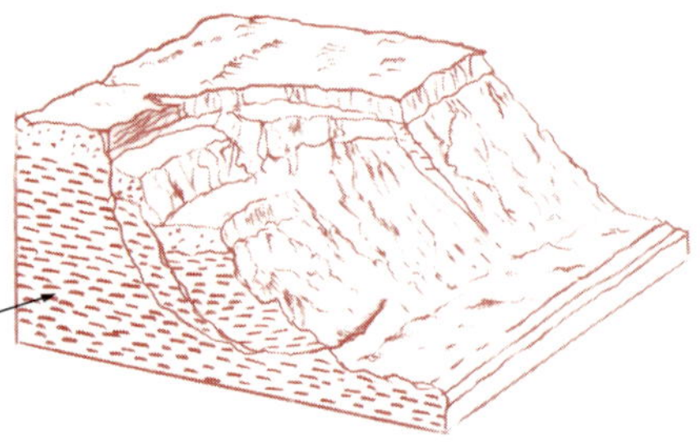

Clay

**Glide** movements are along more or less planar or undulatory surfaces.

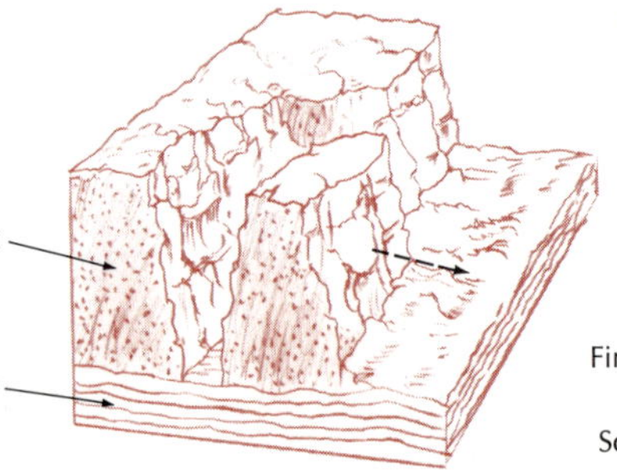

**Glides**

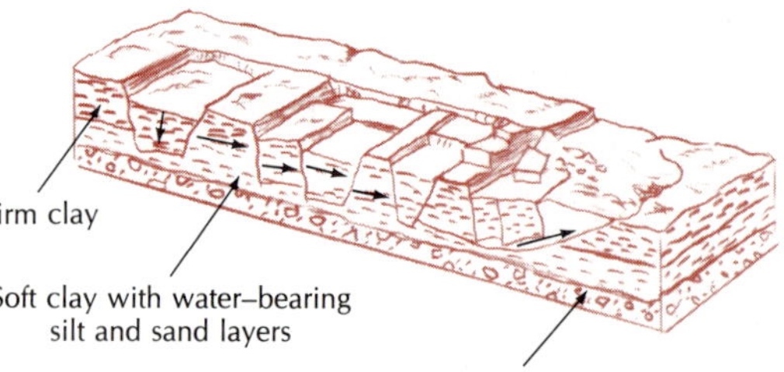

Massive rock

Shale

Firm clay

Soft clay with water–bearing silt and sand layers

Firm clayey gravel

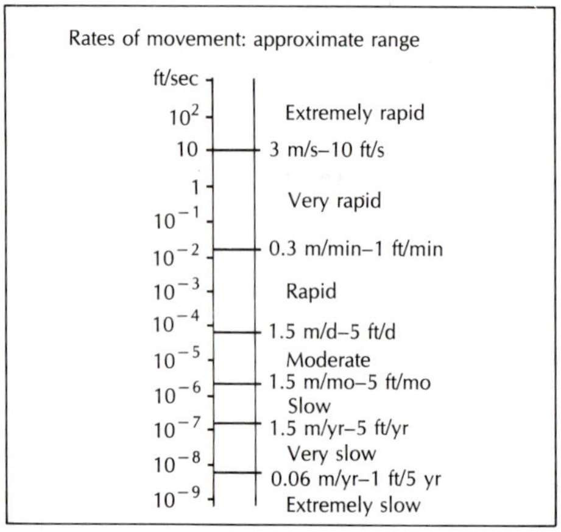

Rates of movement: approximate range

| ft/sec | |
|---|---|
| $10^2$ | Extremely rapid |
| 10 | 3 m/s–10 ft/s |
| 1 | Very rapid |
| $10^{-1}$ | |
| $10^{-2}$ | 0.3 m/min–1 ft/min |
| $10^{-3}$ | Rapid |
| $10^{-4}$ | 1.5 m/d–5 ft/d |
| $10^{-5}$ | Moderate |
| $10^{-6}$ | 1.5 m/mo–5 ft/mo |
| | Slow |
| $10^{-7}$ | 1.5 m/yr–5 ft/yr |
| $10^{-8}$ | Very slow |
| | 0.06 m/yr–1 ft/5 yr |
| $10^{-9}$ | Extremely slow |

**Fig. 11-4.** Classification of mass movements.

TYPE AND RATE OF MOVEMENT

## FLOWS

In bedrock, rate of movement is extremely slow and discontinuous. In soil and unconsolidated sediments, movement resembles that of viscous fluids and can be very rapid (except for creep).

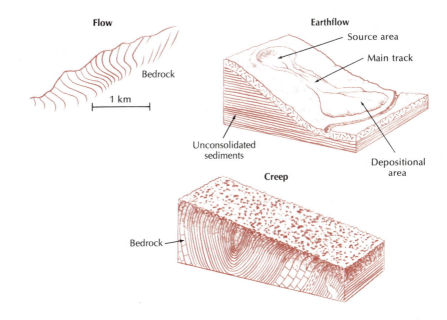

**Flow**

Bedrock

1 km

**Earthflow**

Source area

Main track

Unconsolidated sediments

Depositional area

**Creep**

Bedrock

## COMPLEX MOVEMENTS

A combination of two or more of the above types of movements.

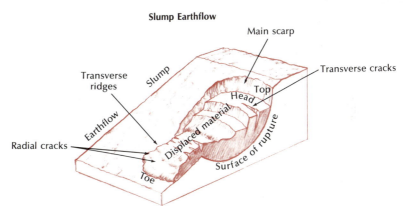

**Slump Earthflow**

Main scarp

Transverse ridges

Slump

Transverse cracks

Top

Head

Earthflow

Radial cracks

Displaced material

Surface of rupture

Toe

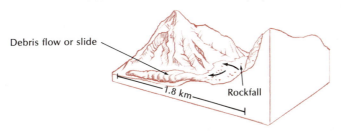

**Rockfall or debris avalanche (extremely rapid rate of movement)**

Debris flow or slide

1.8 km

Rockfall

**Fig. 11-5. A.** *(Left)* Rockfall on the headwall of a cirque in western Colorado. The rock is well jointed, and the frequent rockfalls in the summer may be due to loosening of the blocks by a freeze–thaw process. (Peter W. Birkeland). **B.** *(Right)* Rockfall from the Flimerstein, Switzerland, occurred on April 10, 1939. It buried not only forests and arable land, but also a building and 11 persons. (Swissair-Photo)

another is not always clear cut. ***Landslide*** is a broad term, commonly used for many of these movements.

## Rockfalls

When rock material drops at nearly the velocity of free fall, the event is called a ***rockfall*** (Figs. 11-4 and 11-5). It may range from the plummeting of an individual block to an avalanche of hundreds of thousands of tons.

Should large blocks of rock drop into a standing body of water, such as a lake or fjord, highly destructive waves may be set in motion. Such waves are particularly feared in Norway, where small deltas often are the only flat land at sea level. Should a rockfall-induced wave burst through a village, destruction is likely to be as complete as it is sudden, as the waves often range from 6 to 9 m (20 to 30 ft) in height.

A spectacular wave was set off by a rockfall at the head of Lituya Bay, Alaska, on July 9, 1958. The rockfall, set in motion by an earthquake, began at an altitude of about 900 m (2952 ft) and involved some 31 million m$^3$ (41 million yd$^3$) of rock. Water surged to a maximum height of 530 m (1738 ft) above the level of the bay at its head and moved to the mouth of the bay at about 160 km/hr (100 mph). Don Miller of the USGS has described the experience of a couple who were anchored at locality b (Fig. 11-6):

Mr. and Mrs. Swanson on the *Badger* entered Lituya Bay about 9:00 P.M., first going in as far as Cenotaph Island and then returning to Anchorage Cove on the north shore near the entrance, to anchor in about 4 fathoms of water near the *Sunmore*. Mr. Swanson was wakened by violent vibration of the boat, and noted the time on the clock in the pilot house. A little more than a minute after the shaking was first felt, but probably before the end of the earthquake, Swanson looked toward the head of the bay, past the north end of Cenotaph

Island and saw what he thought to be the Lituya Glacier, which had "risen in the air and moved forward so it was in sight. . . . It seemed to be solid, but was jumping and shaking. . . . Big cakes of ice were falling off the face of it and down into the water." After a little while "the glacier dropped back out of sight and there was a big wall of water going over the point" (the spur southwest of Gilbert Inlet). Swanson next noticed the wave climb up on the south shore near Mudslide Creek. As the wave passed Cenotaph Island it seemed to be about 50 feet high near the center of the bay and to slope up towards the sides. It passed the island about 2 1/2 minutes after it was first sighted, and reached the *Badger* about 1 1/2 minutes later. No lowering or other distur-

**Fig. 11-6.** Lituya Bay, after the giant wave of July 1958 washed over it; the path of forest destruction around the bay is readily seen. (R) marks the rockslide, (D) the maximum altitude [524 m (1720 ft)] of forest destruction, and (B) the location of the *Badger* before it was carried over the spit. (D.J. Miller, USGS)

**Fig. 11-7.** Glide, or translational, slide at Point Fermin, California. Although there are minor slump features at the rear of the slide, most of the movement was along sedimentary strata that dip gently seaward. The glide surface is located just below sea level. The maximum average movement was 3 cm (1.2 in.) per week. (John S. Shelton)

bance of the water around the boat was noticed before the wave arrived.

The *Badger,* still at anchor, was lifted up by the wave and carried across La Chaussee Spit, riding stern first below the crest of the wave, like a surfboard. Swanson looked down on the trees growing on the spit, and believes that he was about 2 boat lengths (more than 80 feet) above their tops. The wave crest broke just outside the spit and the boat hit bottom and foundered some distance from the shore. Looking back 3 to 4 minutes after the boat hit bottom Swanson saw water pouring over the spit, carrying logs and other debris. He does not know whether this was a continuation of

the wave that carried the boat over the spit or a second wave. Mr. and Mrs. Swanson abandoned their boat in a small skiff, and were picked up by another fishing boat about 2 hours later.[*]

## Slides

In a slide movement takes place along one or more surfaces. It may involve the bedrock alone or it may be limited to the overlying soil mantle,

[*]USGS Professional Paper 354–C, pp. 58–59.

especially if the latter is deep and water saturated. Usually, however, it involves both soil and rock, and it can remain more or less intact during movement or it can break apart (Fig. 11-4).

Two main categories of slides are recognized: (1) **glides (translational slides)** and (2) **slumps (rotational slides).** In a glide slippage is dominantly planar, that is, a large mass of rock may become separated from other rocks and glide outward and downward along the surface of an inclined bedding plane (Fig. 11-7). In contrast the motion of a slump is rotational—usually along a concave-upward slip plane—so that the upper part of the landslide is dropped below preexisting ground level and the lower part bulges above it (Fig. 11-8).

Slumps have a very characteristic form. Most of them start abruptly, with a crescent-shaped scarp, or cliff, at their head. Lower down there may be a number of lesser scarps, which on a plan view almost always appear concave downslope. Between the individual scarps the surface of the slide customarily is tilted or rotated backward against the original slope of the ground. The concave-upward slip plane (surface of rupture), which the mass of soil and rocks moves down, may approximate a cross section of a cylinder with an axis that parallels the contour lines of the ground surface if the slide is sufficiently broad. Otherwise, the surface of rupture is likely to be spoon-shaped. The inexorable thrust of the foot of a slump against a building or other structure almost invariably

**Fig. 11-8.** Large slump along the shore of Lake Powell, Arizona. The undisturbed sediments are nearly horizontal, whereas the sediments in the slump block dip downward toward the scarp. The scarp is about 47-m (155-ft) high. (Dennis Netoff)

leads to its collapse, and the foot commonly is responsible for blocking canals, highways, and railroads as well as for engulfing other types of excavations.

Among such slumps, the immense ones that closed the Panama Canal at Gaillard Cut (formerly Culebra Cut) shortly after it was opened in 1914 and that kept it closed more or less continuously until 1920 are impressive examples. Of the 128 million m³ (167 million yd³) excavated in the Gaillard Cut, landslides made necessary the removal of at least 55.5 million m³ (73 million yd³). Great masses of loose, unstable volcanic ash, shale, and sandstone slid on a gently inclined rupture surface toward the canal excavation. One unexpected result was that the bottom of the canal was heaved upward—once as much as 9 m (30 ft)—until what had been the canal bottom appeared as an island in midchannel.

Another rather common kind of slide is the **debris slide** (Fig. 11-4). Here a relatively thin layer of loose material (perhaps the soil) breaks loose and slides over the underlying bedrock surface (Fig. 11-9). Tens of thousands of such slides formed during a 1967 storm near Rio de Janeiro, Brazil (Fig. 11-10). Heavy, pounding rainfall triggered

**Fig. 11-9.** Head of a debris slide scar in the Blue Ridge area of North Georgia. (G.M. Clark)

**Fig. 11-10.** Debris slides that formed during a single storm near Rio de Janeiro, Brazil. Notice the debris flow in the valley. (F.O. Jones, USGS)

the slides, many of which formed mud and debris flows on reaching the valley floors.

## FLOWS

Flows—a very common mass movement phenomenon—differ from slides in that their movement resembles that of a viscous fluid, such as hot tar. It is not unusually for flows in unconsolidated materials to be broken up internally, whereas many slide blocks retain their internal character, such as bedding, intact. However, the boundary between the two is hazy, indeed, as ordinarily a slide grades into a flow downslope. The rates of movement in flows vary from imperceptible to very rapid.

Several kinds of flows are recognized, and all are destructive. Those with the most rapid movement are *mudflows* and *debris flows.* Both are made up of relatively wet, viscous materials, the main difference between the two being that mudflows consist of 80 percent or more of sand or finer particles, whereas debris flows consist of 20 to 80 percent of particles greater than sand size. There is a gradation between these flows, which incorporate water, and water flows, which transport sediment. Features common to both kinds of flows are a bowl-shaped source area; a U-shaped, scoured channel downslope, commonly with levees along the periphery; and despositional lobes at the lower end.

**Fig. 11-11.** Cabin buried to the eaves by a debris flow in 1941 at Wrightwood, California. The flow resulted from rapid snowmelt in the headwaters of the drainage. Cement-like materials surged down the valley some 24 km (15 mi) at velocities averaging close to 3 m/sec (10 ft/sec) in the more fluid parts of the flow. (R.P. Sharp)

Mudflows are an impressive feature of many of the world's deserts. In arid or semiarid lands, normally empty stream courses may fill almost at once with a racing torrent of chocolate-colored mud following a cloudburst. Where arroyos are shallow, the flow may exceed the channel's capacity and spill out over the desert.

Debris flows and mudflows not only are capable of transporting large natural objects, such as house-size boulders, but may trap and sweep along trucks, buses, or even locomotives. Houses inundated by such flows have been buried to the eaves (Fig. 11-11).

Arid or semiarid lands are by no means the only regions where debris flows and mudflows may be seen. They are characteristic of alpine regions, too, and are likely to be exceptionally destructive in areas where a combination of steep slopes, a large volume of water freed by melting snow, and a great mass of loose debris prevails. Geologist Robert R. Curry described such a debris flow in the Colorado mountains in 1961:

Direct observations of the mudflows were hampered by very intense rain and by the fact that the author was about 900 m from the cirque headwall, at the rain gauge, at the time the flows began. At about 4 P.M. on August 18 a loud roar became clearly audible above the thunder. A series of what appeared to be rockfall avalanches were noted in four different localities around the cirque headwall. These appeared confined to areas previously covered with talus cones and, even though the talus had been soaked by 48 hours of intense rain, large rock-dust or water-vapor clouds accompanied the disturbances.

. . . Individual flows occurred as a series of lobate pulsations which, in the case of the largest unit, lasted for 1 hour. The unit was made up of ten more or less distinct flow pulses, each traveling with a maximum surface velocity of 915–980 m/minute near the center of the flow at an elevation of 3750 m. The velocity of the flow pulses dropped to 1 m/minute or less where the flow went out onto the valley floor beyond the base of the talus slope or where a relatively small pulse breached the side of the 0.6–0.8 m high natural levees and slowly flowed out over porous talus. Intervals of 4–15 minutes of relative quiescence elapsed between the flow pulses. Velocity measurements were made by timing the travel of a given flow front between two reference points slightly less than 300 m apart and by analysis of 8 mm motion pictures.

Curry was interested in obtaining samples of the moving material and did so

by forcing wide-necked glass liter-sized jars into the slowly moving side of the flow. The jars were inserted 30–45 cm into the flowing debris about 1 m above the base of the flow. Some bottles were broken and it was generally difficult to force the bottles through the ar-

**Fig. 11-12.** Deposits of talus below bedrock outcrops, Southern Alps, New Zealand. (Peter W. Birkeland)

moring surficial boulders moving up to 150 m/minute, but once beneath the outermost boulders the flow was liquid enough to fill most of the jars. A total of almost 3 liters of matrix of a median diameter of 5 cm or less (a limit imposed by the size of the jars) was collected.*

*R.R. Curry, 1966, "Observation of Alpine Mudflows in the Tenmile Range, Central Colorado," *Geological Society of America Bulletin*, vol. 77, pp. 771–776.

Curry's observations were important to the understanding of these processes. He was also fortunate in having been at the right place at the right time. How many of us, though, would have rushed *to* the mudflow rather than *away* from it on that rainswept August afternoon?

Debris flows and rockfalls commonly combine to form loose rocky ramparts along the bases of cliffs in most mountainous areas (Fig. 11-12). These are called **talus** deposits, from the term used to describe the fortifying slope at the base of a rampart in medieval times.

Mudflows are a common occurrence on the steep slopes of andesitic volcanoes because loose material is abundant, steam can turn the volcanic rocks into a clayey goo, and rainfall or rapid snowmelt can set huge masses in motion. Volcanoes in the Pacific Northwest are no exception. One very real hazard on any of those volcanoes is the rapid melting of snow or ice that would be caused by the eruption of steam or lava. Vast amounts of water would be unleashed, which could pick up surface debris and turn into a mudflow downvalley. More than 55 mudflows have originated at Mount Rainier in the last 10,000 years (Fig. 11-13), and future occurrences are a major threat. Because mudflows can travel so far and with such high speed, it has been recommended that valley floors within 40 km (25 mi) of Rainier should be evacuated in the event of an eruption. Residents of those areas would be hardpressed to get out if a mudflow were already in motion. For example, one such flow in Japan traveled at a velocity of about 90 km/hr (56 mph). In Indonesia volcanic mudflows are called lahars, a term now accepted and used in many parts of the world.

Another associated hazard is that mudflow debris might quickly fill in reservoirs in valleys flanking a mountain, displacing the water and producing downstream floods. An ingenious way of controlling smaller mudflows is to empty the reservoir, trapping the mudflow in it. There is no such hope of containing the large flows, however.

**Earth flows** move slower than debris flows and mudflows. They usually have a spoon-shaped sliding surface with a crescent-shaped cliff at the upper end and a tongue-shaped bulge at the lower end (Figs. 11-4, 11-14).

Some earthflows move quite rapidly, but only under particular circumstances. They involve special clays, aptly called **quick clays.** Quick clay is composed primarily of flakes of clay minerals, and it has a water content that can often exceed 50 percent by weight. It is commonly part of the

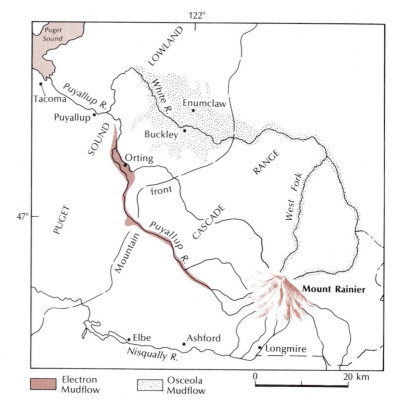

**Fig. 11-13.** Extent of two mudflows that originated on the flanks of Mount Rainier, Washington. The Osceola Mudflow, which is 5800 years old, extended far from the mountain and buried the area now occupied by Enumclaw under 21 m (70 ft) of debris. Five hundred years ago, the Electron Mudflow moved downvalley and deposited 5 m (16 ft) of debris near the present site of Orting. (From D.R. Crandell and D.R. Mullineaux, 1967, "Volcanic Hazards at Mount Rainer, Washington," *U.S. Geological Survey Bulletin 1238*)

**Fig. 11-14.** Slumgullion earthflow in southwestern Colorado. Derived from highly altered volcanic rocks at 3500 m (11,480 ft), the flow has descended to 2500 m (8200 ft), where it has dammed the valley and formed a lake. A younger, active earthflow may be seen advancing over the older, stable one. The rate of movement varies from 6 m (20 ft) per year at the midpoint on the flow to less than 1 m (3 ft) per year at its lower end. (C.W. Cross, USGS)

debris deposited on the seafloor adjacent to glaciers, so it is not surprising that in Norway, Sweden, and parts of eastern Canada several earthflows involving quick clays take place every year. Quick clay has a most amazing and treacherous quality: Ordinarily it is a solid capable of supporting 1 kg/cm$^2$ (14 lb/in.$^2$) of surface, but the slightest jarring motion immediately turns it into a flowing liquid. One hypothesis for this behavior is that when quick-clay layers were originally deposited, usually in saltwater, they contained sodium ions that kept the fine particles together. But when clays are exposed to weathering, the sodium ions are leached out by rainwater and their cohesive effect is lost. Another is that weak attraction forces between grains hold the water-saturated mass together.

Any sudden shock can disturb the mass and produce liquefaction. In one slide in Sweden, the trigger is believed to have been the hammering of a pile driver. The result was that 32.3 million m$^3$ (42 million yd$^3$) of soil and gravel slid into the nearby river, carrying with it 31 houses as well as a paved highway and a railroad it picked up along the way. In less than 3 minutes 1 person was killed, 50 people were injured, and 300 homes were destroyed.

The most damaging quick-clay earthflow on record occurred in 1893 in Norway where a 9-km$^2$ (3.5-mi$^2$) area was wrecked, killing 120 persons. And extensive damage resulted from the 1964 Alaska earthquake in which the city of Anchorage and other coastal towns were especially hard-hit. Much of the damage was due to liquefaction of a quick clay (Fig. 11-15). In that case the liquefying shock was the earthquake.

### Creep

This descriptive word is used for the slow movement downslope of shallow soil material (Fig. 11-4). We are likely to be oblivious of such movement, although we may observe building founda-

tions thrown out of line, power and telephone poles tilted, bowed tree trunks (Fig. 11-16A), and cracked sidewalks and retaining walls. Cuts made in hillsides generally will reveal active creep, commonly shown by bent-rock layering—look carefully for such layers should you ever think of building a hillside house (Fig. 11-16B).

Several mechanisms account for creep. One is subtle expansion and contraction of the ground surface, which results in an imperceptible movement of material downslope (Fig. 11-17). Expansion and contraction can result from a wet–dry cycle in clayey material or from freezing and

**Fig. 11-15.** Turnagin Heights, Anchorage, Alaska, after the 1964 earthquake. During the quake, a bed of quick clay close to the surface liquified and flowed, causing the massive breakup. (U.S. Army)

**Fig. 11-16. A.** Bent or bowed tree trunks are often cited as evidence of active creep. The trees are rooted at some depth in material that is not moving or is only slowly moving downslope. Closer to the surface, however, the soil moves more rapidly, pushing against the trunks and bowing them downslope. The trees respond by maintaining vertical growth in their upper parts—hence the curved trunks. However, snowpack pushing against trees may give the same result. (Peter W. Birkeland) **B.** *(Below)* A common example of creep as shown by bent strata in sedimentary rocks. The effects are more pronounced toward the surface because creep is mainly a near-surface phenomenon. (W.C. Bradley)

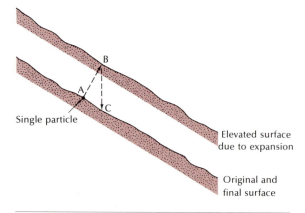

**Fig. 11-17.** The mechanism of creep owing to expansion and contraction. Before an expansion–contraction cycle, particles on the original surface rest at (A). On expansion the surface is elevated and all particles move upward at right angles to the original surface (A→B). On contraction the surface more or less assumes its original position as all particles move vertically downward under the influence of gravity (B→C).

Single particle

Elevated surface due to expansion

Original and final surface

thawing. Another important mechanism is the saturation of the ground with water, such as occurs during rainstorms. Water adds weight to the slope and causes it to lose some of its cohesiveness. Gravity then takes over.

Fault scarps modified mainly by creep show promise in dating fault movements. In parts of the western United States, recent fault scarps are fairly common; they appear as abrupt rock walls, are several meters high, and consist of rock or unconsolidated debris that zigzag across the landscape (Fig. 11-18). Over time, however, creep smooths out the sharp top edges of the scarp, and its slope angle is reduced. Geologists use scarp modification to help determine the age of faults and the frequency of faulting. If the method is carefully applied, these data can then be used to help to assign a hazard risk to various construction sites. This is just another example of applying basic

**Fig. 11-18.** Movement along a fault at the base of the Stillwater Range in Nevada produced this scarp between the tilted cabin and the toppled outhouse. At the time of formation, most fault scarps are steep and the junction between the upper part of the scarp and the land surface is sharp. Erosion or deposition mute the angle and the scarp with time. (Perry Byerly, from A.D. Howard and I. Remson, 1978, *Geology in Environmental Planning*, McGraw-Hill)

research, here on the laying back of a slope with time, to solve a practical problem.

## COMPLEX MOVEMENTS

It has been stressed that there is no sharply defined boundary between types of movements. Where one or the other clearly dominates, the classifications described thus far can be made. Often, however, several types are involved and they are classified as *complex* movements (Fig. 11-4).

Perhaps the most common complex movement is the combination of slump and earthflow. Here the upper part has all the attributes of a classical slump block, but beyond the toe of surface rupture flowage takes place (Fig. 11-19).

Some large rockfalls grade downvalley into very rapidly moving debris flows, the result being stupendous mass movements called *rockfall (debris)*

**Fig. 11-19.** Complex slump/earthflow blocking Highway 24 near Oakland, California. (USGS)

*avalanches,* depending on the predominant material involved. Such events start as rockfalls high on a mountain face, plunging down slopes and sweeping far across valleys with a velocity of 160 km/hr (100 mph) or more.

Several such avalanches have occurred in historic times in the western United States. At the base of Mount Shasta, an active volcano in California (Fig. 11-20) is evidence of the largest known landslide on earth in the last 2 million years. This huge debris avalanche slopes to the northwest for about 43 km (27 mi) on slopes as gentle as 5m/km (26 ft/mi). An area of at least 450 km$^2$ (176 mi$^2$) is covered by the ancient debris; the volume is about 26 km$^3$ (6.2 mi$^3$). Radiometric dating puts its age at between 300,000 and 360,000 years, meaning that the avalanche came from the slopes of an ancestral Mount Shasta, not the present volcano. The cause of the debris avalanche is unknown, but they do occur most often on steep-sided andesitic volcanoes, triggered either by volcanic activity or by earthquakes.

The most tragic modern avalanche was related to the 1970 earthquake in Peru. About 15 km (9 mi) east of Yungay, a town of 20,000 inhabitants, rise the lofty Peruvian Andes and the snow-covered peak of Nevado Huascarán, which soars to 6663 m (21,855 ft). Ground motion from the quake shook the mountain and broke loose a huge block of ice, snow, and rock from the top of the precipitous north face (Fig. 11-21). The mass fell in free-flight, landing 1000 m (3280 ft) below, where it knocked loose a large volume of rock and burst into thousands of smaller fragments. Boulders— one weighting 65 tons—were airborne for as much as 4 km (2.5 mi), and some of the impact velocities exceeded 1000 km/hr (620 mi/hr). All the debris continued to cascade down the north face. Frictional heating of the ice during impact caused some melting, so that the entire mass was transformed into a rapidly moving debris flow. It raced along previously established avalanche and stream channels; by the time it had moved from the mountains, it contained an estimated 50 to 100 million m$^3$ (65 to 130 million yd$^3$) of water, mud, and rocks. Its speed has been estimated at about 210 to 280 km/hr (175 to 210 mph). Just upstream from Yungay, the deep valley bends sharply. Most of the surging mass of mud and boulders made

**Fig. 11-20.** View about 40 km (25 mi) southward to Mount Shasta, California, across the largest landslide deposit on earth over the last 2 million years. The slide came from a volcano ancestral to the present Mount Shasta. Intact blocks of the old volcano underlie many of the mounds, and individual blocks range in size from tens to hundreds of meters in maximum dimension. (D.R. Crandell, USGS)

the turn. But some rode up the valley wall and over the ridge, within seconds the entire city was buried. About 18,000 residents were interred almost instantly. The main mass of the mudflow continued downvalley, where it buried another town—Ranrahirca (1800 people)—and extensively damaged road and rail routes, power and communication lines, and a hydroelectric plant. Eventually it reached the bottom of the valley and the Río Santa, where its momentum carried it across the river as far as 60 m (197 ft) up the opposite bank. It finally came to rest far downstream. The total vertical fall was almost 4000 m (13,120 ft).

A Peruvian geophysicist gives this eyewitness account of the event:

As we drove past the cemetery the car began to shake. It was not until I had stopped the car that I realized that we were experiencing an earthquake. We immediately got out of the car and observed the effects of the earthquake around us. I saw several homes as well as a small bridge crossing a creek near Cemetery Hill collapse. It was, I suppose, after about one-half to three quarters of a minute when the earthquake shaking began to subside. At that time I heard a great roar coming from Huascarán. Looking up, I saw what appeared to be a cloud of dust and it looked as though a large mass of rock and ice was breaking loose from the north peak. My immediate reaction was to run for the high ground of Cemetery Hill, situated about 150 to 200 m away [Fig. 11-22]. I began running and noticed that there were others in Yungay who were also running toward Cemetery Hill. About half to three-quarters of the way up the hill, the wife of my friend stumbled and fell and I turned to help her back to her feet.

The crest of the wave had a curl, like a huge breaker coming in from the ocean. I estimated the wave to be at least 80 m high. I observed hundreds of people in

Yungay running in all directions and many of them towards Cemetery Hill. All the while, there was a continuous loud roar and rumble. I reached the upper level of the cemetery near the top just as the debris flow struck the base of the hill and I was probably only 10 seconds ahead of it.

At about the same time, I saw a man just a few meters down hill who was carrying two children toward the hilltop. The debris flow caught him and he threw the two children towards the hilltop, out of the path of the flow, to safety, although the debris flow swept him down the valley, never to be seen again. I also remember two women who were no more than a few meters behind me and I never did see them again. Looking around, I counted 92 persons who had also saved themselves by running to the top of the hill. It was the most horrible

thing I have ever experienced and I will never forget it.*

One of the authors visited the site in 1986. The new town of Yungay is located north of the cemetery outside of the 1970 deposit, but Ranrahirca has not been rebuilt. People now farm the land below the cemetery and along the ridge that was overtopped (B and C in Fig. 11-21).

How do we account for such awesome velocities and such long distances of transport over relatively flat surfaces, as evidenced by the events just described? Some believe that the high veloc-

*B.A. Bolt et al., 1977, *Geological Hazards,* Springer-Verlag, N.Y., pp. 37–39.

**Fig. 11-21.** Area affected by the rock and debris avalanche of May 1970 that originated on the north peak of Nevado Huascarán (A). Before descending on Yungay, the avalanche rode over a ridge 180 to 240 m (590 to790 ft) high (B). The only place spared in Yungay was the cemetery (C). Debris reached the banks of the Río Santa (D) 14.5 km (9 mi) from the source in about 4 minutes. Part of it flowed more than 2 km (1.2 mi) upstream (E), but most of it flowed downstream toward the Pacific, causing more destruction in the flood plain of the river. (After R.L. Schuster and R.J. Krizek, eds., 1978, *Landslides,* National Academy of Sciences Special Report 176, p. 22 *or* Fig. 2-27)

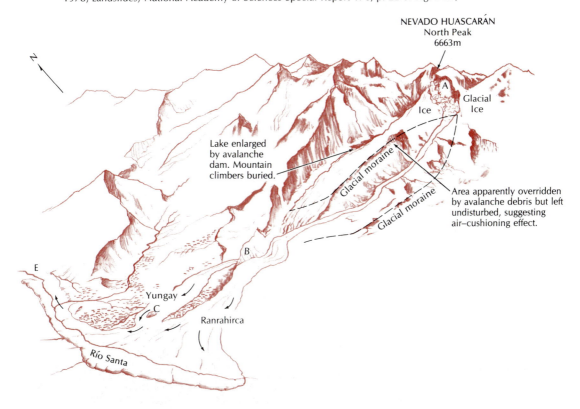

**Fig. 11-22.** The statue of Christ at Cemetery Hill was all that remained of Yungay. (USGS)

ities are due to air entrapped and compressed beneath the falling mass of debris. Material is temporarily buoyed up in much the same manner that air temporarily buoys up a sheet of plywood dropped onto a flat surface. When the falling debris is pitched into the air, a compressed air cushion forms beneath it, and some geologists have noted the occurrence of such geological "ski-jumps" in the field. The eventual loss of the air cushion ends this remarkable transport mechanism. Another theory is that countless semielastic collisions of blocks with each other may keep the debris acting as a fluid, causing it to move great distances beyond the cliffs and slopes from which it was derived.

## CONDITIONS FAVORING LANDSLIDES

Most slides and flows are the result of the forces of gravity acting upon earth materials in an un-

stable condition or position. Although the movement itself may be extremely rapid or imperceptibly slow, a landslide usually does not suddenly spring into being but rather develops gradually over time.

Engineers and engineering geologists have a way of predicting the stability of a slope. They calculate a safety factor (N) according to the following:

$$N = \frac{\text{resisting forces}}{\text{driving forces}}$$

The forces are calculated for the potential slide surface, and the equations are complex. An N value less than or equal to 1 denotes a slope likely to fail; between 1 and 1.25, a slope that can be made safe if remedial measures are taken; and greater than 1.5, a safe slope.

Examples of common *resisting forces* are changes in pore-water pressure and the orientation of geological structures. Water has been called the "hidden devil" in the ground—the cause of many landslides. When more water enters the ground, the water pressure in the pores between grains increases, the grains are floated farther apart, and the cohesion of the material is weakened.

Prolonged or intense rainfall most commonly adds water to the ground, but a simple lawn watering can also create unstable conditions. Clearcutting forests on slopes can have a similar effect. Trees might be thought of as water pumps, returning the water to the atmosphere by transpiration. Remove the pumps, however, and the slopes probably will retain more moisture—an invitation to instability. An artificial reservoir can also be the cause of a landslide; if enough rising water soaks into the banks, they lose much of their strength. Of course, a natural cause, such as climatic change involving greater precipitation, can also set landslides in motion. We should be aware of both the natural and artificial causes of such events; if personal property is damaged, the case might easily find its way to court. Finding the real culprit, however, is no easy task.

Within a rock mass there are often planar surfaces along which sliding can take place. The bedding planes of sedimentary rocks, cleavage or foliation of metamorphic rocks, and joints within igneous rocks are examples. If any of these planes

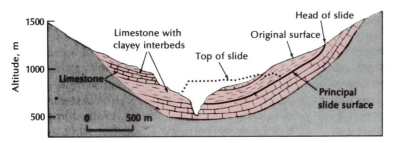

**Fig. 11-23.** Cross section of the valley in which the Vaiont Reservoir is situated, showing the geological setting and the pre- and postlandslide topography. (After G.A. Kiersch, "The Vaiont Reservoir Disaster," *Mineral Information Service*, vol. 18, no. 7)

slant toward a valley wall or an artificial cut, movement along them can take place. In contrast, if the planes slant into a hillside or an artificial cut, the chances of sliding usually are minimal.

Another important resisting force are tree roots, for they increase the strength of shallow slope materials severalfold. One study in the Cincinnati, Ohio, area suggested that 30° slopes under tree vegetation were stable and that slopes of about half that steepness could slide if trees were removed.

Common **driving forces** in landslides are the steepening of a slope, the removal of lateral support, and the addition of mass. A slope can be steepened by natural causes, such as the lateral erosion of rivers or the sea or the rapid vertical erosion by rivers. The material making up the upper part of the slope is held in place by the weight of the material at the base of the slope; if this anchor is removed, the slope is likely to fail.

Highway engineers are particularly sensitive to the problems of oversteepened slopes. In making highway cuts, they are called on to create slopes with a sharper pitch than the original ones—rather perilous business, for the slope must not only be steep to avoid costly removal of material, but also stable. Improperly designed highway fills can also fail, as can cuts in hillslopes for house foundations.

By adding mass to a slope, we mean loading the upper parts of the slope with new buildings or piles of artificial fill. The introduction of water also adds considerable mass to a slope.

Earthquakes, of course, also are a major driving force, for rare is the large earthquake that does not set some rocks in motion. In fact, most of the great catastrophies associated with landslides are related to earthquakes. It follows that the greater the magnitude of the event, the greater the possibility of landslides. Earthquakes of less than magnitude 4 commonly do not result in landslides; one of magnitude 9.2 could set off landslides over an area of some 500,000 km² (195,000 mi²). The correlation of earthquake magnitude and type of mass movement is complex, but it is clear that major rock and debris avalanches are associated with larger-magnitude earthquakes. It should come as no surprise that the greatest loss of life is also during earthquake-related landslides. The devastating avalanche in Peru took the lives of almost 20,000 people but the greatest loss is put at about 1 million lives during an earthquake in China in 1556 (see chap. 5).

With these forces in mind, what then caused the Vaiont Reservoir disaster (Fig. 11-2)? Well, about everything went wrong that could. In the first place, the valley is steep-sided because of a relatively recent river downcutting (Fig. 11-23). Engineers commonly seek such narrow gorges for dam sites because less concrete is needed to plug the valley. Second, at Vaiont the bedrock is made up of limestone and layers of slippery clay, and the bedding planes are in the worst possible configuration—toward the valley axis. Further, the limestone itself is rife with underground solution caverns that serve as collection basins for water that helps to saturate the ground. The valley has also been the site of other landslides; witness the ancient one near Casso and the 1960 slide into the reservoir along its south side. Ultimately, it was excess water introduced into an already unstable system that triggered the slide. Some seeped in laterally from the reservoir and some was contributed by the downpours. Creep in the slide area had increased from about 1 cm/wk (0.4 in./wk), the average rate after the reservoir was built, to 80 cm/day (31 in./day) on the night of the failure.

Even animals grazing on the slopes sensed the danger and moved away in time. What could have been done to prevent the disaster? The only thing that comes to mind is that the geological setting of the area could have been studied more carefully. Certainly, sites that in a geological sense have so many things wrong with them should be avoided. No amount of engineering could have saved the reservoir, at least within acceptable costs.

## STABILIZING SLIDES

Geologists and engineers have devised many ingenious methods of controlling unstable slopes, and perhaps not enough people are aware that it is easier, cheaper, and safer to apply the methods of stabilization before, not after, mass movements take place.

One common way of stabilizing a landslide area is to regrade it to a low angle—one that is stable under the newly imposed conditions, natural or artificial. At the same time the amount of water in the slope should be reduced drastically, either by surface or subsurface drainage or by preventing its entry in the first place, that is, covering the surface of the slide with impermeable material. An example of slope treatment and drainage is provided by the history of the large slides—some covering as much as 64 hectares (158 acres)—that interrupt the hilly terrain of the Ventura Avenue oilfield in southern California. In the rainy winter of 1940–41, one 24-hectare (59-acre) block slid as a single unit for a distance of about 30 m (100 ft). Because the slides sheared oil wells as they moved [in the 1940–41 episode 23 wells were cut off as deep as 30 m (100 ft) below the ground surface], unusually extensive and expensive efforts to curb the movement were made later. Partial success was achieved by spreading tar over the surface and by drilling horizontal drainage holes into the slides—64 km (40 mi) of them. Vertical wells were also drilled through the slides to a porous layer of sandstone, which served as a conduit to carry water away from the slides and into adjacent solid ground. There the excess flow could be pumped out.

Large corporations can afford such expensive remedial measures, but most homeowners cannot. One recourse a homeowner can take is to cover the potential slide area with plastic sheets—a tactic that works, but one that hardly enhances the beauty of the homesite. Planting trees is another solution. The roots add strength to surface materials and foliage transpires a considerable amount of water from the soil zone.

Another common stabilizing method is to strengthen the toe of small slides. Steel or concrete walls are built to hold back the force of the slide, or, in places where bedrock is easily available, the toe is weighted down with large blocks of rock.

The best preventative measure, however, is foresight. Geologists are now able to examine terrain and assess with some degree of accuracy landslide probability. Maps of some urban areas show landslides and the approximate time they last moved are also available to planners and builders. A landslide-potential map has been compiled by the USGS for the coterminous United States. Because it is a small-scale map, it can serve only as a warning to regions rather than to specific sites.

All too often it takes a disaster to move legislators to appropriate funds for research on hazardous environments. Such was the case in Utah when debris flows wreaked havoc on steep mountain slopes during the storms in the spring of 1983. Only after the fact were attempts made to predict the areas of future debris flows. Funding was obtained from both national and state agencies.

## PERMAFROST

Mass wasting is particularly active in high northern and southern latitudes and at high altitudes. The reason is that close beneath the surface the ground is frozen solid; in places, animals have been preserved in near-perfect condition for many thousands of years. We now examine frozen ground, some of the landscape features associated with it, and a few of the engineering problems that result from building on such a precarious surface.

Ground that remains frozen from one year to the next was first called **permafrost** during World War II. Permafrost is defined only on the basis of temperature, it being ground that remains below 0° C (32° F). Hence, the proportion of ice in permafrost can vary from small to large. A knowledge of how much ice is present and where it is located is extremely important, as we will see, to engineers and builders working on such terrain.

Permafrost, which is more widely distributed than many people realize, underlies almost 20 percent of the earth's land surface, including about 85 percent of the land area of Canada and the USSR (Fig. 11-24). It reaches its maximum thickness around the margins of the Arctic Ocean in Alaska, Canada, and the Soviet Arctic. Siberia holds the record for maximum thickness: 1500 m (4920 ft). Sites of maximum thickness in Alaska

and Canada are about half that amount. In a general way the thickness decreases southward, thinning to nearly zero at about the southern boundary shown on the map.

Two major regions of permafrost are recognized (Fig. 11-24). In the north the distribution is continuous, whereas to the south it occurs only in patches. It is hard to predict where those patches are located, as a result land-use planning in the south can be more difficult than in the north. In general, however, the −6° C isotherm coincides with the

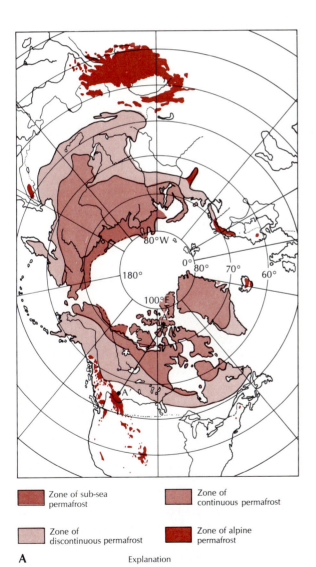

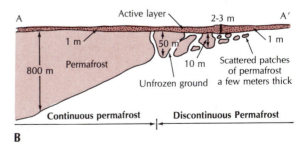

**Fig. 11-24. A.** Extent of permafrost in the Northern Hemisphere. (After Péwé, 1983, Arctic and Alpine Research, vol. 15, p. 145) **B.** Cross section of permafrost along the line A–A'. (From R.J.E. Brown, 1970, *Permafrost in Canada*, University of Toronto Press, Fig. 4. By permission of the University of Toronto Press)

Zone of sub-sea permafrost

Zone of continuous permafrost

Zone of discontinuous permafrost

Zone of alpine permafrost

**A**        Explanation

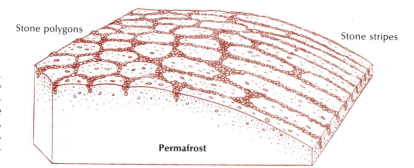

**Fig. 11-25.** Patterned ground comes in many shapes and sizes. On level terrain stone polygons form, whereas on sloping terrain, where creep occurs, the polygons are stretched out, forming stone stripes. (After C.F.S. Sharpe, 1938, *Landslides and Related Phenomena*, Columbia University Press, Fig. 5)

**Fig. 11-26.** Sorted circles formed in old, raised beach material in the periglacial environment of Spitsbergen, Norway. The circles (dark) consist of fine-grained material; the border materials (light) consist of coarse-grained material. (J.L. Sollid)

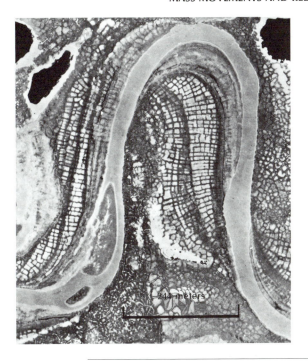 

**Fig. 11-27. A.** *(Left)* Patterned ground along a river southeast of Barrow, Alaska. The pattern of interconnecting ice wedges is formed in extremely cold climates when ice-cemented permafrost cracks. The wedges are widest at the top and narrow downward. Evidence for former ice wedges may be seen in many countries just south of the Pleistocene glacial boundary. (O.J. Ferrians, Jr., USGS) **B.** *(Right)* Foliated ice wedge in late Quaternary silt along the Aldan River, central Yakutia, USSR. (T.L. Péwé)

southern boundary of continuous permafrost, and the −1° C isotherm with the southern limit of discontinuous permafrost. Permafrost also occurs in high mountains south of the main permafrost areas (Fig. 11-24). It has been reported, for example, on the summit of Mount Washington in New Hampshire, in Colorado's Rocky Mountains, and even on Mauna Kea, Hawaii.

Overlying the permafrost is a thin layer of soil in which **ground ice** (subsurface ice) thaws in the spring and freezes in the fall, remaining frozen throughout the winter. That soil is the active layer, and it varies from less than 1 to about 3 m (3 to 10 ft) in thickness. The upper surface of permafrost is called the **permafrost table;** below that the pore spaces are filled with ice. Hence water in the active layer cannot sink underground—one reason why much of the Arctic tundra is so boggy and water soaked. Precipitation over large segments of the Arctic is very slight, although the many lakes and muskegs seem to belie that fact. Water re-

mains at the surface because it cannot sink underground. Further, the evaporation rate is relatively low.

### Patterned Ground and Solifluction

A bizarre, but common manifestation of ice-churned ground in permafrost areas is a curiously regular patterned surface (Fig. 11-25). From the air the ground resembles a gigantic tiled floor (Fig. 11-26). Some geometrically shaped polygonal areas are thought to be the result of **frost heave,** which is much more prominent in fine-grained soils than in coarse. When frost heaving takes place year after year in a soil of mixed composition, coarse materials, such as boulders and gravel, are gradually shoved radially outward from the central area and finer materials remain behind and become concentrated. Other patterned ground results from a network of vertical ice wedges (Fig. 11-27).

*Gelifluction* is an extreme type of creep of water-saturated surface material that develops in the presence of permafrost. Hilly terrain underlain by permafrost exemplifies it best. As we have noted, surface water cannot sink into permafrost, so that water that would normally percolate far beneath the surface is concentrated in the active layer. This active layer, being saturated, is far more susceptible to creep than similar terrain with lower water content would be.

Active gelifluction produces a landscape that bears some resemblance to the wrinkled hide of an aged elephant. Different parts of the water-saturated surface layer creep downslope at different rates, so that hillsides where these processes are active are festooned with soil lobes, or tongues, some of which advance rapidly and some slowly (Fig. 11-28). Maximum rates of lobe movement approach 40 m/1000 yr (131 ft/1000 yr). A curious aspect of gelifluction is that it tends to produce a rounded, smooth terrain that stands in strong contrast to the rugged terrain of glaciated valleys.

A unique landform found in some permafrost terrain is the **pingo,** from the Eskimo word for "conical hill" (Fig. 11-29). Pingos are near-circular, turf-covered ice mounds, some as high as 70 m (230 ft) and 600 m (1970 ft) in diameter. The core of the mound is nearly pure ice, formed during the freezing of water-rich sediment.

Some permafrost features were formed in the geological past and, if correctly identified, can serve as indicators to temperatures during the glaciations. Mean annual air temperatures no higher than the following are associated with these features: 0° C (32° F) for certain forms of patterned ground, −5° C for ice-wedge polygons, and −2° to −6° C for pingos. During the last glaciation, permafrost may have spread over twice the area that it does today.

Another unusual feature found in permafrost

**Fig. 11-28.** Gelifluction lobes on a 14° slope in the Ruby Mountains, Yukon Territory. (Larry W. Price)

endless, but the degree to which Arctic pioneers have overcome them is a testimonial to their ingenuity and perseverance. Even such a simple thing as developing a water supply in a permafrost area can become a major frustration. Groundwater in the active layer is available only during the summer and is usually at so shallow a depth that it is readily contaminated by surface wastes. There may be groundwater below the permafrost, but it is deep, and well sections drilled through the frozen ground are almost certain to freeze. Delivery of water also poses a problem. If water pipes are buried underground, they freeze; if placed above ground, they freeze, too, and are likely to be thrown out of line as the ground heaves when it freezes and sinks when it thaws. Expensive insulation is the only solution.

One solution to water supply in permafrost areas is to build dams and collect the summer snowmelt for year-round use. Yet climatic conditions can create difficulties for dams and reservoirs. The Russians, for example, ran into trouble when a dam in the far north started to leak shortly after the reservoir behind it was filled. The dam was built on volcanic rock with tiny cracks that were permanently filled with veinlets of ice. Ordinarily, ice-filled rock below the permafrost level can be treated as solid rock. In the Russian case, however, the filled reservoir with its insulating layer of ice on the surface acted as a heat trap. The ice veinlets melted and the bottom of the dam became virtually a sieve. Newer dams built in similar cold areas are refrigerated by pumping cold air into them to prevent such melting.

Sewage disposal is perhaps the ultimate problem. Septic tanks and leach fields freeze, and in the absence of bacteria, waste does not decay and disappear as it does in warmer climates. At Point Barrow in Alaska one unsightly (but practical) solution was to heap everything atop an ice floe during the winter. In the summer the ice cake floated out into the Arctic Ocean, melted, and the waste sank. Waste disposal is even more of a problem now that oil has been discovered in northern Alaska.

Permafrost and the discovery of oil in the North Slope of Alaska also led to a heated controversy over the Alaska pipeline, which extends from the petroleum fields southward to the ice-free port of Valdez, a distance of about 1300 km (808 mi). The oil flowing through the pipeline, which is more than 1 m (3 ft) in diameter, has a temperature of 70 to 80° C (158° to 176° F). It was argued that if the pipe were buried, difficult problems would arise from melting of the permafrost. It was calculated, for example, that within the first decade after burial a cylindrical area 6 to 9 m (20 to 30 ft) in diameter would be thawed around the pipe. In successive decades the thawing would continue, but at a diminishing rate.

The major construction problem was the condition of the permafrost before thawing. If it were dry, thawing would have only a slight effect. However, permafrost is in large part composed of fine-grained sediments with a high ice content, and thawing of the material forms a water-saturated slurry into which the pipe could sink. If the pipeline were on a slope, the slurry could flow out onto the landscape, allowing the pipe to settle even deeper—into still-frozen ground—and melting would continue. Stress caused by such processes could cause the pipe to rupture, leading to oil spills that would rival in seriousness those that occur at sea. It was vital, therefore, to identify all potential problems before the pipeline was constructed.

After detailed investigations were made, it was found that the pipeline could be placed below ground for about one-half its length without serious consequences. Conventional burial procedures were used. Because of the problems associated with the melting of the permafrost and to insulate it against the arctic cold, the other half of the pipeline was built above ground on platforms about 15 to 21 m (50 to 70 ft) apart (Fig. 11-33). Areas disturbed during the construction were revegetated, returning the environment to a stable condition in which the permafrost, protected by the vegetative cover, is prevented from melting.

# SUMMARY

1. Gravity operates on all sloping ground, so that material moves to lower positions on the slope.
2. The transfer rate of material on a slope varies from millimeters per year to meters per second. Similarly, the amount of material involved in the transfer varies from small amounts of material to entire mountainsides.
3. *Slides* and *flows* are the mass movements that most often destroy property and dwellings. They can occur naturally or be produced by human activities. Once set in motion, they are difficult to stop. Slopes prone to slides or flows should not be used as building sites.

4. A rock avalanche is the most spectacular kind of mass movement because so much material is transferred downslope so quickly.
5. A safety factor can be calculated for many slopes by taking into account the resisting and driving forces. The resulting number is indicative of the stability of the slope.
6. Mass movement takes place at a fairly rapid rate in areas of permafrost. Such terrain is extremely sensitive to human manipulation; any engineering projects on it must be undertaken with great care.

# QUESTIONS

1. Discuss the role of water, rainstorms, and mean annual rainfall in initiating slides and flows.
2. What kinds of mass movements can be triggered by earthquakes? Discuss the processes involved in such movements.
3. What are the geological factors involved in slides and flows in an area of tightly folded, interlayered sandstones and shales?

4. Are mudflows always an arid region phenomenon?
5. Discuss the geological environmental problems of clearcutting large tracts of rain forest on steep slopes with lateritic soils. A dense network of dirt roads would be used to transport the logs.
6. Discuss the problems of building houses and roads in permafrost terrain underlain by (1) granite and (2) interbedded limestone and shale.

# SELECTED REFERENCES

Costa, J.E., and Baker, V.R., 1981, Surficial geology, Wiley, New York.

Crandell, D.R., and Mullineaux, D.R., 1967, Volcanic hazards at Mount Rainier, Washington, U.S. Geological Survey Bulletin 1238.

Crandell, D.R., et al., 1984, Catastrophic debris avalanche from ancestral Mount Shasta volcano, California, Geology, vol. 12, pp. 143–46.

Ericksen, G.E., and Plafker, G., 1970, Preliminary report on the geologic events associated with the May 31, 1970, Peru earthquake, U.S. Geological Survey Circular 639.

Ferrians, O.J., Jr., Kachadoorian, R., and Greene, G.W., 1969, Permafrost and related engineering problems in Alaska, U.S. Geological Survey Professional Paper 678.

Fleming, R.W., and Taylor, F.A., 1980, Estimating the costs of landslide damage in the United States, U.S. Geological Survey Circular 832.

Hays, W.W., ed., 1981, Facing geological and hydrologic hazards, U.S. Geological Survey Professional Paper 1240–B.

Kiersch, G.A., 1965, The Vaiont Reservoir disaster, Mineral Information Service, vol. 18, no. 7, California Division of Mines and Geology, Sacramento.

Lachenbruch, A.H., 1970, Some estimates of the thermal effects of a heated pipeline in permafrost, U.S. Geological Survey Circular 632.

Péwé, T.L., 1983, The periglacial environment in North America during Wisconsin time, *in* Late-Quaternary environments of the United States, vol. 1, The late Pleistocene, S.C. Porter, ed., University of Minnesota Press, Minneapolis, Minn.

Price, L.W., 1981, Mountains and man: University of California Press, Berkeley.

Ritter, D.F., 1986, Process geomorphology, W.C. Brown, Dubuque, Iowa.

Robinson, G.D., and Spieker, A.M., 1978, "Nature to be commanded . . .," Earth science maps applied to land and water management, U.S. Geological Survey Professional Paper 950.

Schuster, R.L., and Krizek, R.J., eds., 1978, Landslides, National Academy of Sciences, Washington, D.C.

Selby, M.J., 1985, Earth's changing surface, Oxford University Press, New York.

Washburn, A.L., 1980, Geocryology, Wiley, New York.

# 12

# RIVER SYSTEMS AND LANDFORM EVOLUTION

Few natural phenomena are more intimately involved with human affairs than rivers. In past centuries, the Nile, Tigris, and Euphrates were literally givers of life as they threaded their way across a desert land. Ancient civilization depended on these sources for irrigation, a communal enterprise that contributed to the development of modern urbanized society. Mathematics, surveying, and hydraulics began with the designing of dams and canals. One of the earliest projects was a long dike built about 3200 B.C. on the west bank of the Nile, with cross dikes and canals, to carry floodwaters into basins adjacent to the river.

Rivers have long played a role as natural barriers (e.g., of decisive importance in Roman times were the Rhine and the Danube) and served as routes from the sea to the interior—explorers have traditionally followed rivers. Most of the world's leading cities are built on riverbanks. And rivers are identified indissolubly with the history and national aspirations of almost all the lands that border them. It would be difficult to conceive of China without the Chang Jiang (formerly Yangtze River), Brazil without the Amazon, or India without the Ganges.

To devout Hindus, the Ganges is sacred. The faithful follow its course in pilgrimage to the headwaters at Gangotri before they die (Fig. 12-1). Their ashes are cast into its waters at burial. But the river can also be the scene of devastation during the monsoon season, flooding the great cities of Calcutta and Dacca in its lower reaches.

In this chapter we first discuss stream flow—how water flows down a valley—and then relate it to the geological work that rivers perform, most of which takes place during major flooding. We then present ways of deciphering river history and the role of rivers in landscape evolution. Throughout the chapter the environmental problems involved in manipulating rivers are underscored. We have become aware of such problems through mistakes, and only by recognizing why we erred can we avoid future problems.

## THE HYDROLOGIC CYCLE

Before discussing stream flow, we should consider the amount of water on earth and how it moves

**Fig. 12-1.** The Ganges River in the Himalaya, near the town of Tehri. (Peter W. Birkeland)

347

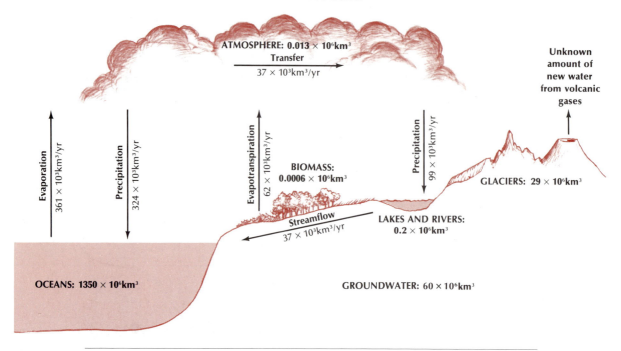

**Fig. 12-2.** The amounts of water contained in various natural reservoirs and the amounts transferred annually from one reservoir to another. (After A.L. Bloom, 1969, *The Surface of the Earth,* Prentice–Hall)

from place to place (Fig. 12-2). There are several water reservoirs—the oceans, the rivers and lakes, the glaciers, voids in underground rocks, the atmosphere, and the biomass. Of them, the oceans contain by far the most water. Although the volume content of these natural reservoirs has changed, as during major glaciations when large amounts of ocean water were transferred to glaciers, most geologists believe that the total amount of water on earth has been more or less constant for the past billion years. New water may have been added to the earth's surface by steam condensation during volcanic eruptions, but the amount is insignificant overall.

Water is transferred from reservoir to reservoir each year, but the entire budget is in balance. The cycling process is known as the **hydrologic cycle.** The water that is annually evaporated from the ocean [approximately equivalent to 1 m (3 ft) in depth] exceeds precipitation over the oceans, and the excess is transferred to land via the atmosphere. On land the opposite occurs—that is, more

water reaches the ground as precipitation than leaves through **evapotranspiration.** The latter term encompasses water losses through evaporation as well as through transpiration from plants and animals. In fact, in well-vegetated areas, the main loss of water to the atmosphere is through transpiration from plants. The cycle is completed and balanced as stream flow removes the excess from the land.

## STREAM FLOW

Stream flow performs work that continually changes the shape of the land. The geologist Arthur L. Bloom expresses the amount of available stream energy in a down-to-earth way:

The average continental height is 823 m above sea level. If we assume that the 37,000 cubic kilometers of annual runoff flow downhill an average of 823 m, the potential mechanical power of the system can be calculated. Potentially, the runoff from all lands would continuously generate over 12 billion horsepower. If all this power were used to erode the land, it would be comparable to having one horse-drawn scraper or scoop at work on each three-acre piece of land, day and night, year around. Imagine the work that would be accomplished! Of course, a large part of the potential energy of the runoff is wasted as frictional heat by the turbulent flow and splashing of water, but we will see that the "geomorphology machine" is really quite efficient, and in fact does erode and transport rock debris down to the sea almos᛫ as fast as if horse-drawn scrapers were hard at work on every small plot of land, over all the Earth.*

Great impetus has been given to the study of stream flow because of its importance in the design of hydroelectric plants, dams, spillways, and increasingly complex irrigation systems. Every industrialized nation is actively engaged in research into the nature of stream flow (Fig. 12-3). The largest laboratory in the United States is the U.S. Waterways Experiment Station, operated by the U.S. Army Corps of Engineers at Vicksburg, Mississippi. There, elaborate models of the Mississippi have been constructed and an immense amount of data collected and analyzed to find ways to bring that unruly river and its tributaries under control.

As anybody who has rafted a wild river will testify, water flow is mainly turbulent (Fig. 12-4). In turbulent flow the water particles go every which way. Sometimes the particles swirl upward like autumn leaves or like dust devils in the desert; at other times they descend just as violently in the vortices of whirlpools and eddies. Despite the random paths of individual water particles, the main thrust of the water is forward, downslope in the direction the stream is flowing.

In general terms, the velocity of a stream can be defined as the direction and magnitude of displacement of a portion of the stream per unit of time. Customarily, we measure it in meters per second (m/sec) [feet per second (ft/sec)] or kilometers per hour (km/hr) [miles per hour (mph)]. Velocities of less than 6 km/hr (4 mph) are the most common; few streams attain velocities in excess of 30 km/hr (19 mph). As much as any single factor velocity is responsible for determining

*A.L. Bloom, 1969, The Surface of the Earth, Prentice-Hall, Englewood Cliffs, N.J., p. 15.

**Fig. 12-3.** Car, suspended from a cable, allows geologists to measure the discharge and velocity of the waters and to take samples. Photograph taken in 1890 on the Arkansas River near Canon City, Colorado. (USGS)

the size of the particles a stream can transport as well as the way in which it carries the particles, or load.

At what point is an "average velocity" likely to be located within a stream? Different parts of the water in a stream move at different rates and, as in glacier flow, the center moves faster than the

**Fig. 12-4.** Turbulent flow of the muddy Little Colorado River at Grand Falls, Arizona. (Curt Smith)

sides and the top faster than the bottom because of friction along the channel sides and bottom, or perimeter (Fig. 12-5). Data show that the average velocity of a river is approximated by that velocity at 0.4 of the distance above the bed of the river, about in midstream. How, then, do we obtain this value short of swimming? Actually, a reasonable figure can be obtained by throwing a stick into the middle of the stream, timing its travel over a known distance, and multiplying that velocity value by 0.8.

**Discharge** is the quantity of water that passes a point in a given interval of time as expressed in cubic meters or feet of water per second. It is calculated by multiplying average velocity by the cross-sectional area of the channel at that point. However, discharge of most rivers is far from constant. In northern rivers, it fluctuates with the melting of snow and ice. Rivers in tropical monsoonal regions show large seasonal variations. Streams of the arid southwestern United States show as great a range as any. Throughout most of the year they may have no surface flow at all, but during a sudden cloudburst they can become raging torrents, filling their channels from bank to bank.

The Amazon is by far the world's largest river.

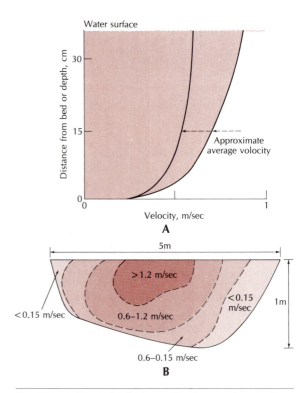

**Fig. 12-5. A.** Vertical velocity profiles for two rivers in Wyoming. Notice that the velocity is zero at the bed of each stream. **B.** Velocity distribution in a cross section of a river channel in Wyoming. (After L.B. Leopold et al., 1964, *Fluvial Processes in Geomorphology,* W.H. Freeman, Figs. 6–1, 6–9)

350

It rises high in the Andes, about 160 km (100 mi) from the Pacific Ocean, and it flows eastward some 6500 km (4030 mi) to the Atlantic Ocean. Where the river channel is deepest, the discharge averages 200,000 m³/sec (7 million ft³/sec) or about one-fifth of the water flowing on the entire earth at that second! The volume is so great that water 160 km (100 mi) offshore at the mouth is still fresh. To handle this amount of water, an enormous channel is required; its width at the mouth is equivalent to the distance from Paris to London, and some islands in the estuary rival Switzerland in area. The world's second largest river, the Zaïre in Africa, has only one-fourth the discharge; our Mississippi–Missouri system ranks eighth, with only one-tenth the discharge.

The average velocity of a stream depends on several factors: (1) the gradient (or downvalley slope), (2) the cross-sectional shape of the channel, (3) the roughness of the sides and bottom of the channel, (4) the discharge of the stream, and (5) the amount of sediment the stream is carrying. We will look at the effect of each factor separately, and then at how each can vary along the length of a river.

1. Increased gradient obviously speeds up a stream's flow. Where the gradient is low, its velocity is low. When slopes are vertical, as in a waterfall, the velocity approaches that of free-fall.

2. The channel that allows for the greatest velocity resembles a semicircle—a shape that gives the maximum cross-sectional area for the lowest channel perimeter. The lower the perimeter length, the less the frictional retardation and the greater the velocity. In contrast, in wide, shallow rivers much of the water is in contact with the channel perimeter; hence frictional retardation is great and velocity relatively low.

3. The rougher the channel, the lower the velocity. A smooth, clay-lined channel will promote a higher velocity than one lined with large boulders or dense vegetation.

4. Velocity varies in proportion to discharge. To illustrate this, consider a river with fluctuating discharge. During a low-discharge stage, a river is shallow, much of the water is close to the channel perimeter, and the velocity is low. Increase the discharge, though, and less water per unit volume comes in contact with the perimeter. In such a situation frictional retardation is less and the river flows downvalley at a higher velocity.

5. An increase in sediment load with a corresponding decrease in the percentage of water has a strong breaking effect on velocity. This is readily understandable because the more sediment a stream receives, the muddier it becomes, until finally the viscosity can increase to the point where the stream can no longer flow. Such an effect is sometimes demonstrated by the ephemeral streams produced by short-lived thundershowers in the desert.

All the factors affecting velocity must vary in a downvalley direction and in a fairly systematic way. For example, if the velocity of the upper part of a river were much faster than the lower part, it would overtake the lower part. Or, if the lower part were to run much faster than the upper part, it might run away from the latter. Of course, this does not happen, for rivers continually adjust themselves in order to remain intact ribbons of water flowing toward the sea.

Consider, for example, a river heading in high mountains and flowing across vast lowlands. In the mountains, the gradient is steep, but the velocity is slowed by a boulder-strewn rough channel and a relatively low discharge. Progressing downstream, the gradient may lessen, but not necessarily the velocity. Indeed, it may remain constant or even increase. Commonly, discharge increases as more tributaries join the main stream and as the channel becomes smoother—both contribute to a greater velocity. The tendency for greater velocity, however, could be offset by a lesser gradient, and the result could be constant velocity. The point to be made here is that determining the velocity of a stream is a complex matter, depending on many variables, none of which is easily measured. The interplay of these factors commonly results in a downvalley decrease in stream gradient (Fig. 12-6).

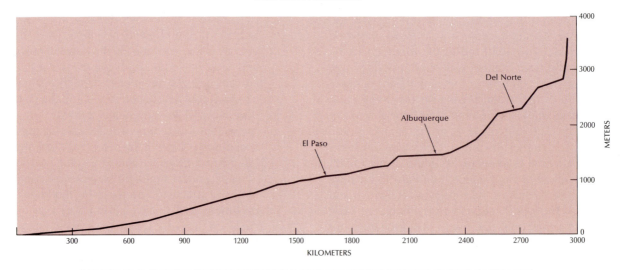

**Fig. 12-6.** Gradient of the Rio Grande from its headwaters in Colorado to its mouth at the Gulf of Mexico. (From U.S. Geological Survey Paper no. 44)

## STREAM TRANSPORTATION

The roiled sediment brought down by the Mississippi River clouds the waters of the Gulf of Mexico far seaward of the river's mouth. In the Southwest, within a generation, some fairly large reservoirs have silted up completely, and the lakes behind the dams have been converted into dreary expanses of muddy or dusty silt, depending on the season. Still other examples could be cited to show that the land is inevitably wasting away and that much of it is being carried to the sea.

That annual wastage can be an imposing amount is demonstrated by a single river, the Mississippi, which every day carries slightly less than 1 million tons of sediment to the Gulf of Mexico—even more when in flood. This colossal drain of the U.S. central lowlands removes 4 cm (1.6 in.) of soil every 1000 years and has resulted in the construction, within the past million years or so, of a broad platform of sand, silt, and clay that covers an area of around 31,000 km$^2$ (12,000 mi$^2$) at the river's mouth, with a central thickness of at least 1.5 km (1 mi).

Most people know that streams carry a heavy burden. A problem in determining how the load is carried is that most rivers transport most of their load during flood stages and that sediment sampling at such times can be rather perilous.

A river moves its load in three major ways—in part by solution, in part by suspension, and in part by bodily movement (rolling and sliding) along the bottom of the channel.

### Dissolved Load

The dissolved material rivers carry is supplied largely through the leaching out of ions derived from the weathering of minerals and rocks in soils. It is this invisible dissolved load that gives some river water—especially the water in western U.S. rivers that cross arid or semiarid regions—its distinctive taste. When such water evaporates, it leaves behind a white residue of alkali salts. Such accumulations are fatal to nearly all plants.

Dissolved loads go wherever river water goes and only precipitate out if conditions permit. Total dissolved load may be determined through an analysis of the water. Error can creep into the interpretation of data, however, because pollution can account for as much as 50 percent of the dissolved load of some rivers.

### Suspended Load

Muddy, roiled water sluicing through an arroyo after a desert cloudburst is largely a function of the suspended load: the cloud of sediment in the water. How long such material stays in suspension depends on several factors: (1) the size, shape, and specific gravity of the sediment grains; (2) the velocity of the current; and (3) the degree of turbulence.

Finer particles, such as silt and clay are commonly carried in suspension; in contrast, larger sand grains usually are carried only briefly in suspension by strong currents before they sink to the bottom of the channel. Flat mineral grains, such as mica flakes, will sift down through the water much like confetti when compared to the more direct way in which nearly spherical grains settle out. Specific gravity is also important because denser substances (e.g., gold nuggets) with specific gravity of 16 to 19 are deposited far more rapidly than feldspar grains of the same dimensions that have a specific gravity of about 2.7.

One of the most important processes that keep particles in suspension is turbulence. For example, if a particle is settling, it may in no time be caught in an upward swirl of water. So it is unlikely that individual grains of sediment will settle out at a uniform velocity along the entire course of a river. Rather, each particle follows a complex path, drifting down the river with the moving current, here and there swirling erratically—much like a sheet of paper caught in a vagrant wind.

Measurements of suspended sediments are fairly easily taken. All that is needed is a bottle of river water from which the solid sediment is separated out and weighed. If the average discharge is known, the annual suspended load can be calculated.

## Bed Load

The **bed load** moves along the bottom either as individual particles or as aggregates of particles (Figs. 12-7). Individual particles may move by sliding, rolling, or **saltation** (from the Latin word *saltare*, "to jump"). Saltation could be compared to the game of leapfrog. A sand grain may be rolling along the bottom, or may even be stationary, when it is caught up by a swirling eddy. Then it bounds through the water in an arching path. Should the velocity be great enough, it may be swept upward to become part of the stream's suspended load temporarily; if not, the particle

sinks again, either to remain stationary or perhaps to continue downstream by leaps and bounds.

It is more difficult to measure the bed load of a natural stream than the dissolved or suspended load. The latter two loads are diffused throughout the main body of the river, but the bed load moves along the most inaccessible part of the stream—the bottom—mainly during major floods, a time when the river's bed is least accessible for study. Bed loads can be trapped in the deltas of downstream reservoirs, and the simplest way to estimate annual bed-load transport is to measure the volume of the delta.

## Amount of Dissolved and Solid Load

The ratio of dissolved load to **solid load** (suspended load plus bed load) varies from river to river for several reasons. It might come as a surprise, however, to hear that about one-half the total worldwide load transferred to the oceans is in the dissolved form. In short, the continents are continually being dissolved through the effects of rainfall and weathering.

Rivers show a complete range in the solid-load to dissolved-load ratio (Fig. 12-8). In areas of high relief, such as the Colorado River basin and parts of Southeast Asia, erosion is rapid and rivers carry a high solid load. On the other hand, areas of low relief, such as the Amazon basin, the Columbia River Valley, and parts of the southeastern United States, are characterized by a low erosion rate and low solid load. Chemical weathering effectively controls the amount of dissolved load; we see relatively little in semiarid areas but a considerable amount in humid and well-vegetated areas. Only a few rivers have a higher dissolved load than a solid load; the Volga in the USSR is one of these.

An extremely important factor affecting the amount of solid load delivered to a stream is the interaction between vegetation and precipitation. This, in turn, affects the **sediment yield**—the amount of eroded material transferred from the land to a stream system. In arid and semiarid regions with amounts of precipitation up to about 38 cm (15 in.), the sediment yield progressively increases. Beyond 38 cm, however, the grass cover thickens, protecting surfaces from erosion and the sediment yield starts to decrease. At still higher rates of

**Fig. 12-7.** Boulders near Manzanar, California, once swept along as bed load by flash torrents from the distant canyon. (Ansel Adams)

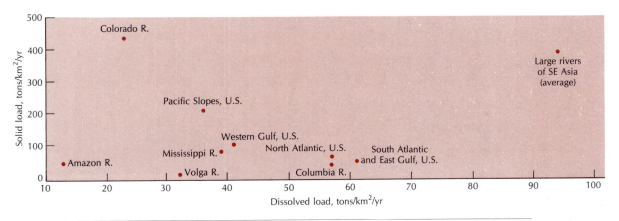

**Fig. 12-8.** Amounts of dissolved and solid load of various rivers and regions. (After Judson and Ritter, 1964, "Rates of Regional Denudation in the United States," *Journal of Geophysical Research,* vol. 69, pp. 3395–3401; after N.M. Strakhov, 1967, *Principles of Lithogenesis,* vol. 1, Oliver and Boyd, Edinburgh)

precipitation, forest vegetation becomes dominant and may retard erosion and sediment yield even further. The curve obtained by plotting these relationships (Fig. 12-9) is a valuable tool for predicting the responses of a stream to climatic change or to changes effected by humans, such as forest clearcutting. Although the curve is derived from studying parts of the United States, it may be applicable in other countries.

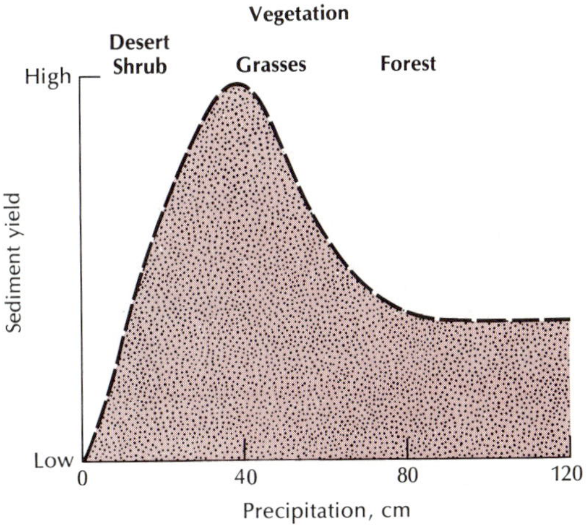

Putting it all together, we can estimate how long the soil in the United States and the rest of the world will last. The average rate of erosion of the United States, before people began to disrupt the landscape, has been estimated at 3 cm/1000 yr (1.2 in./1000 yr). Through construction and agricultural practices, humans have caused a dramatic increase in places—of one order of magnitude or greater (Fig. 12-10). Taking the United States as a whole, however, the present rate is about double the prehuman rate given above.

The prehuman worldwide rate of erosion has been calculated at slightly over 9 billion tons/yr, equivalent to a lowering of the total land surface at a rate of about 2.4 cm/1000 yr (1 in./1000 yr). Human intervention has increased the rate to some 24 billion tons/yr by one estimate, whereas another estimate (chap. 10) is 25 billion tons/yr from croplands alone.

At the natural rate the continents would be swept to the sea in a little over 300 million years; humans are merely hastening the process. It is doubtful that all the lands would be reduced to sea level, however, because the geological record clearly shows that such an event has never happened in the past. Highland areas are surely worn

**Fig. 12-9.** Generalized relationship of annual sediment yield, mean annual precipitation, and vegetation. (After W.B. Langbeim and S.A. Schumm, 1958, "Yield of Sediment in Relation to Mean Annual Precipitation," *American Geophysical Union Transactions,* vol. 39, pp. 1076–84)

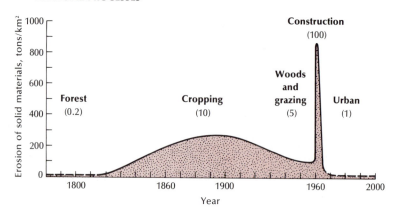

**Fig. 12-10.** Erosion rate related to land use in the Middle Atlantic region of the United States. The numbers in parentheses are the approximate erosion rates in cm/1000 yr for the various time periods. The first increase in the erosion rate coincided with the clearing of forests for agriculture in the early 1800s. The rate decreased in the early to mid-1900s when the land was used for grazing and partly returned to forests. Construction in the 1960s bared much of the land to runoff and erosion; that was soon halted, however, as paving, lawns, and other covers that inhibit erosion were added to the landscape. (M.G. Wolman, 1967, "A Cycle of Sedimentation and Erosion in Urban River Channels," *Geografiska Annaler*, vol. 49, Series A, 2–4, p. 386)

down, but other areas, formerly low, are pushed up to form highlands and become good sources of sediment.

## Competence

The **competence** of a stream refers to the size of sedimentary particles a stream can transport, and depends primarily on velocity. At low velocities many streams run clear, and the sediment grains on their beds rest relatively undisturbed. With increasing velocity, the water becomes more and more roiled and larger and larger particles are picked up.

The failure of the Saint Francis Dam, Santa Paula, California in 1928 is a prime example of the transporting power of running water. When the 62.5-m (205-ft) high concrete structure collapsed, a wall of water 38-m (125-ft) high surged down the canyon with a velocity of perhaps as much as 80 km/hr (50 mph), enough to move 18-m (59-ft) blocks of concrete weighing as much as 10,000 tons, almost 1 km (0.6 mi) downstream.

The relationship of load size to stream velocity is not a simple one (Fig. 12-11). Research has shown that in order to initiate particle motion in sand or larger bed-load sizes, a greater velocity is required than that needed to keep the particles in motion. It is the added force of momentum that helps to keep the particles moving. A somewhat surprising result of the research is that the velocity required to initiate motion of clay is similar to that

required for sand. The reason for the apparent anomaly is twofold: (1) clays rest at the bottom of the stream where velocities are nil and (2) clay particles are held tightly together by cohesive forces. High velocities are required to produce the conditions needed to disrupt those forces, setting the fine-grained material in motion. Once in motion

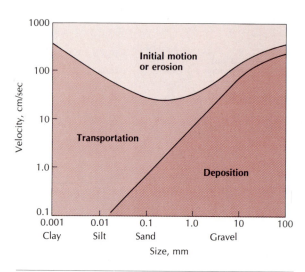

**Fig. 12-11.** Relationship of stream velocity to initial motion, continuing transportation, and deposition of particles of various sizes. (After F. Hjulstrom, 1935, "Studies on the Morphological Activity of Rivers as Illustrated by the River Fryis," *University of Upsala Geological Institute Bulletin* no. 25, pp. 221–527)

**Fig. 12-12.** Hydraulic mining for gold in Nevada County, California, in the late nineteenth century. The excess sand and gravel were washed to the rivers, thus greatly increasing their loads. (Bancroft Library, University of California)

the fine particles stay in suspension over a wide velocity range, quite unlike the narrow velocity range for bed-load material.

## Graded Streams and Their Disruption

Every stream has the capacity to transport material. A stream with a load that exceeds its capacity will drop its overload as abruptly as a Peruvian llama deposits its burden if it is convinced that it is beyond its carrying capacity. When a stream deposits its excess load on the channel bottom, we say that it is **aggrading** its channel. This may happen when too much sediment is supplied or when the particle size exceeds the stream's competence. Conversely, an underloaded stream—one in which the energy to transport exceeds its load—is likely to pick up an additional quantity of material by **degrading,** or eroding, its channel.

When a stream is balanced between the two extremes and has achieved equilibrium so that its slope and discharge give it sufficient velocity to handle the load, it is **at grade,** or in equilibrium.

a **graded stream** was aptly defined by the geologist J. Hoover Mackin:

A graded stream is one in which, over a period of years, slope is delicately adjusted to provide, with available discharge and with prevailing channel characteristics, just the velocity required for transportation of the load supplied from the drainage basin. The graded stream is a system in equilibrium; its diagnostic characteristic is that any change in any of the controlling factors will cause a displacement of the equilibrium in a direction that will tend to absorb the effect of the change.*

A brief discussion of Mackin's statement is in order. First, it is important to point out that equilibrium is reached only over a period of years. A river might carry a maximum load only a few days or weeks of the year, during times of flood. The rest of the time it might display little tendency to work.

Another important aspect is that if the equilib-

*J.H. Mackin, 1948, "Concept of the Graded River," *Geological Society of America Bulletin,* vol. 59, p. 471.

rium of a graded stream is disrupted, the stream will react quickly to counteract the disturbance. For example, if more load or larger particles are imposed on a stream, the stream might not be able to handle it. It will deposit the load in the streambed where it will remain until a steeper slope is built that can handle the load and a new grade attained. A classic example occurred in the gold rush days in California when hydraulic mining in the mountains greatly increased the loads streams had to carry (Fig. 12-12). Far downstream in the Great Valley rivers filled in, or aggraded, part of their channels, resulting in frequent flooding. In contrast, the removal of part of a load from a river can bring about downcutting, or degradation.

Altering discharge also can greatly disrupt the equilibrium of a stream because discharge and velocity are closely related. A decrease in discharge results in aggradation, whereas an increase results in degradation. Discharge can be altered through climatic change or water-diversion projects.

The side effects of damming a river vividly point out what could happen when stream equilibrium is disrupted. Above the reservoir the stream carries its usual load, but on meeting the quiet reservoir waters, dumps its load to form a delta (Fig. 12-13). Because the delta has a low slope, some material is deposited on its surface and some is swept out and deposited on the delta front. As the delta continues to build out into the reservoir, the valley upstream must aggrade to maintain the equilibrium slope.

Things are not much better downstream from

**Fig. 12-13.** Delta formed by a tributary to the Snake River, near Huntington, Oregon. (Omar Raup)

the dam. There the stream has been deprived of its load—part of which now forms the delta—and the river degrades. Such behavior could lead to the undermining of structures along the banks of the stream.

Another common cause of stream disequilibrium is a change in **base level.** Base level is the low point to which most streams flow—the ocean. If the ocean level rises or falls, the streams will aggrade or degrade, respectively. Many coastal streams responded in such a way during the rapid fluctuations in sea level that accompanied the major Pleistocene glaciations.

## FLOODS

Worldwide, in the period 1947–80, floods have claimed almost 200,000 lives, thus ranking third behind cyclones and earthquakes in fatalities.

When rains are heavy, and fall day after day, floods begin to threaten those who live close to a river. The ground will soak up a certain amount of water, but if the rainfall continues, the accumulated rain exceeds the capacity of the ground to absorb more, just as a sponge held under an open faucet will cease to hold water. At such a point runoff increases and the water in a river channel can swell to flood proportions (Fig. 12-14). Increased discharge means an increase in velocity, so that the riverbed is scoured and the channel is enlarged. Some of the floodwaters are thus accomodated. But if discharge continues to mount, the rampaging river will spill out over the **flood plain**—that low-lying ground that periodically is inundated. The force of these floodwaters can cause considerable damage (Fig. 12-15).

Today it is possible, especially in extensively studied flood-prone areas, to warn people before a flood strikes. Those living in other areas, however, are not so fortunate. The Mississippi River floods have been tracked for decades. In one recent flood, higher-than-normal rainfall was first recorded in the fall of 1972, yet floodwaters reached their maximum depths in the spring of the following year (Fig. 12-16) when record consecutive days above flood stage were set in Saint Louis (77 days) and Vicksburg (88 days). Damage was estimated at over $400 million, or $33 per acre of inundated land. Yet flash floods, such as those that

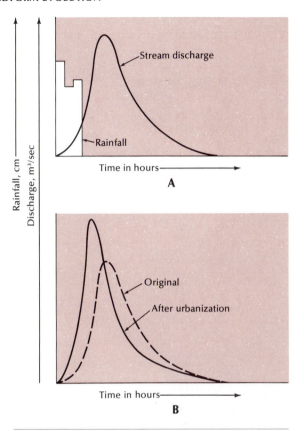

Fig. 12-14. **A.** Hypothetical relationship between the rainfall aociated with an actual storm and the discharge of a stream. The lag between peak rainfall and peak discharge occurs because of the time required for the ground to become saturated. **B.** Alteration of the storm-discharge curve aftrer urbanization. Pavements and other impervious ground result in greater run-off and an earlier arrival of peak discharge, and floods may occur more frequently. (From L.B. Leopold, 1968, *Hydrology for Urban Land Planning—A Guidebook on the Hydrologic Effects of Urban Land Use,* U.S. Geological Survey Circular no. 554, Fig. 1)

characterize large areas of the western United States, are unpredictable, arriving with little warning and lasting only a few hours.

In 1976 a tragic flood took place in the Rocky Mountains. During the evening and night of July 31–August 1, as much as 51 cm (20 in.) of rain fell in the canyon of the Big Thompson River, Colorado. The soil could not absorb that amount of water and the river flooded. The amount of water was two hundred times normal flow, reaching a maximum depth of almost 6 m (20 ft). At

**Fig. 12-15. A.** The flood of March 12, 1963, inundated the floodplain of the North Fork of the Kentucky River at Lothair and Hazard, Kentucky. (Reprinted with permission from The Courier-Journal)

least 139 persons died, many of whom were attempting to flee the area by driving down the road that paralleled the river (Fig. 12-17). Many lives could have been saved had people parked their cars and climbed a short distance up the valley sides. Damage was put at $35.5 million.

Earth scientists have estimated how often floods of certain discharges will recur and use terminology that reflects this, for example, a *5-year flood,* a *25-year flood,* and so on. Flood maps, showing to what extent floods of various sizes might inundate a flood plain are then made up. Such maps

are an integral part of wise land-use planning, and should be consulted if construction close to a river is being considered.

Flood maps must be updated fairly often. One reason is that with each new flood there is new data that refines the statistics on flood-recurrence intervals. The data base, however, is short, commonly less than a century. Flood-frequency estimates over a much longer time span can be obtained by dating old flood deposits on the flood plain or even those deposits isolated in rock shelters high above present river level. In this way an extraordinary flood in 1954 on the Pecos River of Texas was shown to have a recurrence interval of more than 2000 years.

Another reason for updating is that urbanization tends to increase the incidence of floods of a particular size. The reason is simple: Roads and houses and shopping centers cover ground that once soaked up rainwater. The consequence is, of course, that the runoff for a given-size storm is increased, which, in turn, increases the incidence of floods of a given size (Fig. 12-14B).

Few people driving through the desert of eastern Washington State realize that part of their route is along the course of an ancient flood, quite likely the largest on this planet. Called the Lake Missoula flood, it left such a clear mark on the Columbia Plateau that it appears in photographs taken from orbiting satellites, even though it took place thousands of years ago (Fig. 12-18). Geologist J. Harlan Bretz, more than anyone else, put together the geological pieces on which this fascinating story is based.

The story begins less than 20,000 years ago when an enormous glacier pushed south along a wide front from Canada into northeastern Washington, Idaho, and Montana. Ice blocked major

**Fig. 12-15. B.** Helixed railroad track along the south branch of the Potomac River, West Virginia. Floodwaters discharged from a two-day storm in November 1985, locally exceeded the 500-year-recurrence event and had enough force to deform these tracks. (G. Michael Clark)

**Fig. 12-16.** The Mississippi River flood of spring 1973. (J. Skelton, USGS)

drainages and formed a large lake known as Lake Missoula. Ice can hardly be said to make a stable dam, for it floats on water. Conditions changed, the dam burst, and an amount of water equivalent to about 20 times the worldwide average discharge was unleashed on the basaltic plaeau. Peak discharge lasted a day, and in a week or two the lake had drained. Maximum water depths exceeded 200 m (660 ft), and velocities up to 75 km/hr (47 mph) seem possible. Researchers are now arguing about the number of these floods, with over 40 being suggested.

The ancient flood of Lake Missoula devastated the area. Many square kilometers of basalt were reamed out, creating a landscape appropriately called the channeled scablands. One of the most spectacular erosional features is Dry Falls, a cataract twice as high and nine times wider than Niagara Falls. Boulders up to 30-m (100-ft) long were moved as far as 3 km (2 mi) by the flood-

**Fig. 12-17.** Drake, Colorado, before and after the Big Thompson River flood of 1976. (Hogan and Olhausen)

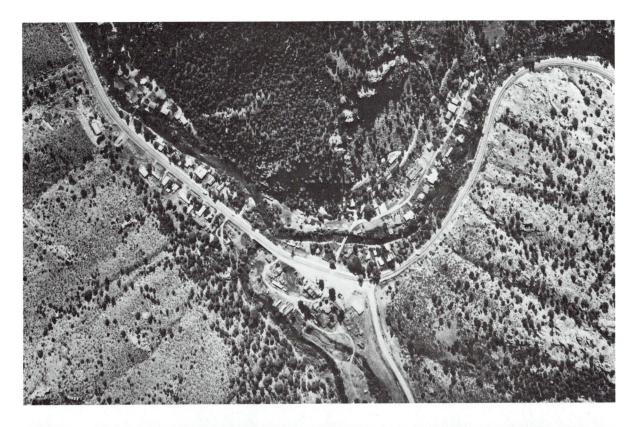

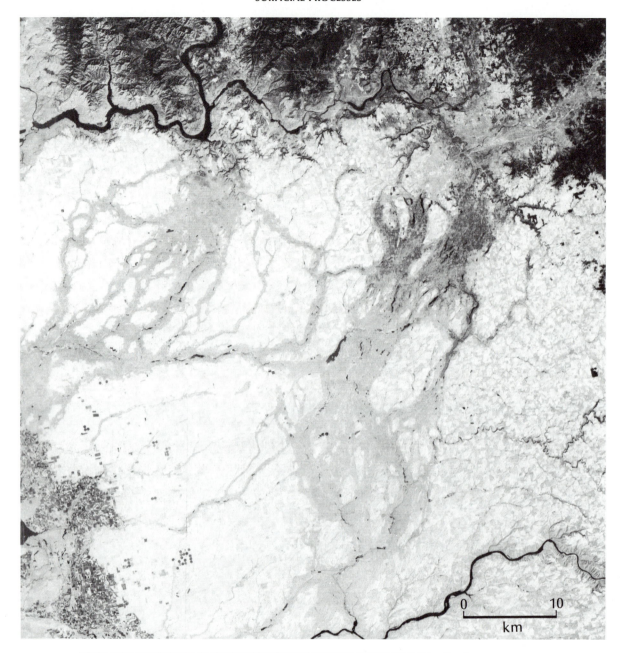

**Fig. 12-18.** Satellite photograph of the channeled scablands of eastern Washington carved out by the ancient Lake Missoula flood. The path taken by the flood is shown by the gray braided pattern, whereas the present course of the Columbia River can be seen in the upper part of the photograph. (Eros Data Center, USGS)

waters. Gravel bars formed, and some of the larger ones were more than 3-km (2-mi) long and 20-m (100-ft) high. Streams commonly form small sand ripples on their beds, but the Lake Missoula flood formed giant gravel ripples up to 7-m (23-ft) high, spaced more than 100 m (330 ft) apart.

The scabland features are so different from the expected norm that it is no wonder that in the 1920s Bretz had trouble convincing his colleagues of a great prehistoric flood out West. He persevered, gathered more data, and eventually won them over. Today few geologists doubt his interpretation. In fact, when a group of geologists from many nations were taken through the flood area in 1965, they wired Bretz a message that concluded: "We are now all catastrophists."

Large-scale channels on Mars have features that suggest to some geologists ancient and catastrophic flooding. Although formed over 1 billion years ago, the channels are still well preserved. Atmospheric weathering, erosion, and other complex events on earth eradicate most topographic features in far less time than that. If, indeed, the features on Mars are the result of flooding, the discharge was 50 times greater than the Missoula flood according to one estimate.

Similar floods, but on a much smaller scale, occur when a dam fails. On June 5, 1976, the Teton Dam in southern Idaho failed, releasing 3.4 billion m³ (4.4 billion yd³) of water in 8 hours (Fig. 12-19). A total area of 480 km² (185 mi²) was inundated and 11 persons perished. Damage approached $1 billion. No one knows exactly what happened. It is thought that the highly fractured and porous bedrock probably was not adequately plugged with concrete, thus allowing water to flow from the reservoir to the core of the earthfill dam, where it rapidly eroded the structure away.

## STREAM DEPOSITION

The broad plains bordering many of the world's large rivers have been tempting sites for settlement since the beginning of history. Both the Egyptian

**Fig. 12-19.** Looking upstream at the remaining section of the Teton Dam, southern Idaho, one day after it failed. To the left is the spillway, and the debris adjacent to the stream below the dam was deposited by the floodwaters. (U.S. Army Corps of Engineers)

and Babylonian civilizations were bound to the rivers that shaped them—the Nile and the Euphrates, respectively. The Nile spreads across its bordering broad lowland in time of flood. In fact, the annual flood was an event of such importance to the survival of Egypt that a whole pantheon of deities was given credit for the phenomenon. For centuries, the Nile floods laid down sediment—natural fertilizer—on adjacent farmlands. But the building of the Aswan Dam in the mid-twentieth century cut off that natural process. Even along the coast, where the Nile meets the Mediterranean, a decline in fish harvests and increased erosion may be tied to the fact that the dam checks not only the floods, but the nutrients and sediments that the floodwaters once carried to the sea. Of equal economic importance is the increased incidence of the parasite *Schistosoma*, which causes a debilitating, sometimes fatal, disease in humans.

In our discussion, we will look at the characteristic features of flood plains, mention some problems in flood plain manipulation, and conclude with a description of sedimentation and landform patterns where rivers meet the sea.

## Flood Plain Features

From the term itself, we expect the surface of a flood plain to be covered with deposits from a flooding river. A prominent feature of any flood plain is the river-channel pattern. Two principal patterns are recognized. Rivers with single channels and large curving bends are said to have a **meandering pattern** (Fig. 12-20); the word is appropriately from the Latin, meaning "to wander."

If rivers do not have a meandering pattern, chances are that the pattern will be **braided,** the other major river pattern. Rather than flowing in a single rather narrow but deep channel, the river follows a braided pattern—a series of wide, shallow anastomosing channels that continually join and part from one another in a downstream direction (Fig. 12-21). The individual channels may change position hourly, daily, or seasonally—they are quite active.

Some rivers are braided and others meander mainly as a result of the type of sediment they carry. Meandering streams require a rather tough bank material to restrict the river to a single chan-

nel. The toughest material is cohesive silt and clay, thus meandering rivers are those with a relatively high suspended load. In contrast a stream can easily spread into many shallow channels if the bank material is loose and noncohesive. The weakest bank materials are sand and gravel, hence braided streams are characteristically bed-load streams.

Stream pattern is a part of the equilibrium of a stream, which is why J. Hoover Mackin refers to "prevailing channel characteristics" in his definition of a graded stream. A high-gradient, boulder-carrying braided stream can be at grade just as

**Fig. 12-20.** River meandering across the plains of southern Wyoming. (Janet Robertson)

**Fig. 12-21.** Braided channels of the Muddy River near its junction with the McKinley River, Alaska. (Bradford Washburn, Boston Museum of Science)

can a low-gradient, low-velocity meandering stream burdened with silt.

The flood plains of meandering rivers are often flooded—the lower Mississippi is typical—so let us discuss some of their flood-plain features (Figs. 12-22, 12-23). Low embankments, or natural **levees,** border and slope gently away from the river. Floods build the levees, for as floodwaters surge over the banks, the velocity quickly decreases and coarser sediments—usually sand and silt—are deposited. Levees can be the higher, firmer ground of some flood plains and thus are the favored site for roads, farms, and settlements. Between the levee and the bluffs bordering the flood plain is the **backswamp.** This is a boggy section that holds much of the floodwaters; sediments commonly are clayey, derived from the finer suspended load typical of a quiet water environment.

In some places tributary rivers are trapped in the backswamp and run parallel to the main stream for many kilometers before they are able to join it. Geologists have termed such rivers **yazoos** after a stream of that type—the Yazoo River in Mississippi (Figs. 12-22, 12-23).

Some meandering rivers are known for rapidly changing their course, and this leads to some characteristic features. Both velocity and the shape of the channel cross section vary downstream (Fig. 12-24). Between bends the channel is quite symmetric and the maximum velocity is near mid-

**Fig. 12-22.** Characteristic features of the flood plain of a meandering river.

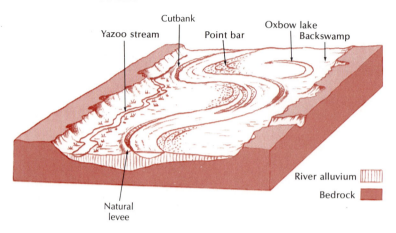

**Fig. 12-23.** The Yazoo River, a tributary of the Mississippi, showing point bars and the position of a possible future cutoff if the river is left to its own devices. (Frank Beck)

368

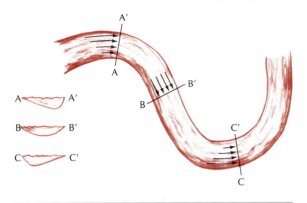

**Fig. 12-24.** Plan view and cross sections of a meandering river, showing the change in channel shape with position along the river. The length of the arrows depicts relative river velocity. (After D.J. Easterbrook, 1969, *Principles of Geomorphology*, McGraw-Hill, Fig. 6–10)

stream. In contrast the maximum velocity in bends is at the outer edge of the bend where the water is deepest. This results in the often-rapid undercutting of the outside bank, hence the term **cut bank.** Materials eroded from one cut bank usually are deposited on the inside of the next bend downstream to form **point bar** deposits (Figs. 12-22, 12-23).

Because river bends migrate in the direction of the cut bank, the neck of land between the bends can be eroded and a new, more direct channel—called a **cutoff**—can form (Fig. 12-25). All told, there have been about 20 such naturally occurring cutoffs on the lower Mississippi since 1765. About 15 have been made artificially since 1932 by the Mississippi River Commission in order to straighten the river's course, thereby increasing the gradient

**Fig. 12-25.** Major changes in the courses of the Mississippi and Yazoo rivers over a 16-year period. Historically, one of the more interesting cutoffs of the Mississippi occurred at Vicksburg in 1876. Before that date the river made a broadly sweeping curve past the city. In 1876, however, the river formed a cutoff south of Vicksburg that isolated that town from the river. Ironically, in 1862 General Ulysses S. Grant (1822–85) had tried unsuccessfully to construct an artificial cutoff at about the same place so that Union river traffic could bypass Confederate guns at Vicksburg.

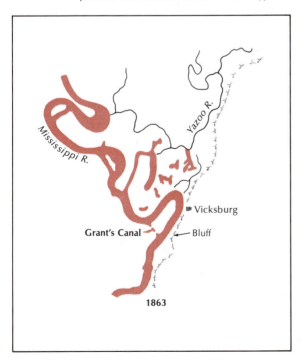

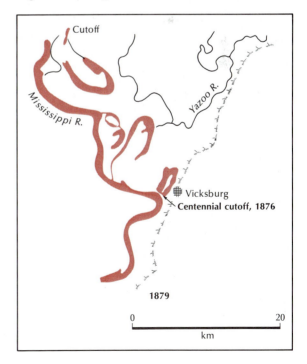

and, thus, the river's velocity: as a consequence theoretically diminishing the flood hazard by improving the hydraulic efficiency of the channel. As we will see later, however, straightening channels can have some disastrous side effects.

Let Mark Twain (1835–1910), in his role of steamboat pilot, have the last word on meandering streams. In *Life on the Mississippi* (1883) he writes about the river meanders and how the river is shortened when it cuts through the narrow neck of a meander. He grossly misuses the principle of uniformitarianism and is not very kind to science in general. But who will argue?

Therefore, the Mississippi between Cairo and New Orleans was 1215 miles long 176 years ago. It was 1180 after the cutoff of 1722. It was 1040 after the American Bend cutoff. It has lost 67 miles since. Consequently its length is only 973 miles at present.

Now if I wanted to be one of those ponderous scientific people, and "let on" to prove what had occurred in the remote past by what had occurred in a given time in the recent past, or what will occur in the far future by what has occurred in late years, what an opportunity is here! Geology never had such a chance, nor such exact data to argue from! . . . Please observe:

In the space of 176 years the Lower Mississippi has shortened itself 242 miles. That is an average of a trifle over one mile and a third per year. Therefore, any calm person, who is not blind or idiotic, can see that in the Old Oolitic Silurian Period, just a million years ago next November, the Lower Mississippi River was upwards of 1,300,000 miles long, and stuck out over the Gulf of Mexico like a fishing rod. And by the same token any person can see that 742 years from now the Lower Mississippi will be only a mile and three-quarters long, and Cairo and New Orleans will have joined their streets together, and be plodding comfortably along under a single mayor and a mutual board of aldermen. There is something fascinating about science. One gets such wholesale returns of conjecture out of such a trifling investment of fact.*

### Deltas

Herodotus [c. 484–430? (420?) B.C.] impressed by the branching pattern of the distributaries of the Nile, compared the form of the watery, muddy

*Mark Twain, *Life on the Mississippi*, 1939, in *The Favorite Works of Mark Twain*, Garden City Publ. Co., Inc., New York, pp. 86–87.

region between Cairo and Alexandria to the Greek letter delta (Δ). The comparison is so apt that it has won general acceptance.

The Nile Delta is a nearly ideal example of that particular landform, so much so that few others measure up (Fig. 12-26). A high-altitude photograph shows how the main channel of the river separates into a host of branching, lesser arms called, quite appropriately, **distributaries.** Another feature of the Nile Delta is the bordering bays and lakes, of which Abukir Bay is a good example. Abukir was the site of the Battle of the Nile in which the French fleet was destroyed by Nelson, thus ending Napoleon's hopes for an Eastern empire. Similar water bodies border many of the other deltas of the world. Well-known ones are Lake Pontchartrain connected by canal to the Mississippi Delta, the Zuider Zee (IJsselmeer) and marshes of Zeeland adjacent to the Rhine, and the lagoon surrounding Venice at the mouth of the Po River.

The best-developed deltas—those with the ideal Δ form—are most likely to develop be built up where a river moves a large load of sediment into a relatively undisturbed body of water. Examples of such impressive accumulations of riverine deposits are the great deltas at the mouths of the Ganges–Brahmaputra in India and Bangladesh, the Niger in Nigeria, and the Volga in the USSR.

Several factors combine to alter the conditions at the mouth of a river, so that many deltas are far from the ideal shape. River hydrology, the amount and size of river load, the geometry of the coast including the seaward slope, and the tectonic stability of the coast are factors to consider. In fact, some rivers, such as the Columbia, have no delta at all.

In North America the Mississippi is by far the best-known delta, not only for its long record of channel changes but for the thousands of oil wells drilled in its sediments and the repeated geophysical surveys made up and down its length. This information gives us a uniquely detailed, three-dimensional picture, not only of the 29,000 km$^2$ (11,200 mi$^2$) of delta surface, but of the over 1.5-km (0.9 mi) thick sediments below the waters of the Gulf of Mexico as well.

The Mississippi Delta is called an **elongate delta** because it extends far into the Gulf of Mexico. It is also called a **bird'sfoot delta,** as suggested by the outline of its area (Fig. 12-27). Because of a high sediment load, the delta builds out fairly rapidly on a flat slope with a channel more or less confined within levees. In time the channel shifts, taking a shorter and steeper route to the sea. The new channel ends in a subdelta that extends seaward until it too is abandoned. Mapping and dating reveal the times various parts of the delta have been the depository for sediment from the midcontinent.

The delta that is seen now represents only about the last 10,000 years. Before that, during the last glaciation sea level was low, and the river near New Orleans would have been deeply incised, ending in a delta far south of the present one.

The hazards of living on deltas are several. Not only are residents' homes prone to flooding, but subsidence is a common problem. In New Orleans, for example, long-term subsidence is measured at 12 to 24 cm (5 to 10 in.) per 100 years. This results from the compaction of delta sediments, tectonic downwarping, and groundwater withdrawal. The consequence is that 45 percent of the urbanized area is at or below sea level, with the lowest spot some 2 m (6.5 ft) below sea level. Thus the city is prone to flooding from river and

**Fig. 12-26.** Satellite mosaic photograph of the Nile Delta. The irrigated delta contrasts sharply with the surrounding desert. (NASA)

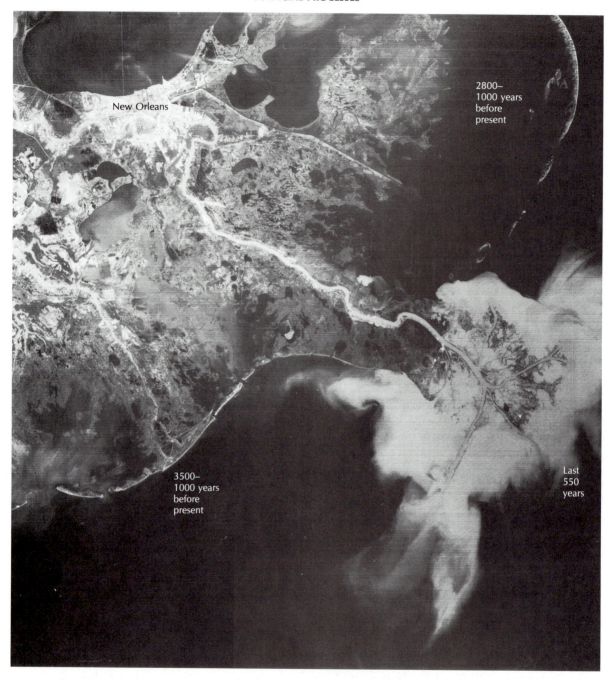

**Fig. 12-27.** Bird'sfoot delta of the Mississippi River, showing the approximate times various parts of the delta were formed. (NASA)

storm surges during hurricanes; its main defense has been to construct levees up to 5-m (16.5-ft) high that encircle the city. Venice, Italy, is another city with problems. There, too, groundwater withdrawal has increased the rate of subsidence (see chap. 18).

### Stream Terraces—a Case of Disequilibrium

Extensive flat areas slope downstream more or less parallel to the slope and flood plain of many streams (Fig. 12-28). If the "flats" are underlain by river deposits, the landform is called a *river terrace.* In essence such terraces are ancient flood plans, long abandoned and stranded high enough, so that present-day floods no longer surmount them. River terraces and their associated deposits are important in deciphering the history of a river—a task, as we will see, that is far from easy.

Two major kinds of terraces are recognized (Fig. 12-29). One is a **cut terrace,** so named because it consists of a thin veneer of gravel resting on a fairly smooth surface cut from bedrock (Fig. 12-30). How thin is thin? Geologists have their opinions, but most would say that the layer of gravel can amount to no more than the thickness moved during the deep scour that accompanies major floods. For small rivers that means a depth of about 5 m (16 ft) or less; for large rivers, about 10 m (33 ft). The other kind of river terrace is a **fill terrace.** It, too, rests on bedrock, but the river deposits are relatively thick. It is difficult to generalize, but the

**Fig. 12-28.** River terraces along the Rangitata River, Canterbury, New Zealand. Most of the terraces were formed during the time of the last glaciation and subsequent to it. (New Zealand Geological Survey)

deposits are thicker than the depth of scour and hence thicker than the gravels of cut terraces.

We can learn something about a river's behavior in an earlier age from studying its terraces and deposits. First, consider cut terraces. In some places they might be 1 km (0.6 mi) or more wide. In order to have developed such an extensive former flood plain, a river must have been in near-perfect equilibrium for a long time, meandering across the flood plain and nipping away at valley walls. Imagine a horizontal saw cutting through the landscape. Although the rates at which rivers cut through bedrock are hard to determine, we can estimate that surfaces of 1 km (0.6 mi) or more wide found in many parts of the western United States may have required hundreds of thousands of years.

Something then changed the equilibrium of the river, so that it subsequently downcut, leaving the abandoned flood plain as a cut terrace.

Fill terraces have a more complicated history. Initially, a river flows at a low level in its valley, perhaps swinging back and forth on a flood plain. Aggradation then follows as river sediment is deposited on the flood plain rather than being carried farther downstream. Next, the flood plain becomes a terrace as the river downcuts. The sequence of events, therefore, is aggradation followed by degradation. Some valleys have many fill terraces, one inset into another (Fig. 12-31)—indicating that the aggradation–degradation cycle has taken place many times.

It takes nothing short of a good detective to

**Fig. 12-29.** The formation of a cut terrace (**A**) and a fill terrace (**B**).

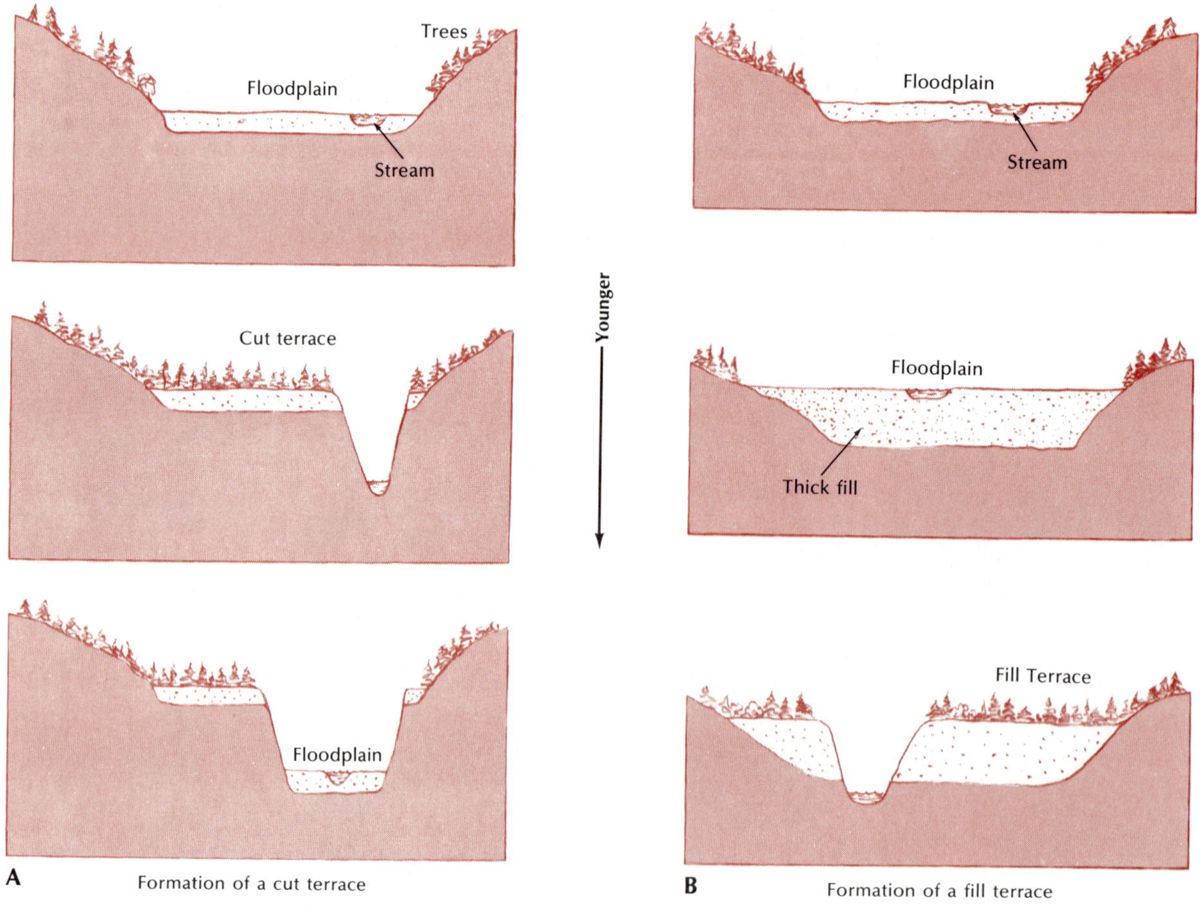

A    Formation of a cut terrace

B    Formation of a fill terrace

**Fig. 12-30.** Rounded river gravel overlying light-colored bedrock in a cut terrace of the Awatere River, New Zealand. Finer-grained deposits, either of flood plain or aeolian origin, overlie the gravel. Before a flat area adjacent to a river can be called a river terrace, deposits such as these must be present. (Peter W. Birkeland)

come up with a reason for terrace formation, and like so many other geological solutions, the best one usually seems to be a list of possibilities. In general, geologists look for downstream or upstream causes. Downstream variation in base level always is a possibility because a river responds quickly to variations in sea level at its lower end, aggrading as the sea level rises and degrading as it falls. An upstream cause could be the reaction of a river to changes in the size and amount of load or in the volume of discharge—changes that might have resulted from a climatic change. The slope of a terrace, the characteristics of its deposits, and any remaining ancient channel patterns etched on its surface all must be compared with

**Fig. 12-31.** Three fill terraces, each underlain with stream deposits of a different age. The highest terrace is the oldest, the lowest is the youngest. (Adapted from L.B. Leopold and J.P. Miller, 1954 "A Postglacial Chronology for Some Alluvial Valleys in Wyoming," U.S. Geological Survey Water Supply Paper no. 1261)

similar features of the present-day river before any one hypothesis is chosen over others.

Tectonism also can affect river-terrace formation. For example, rivers usually entrench in areas undergoing uplift but aggrade in areas undergoing tectonic flattening. Some thousands of years later, an uplifted area might be characterized by cut terraces and a flattened area by fill terraces. Furthermore, because faults might not be recognizable in the loose gravels, river terraces may be the only clue to earlier tectonic activity.

## Artificial Stream Disequilibrium

Rivers may be in or out of equilibrium naturally or they can be helped along by humans. The problem is that people choose to live along rivers but do not always build in the right places—suitable sites can only be chosen after long-term study of the behavior of the river. So when catastrophic events take place—catastrophic to landowners but run-of-the-mill events in the life of the river—the people affected want somebody to do something. In many places somebody did do something and the results have sometimes had unforeseen side effects.

Examples of river disturbances caused by humans abound. Because there are many, we will cover only one fairly "innocent" one here, innocent at least at first glance. It should be pointed out that as with any patient, before remedial measures are prescribed, it is not a bad idea to try to find out how a river reacted to natural changes in the past before imposing new changes on it. As was emphasized earlier, however, many rivers, like many people, do not divulge their past histories readily.

An example is that of the meandering Blackwater River in Missouri. Local flooding had been a problem before 1910, and measures to control it seemed in order. The parameters of the original river in the headwaters' reach were 54-km (34-mi) long, 1.7 m/km (9.3 ft/mi) gradient, with 1.8 meanders/km (3 meanders/mi). Bridges 15- to 30-m (50- to 100-ft) wide spanned the river. Because meandering rivers are slow moving, they cannot get rid of the rapidly increasing discharge that accompanies storms—the result is flooding. The common remedial practice is to cut a straighter channel, which increases both stream gradient and velocity and disperses the floodwaters more quickly.

In 1910 the Blackwater was straightened and channelized. As Mark Twain points out so vividly, straightening also results in shortening. The river was shortened by about half at the expense of increasing the gradient to 3 m/km (17 ft/mi). The artificial channel was cut 9-m (30-ft) wide at the top, 1-m (3-ft) wide at the base, and 3.8-m (12-ft) deep, for a cross-sectional area of 19 m$^2$ (23 yd$^2$).

The floods were contained in the channel, but the channel grew larger. Over the next 60 years, the river sector that enlarged the most measured 71-m (233-ft) across at the top, 13-m (43-ft) across at the bottom, and 12-m (40-ft) deep, for a cross-sectional area of 484 m$^2$ (580 yd$^2$). In any river system the tributary streams follow the lead of the main stream; this drainage was no exception. The tributaries of the Blackwater downcut just as fast as the main stream and gullies began to spread over the landscape.

Bridges were built across the straightened channel, but as the channel grew wider and deeper, the bridges had to be lengthened and the vertical pilings extended. In spite of those efforts, many bridges could not withstand the stress and collapsed.

The effects of channel straightening were also felt downstream beyond the limits of the artificial channel as sediment derived from the enlarging channel was deposited. Not only was the channel filled, but two successive generations of fenceposts were buried in the flood debris that covered the landscape.

The story of the Blackwater is an excellent small-scale example of what can happen when an attempt is made to regulate a river. Once the forces are set in motion, they are difficult to check. We now have enough case histories to predict intelligently the side effects of most regulatory measures. We can, then, only hope that the arguments of cost versus benefit can be properly laid before those individuals charged with making decisions.

## DRAINAGE PATTERNS AND THEIR INTERPRETATION

The patterns of river systems are varied. In some, tributaries all flow more or less parallel to geological structures. In others rivers cut directly through mountain ranges. All patterns have a geological explanation; here we will discuss the more obvious and typical relationships found in nature (Fig. 12-32).

In landforms where an original sloping surface is formed, all streams flow down the surface. If the landform is a volcano, the drainage pattern radiates from the central highland; a tectonic dome may give a similar pattern. In tilted mountain ranges, on the other hand, the streams and tributaries parallel one another. Drainage systems are more complicated in areas of folded sedimentary rocks of varying resistance to erosion. The smaller tributaries may flow along the more easily eroded rocks and the larger streams might cross the structural grain of the region. The result is a network of streams that meet at right angles, in a *trellis pattern.* Where there is no structural or rock-layering control, a *dendritic pattern* forms, so named for its similarity to the way in which branches join the trunk of a tree.

River patterns can change with time. Usually the uppermost tributaries extend headward. As this happens, they capture parts of the adjacent drainage system, and with it perhaps a pattern with different controls.

One extraordinary river pattern is that which cuts directly across a high mountain range, seemingly showing little respect for the barrier. This can happen in one of two ways. One is when tectonics raise a mountain range across part of a river's course. If the power of the river is sufficient, it can maintain its course across the rising barrier. An outstanding example is the Brahmaputra River (Fig. 12-33), which rises on the northern slopes of the Himalayas and collects drainage from both Nepal and Tibet as it flows east for about 1400 km (870 mi). It then turns south, crosses the Himalayas, and exits to the Bay of Bengal. Such streams, established in their course before a topographic barrier is raised across the course, are called **antecedent streams.** A similar origin can be ascribed to the Arun River, which rises in Tibet east of Mount Everest and flows south across some of the higher reaches of the range (Fig. 12-33).

The second way that this unusual pattern can develop is when a stream downcuts and encounters different structures, yet maintains its course.

Consider the rivers that cross the Appalachians (Fig. 12-32C). Although the smaller tributaries flow parallel to the rock structures, some of the larger streams cut directly through the anticlines; the Susquehanna River is one of the best examples of this behavior. Such rivers attain their courses long after the time of folding as they meander across a low-relief erosion surface, cutting across the folds or across younger sediments that overlie the folded sediments. If regional downcutting follows, the river can encounter and cut across any previously buried structure. In the case of the Appalachian rivers, the larger streams maintained their courses across the anticlines, whereas the less powerful tributaries cut down and expanded headward along the more easily eroded rocks, thereby etching out the regional structural patterns. Streams with this history are called *superposed* **streams.**

## LANDSCAPE EVOLUTION IN HUMID CLIMATES

In most climatic regimes, streams play a major role in the evolution of the landscape. Here we focus our attention on the relatively humid climates (in chap. 13 we will look at the arid climates). During the nineteenth century the role of streams in landscape evolution was widely debated. As late as 1880, although no one doubted that streams were capable of downcutting, some geologists ascribed such large-scale gorges as the Grand Canyon (Fig. 12-34) to violent sundering of the earth's crust. Following this reasoning, the presence of rivers in such canyons meant that they followed the cataclysmic troughs, not created them. Yet long before that time a Scottish geologist eloquently made the case that most streams excavate the valleys they occupy:

If indeed a river consisted of a single stream, without branches, running in a straight valley, it might be supposed that some great concussion, or some powerful torrent, had opened at once the channel by which its waters are conducted to the ocean; but when the usual form of a river is considered, the trunk divided into many branches, which rise at a great distance from one another, and these again subdivided into an infinity of smaller ramifications, it becomes strongly impressed upon the mind, that all these channels have been cut by the waters themselves; that they have been slowly

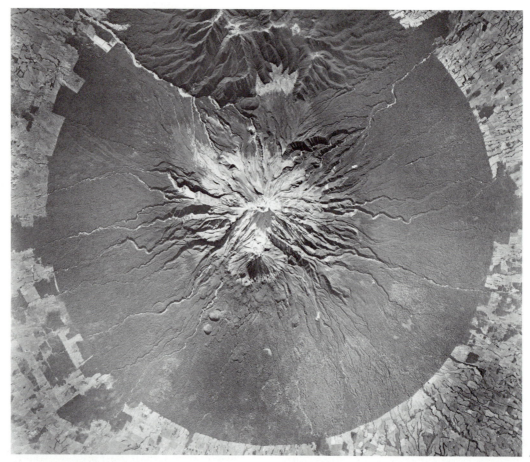

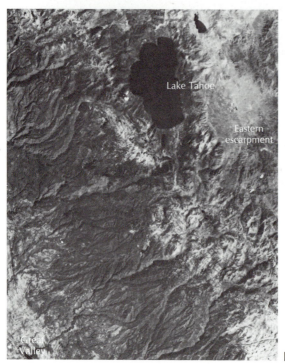

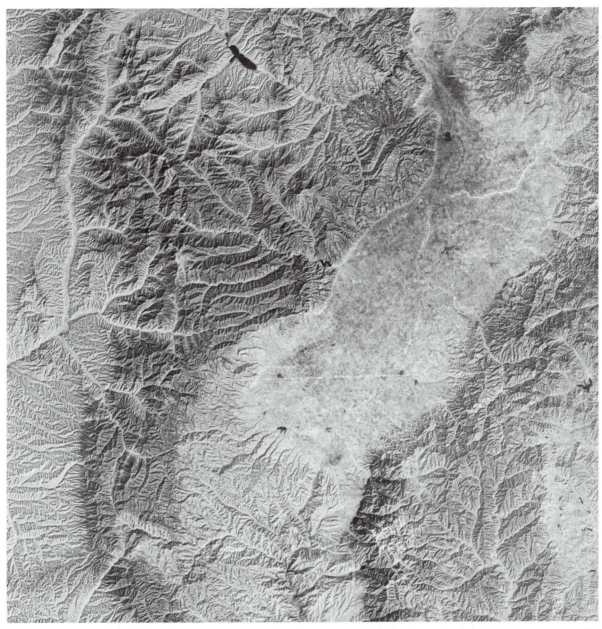

**D**

**Fig. 12.32.** Various drainage patterns. **A.** Radial drainage on Mount Egmont, a volcano in New Zealand. (Surveyor General, New Zealand Department of Survey and Land Information) **B.** Parallel drainage flowing southwest down the regional dip of the northern Sierra Nevada (California and Nevada), a range bounded by a zone of normal faults on the eastern side. Lake Tahoe lies in a fault-bounded basin. (NASA) **C.** Trellis drainage pattern in folded rocks of the Appalachian Mountains near Harrisburg, Pennsylvania. (EROS Data Center, USGS) **D.** Dendritic drainage formed in loess deposits, China. (NASA)

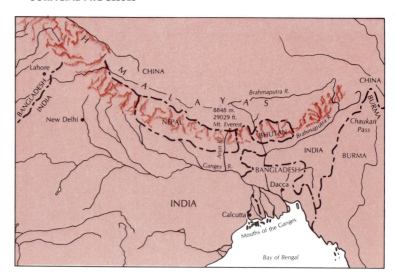

**Fig. 12-33.** Antecedent courses of the Arun and Brahmaputra rivers as they cross the Himalayas.

**Fig. 12-34.** The Grand Canyon at the foot of the Toroweap. (Drawing by William H. Holmes, from J.W. Powell, The exploration of the Colorado River, 1875)

dug out by the washing and erosion of the land; and that it is by the repeated touches of the same instrument that this curious assemblage of lines has been engraved so deeply on the surface of the globe.*

Let us look at the Colorado River again to review the role of streams in cutting and shaping canyons. The Grand Canyon of the Colorado flares upward in a series of gigantic steps and is 21-km (13-mi) wide at the rim. But the river shaped only a part of the canyon—that occupying a narrow vertical slot—as it cut down into the plateau. Other processes—weathering, mass movements, and slope wash—moved the materials making up the valley walls to the river. The river served as a conveyor belt to remove material from the region. Thus, the volume of the excavated canyon is much greater than that volume carved by the river alone.

Now if we move on to imagine a region dissected by large canyons, with erosion playing a role for millions of years, what can we surmise the region eventually will look like? Obviously, the land cannot be eroded below the position of sea level. This prompted Major John Wesley Powell (1834–1902), a pioneering geologist and the leader of the first party to explore the Grand Canyon (Fig. 12-35), to call sea level the base level of erosion.

The long-term effect of erosion is to wear the land down to an almost featureless plain near base level. In Powell's view, the formation of such a plain included not only a narrow stream channel carving the landscape but the wearing down of all interstream areas until an entire region was nearly at sea level. Running water and mass wasting processes combine to achieve this landscape, one that cuts across all geological structures. There would be some relief to the erosion surface, but all slopes would be quite gentle.

It is difficult to point out present-day low-relief erosion surfaces. Much of the world's land surface has stronger relief, an indication in itself of the recency and the continuing constructional forces of deformation. Add to this crustal unrest the fluctuating level of the oceans during the ice ages as well as the reaction of rivers to this fluctuation

and it is no wonder that low relief erosion surfaces are so rare.

Low-relief erosion surfaces probably have formed again and again throughout geological time. Some dissected surfaces are thought to have formed during times of crustal and base-level stability (Fig. 12-36). And some ancient and buried unconformities could be low-relief landscapes graded to an unknown base level (see Fig. 17-33).

Because low-relief erosion surfaces probably require millions of years to form and few modern analogs exist, there is disagreement as to how they evolve. Two main hypotheses have been suggested: ***downwasting*** and ***backwasting.***

In downwasting, slopes gradually diminish through time. Let us imagine an undissected landscape, such as a recently uplifted mountain block, as starting point and extrapolate the development of a low-relief erosion surface. Streams in the mountain block are widely spaced and of high gradient and erosive power. (An example would be the canyonlands of the western United States.) Eventually, the landscape is totally dissected and all slopes are more or less straight (Fig. 12-37). The streams occupy narrow valley floors and are still eroding on high gradients. The outline of the original mountain block has been eradicated. In time the rate of stream downcutting decreases, floodplains form, and the streams may be graded. Ultimately, the valley slopes take on a sigmoidal form with the upper convex slope formed by creep and a lower concave slope formed by wash processes. As erosion continues, slopes, relief, and river gradient all decrease, but because they are all closely tied together, they do so in a mutually compatible way. The lower slopes become, the more time is required to lower the landscape a fixed amount. The eventual form is called a ***peneplain.*** According to the downwasting hypothesis, this would be the idealized landscape evolution for a relatively humid region.

Low-relief erosion surfaces are also thought to form by backwasting. Here the slopes retreat essentially parallel to themselves. As in the downwasting model, we can think of an uplifted mountain block as a starting point. In time floodplains develop, the rivers become graded, and the landscape is dissected. Thereafter the slopes hold the same angle and slowly retreat parallel to them-

---

*From Illustrations of the Huttonian theory of the earth, by John Playfair, Edinburgh, 1802.

**Fig. 12-35.** Major John Wesley Powell, shown here with his companions before the journey, led the first expedition into the Grand Canyon in 1869. (J.K. Hillers, USGS) Powell's own words describe well the mood of the small band as it moved into the chasm:

*August 13.* We are now ready to start on our way down the Great Unknown. Our boats, tied to a common stake, are chafing each other, as they are tossed by the fretful river. They ride high and buoyant, for their loads are lighter than we could desire. We have but a month's rations remaining. The flour has been resifted through the mosquito-net sieve; the spoiled bacon has been dried, and the worst of it boiled; the few pounds of dried apples have been spread in the sun, and reshrunken to their normal bulk; the sugar has all melted, and gone on its way down the river; but we have a large sack of coffee. The lighting of the boats has this advantage; they will ride the waves better, and we shall have but little to carry when we make a portage.

We are three quarters of a mile in the depths of the earth, and the great river shrinks into insignificance, as it dashes its angry waves against the walls and cliffs, that rise to the world above; they are but puny ripples, and we but pigmies, running up and down the sands, or lost among the boulders. . . .

With some eagerness, and some anxiety, and some misgiving, we enter the canyon below, and are carried along by the swift water through walls which rise from its very edge. They have the same structure as we noticed yesterday—tiers of irregular shelves below, and, above these, steep slopes to the foot of marble cliffs. We run six miles in a little more than half an hour, and emerge into a more open portion of the canyon, where high hills and ledges of rock intervene between the river and the distant walls. Just at the head of this open place the river runs across a dike; that is, a fissure in the rocks, open to depths below, has been filled with eruptive matter, and this, on cooling, was harder than the rocks through which the crevice was made, and, when these were washed away, the harder volcanic matter remained as a wall, and the river has cut a gateway through it several hundred feet high, and as many wide. As it crosses the wall, there is a fall below, and a bad rapid, filled with boulders of trap; so we stop to make a portage. Then we go, gliding by hills and ledges, with distant walls in view; sweeping past sharp angles of rock; stopping at a few points to examine rapids, which we find can be run, until we have made another five miles, when we land for dinner.

Then we let down the lines, over a long rapid, and start again. Once more the walls close in, and we find ourselves in a narrow gorge, the water again filling the channel, and very swift. With great care, and constant watchfulness, we proceed, making about four miles this afternoon, and camp in a cave.

**Fig. 12-36. A.** *(Left)* Low-relief erosion surface (peneplain) cut across schists and subsequently was dissected by streams; site west of Dunedin, New Zealand. (W.C. Bradley) **B.** *(Below)* Flat erosional remnants form the summit plateau of the northern Wind River Mountains, Wyoming. These are interpreted by some to be the last remaining parts of an ancient peneplain. (Janet Robertson)

selves. Eventually, a low-gradient erosion surface develops at the base of the parallel-retreating slopes, widening as the slopes retreat. Ultimately, all land above the level of the widening platform will be stripped off as the separate retreating slopes meet, and an entire region will have been worn down to a base level of erosion with a gentle gradient. Such a surface is called a *pediplain.* Because this is the envisioned landscape evolution for arid areas, more will be said in chapter 13.

Because the final product of long-term erosion in humid and arid areas is a similar low-relief erosion surface (peneplain and pediplain, respectively), if all that remains are dissected remnants of the surface (Fig. 12-36B), it is difficult to say whether that surface was the result of downwasting or backwasting.

## RATIONAL USE OF RIVERS

In the early days of our country before the transcontinental railroad was completed and before the present web of highways was constructed, rivers provided the principal path of transportation for both goods and people. Now, of course, one of their most important uses is as a water supply.

In an attempt to harness the forces of nature, we have built and propose to build a great many dams across our rivers. The water stored behind the dams is to be used for a number of purposes: production of electric power, a steady supply of water for irrigation and other purposes, flood control, among other things. At the same time the number of rivers now preserved in national parks and monuments, or designated as wild rivers, in-

**Fig. 12-37.** The area around the Great Wall of China is representative of erosional landscapes, in which valley bottoms are narrow and most of the landscape is steeply sloping. (Wes LeMasurier)

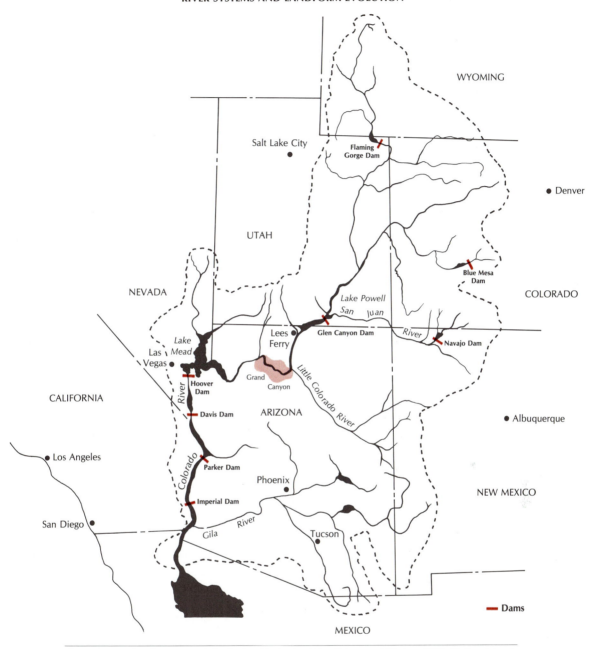

**Fig. 12-38.** Colorado River and the limits of its basin. (After *Water and Choice,* 1968, National Academy of Sciences, Publication 1689)

dicates that their aesthetic and recreational value is receiving more and more recognition.

That there is a conflict among users is clear from the minimum attention given to this topic even by the media. Choices must be made, but too often they are made by a small group of persons whose views do not necessarily represent those of the majority affected.

Even when plans for development of a river are disclosed, the proposers are often remarkably stubborn, refusing to listen to the opposition's arguments or to consider alternatives.

Luna Leopold, a prominent hydrologist, believes that a reason for such behavior is that although proposed benefits can be *quantitatively* stated, the "non-monetary values are described either in emotion-laden words or else are mentioned and thence forgotten." In an effort to remedy such situations, Leopold devised a chart to evaluate quantitatively some of the aesthetic factors of river sites. Disclaiming any personal bias, he applied his method to 12 river sites in Idaho in the vicinity of Hells Canyon along the Snake River, an area in which the Federal Power Commission wanted to construct one or more hydroelectric dams.

After comparing the Hells Canyon site with other sites in Idaho capable of hydropower development, he found Hells Canyon the most worthy of preservation. Leopold then compared the site to other river valleys that lie within national parks: (1) the Merced River in Yosemite, (2) the Colorado River in the Grand Canyon, (3) the Yellowstone River near Yellowstone Falls, and (4) the Snake River in Grand Teton National Park. His conclusion was that "Hells Canyon is clearly unique and comparable only to the Grand Canyon of the Colorado River in these features." It remains to be seen whether his method will be followed in determining the rational use of our river resources.

The Colorado River is an example of the myriad problems facing river planners. Although its headwaters are in the Colorado Rockies, the river basin encompasses parts of seven western states and Mexico before the river empties into the Gulf of California (Fig. 12-38). Each of these states has been allotted a portion of the water, and a certain amount is guaranteed to Mexico. Cities in the basin—which is part of the Sunbelt—have grown rapidly in recent years, and the demand for water has risen greatly. Even Denver and Salt Lake City, which lie outside the watershed, tap the Colorado, diverting water through tunnels. As other cities in the area grow, agriculture and industrial needs will also increase. In addition, if oil shale deposits in the northern part of the basin are ever developed, large amounts of water will be required. Increased pollution, too, is a threat; much of this consists of increased salinity resulting from agricultural uses along the river. As a result the water that reaches Mexico is of very poor quality.

Dams are controversial for a number of reasons. They flood farmlands and areas of scenic beauty and displace people. Evaporation losses from reservoirs are very high, so that much precious water is wasted. Lake Powell, for example, loses about 2 m (6.6 ft) of water per year through evaporation; one study estimates the total evaporation losses of the Colordao basin at nearly one-tenth of the river's discharge. Proposals for future dams will meet increasing opposition from conservation groups and local residents, who argue that there should be no further interference by developers in planning the fate of the Colorado or other rivers.

## SUMMARY

1. Streams are the main agents that transport material from land to sea.
2. Water in a stream flows in a turbulent fashion, and turbulent flow aids greatly in the transportation of the stream's *solid load (bed load* and *suspended load).* The *dissolved load* is derived from chemical weathering and quantitatively makes up an important percentage of the total load carried by a stream.
3. Stream velocity is determined by channel shape and roughness, discharge, and stream gradient. The greater the velocity, the greater the size of the material a stream is capable of moving.

4. Streams will reach an equilibrium in which the load added to the stream is in balance with the material carried by the stream. A stream that has reached equilibrium is called a *graded stream.*
5. Stream terraces reflect former stream disequilibrium during which the stream either *aggraded* or *degraded.* The two main kinds of stream terraces—*cut terraces* and *fill terraces*—each reflect a different stream history. Human intervention can bring about stream disequilibrium; often the effect is detrimental to people and to structures along the river.
6. Floods are natural to all streams; ascertaining the

extent of floods of various magnitudes is an important part of land-use planning. Flooding that occurs more frequently than the norm can be the result of building practices.

7. The two main kinds of stream patterns are largely related to the kind of load. *Braided* patterns reflect a high bed load, *meandering* patterns a suspended load.

8. High-relief landscapes, given sufficient time and tectonic stability, can evolve into a low-relief rolling erosion surface. In a humid climate slope processes are mainly responsible for the lowering of the original landscape, whereas the role of rivers is to transport material out of the drainage basin.

## QUESTIONS

1. Discuss the factors that would contribute to (a) an increase in velocity downstream and (b) a decrease in velocity downstream in an imaginary river.

2. In a particular river how would climatic change influence the amount of size of solid load? The amount of dissolved load? What are some of the factors that might help explain the ratio of solid-to-dissolved load in the Amazon River (Fig. 12-8)?

3. Keeping costs and benefits in mind, what factors must be considered before damming a river?

4. Would the recent geological history of a river be the same if it were lined with cut terraces or with fill terraces?

5. If low-relief erosion surfaces can be formed in both arid and humid climates, how could one determine the climatic conditions that prevailed during the formation of an ancient surface some 100 million years old?

## SELECTED REFERENCES

Baker, V.R., and Bunker, R.C., 1985, Cataclysmic late Pleistocene flooding from glacial Lake Missoula: a review, Quaternary Science Reviews, vol. 4, pp. 1–41.

Baker, V.R. and Milton, D.J., 1974, Erosion by catastrophic floods on Mars and Earth, Icarus, vol. 23, pp. 27–41.

Bloom, A.L., 1978, Geomorphology: a systematic analysis of late Cenozoic landforms, Prentice-Hall, Englewood Cliffs, N.J.

Chin, E.H., Skelton, J., and Guy, H.P., 1975, The 1973 Mississippi River Basin flood: compilation and analyses of meteorologic, streamflow, and sediment data, U.S. Geological Survey Professional Paper 937.

Chorley, R.J., Schumm, S.A., and Sugden, D.E., 1984, Geomorphology, Methuen, New York.

Costa, J.E., and Baker, V.R., 1981, Surficial geology, Wiley, New York.

Czaya, E., 1981, Rivers of the world, Van Nostrand Reinhold, New York.

Dunne, T., and Leopold, L.B., 1978, Water in environmental planning, W. H. Freeman, San Francisco.

Emerson, J.W., 1971, Channelization: a case study, Science, vol. 173, pp. 325–26.

Hays, W.W., ed., 1981, Facing geological and hydrologic hazards, U.S. Geological Survey Professional Paper 1240–B.

Leopold, L.B., 1969, Quantitative comparison of some aesthetic factors among rivers, U.S. Geological Survey Circular 620.

Mackin, J.H., 1948, Concept of the graded river, Geological Society of America Bulletin, vol. 59, pp. 561–88.

National Research Council, 1968, Water and choice in the Colorado Basin, National Academy of Sciences Publication 1689, Washington, D.C.

Ritter, D.F., 1986, Process geomorphology, W.C. Brown, Dubuque, Iowa.

Schumm, S.A., 1977, The fluvial system, Wiley, New York.

Selby, M.J., 1985, Earth's changing surface, Oxford University Press, New York.

**Fig. 13-1.** Sand dunes encroaching on the barren slopes of the Sierra del Rosario Range, Sonora, Mexico. (Peter Kresan)

# 13

# DESERT LANDFORMS AND DEPOSITS

The world's deserts are some of the least familiar of land areas (Fig. 13-1). Perhaps their seeming mystery lies in their distance from lands such as those of western Europe and the Atlantic coast of North America, where modern Western civilization became industrialized. Had Western life remained centered on the Mediterranean, deserts would have been much closer to our daily lives because the limitations imposed by aridity bear heavily on such desert-bordering countries as Spain, Morocco, Algeria, Libya, Egypt, and Israel.

In earlier days much of the southern shore of the Mediterranean was the granary of Rome, but with time once-flourishing cities, such as Leptus Magna in Libya, became stark ruins, half buried in the sand. One of the problems in studying deserts is that the boundaries are not fixed but may change through the centuries. Deserts require a rather special set of circumstances for their formation and existence. Our focus will be primarily on the action of wind and water in shaping desert terrain. We will show that deserts are not fixed areas, that their boundaries shift with time—sometimes helped along by the activities of humans—and that major climatic changes have taken place in arid areas in the geological past.

## CHARACTERISTICS OF DESERTS

First, we must agree on what constitutes a desert. Temperature is not the only factor; some are hot most of the time, others may have hot summers and cold winters, and some are cold throughout much of the year. Drought is their common characteristic; in a general way deserts are those areas where more water is potentially lost through surface evaporation or transpiration from vegetation than actually falls as rain.

Desert regions are classified according to their degree of moisture deficit as extremely arid, arid, or semiarid (Fig. 13-2). Extremely arid areas are those where no rain has fallen for at least 12 consecutive months. Arid and semiarid regions occupy one-third of the earth's land area and are concentrated in subtropical and in middle-latitude regions. For example, nearly continuous desert runs from Cape Verde on the west coast of Africa, across the Sahara, the barren interior of Arabia, the desolate mountains of southern Iran, and on to the banks of the Indus in Pakistan. The preponderance of the dry areas of the earth—exclusive of the Arctic—are on either side of the equator, chiefly around latitude 30°, and they tend to occur

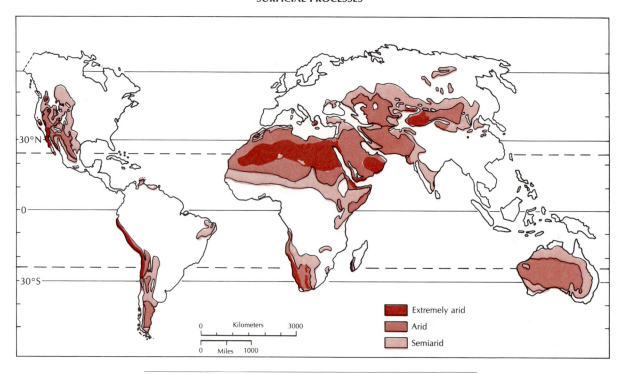

**Fig. 13-2.** Distribution of the earth's deserts, exclusive of the polar deserts.

on the western side of continents. The polar deserts are dry and extremely cold. Most occur above latitude 70° N in the Arctic or in the dry valleys of the Antarctic. Many landforms, weathering features, and soils are common to both kinds of deserts, but our focus will be on the hot deserts.

Desert climates are characterized by great extremes. Temperatures as high as 57° C (135° F) have been recorded in Algeria and as low as −20° C (−4° F) in Arizona. Precipitation at one station in Chile ranges from 0 to 390 mm (16 in.) per year, but the lowest mean annual precipitation is at an Egyptian station, where only 0.4 mm (0.02 in.) per year has been recorded. The record number of years without rainfall stands at 15 years for a station in Namibia. However, when the rains do come they can be intense; 1 mm (0.04 in.) per minute falls in some Saharan storms.

Because drought is their prevailing characteristic, deserts notably are regions of sparse vegetation. Only extremely arid deserts are completely devoid of plants, except for the few scattered ones along the dry riverbeds, which almost always hold some moisture. In more typical deserts plants are widely spaced. Their colors tend to be subdued and drab, blending with their surroundings; but in the spring even the cactus blooms with colorful flowers. Their leaves may be small and leathery to reduce evaporation. In fact, some, such as the saguaro or the barrel cactus do not have true leaves. Other desert plants, such as the ubiquitous sage, may develop deep root systems in proportion to the part of the plant that shows above ground.

Animals, although few in number, are also found in deserts. Prehistoric natives in North America hunted mammoth and bison in the western deserts; today in the extremely arid desert of the Mideast gazelle, ibex, and even wolves can be seen.

Contrary to the popular image, most deserts are not vast shimmering seas of sand. More likely they are broad expanses of barren rock or stony ground with shallow soil profiles. Ground colors are largely those of the original bedrock. For example, the bright red color that we associate with such places as the Grand Canyon and Monument Valley comes in large part from iron oxides within the rocks themselves. In some deserts, however, red colors can also be contributed by soils.

It is typical of many arid regions, especially those in continental interiors, that streams originating within the desert often falter and disappear within the desert's boundaries. The term given to a pattern of streams that do not reach the sea is *interior drainage.* Some streams simply shrink and sink into the sand, whereas others may carry enough water to maintain a temporary or permanent lake in a structural basin at the end of their course. Such an area is the Great Basin of the western United States (Fig. 13-3), a vast desert region

**Fig. 13-3.** Satellite image of the desert in the southwestern United States, along the California-Nevada border. Death Valley is near the center of the image. The topography is typical of the region with ranges flanked by smooth gravel plains (dark tones), separated by basins with lake deposits (light tones). (NASA)

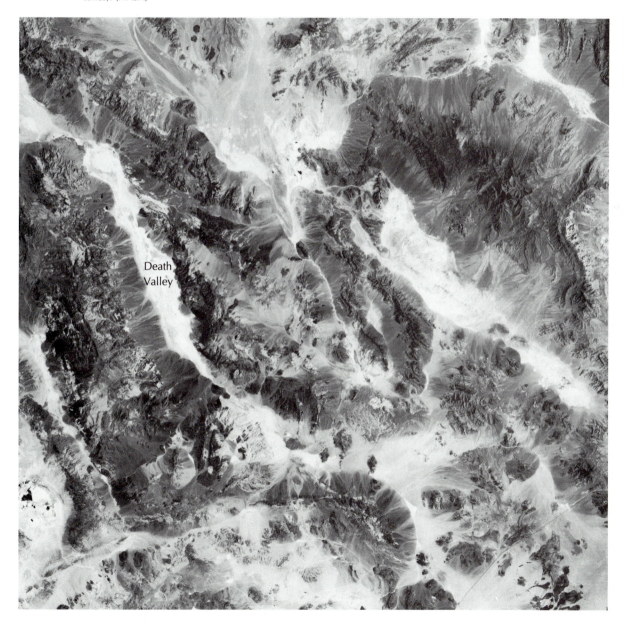

Death Valley

stretching from the Sierra Nevada eastward to the Wasatch Range in Utah and from the area north of the Colorado River into eastern Oregon and southern Idaho. Early pioneers often followed rivers as they moved westward across this territory, only to find that the rivers drained into lakes with no outlet or that they simply vanished in the desert. Most lakes in a basin are highly saline and fluctuate in depth with climatic variations.

Most desert rocks have a dark surface coating—a shiny bluish-black substance called **desert varnish.** This thin coating can cover entire stones and form patterns on high vertical cliffs. Native North Americans scraped off the varnish to make their picturesque petroglyphs (Fig. 13-4).

Chemical analysis indicates that desert varnish is a mixture of iron and manganese oxides and clay minerals. Recent research suggests that much of the varnish is from airborne substances and that microorganisms played a role in the precipitation of the oxides. Archaeological and geological evidence indicates that several thousand years must elapse before desert varnish becomes a prominent feature.

Another unique feature of deserts is **stone pavement**—a thin veneer of stones one to two stones thick. The stones form a tightly packed mosaic as if put together by a mason—hence the term (Fig. 13-5). Typically, the layer of soil (the A horizon) directly beneath the surface is light-colored. Early inhabitants of the Peruvian Desert made patterns by selectively removing the stones and exposing the light-colored soil layer (Fig. 13-6).

Three hypotheses have been advanced to explain the origin of most stone pavements. One is wind erosion. Consider a sand and gravel deposit exposed at the surface with little vegetation for cover. Winds sweeping across the surface can pick up sand and finer-grained materials and carry them away, but the gravel will remain. If the process continues, eventually the gravel and larger stones will form a tightly packed lag concentrate. A second hypothesis states that the stones move from the shallow depth to the surface. If the gravels were originally scattered throughout a somewhat clayey matrix, the shrinking, cracking, and swelling that accompanies drying and wetting cycles could move gravel toward the surface and there concentrate it into a pavement.

**Fig. 13-4.** Desert varnish on sandstone in the Glen Canyon of the Colorado River. Prehistoric Indians inscribed petroglyphs by scraping away the dark coatings. Notice the bighorn sheep, important prey of the Indians, and the self-portrait with a headdress of bison horns. (Utah Museum of Natural History)

A third hypothesis—formulated only recently—considers the deposition of atmospheric dust that is so common to deserts. Once a surface (alluvial fan, basalt flow, etc.) is formed, dust begins to settle on it. The dust infiltrates the surface layer of rocks and accumulates just beneath it. Eventually, as more dust comes in, a desert pavement is formed, overlying nongravelly fine-grained material. This theory could fit the field evidence best.

Whatever their origin, stone pavements form an armor that protect the surface from further wind erosion. And, like desert varnish, they take a long time to form; commonly, those gravel surfaces with the best development of stone pavement also are well coated with varnish.

**Fig. 13-5.** Stone pavement in southern Israel. The material beneath the pavement is relatively free of stones. (Peter W. Birkeland)

## CONDITIONS CAUSING DESERTS

What special circumstances are responsible for the aridity of some parts of the earth's surface? If we omit the polar regions, there are four major types of arid regions: (1) horse latitude deserts, (2) continental deserts, (3) rain shadow deserts, and (4) deserts produced by cold coastal currents in tropical and subtropical regions.

Most of the world's deserts are located about 30° north and south of the equator, as noted earlier, in the so-called horse latitudes. Such deserts owe their origin, in part, to global atmospheric circulation (Fig. 13-7). Close to the equator, atmospheric pressure is low because of the rising hot, moist air. As air rises it cools and loses its ability to hold moisture, thus producing the high rainfall of the equatorial regions. The cooled dry air then spreads northward and southward from the equator into the horse latitudes, where it descends and is warmed. The lands of the horse latitudes, then, are deserts because they are continually parched by warm, dry winds that evaporate any moisture present. The three largest deserts on earth, the Sahara and the Australian and Arabian deserts (in order of decreasing size) all are of this origin.

In contrast continental deserts are so named because they lie in the interior of large continents, far from the main moisture source—the oceans. A familiar example is the Gobi Desert of northern China and Mongolia.

A third type of desert is caused by the interposition of a mountain barrier in the path of a moisture-bearing air current. A striking example is found in the western United States where the desert stretches eastward in the rain shadow of the Sierra Nevada of California. As moisture-laden air from the ocean rises along the west side of the Sierras, it is cooled, and rain falls; maximum rainfall actually occurs west of the divide. The now-dry, cooler air flows down the eastern slope, is warmed again, and evaporates any water in its path.

Least familiar to inhabitants of the Northern Hemisphere are the coastal deserts of the mid-latitudes. They owe their origin to cool coastal currents and cool air masses that move inland from the oceans and, therefore, have little capacity to hold water. Perhaps the best known are on the west coast of two Southern Hemisphere continents. One is the desert of southern Peru and northern Chile formed by the Humboldt Current; the other is the Namib Desert of Namibia formed by the Benguela Current. Both currents run north-

**Fig. 13-6.** Geometrical patterns—created by piling up rocks coated with desert varnish—stretch across the desert in the Nazca Valley, southern Peru. The light color of the pattern is the exposed surface of the soil. The patterns date from about 100 B.C. to A.D. 700, but no one knows why they were made. The size of many of the designs suggests that they were meant to be visible only from the air, or, possibly, that they were monuments to important gods, hence their size. (Copyright © Marilyn Bridges)

ward along the coasts, carrying cool antarctic waters with them.

Coastal deserts are impressive because of the dramatic climatic contrasts encountered within extremely short distances. The Atacama Desert, for example, is one of the driest lands on earth, yet its seaward margin is concealed in a virtually unbroken gray wall of fog. Air masses moving across the cold waters of the coastal current are chilled, so the little moisture that does exist is condensed to form a seemingly eternal blanket of fog over the sea. Once the fog drifts landward,

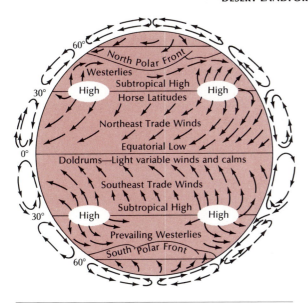

**Fig. 13-7.** Schematic diagram of atmospheric circulation. Because of the rotation of the earth, the winds are not north–south but at an angle, thus producing the equatorial easterlies and the midlatitude prevailing westerlies.

where air temperatures are higher, it burns off, and the capacity of the air to hold moisture increases rather than decreases as it moves across the heated land. In southern Peru vegetation and cultivated fields are found only a short distance from the barren coast.

About once every decade the coasts of Ecuador and Peru are drenched by torrential rains during a heat wave called *El Niño* (Spanish for "The Child") as this phenomenon appears around Christmastime. For complex reasons that involve the earth's atmospheric and oceanic circulation, the cool oceanic currents weaken, the coastal waters become noticeably warmer, and severe storms batter the coastal region. During the last episode of El Niño in 1982–83, parts of the desert received between 3 and 4 m (10 and 13 ft) of rain (Fig. 13-8). Total rainfall decreases to the south. The floods can be localized, devastating some areas but not others.

## EROSION AND DEPOSITION—STREAMS AND LAKES

Paradoxical as it may seem, running water is the agent most responsible for sculpturing desert landforms, perhaps more so than in humid regions. It is, indeed, puzzling how the enormous volumes

**Fig. 13-8.** Floodwaters engulf school buses during the El Niño of 1983, Río Seco, north of Lima, Peru. Many rivers, such as this one, flow only during the heavy rains that accompany El Niño phenomena in the arid Peruvian desert. (Pablo Lagos)

of rock that once filled desert canyons or shaped the mountains themselves were removed or worn down to bare rock plains in a land where there appear to be no streams.

It may be that not all landforms in all deserts were formed under the climatic regime we see today; they may be fossil landscapes, in a sense, that is, survivals of erosional patterns carved during a more humid time. We will see one result of a recent climatic shift a little farther on when we discuss the lakes that formed in desert regions during times of recent glaciation. Not all the ero-

sion of desert landscapes, however, can be attributed to an earlier, wetter climate. Almost all deserts have some rainfall, even though 10 to 15 years may elapse between showers. So it could be that the infrequent storms over a long enough period of time provide enough water for erosion to take place.

## The Nature of Runoff

Contrary to popular belief, cloudbursts are relatively rare in arid lands—they are much more

**Fig. 13-9.** Badland topography, formed in shale, in the arid region of southern Israel. (Peter W. Birkeland)

**Fig. 13-10.** Arroyo incised in a river valley floor in Nevada. The flat riverbed and steep walls composed of young, fine-grained river deposits are typical of many arroyos. (Karl Birkeland)

common in areas where rainfall is greatest, for example, in the rainy tropics or the southern Atlantic coastal states. But a moderate rain in the desert can assume the proportions of a cloudburst in a more humid region and also do a very effective job of erosion. The reason is that little or no vegetation protects the slopes from the spattering effect of raindrops or from water running across the surface, rapidly cutting gullies. This can lead to the formation of **badlands,** or slopes scored with great numbers of gullies, large and small. These are characteristic desert landscape elements where the bedrock is mainly impermeable shale (Fig. 13-9).

In many deserts river channels, although they seldom carry much water, are incised in the valley bottom. These steep-sided, flat-floored river washes are called **arroyos** (Fig. 13-10), Spanish for "stream" or "brook." In the West accelerated erosion in the late 1800s caused many of these features. It may

be that a subtle climatic change or overgrazing by livestock recently introduced, or both, were responsible for the downcutting.

In a sudden desert downpour, it takes but moments for a dry, sandy arroyo bordered by low but steep cliffs to be flooded with surging, mud-laden water. The stream swirls and churns violently forward, sweeping along a great mass of debris. Boulders of all sizes can be moved in the torrents. Such flash floods make the desert impassable until the arroyos drain. And they drain almost as rapidly as they fill because there is no continuing source of water. After only a few hours beneath the desert sun, an arroyo floor earlier covered by water may be dry sand again, with only an occasional pool of muddy water. Evidence of the event, however, will remain in the form of transported rocks and other debris.

Occasionally, the short-lived torrential floods overflow the low banks of dry washes, spreading

1 km

N ← North

**Fig. 13-11.** Vertical air photo of alluvial fans where streams debouch onto the valley floor—east side of Death Valley, California. (Fairchild Aerial Photography Collection, Whittier College)

a sheet of muddy, turbulent water over the desert floor. Such a sheet flood will pick up loose sediments and shift them around the landscape.

Rates of erosion in deserts vary with runoff and mean annual rainfall. In small drainage basins in the United States, the maximum amount of erosion seems to occur in semiarid regions, with lesser amounts in arid regions and in humid regions. There is less erosion in more-arid regions because rainfall, the main cause of erosion, is scant, whereas in humid regions the stabilizing effect of vegetation lowers the rate of erosion (see Fig. 12-9).

### Depositional Landforms

Wherever erosion occurs, deposition takes place nearby. Deposition is especially prominent in arid regions because ordinary streams cannot escape beyond the confines of the desert. Water evapo-

rates or sinks underground into the sandy stretches of normally dry stream beds or it may be taken up by fiercely competitive water-seeking plants, such as the mesquite and tamarisk, that line the banks of many desert watercourses. Much of the desert landscape is dominated by stream deposits, in large part because there is not enough running water to move debris out of a desert basin and on to the sea.

The main depositional desert landform is the **_alluvial fan_** (Fig. 13-11). Its name comes from its shape, which approximates a portion of a cone that enlarges downslope from the point where a stream leaves the mountain front. Fans commonly are composed of accumulations of gravel and sand; deposition on their surfaces is attributed to two factors, both of which reduce the velocity of the stream: (1) much of the water sinks into the porous, sandy subsurface layers of the fan, thereby decreasing the discharge, which helps to control the velocity; (2) the main channel, on leaving the mountain front and entering the fan, soon branches into a score of distributaries in a braided pattern. The resulting increase in length of channel perimeter also helps to decrease the velocity and to bring about deposition.

Alluvial fans vary in size and degree of slope, but there is consistency in the variation. In a general way, the area of a fan is proportional to the area of the drainage basin from which the fan debris is derived; thus, small drainage basins produce small fans and large drainage basins, large fans. The slope of a fan is related to the parameters that control the slopes of streams. Fans derived from large drainage basins tend to have relatively low slopes as do those constructed by streams with large discharge. Finally, sediment size helps to control the slope, for larger gravels are usually associated with fans having a steeper gradient.

Streams may be able to cross the fan in time of flood, but under normal conditions they sink into the sandy ground almost as soon as they leave the bedrock of the mountains. Visitors to desert country almost always are surprised to see a stream waste away, growing thinner and thinner downstream and then vanishing completely—quite the reverse of humid-climate streams, in which volume commonly increases downstream. The water does not actually vanish but percolates slowly through pore spaces in the fan. Far down the fan, the same water may seep out in springs—commonly marked in the desert by clumps of mesquite trees, one of the best guides to water in the arid Southwest.

Given enough time, the alluvial fans flanking most mountain ranges will coalesce with adjacent fans. Overlapping like palm fronds, the fans form a nearly continuous apron sloping away from the mountain front to the basin floor. The apron is called a **bajada,** from Spanish, meaning "a gradual descent" (Fig. 13-12).

**Fig. 13-12.** Bajada on the west side of Death Valley at the foot of the Panamint Range. Each fan has deposits of several ages as shown by the position of stream (the higher fans are older) and tonal differences caused by desert varnish (the darker the older). (H.E. Malde, USGS)

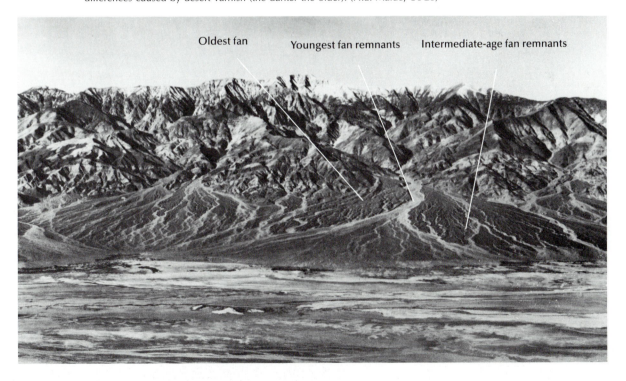

Many desert basins are closed areas with interior drainage. During a downpour water rushes down the bedrock canyons out onto the bajadas and, if rainfall is sufficient, collects to form a shallow lake in the lowest part of the basin. High evaporation rates quickly remove the water. At times, however, no water reaches the lower part of the basin. Contrariwise, in a humid region, where precipitation exceeds evaporation, such areas are almost always occupied by a lake with waters that rise until they spill over the lowest part of the basin rim. This latter process continues basin by basin until streams downcut, drain the lakes, and form an integrated regional drainage system.

Lakes in arid regions are salty or brackish—the result of high evaporation losses. The Dead Sea, about 396 m (1300 ft) below sea level, into which the River Jordan flows, is a renowned example. The Dead Sea waters contain so much salt that swimmers float high in the water and find that swimming on the back is the most stable position (Fig. 13-13).

A larger body of water without an outlet is the Caspian Sea, and even though it is supplied by the mighty Volga, not enough water reaches it to allow the lake to spill over the low divide separating it from the Don River and the Black Sea.

A desert lake familiar to North Americans is the Great Salt Lake, which covers some 3800 km$^2$ (1467$^2$ mi) in Utah. Similar to the Dead Sea, parts of the lake have a salt content that approaches 27 percent, or nine times that of ocean water.

In many parts of the world, however, where the water brought in by streams evaporates, desert lakes may be only short-lived seasonal affairs and may be completely dry for decades. Such ephemeral lakes are called **playa lakes;** they evaporate quickly, leaving a **playa** (Fig. 13-14).

Playa lakes vary in salt content and in the amount of fine-grained (silt and clay) particles they carry. Accounting for the high salinity for some lakes are salts brought in as dissolved stream load that become concentrated as pure water evaporates from the lake surface. Because soluable constituents in desert rocks may be carried into such a lake, it is no wonder that some desert lakes as they evapo-

**Fig. 13-13.** Swimmers floating in the highly saline waters of the Dead Sea, Israel. (Peter W. Birkeland)

**Fig. 13-14.** Laguna Salada, a shallow lake in an extensive playa at the base of the Sierra los Cucapas Mountains, northeastern Baja, Mexico. (Peter Kresan)

**Fig. 13-15.** The Devil's Golf Course, Death Valley, California, is composed of a variety of salts that were concentrated in ancient Lake Manley at the end of a long drainage system. The salts were precipitated as the lake disappeared. The reason for the depressions and jaggedness is both the dissolution of preexisting salt and the formation of new salt crystals. (Ansel Adams)

**Fig. 13-16.** Remnants of a pediment cut on granite that slopes away from steep mountain fronts in the southern Sinai Desert. This pediment has been dissected and the underlying bedrock exposed. In the same area is the famous sixth-century monastery of St. Catherine as well as Mt. Sinai, the latter a landmark since the time of Moses. (Ran Gerson)

rate precipitate such minerals as potash, potassium salts, and boron compounds. A well-known example of a salt playa is the Devil's Golf Course, a jagged terrain of salt pinnacles on the floor of Death Valley (Fig. 13-15), nearly 90 m (295 ft) below sea level. It was a fearsome stretch for the first party of emigrants to cross. That playa receives some of its water and salt from the Amargosa River, which name appropriately enough means bitter. If, however, the streams are choked with fine-grained sediments, the result is a clay playa.

Most salt and clay playas are expansive and dead flat. Automobile and bicylce and speed records are set on the Bonneville Salt Flats on the edge of the Great Salt Lake, and the hard clay playas in California serve as landing strips for space shuttles.

### Erosional Landforms

In addition to gullies, valleys, and canyons, there is in the desert another erosional landform, the **pediment**. Pediments are bedrock surfaces stripped bare, or with a thin veneer of sediment, that slope gently away from desert mountains toward the lower part of an intermontane basin (Fig. 13-16). The surface usually meets the mountain front at a sharp angle. From a distance the broad, encircling surfaces look like a uniformly sloping bajada rather than the product of long-continued degradation involving the removal of large volumes of bedrock. In fact, there is no easy way to visually differentiate fans, which are areas of deposition, from pediments, which are areas of erosion.

Pediments were first described in this country in 1897 by an American exploring geologist, William J. McGee (1853–1912). He was as surprised as anyone by their true nature:

During the first expedition of the Bureau of American Ethnology [1894] it was noted with surprise that the horse shoes beat on planed granite or schist or other rocks in traversing plains 3 or 5 miles from mountains rising sharply from the same plains without intervening foothills; it was only after observing this phenomenon on both sides of different ranges and all around several buttes that the relation . . . was generalized.

The formation of pediments has long been a puzzle to geologists. Some think that they are formed by lateral planation of streams, that is, erosion caused by a stream as it swings back and forth over a rock surface. The materials carried by the stream slowly cut away at the bedrock, and where the stream touches the mountain front, it causes the mountain front to retreat slightly. Other geologists believe that pediment surfaces weather, and that the products of weathering are removed by water sweeping across the pediment during rainstorms. The mountain front also weathers, and various processes later deliver this debris to the upper end of the pediment. Debris of a certain size will continue to be carried across the pediment; all other material will remain at the mountain front until weathering reduces it to a size that can be transported.

Another hypothesis on the origin of pediments was advanced to explain those in the Mojave Desert in southeastern California. Evidence suggests that the Tertiary landscape, say 8 million years ago, was one of soil-covered hills with fronts possibly retreating in a parallel fashion (Fig. 13-17). A climatic change toward greater aridity a few million years ago increased the erosion rate and stripped the land of its soil cover. Bedrock pediments and bouldery mountain fronts are part of the stripped surface. The pediment is inherited from the past, so that one does not have to look to present arid conditions for its formation.

Certainly, this hypothesis has much merit. However, it may not explain the origin of all pediments. One problem in understanding many arid landforms is that in an area with scant precipitation there is little opportunity to observe formative processes if running water is the main formative agent.

### Erosional Cycle in an Arid Region

Geologists have made a considerable effort to determine if landforms evolve through time from one characteristic form to another. Some, for example, envisage the landscape in a humid climate as initially hilly, only to be converted into a peneplain after millions of years of erosion, dominantly backwasting. And, so the argument goes, perhaps an erosion cycle can also be deciphered from desert landforms.

As an example, we will use the Southwest (Fig. 13-18). We start with a recently uplifted mountain range bounded by normal faults—hence it has a relatively straight front. Down-dropped blocks form basins that separate adjacent ranges. As weathering and erosion proceed, streams incise the mountains and the mountain fronts recede nearly parallel to their original slopes. The flattening and rounding caused by soil creep in humid regions is not to be found. Also, debris is not carried out of the area, but accumulates in the basins. Because streams are few and flow short distances only, an integrated drainage system cannot develop. As a consequence the valley floor is occupied by a playa. Alluvial fans lead out to the playa. The basin gradually fills with sediment and drainage eventually can spill into the next basin, becoming "throughgoing."

Provided the area remains tectonically stable for a long period of time, and through-flowing drainage is established, slopes eventually can retreat from the original mountain front to form pediments. The pediment continues to expand at the expense of the mountain until only mere fragments of the mountain—called **inselbergs** (German word for "island mountains")—remain. Finally, the inselbergs are consumed and low bedrock domes are all that remain. These extensive coalescing pediments are called *pediplanes* and correspond to the peneplains of a humid region. Thus it is believed that gently rolling terrain of very low relief is the end result of erosion in any climate. Few areas remain tectonically stable long enough for such landscapes to develop, however.

Active faulting and climatic variation can alter the processes that form alluvial fans and pediments. In the Great Basin, alluvial fans are generally found in the north, where active faulting is taking place, and pediments occur in the south, where faulting processes have long been dormant. So, tectonic stability seems essential to pediment formation.

## WIND

### Wind Erosion

Over many of the dry lands of the world, the wind blows seemingly without restraint, thus adding a note of melancholy to an already desolate terrain. The drier part of Patagonia in the far southeastern reaches of Argentina is such a land. Other deserts are perhaps as windy on occasion, but in most of them times of extreme winds alternate with times of calm.

When the wind blows in the desert, its erosional effectiveness is likely to be much greater than in

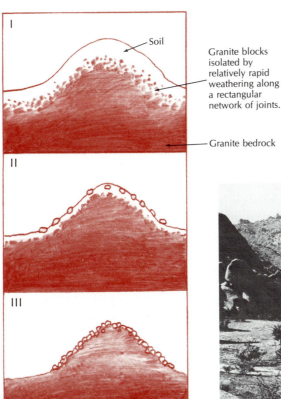

Soil

Granite blocks isolated by relatively rapid weathering along a rectangular network of joints.

Granite bedrock

A

**Fig. 13-17. A.** Possible landform evolution in the Mojave Desert. Stage I is the soil-covered granitic landscape in Tertiary time. The bedrock weathers more toward the surface, and jointing controls the blocky weathering forms. Stages II and III represent progressively greater stripping of the soil and weathered granite to the bedrock that dominates the landscape today. The low-lying flat areas would be our present-day pediments. (After T.M. Oberlander, 1972, ''Morphogenesis of Granitic Boulder Slopes in the Mojave Desert, California,'' *Journal of Geology*, vol. 80, pp. 1–20, Fig. 14) **B.** *(Below)* Mountain front in the Mojave Desert made up of huge granitic boulders similar to those in Stage III of the drawing. The flat area in the foreground is the pediment. (W.C. Bradley)

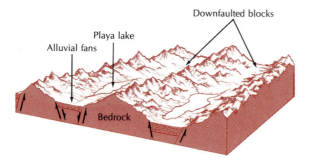

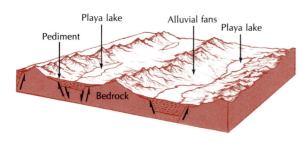

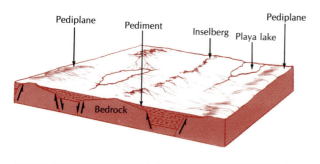

**Fig. 13-18.** The erosional cycle in the semiarid-to-arid Southwest. (After W.K. Hamblin, 1985, *The Earth's Dynamic Systems, Fourth Edition*, p. 221, Fig. 12-16. By permission of Macmillan Publishing Company)

ciers will be moved by the wind—even by a tornado, but small pebbles, about 4 mm (0.16 in.) in diameter, commonly are moved. It is the particles being moved that erode surfaces during high winds.

Sand grains move along the ground in a series of hops—a motion known as *saltation* (Fig. 13-19). Grains may bounce off other grains on the surface or they may stop abruptly, the force of their impact setting other grains in motion. The height to which the grains bound depends on the characteristics of the surface. Grains bound higher when bouncing off a hard surface, such as a pebbly stone pavement or a road, than from a loose sand surface that absorbs much of the impact energy. Saltating sand particles seldom rise to heights greater than 1 m (3 ft) or so above the surface.

Saltating grains have a marked effect on the shape of rock outcrops or gravel. The outcrops are undercut and smoothed. In contrast hard gravel particles or stones will be cut on the upwind side to form a low-angle planar surface called a **facet.** Faceted stones are called **ventifacts;** because the facets form at approximately right angles to the wind, they have been used to help deduce ancient wind directions. Because stones can be shifted about by wind scouring near their bases or by other means, including frost heave or animal disturbance, multifaceted stones are more common than not (Fig. 13-20).

Wind also erodes soil or bedrock, producing **deflation hollows.** Thousands of these small closed basins are found in the Great Plains. In Australia there are similar basins, their wind origin proved by the accumulation of deflated material just downwind from the basin. An exceptional closed basin is the Qattara Depression in the western desert of Egypt, about 300-km (186-mi) long with a floor 134 m (440 ft) below sea level. It is a forbidding quagmire of salt and shifting sand that covers about 19,500 km² (7520 mi²). Whatever the origin of the Qattara Depression, the wind almost certainly played a prominent role in enlarging it and in deepening it to the water table. Evaporation of the water through centuries has led to the accumulation of salt deposits. This immense, impassable saline trough acted as a barrier in World War II, making it impossible for Rom-

humid regions. In the latter the ground surface is protected by vegetation—vegetation that also acts as an extremely effective baffle to decrease wind velocity—and by a more tenacious mantle of weathered soil—soil that also may be damp throughout most of the year. The wind is very effective in transporting certain sizes of particles, but not others. Obviously, there is an upper limit to the size of loose particles that it reasonably can be expected to move. No boulders of the size that are transported by streams, ocean waves, or gla-

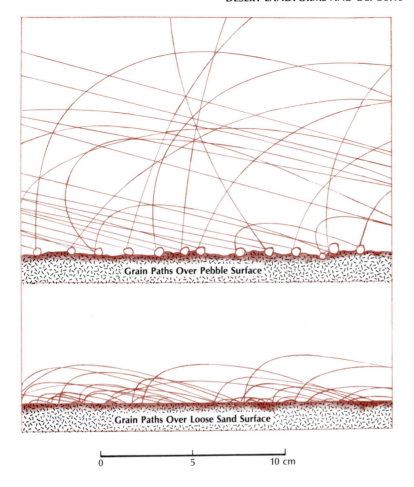

0    5    10 cm

**Fig. 13-19.** The paths that saltating sand grains take across two different surfaces. (After R.A. Bagnold, 1941, *The Physics of Blown Sand and Desert Dunes,* Metheun, London. Halsted, N.Y., 1965.)

**Fig. 13-20.** Faceted fine-grained igneous rocks, Antarctica. Stone **C** is 7-cm (2.8-in.) wide. (Peter W. Birkeland)

mel's Afrika Korps to turn Montgomery's flank. Thus the Germans were compelled in 1942 to attack the grouped British forces at El Alamein on the narrow neck of land separating the Qattara from the Mediterranean.

Groundwater sets a lower limit to wind erosion. Once the desert surface is deflated to the level of the water table and the ground is kept damp, the wind can no longer pick up loose material with the same ease and deflation virtually halts. A beneficial effect of deflation is the appearance of springs around the margins of some depressions. Such springs are the source of water for the true oases, for with only a little groundwater to draw on, date palms flourish, making a startling contrast with their stark surroundings.

## Wind Transportation

Wind can transport enormous amounts of fine-grained sediment (dust) for long distances. It is estimated that about 500 million tons of dust are carried by the wind from deserts each year. One recent storm in southern California moved 25 million tons of sediment in 24 hours, some of which was deposited in northern California (Fig. 13-21). Satellite imagery shows dust plumes 600-km (370-mi) wide that extend 2500 km (1550 mi) from Saharan source areas westward to the Caribbean islands and South America. Dust deposits on the Hawaiian Islands have a source in Central Asia, and sensitive instruments on Mauna Loa even detect dust from spring plowing in China. It has been estimated that Egypt has about 10 dust storms a year, that China has 30, and that Kazakhstan in Soviet Central Asia tops the list with 60.

Such statistics are strong evidence that the wind can be an effective erosional agent and one of great significance in arid regions. In fact, it is the only agent that can transport large amounts of material beyond the confines of a typical desert.

Odd as it may seem, very fine-grained particles, such as the silt or clay of dust, are not easily set into motion by the wind. In general, the same

**Fig. 13-21.** Wavelike cloud of dust in a 1977 windstorm in the southern San Joaquin Valley of California. Maximum altitude reached was about 1600 m (5250 ft). An area of 2000 km² (780 mi²) was affected by winds that may have reached 300 km/hr (186 mi/hr). Drought, overgrazing, and the lack of windbreaks all contributed to the severity of the storm. (H.G. Wilshire, USGS)

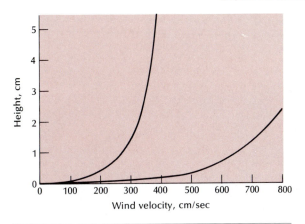

**Fig. 13-22.** For two different events, variation in wind velocity in relationships to ground-surface proximity. (After R.A. Bagnold, 1941, *The Physics of Blown Sand and Desert Dunes*, Metheun, London. Halsted, N.Y., 1965.)

principles apply here that apply to streams. Silt and clay particles, which are small and strongly cohesive, cannot be picked up readily by a moving current of air because wind velocity diminishes close to the ground (Fig. 13-22). This is demonstrated when the wind blows full force across the surface of a clay playa. Very little dust is stirred up as a rule and the hard-packed clay particles hold firm. Some of the looser sand and silt around the margin, however, may be picked up, especially if the playa surface is sun-cracked. If these sand grains saltate across the clay, particles of clay can be kicked into the air and carried aloft.

Once silt-or clay-size particles are picked up by the wind, however, they remain aloft much longer than sand does. There are no restraints imposed on their travel comparable to those of their water-borne equivalents, which are restricted by the drainage pattern of a stream or a glacier's course.

## Wind Deposition

Materials deposited by the wind generally fall into two size groups. Those that are carried as dust in suspension are mainly silt and clay, whereas those that saltate along the ground are sand.

**Loess**   The term *loess* comes from a Swiss-German word meaning "loose"; it originally referred to fine loams in the Rhine and other European valleys. The most renowned dust deposits are those of northeast China, described by the German geologist Baron von Richthofen (1833–1905) while on an exploring expedition to the outermost parts of the Russian and the Chinese empires. The loess of China was transported by the wind out of the Gobi and across the barren Zhangjiakou (formerly Kalgan) range on which the Great Wall was built. Deposited on the North China plain by the dust storms of centuries, loess lies deep on that ancient land, often to thicknesses of hundreds of meters (Fig. 13-23). This tan-colored silt gives both the Yellow River (Huang He) and the Yellow Sea (Hwang Hai) their names. The sea is colored for hundreds of kilometers from shore by suspended dust particles. Loess in southern Israel may have been carried from the Sahara. In other regions loess is closely associated with glaciation (see chap. 14). In the United States and Europe, dust derived from glacial-age floodplains forms extensive loess deposits that produce some of the world's most valuable farmlands. Loess has two useful attributes: it is easily excavated and it can stand in a vertical face without failing (Fig. 13-24).

**Sand Dunes**   When sand is plentiful in arid regions, winds move it and pile it up in characteristic heaps called sand dunes. Such features are not, however, limited to deserts. Many of the larger and more renowned of the world's dunes are along shorelines, such as those on the eastern shore of Lake Michigan, the length of Cape Cod, and the coast of the Somali Democratic Republic. Dunes also border the sandy plains of some large rivers—the Volga is an outstanding example.

Few deserts are completely sand covered. Nevertheless, there are sandy areas in almost all deserts; the south-central part of Arabia and the western part of the Sahara are among the best known. Broad dune-covered areas are called *ergs* (sand seas) because they so resemble a wave-tossed sea (Fig. 13-25).

Sand dunes consist primarily of sand-size grains, which bear testimony to the extraordinary sorting ability of the wind. Finer material, such as silt, may be blown far away, perhaps to form a loess deposit, whereas coarse rock fragments, such as pebbles and gravel, may lag behind the sand.

**Fig. 13-23.** Eroded loess landscape, Shanxi (formerly Shansi) Province. In some parts of China, the oldest part of the loess column has been dated at about 2.5 million years. Numerous buried soils indicate that the deposition was episodic. (Lui Tongsen, Academica Sinica)

Dunes are neither stable nor permanent features of the landscape and may be continuously on the march. Usually, they have a gentle side facing toward the wind and a steep side facing away from the wind. Wind-drifted sand blows up the gentler slope of a dune, and when it reaches the crest, it may be carried a short distance beyond— the tops of dunes sometimes seem to be smoking when the sand is driving across them. Behind the crest the sand drops out of the windstream to accumulate on the steeper slope, the **slip face.**

When the sand is dry and well sorted, the inclination of the slip face may be as much as 34°. Should the slope steepen, the sand becomes unstable and shears along a slightly gentler plane, with the result that a small avalanche of sand glides to the base of the dune. When new sand falls on the slip face, the slope steepens once more, and so the process repeats itself. The net result is a transfer of sand from the upwind to the downwind side of the dune. Thus, the dune slowly migrates—in a sense rolling along over itself.

Dunes have a fascinating variety of shapes and patterns. The eventual shape is a function of many factors, for example, wind strength and direction, amount of sand, distance from the source of sand, and the presence of vegetation.

Where winds blow generally from one direction, as they often do along a seacoast, dunes are likely to have a relatively persistent geometry. For example, they may be aligned at right angles to the wind, in which case they are called *transverse dunes* (Fig. 13-26). Such dunes may be quite short and flourish where sand is abundant and the winds are strong. Typically, many coastal dunes are transverse.

A curious, aesthetically appealing dune is the *barchan* (Figs. 13-26, and 13-27), a sometimes perfectly proportioned crescent. Pointing downwind are the horns of the crescent and the steeper slip face that lies between them. As is typical in dune formation, sand blows up the gentle slope of the crescent and slides down the slip face. Sand is swept around the ends of barchans and tails off downwind to form the horns of the crescent. These dunes, which are migratory and move across the desert at rates of 25 m (82 ft) or more per year, flourish in coastal deserts, such as the Atacama of Chile and Peru. The area surrounding a barchan is likely to be barren bedrock, almost completely sandless. The wind is remarkably efficient, whisking up loose sand from the rocky floor of the desert between the dunes. Grains of sand bounce along much more readily over bare rock than they do over sand. Once they start to accumulate, as in a dune, their free-roving days are over—at least temporarily.

It is not unusual for one type of sand dune to grade laterally into another as conditions change. For example, transverse dunes can grade into barchans in a downwind direction as the supply of sand diminishes.

A *parabolic dune* is roughly similar in shape to a barchan (Figs. 13-26, 13-28). More or less U-shaped, it forms where vegetation has partly stabilized the sand. Sand is blown from the center of the dune and deposited immediately downwind, so that the nose migrates but the arms remain anchored.

Other kinds of dunes occur in areas of ample sand supply as a result of winds blowing, gener-

**Fig. 13-24. A.** *(Above)* An elephant and graffiti carved into loess exposed in a steep road-cut in Atchison County, Missouri. Highway engineers like to work with loess because it usually holds a vertical cut well. (Ardel Rueff, Missouri Department of Natural Resources, Division of Geology and Land Survey) **B.** *(Below)* Because it holds vertical faces well, houses can be excavated in loess, such as these in Shanxi (formerly Shansi) Province, northern China. (H.E. Malde, USGS)

**Fig. 13-25.** Great sand seas, shown here in the eastern Rub' al-Kahli, are typical of the desert of Saudi Arabia. (Aramco)

ally, in several dominant directions. *Longitudinal dunes,* for example, are linear ridges aligned generally in the direction of the prevailing wind, with slip faces on each side of the ridge (Figs. 13-26, 13-29) Wind directions are thought to alternate from one side of the ridge to the other. Such dunes are also common in the Great Sandy Desert of Western Australia, where there are ridges as long as 95 km (59 mi).

More complex are *star dunes* (Figs. 13-26, 13-30); their growth, which involves winds blowing from many directions, is more vertical than horizontal.

The interaction of the factors that control dune form is complex. Figure 13-31 shows how variation in wind strength, sand supply, and vegetation create the various forms.

## CLIMATIC CHANGE IN THE DESERT

The positions of deserts have shifted with time. A good example of progressive shift to desert conditions is the evolution of the desert of North America's Great Basin. About 15 million years ago lakes were scattered throughout the region and a deciduous hardwood forest covered the land-

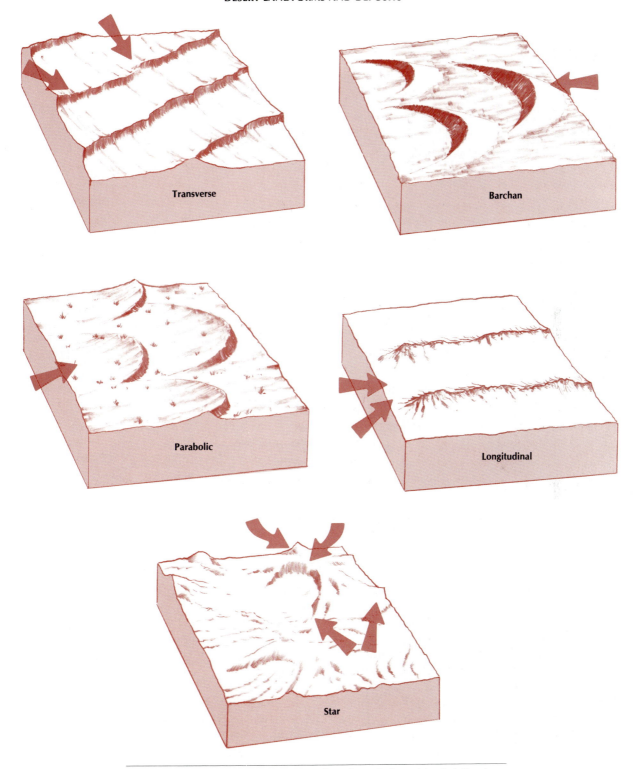

**Fig. 13-26.** Sketches of dunes. The arrows show the main formative wind directions.

**Fig. 13-27.** Barchan dunes along the west shore of the Salton Sea, California. The form of these dunes indicates that they are moving from the upper left to the lower right. (John S. Shelton)

**Fig. 13-28.** A group of sparsely vegetated parabolic dunes in northern Arizona. The wind direction is from right to left. (Tad Nichols)

**Fig. 13-29.** The main linear feature here is a longitudinal dune, in which sand transport more or less parallels the orientation of the dune. Namib Desert, Namibia. (Tad Nichols)

**Fig. 13-30.** Sand dunes in the Sahara. The forms range from relatively small ridges to star-shaped hills (star dunes) that may be as high as 100 m (328 ft). (From H.T.U. Smith, 1963, *Eolian Geomorphology, Wind Direction, and Climatic Change*, AFCRL-63-443, Fig. 16, Geophysics Research Directorate, Air Force Cambridge Research Laboratories, Office of Aerospace Research, U.S. Air Force, Bedford, Massachusetts)

scape. Plant fossils found in desert rocks reveal this history. At that time the Sierra Nevada was a low range of hills and moisture-laden Pacific air masses moving over the region precipitated as much as 1 m (39 in.) each year. Although there followed a gradual drying with some fluctuations, the present desert came into being only about 2 to 3 million years ago. It was then that tectonic

movements pushed up the Sierra Nevada, cutting off the moisture-bearing air masses and folding and faulting the lake beds. Many lakes dried up and the forests were replaced by desert plants.

Even during the past 2 million years we have evidence of climatic change. One problem in unraveling the story is that very little organic material—essential in determining past climates and in dating sediments—has been preserved. Alluvial fans may provide some clues to climatic change as fans can have several surfaces, each of different age (Fig. 13-12). Streams that build fans should respond to climatic changes by degrading at times and aggrading at other times. For example, with a change to a wetter climate and greater discharge, would a stream degrade, only to be followed by aggradation during a drier climate? The link between climatic change and this stream response is not clear, and it is still being debated.

Striking evidence of climatic change may be seen in the ancient deposits and shorelines that rise above present-day lakes or desert basins in which no lake exists. The Great Basin, for example, contained numerous lakes about 20,000 years ago (Fig. 13-32). The most redoubtable of these was Lake Bonneville, the precursor of Great Salt Lake. Ancient shorelines can be seen on the

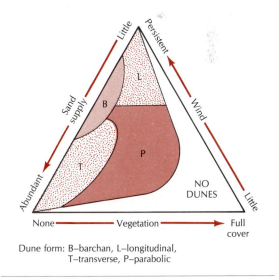

Dune form: B–barchan, L–longitudinal, T–transverse, P–parabolic

**Fig. 13-31.** Relationships among sand supply, vegetation, wind strength, and various dune forms: T = transverse, B = barchan, P = parabolic, L = longitudinal dunes. (Modified by Art Bloom from Jack Hack, 1941)

higher slopes of the Wasatch Mountains, more than 300 m (984 ft) above the modern lake (Fig. 13-33). Lake Bonneville covered an area of more than 51,000 km² (19,690 mi²) and at one time had an outlet at Red Rock Pass, draining north to the Snake River and thence to the Pacific by way of the Columbia River. At times the lake level dropped catastrophically owing to rapid downcutting at the pass. This gave rise to exceptionally large floods moving downstream from Red Rock Pass, with one even approaching depths of 100 m (328 ft).

**Fig. 13-32.** Late Pleistocene lakes in the Great Basin, relative to present lakes. Arrows indicate direction of stream flow from one lake basin to another. In a few places the lakes rose so high that water spilled out of the basin. (From R.B. Morrison, 1968, "Pluvial Lakes," in R.W. Fairbridge, ed., *The Encyclopedia of Geomorphology*, Reinhold, N.Y.)

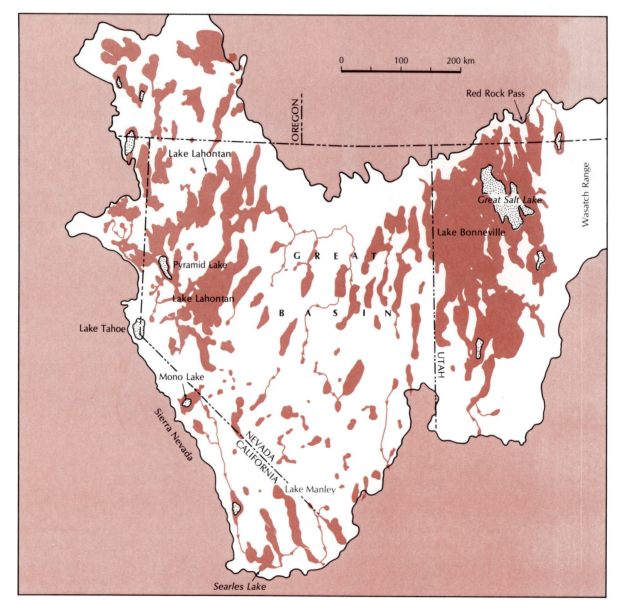

**Fig. 13-33.** Nineteenth-century etching of the Lake Bonneville shorelines on the northern end of the Oquirrh Range, Utah. (G.K. Gilbert, *Lake Bonneville,* 1890, *U.S. Geological Survey Monograph 1*)

In the last several years, Great Salt Lake has been on the rise again. The rate is about 1 m (3 ft) per year. This is enough to cause alarm because an airport, a freeway, and shore facilities are threatened. One beach resort is already under water. Now the excess water is being pumped into a dry basin where it is left to evaporate; this may help to control the lake level.

A contemporary of Lake Bonneville was Lake Lahontan, located mostly in western Nevada, not far from Reno. In that rugged mountainous area, all the intervening valleys were filled with long narrow arms of the lake.

Even Death Valley had a lake at this time. Named Lake Manley (Fig. 13-32)—in honor of Lewis Manley, who saved the first party of pioneers to reach Death Valley from starvation—it was about 145-km (90-mi) long and 183-m (600-ft) deep. Streams flowing southward along the eastern side of the Sierras filled several closed basins to form a string of lakes, the last of which

was in Death Valley. The present-day Mono Lake is a remnant of one such larger lake and the focus of much attention as the city of Los Angeles, in its insatiable quest for water, is removing such large quantities of water that the level of this picturesque lake has been dropping markedly. The shorelines are exposed to erosion and the area's wildlife is endangered.

The obvious recency of the large lakes of the Great Basin coupled with the fact that in a few locations their shorelines cut through deposits laid down by ancient glaciers shows that the last high stand of the lakes generally coincided with the time of the last major ice advance. Furthermore, the record is very clear that such events happened more than once, that all the high stands seem to have occurred during glacial times, and that the low stands (times of near dessication) coincided with the times of glacier disappearance. Climatic change and the increase of meltwater appear to have caused lake expansion and, as the glaciers

**Fig. 13-34.** Lake sediments near the southern end of the Dead Sea, Israel, were deposited when the Dead Sea was much deeper, the result of a late Pleistocene climatic change. (Ran Gerson)

of the world receded, the level of North America's desert lakes fell. Lakes in other climatic belts, such as those in Africa, did not rise and fall at the same time as the North American desert lakes because they were responding to another set of changing climatic conditions.

Desert lakes occur in other arid regions throughout the world. Well-known examples are Lake Chad in Africa, Lake Eyre in Australia, and Lop Nur (formerly Lop Nor), Lake Balkhash, and the Aral and Caspian seas in Central Asia. Not only were the Aral and Caspian seas larger in the recent geological past than they are today, but they were connected with one another as well as with the Black Sea. Many other seas, such as the Dead Sea, are rimmed by older shorelines that scar the barren slopes of the bordering desert hills, much like gigantic flights of steps, and lake sediments partly fill the basins (Fig. 13-34). Some of the Dead Sea Scrolls were found hidden in caves excavated in ancient sediments of the Dead Sea.

There are several other indicators of climatic change, some of them in unexpected places. In areas of desert-derived loess, such as those of China and Israel, layers of buried soils may be found. The presence of a buried soil could mean that there was a relatively sudden climatic change in the area where the loess originated and that the deposition of this airborne material was halted, perhaps for thousands of years. During that time a layer of soil could have formed. This cycle might have been repeated several times. Could vegeta-

tion change in the loess source areas be one reason for the intermittent deposition of loess downwind?

Some artifacts, stone implements, and rock paintings of extraordinary subtlety and sophistication testify to the presence of early man in what today are desolate expanses of the Sahara (Fig: 13-35). Later, much of that region was the granary of Rome. Colonial cities of that day as well as roads along which the Roman legions marched are now covered by drifting sand. The expansion of the desert broke the slender thread of communication linking the cultures of the Mediterranean and the African worlds. Apparently thereafter the two had a vague and uncertain awareness of each other over the centuries; it was not until the introduction of the camel caravan that the land connection was reestablished. By that time the culture of each world had evolved along very different paths.

In parts of the western American desert, trees formerly grew at altitudes several hundred meters below their present limit. The evidence comes from careful examination of tree fragments in abandoned and fossilized rats' nests where today it is too hot and dry for trees to exist.

## DESERTIFICATION—THE MAKING OF A DESERT BY HUMANS

In 1969–73 the effects of a drought in the southern Sahara, known as the Sahel, were catastrophic in terms of loss of humans and livestock. Although the finer points are still debated, it appears that the land had long been misused and the drought made a bad situation even worse. By the time the summer monsoon rains returned, thousands of square kilometers of pasture and farmland had been converted into a brown and barren wasteland that could support little life. From this experience came the term **desertification:** the destruction of the biological potential of land, leading to desertlike conditions.

Examples of land destruction through human misuse are plentiful. Over the past 5000 years the forests of the eastern Mediterranean region have been cut for wood to build ships and temples, to fuel ore smelters, or to convert land to cultivation. The land never recovered, the forests never returned, and the soil in many places was eroded to bedrock. Mesopotamia, the land between the Tigris and Euphrates rivers, was a highly productive agricultural area several millennia ago, but the accumulation of salts in the soils owing to irrigation practices helped bring about its demise.

One of the major causes of desertification is overgrazing—a common problem in many desert fringe areas. The grasses and other plants that livestock feed on eventually are destroyed and desert plants take over. Constant trampling of an-imal hooves can compact the soil. Runoff and water erosion then increase as does wind erosion, and the result is declining productivity of the very lands where a major increase in productivity is most needed.

The pace at which land is turned over to desert can be rapid. A comparison of maps and photographs in southern Sudan documents a southern shift of the desert boundary of about 160 km (100 mi) between 1958 and 1975—in only 17 years!

One can see the enormity of the problem by citing some population figures. It is estimated that over 600 million people—14 percent of the world's population—live in the deserts. More than one-tenth of these people live on land that is classified as severely or very severely affected by desertification. Many may lose their livelihood as the land deteriorates further and thus migrate to already-overcrowded cities. If large-scale calamity is to be avoided, ways to reverse these trends must be found.

## A FRAGILE LANDSCAPE

In many U.S. desert areas population is burgeoning, and with it problems in land use and abuse. In well-watered areas, vegetation usually grows rapidly after any disturbance, helping to hide scars.

**Fig. 13-35.** A Stone Age rock painting in the African desert depicts people gathering grain. This and other similar paintings suggest a wetter climate in the past. (Eric Lessing, Magnum)

**Fig. 13-36.** Erosion of a steep slope caused by the use of off-road vehicles near Gorman, California. Photographs taken in May 1976 *(above)* and in March 1978 *(below)* dramatically show the accelerated erosion. When the riding surface becomes too gullied, vehicle owners seek undisturbed land elsewhere. (H.G. Wilshire, USGS)

In deserts, however, the works of humans remain in plain view for many years.

A recent threat to desert landscapes has been the damage caused by recreational vehicles. The ground is often disturbed to the extent that tire tracks can be seen crisscrossing the desert from miles away (Fig. 13-36). The tracks not only mar the landscape, but also compact the soil and destroy the vegetation, resulting in erosion and gullying. The damage from wind erosion can be extensive. The soil churned up and subsequently picked up by the wind in one cross-country motorcycle race across the Mojave Desert was estimated at 660 tons. Dust plumes originating from disturbed lands are now being identified by satellite imagery. We do know that deserts heal very slowly and that damage done today will affect generations to come. Some areas may never be restored. Even the Peruvian stone patterns (see Fig. 13-6) are threatened by off-road vehicles.

## SUMMARY

1. Hot deserts occupy about one-third of the earth's land area and are characterized by rates of evaporation that greatly exceed the precipitation rate. Sparse, or absent, vegetation cover plus the torrential but localized nature of the rainfall help account for many desert features.
2. Many surfaces in the desert are characterized by *desert varnish,* a bluish-black coating on rock sur-

faces. A tightly packed surface concentration of stones is called a *stone pavement.* Wind action or the upward movement of stones from shallow depths are the cause of pavement.

3. The main depositional landform characteristic of hot deserts is an *alluvial fan,* which is formed from material deposited by a stream as it leaves the mountains and enters an intermontane basin.

4. The main erosional landform in hot deserts is a *pediment.* Its surface form is similar to that of a fan, but it is bedrock rather than deposited material. The origin of pediments is still debated.

5. Wind is a very effective agent in shaping some deserts. The sand it carries cuts away at stones and bedrock alike, and in some places it cuts and excavates bedrock basins.

6. Wind-blown sand can be deposited as a variety of dune forms.

7. Many deserts were areas of higher rainfall in the past, as shown by evidence that lakes existed in many present-day dry basins. Lakes in the southwestern United States were deep and extensive during the last glaciation.

8. Humans can create deserts through the misuse of lands, a process called *desertification.*

## QUESTIONS

1. What conditions are responsible for deserts? Cite the evidence within deserts or areas peripheral to deserts that indicate the boundaries of past deserts were different from the deserts of today.

2. Compare stream erosional and depositional processes downstream in a humid versus an arid environment.

3. How would you distinguish a pediment from an alluvial fan in the field? Would this distinction be important to groundwater exploration?

4. What factors are important in the formation of the main types of sand dunes?

5. What kinds of evidence could be used to suggest that desert lakes grew and dried up repeatedly, just as glaciers advanced and retreated repeatedly?

6. Imagine a grassland adjacent to a desert, one that is at present overgrazed. Using your knowledge of soils and erosional processes, discuss how the area can become a desert with no change in climate.

## SELECTED REFERENCES

Bagnold, R.A., 1941, The physics of blown sand and desert dunes, Methuen, London. (Repr. 1965, Halsted, New York.)

Bloom, A.L., 1978, Geomorphology: a systematic analysis of late Cenozoic landforms, Prentice-Hall, Englewood Cliffs, N.J.

Cooke, R.U., and Warren, A., 1973, Geomorphology in deserts, University of California Press, Berkeley.

Dorn, R.I., and Oberlander, T.M., 1982, Rock varnish, Progress in Physical Geography, vol. 6, pp. 317–67.

Dregne, H.E., 1983, Desertification of arid lands, Harwood Academic Publishers, New York.

Eckholm, E., and Brown, L.R., 1977, Spreading deserts—the hand of man, Worldwatch Paper 13, Worldwatch Institute, Washington, D.C.

Glennie, K.W., 1970, Desert sedimentary environments, Elsevier, New York.

Goudie, A., and Wilkinson, J., 1977, The warm desert environment, Cambridge University Press, London.

Mabbutt, J.A., 1977, Desert landforms: MIT Press, Cambridge.

McGinnies, W.G., Goldman, B.J., and Paylore, P. eds., 1968, Deserts of the world, University of Arizona Press, Tucson.

McKee, E.D., ed., 1979, A study of global sand seas, U.S. Geological Survey Professional Paper 1052.

Morrison, R.B., 1968, Pluvial lakes, pp. 873–83 in The encyclopedia of geomorphology, R.W. Fairbridge, ed., Reinhold, New York.

Oberlander, T.M., 1974, Landscape inheritance and the pediment problem in the Mojave Desert of southern California, American Journal of Science, vol. 274, pp. 849–75.

Péwé, T.L., ed. 1981, Desert dust: origin, characteristics, and effect on man, Geological Society of America Special Paper 186.

Ritter, D.F., 1986, Process geomorphology, W.C. Brown, Dubuque, Iowa.

Short, N.M., and Blair, R.W., Jr., eds., 1986, Geomorphology from Space, a global overview of regional landforms: NASA AP-486.

Wilshire, H.G., 1977, Study results of 9 sites used by off-road vehicles that illustrate land modifications, U.S. Geological Survey Open-file Report 77–601.

# 14

# GLACIERS
# AND THE EFFECTS
# OF GLACIATION

Scenically, the world is more indebted to glaciation than to any other process of erosion. Without glaciation there would be few of the jagged peaks that stand in isolated splendor along the crest of many of the world's lofty mountain ranges. Nor would there be spectacular valleys whose outlines have been sharpened by the glacial file—such as Yosemite Valley, the Lauterbrunnenthal of Switzerland, and the Norwegian fjords. Because a glacier can erode more deeply in some parts of its channel and less deeply in others and because the material it deposits has an irregular surface, it can form lakes and add both beauty to the landscape of alpine regions throughout the world, and drama to such lower-lying areas as the Great Lakes region, northeastern Canada, and Scandinavia.

**Fig. 14-1.** Columbia Glacier, Alaska, as it appeared in 1979, a position close to the maximum for the previous 80 years. The glacier front was grounded on Heather Island and a terminal moraine in shallow water to the left of it. In the early 1980's the front of the glacier retreated from the grounded position, and because the front was now in deep water, calving of icebergs caused drastic retreat. By 1986, the front has retreated 2 km (1.2 mi), and about 30 km (18 mi) of retreat are expected in the next several decades. (Austin Post, USGS, Tacoma, Washington, photo no. 79L3-028)

The Ice Age just ended has affected the lives of us all to a greater or lesser degree. Soil and loose rocks were stripped from vast land areas, leaving barren rock behind. The load of stripped material was then deposited at the glacier margins. And glacially produced silt and clay were blown off the flood plains of the world's major glaciated rivers and deposited across the landscape for many kilometers downwind. These deposits constitute some of the world's finest agricultural lands. Today both geologists and archaeologists are studying the placement and timing of the ice sheets that joined in western Canada because this glacial barrier prevented the migration of humans and animals southward from Alaska.

Many of the areas covered by glacial ice were bowed down under its weight; when the ice disappeared, the land rebounded. In Canada uplifted wave-cut features show that the Hudson Bay region has risen 300 m (984 ft) or more (see Fig. 6-49). Historic records and various shoreline structures show that the land is still rising. It rises at the unusually high rate of about 2 m (6.5 ft) per century at the southern end of Hudson Bay and decreases to about zero in the vicinity of the southern Great Lakes.

Worldwide, the sea level reflects the waxing and waning of the ice sheets. When the ice sheets expanded, tens of millions of cubic kilometers of water were withdrawn from the sea via the atmosphere and locked up on land as ice. Sea level was lowered as a consequence, perhaps by as much as 140 m (459 ft). Although this may not seem like much, it was enough to alter world geography profoundly. Land areas now separated were then connected and migrations of both people and animals occurred across the newly exposed lands.

There are many reasons why geologists study glaciers and the effects and deposits of former glaciations (Fig. 14-1). Perhaps the foremost reason is that they indicate rather dramatic past climatic changes. If we can determine when such changes occurred, we might be able to predict the frequency of climatic change. Another reason for studying glaciers is that glacial deposits often create special engineering problems. Furthermore, monitoring the amount of ice on land helps us to determine if the rise and fall of sea level along some coasts is caused by worldwide changes in the volume of ice or by local tectonic changes.

**Fig. 14-2.** Louis Agassiz (1807–73). (Harvard University Archives)

## GLACIAL THEORY

It is no wonder that evidence of past glaciation attracted the attention of observant inhabitants in Europe and New England in centuries past. The lavish supply of boulders on New England farms was a source not only of wonderment to the early settlers, but also of wearisome, backbreaking toil. So much labor was involved in clearing fields strewn with glacially transported stones that more than one young man was readily convinced that a life at sea could be no harsher—even on a New Bedford whaler.

For many, the presence of those strange stones, found far from their place of origin—for example, granite blocks resting on a limestone terrain—could be explained as the work of the Great Flood endured by Noah.

In Great Britain widely scattered glacial deposits were called drift because it was believed they were left behind by ancient seas. It was a difficult task to persuade English geologists that glaciers had scoured the land and transported rocks, some the size of small houses, for scores of kilometers. The problem was that there were no existing glaciers to serve as models. In the Alps, however, glaciers are common. Through the centuries their snouts have advanced or retreated and alpine passes have been alternately ice free or ice blocked. Some villages occupied in medieval times are now buried by ice. The silver mine of Aiguille d'Argentière, active in the Middle Ages, is now covered by a glacier of Mont Blanc in the French Alps. Many alpine villagers must have been aware that when a glacier receded it left behind a trail of barren, stony ground interrupted by low rocky ridges and diversified by lakes and ponds.

Not until the early nineteenth century were there Europeans convinced that an ice sheet had covered much of northern Europe, an idea that had aroused the curiosity of one of the leading Swiss naturalists of the day, Louis Agassiz (Fig. 14-2). After an expedition in the Alps with a friend to see the evidence firsthand, Agassiz became a believer. In 1846 he arrived in the United States with a professorship at Harvard University, and he spread the word of a Great Ice Age that had once refrigerated most of the Northern Hemisphere. A concept so novel, however, aroused opposition, and it was some time before it was accepted by the scientific community.

## DISTRIBUTION AND FORMATION OF GLACIERS

Glaciers today occupy about 15,000,000 km² (5,791,500 mi²) of the earth's surface—about 10 percent of the total land area. Most of the ice is locked up in two ice caps: Antarctica, which accounts for about 84 percent of the ice in the world, and Greenland which accounts for 11 percent. The rest of the ice is scattered around the mountains of the world. Glaciers are especially prominent features in the ranges that parallel the coasts of Alaska, British Columbia, and Washington; in the Rocky Mountains of Canada and the northern United States; and in Scandinavia, the European Alps, the Southern Alps of New Zealand, and the Andes.

The amount of water locked in glacial ice can be converted to a depth of ocean water; thus we can appreciate its quantity and gain some idea of what might happen if all the ice on earth were to melt. It is no simple task to measure the amount of existing ice because the configuration of the bases of glaciers are not well known. The best estimates, however, are that if all the present ice were to melt, sea level could rise as much as 70 m (230 ft) above its present level. A glance at a map of North America shows that many coastal cities would lie under water, whereas many inland cities would become seaports (Fig. 14-3). But there is no need for alarm because the major ice caps are fairly stable; any changes that might take place probably would do so gradually, over a thousand or so years. The glacial recession since the 1890s, for example, has resulted in a general sea level rise of 12 to 30 cm (5 to 12 in.). In the low-lying areas of Holland, dikes have been built to keep the rising sea from submerging the land. But a rising sea level is not the only reason for such flooding. Tectonic subsidence and the compaction of sediments that make up the land can also cause seawater flooding.

Glaciers may develop in those parts of the world where the combination of sufficient winter snowfall and low summer temperatures allows some snow to remain the year around. The snow is converted to ice, which eventually displays evidence of movement. Thus, glaciers are active today in high mountains over much of the globe and in the farther reaches of the Northern and Southern hemispheres. They occur near sea level in polar regions, but equatorial mountains must be at least 5000-m (16,400-ft) high to harbor glaciers.

**Fig. 14-3.** Changes in the shoreline of North America caused by changes in ice volumes. The solid black line is the present coastline; the outer border is the coastline during the last glacial maximum about 18,000 years ago; and the inner border represents the coastline if the present ice of Greenland and Antarctica were to melt. (After R.H. Dott, Jr., and R.L. Battan, 1971, *Evolution of the Earth,* New York, McGraw-Hill. Reprinted by permission of McGraw-Hill.)

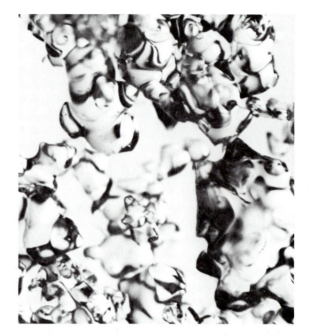

More than cold temperatures are needed for snow to accumulate and remain from year to year. This is demonstrated by the vast areas of the cold Arctic that are not covered with glaciers. The exact amount of snow required to maintain glaciers will vary from place to place. In maritime regions with high summer snowmelt, about 4 m (13 ft) of snowfall (on a water-equivalent basis) are required, whereas only a fraction of a meter is required in the polar deserts of the northern Arctic and Antarctica, where summer snowmelt is at a minimum.

Local environment is significant in determining the position of glaciers. In the Northern Hemisphere glaciers are usually larger or restricted to the more shaded, north-facing slopes. Wind can also be a determining factor. In parts of the Rocky Mountains of Colorado and Wyoming, dry snow falling on the alpine tundra slopes is picked up by winds from the west and redeposited on the heads of east-facing valleys, eventually it is converted to glacier ice.

**Fig. 14-4. A.** *(Above left)* Newly fallen snowflake. (CIRES/National Snow and Ice Data Center, Boulder, Colorado) **B.** *(Above right)* With time snowflakes melt and refreeze into rounded snow particles. (E.R. LaChapelle) **C.** *(Below)* Eventually, interlocking crystals of glacier ice form. As a glacier moves down-valley, crystal size can reach as much as 10 cm (4 in.). (Chester Langway, Jr.)

## Change of Snow to Ice

The conversion of snow to ice is not a simple process. Snow is water that has crystallized directly from water vapor in the atmosphere. Because snow is a crystalline substance, snowflakes grow in regular geometric patterns (Fig. 14-4)—just as the mineral quartz does. The specific gravity of snow is much less than that of water, so that 1 cm (0.4 in.) of snow is commonly equal to 1 mm (0.04 in.) of rainwater.

After snowflakes lie on the ground for a short time, they change form (Fig. 14-4). Individual flakes may **sublimate** (pass directly from a solid to a gaseous state) or they may melt and refreeze into granules of ice. That gritty, granular snow, with a texture much like coarse sand, is a familiar phenomenon in snowbanks and is called **firn** (from a German adjective, meaning "of last year").

Firn, which typically accumulates on the upper slopes of alpine mountains, goes through a gradual transition into glacier ice. Firn is usually white or grayish-white, and the spaces between the granules are filled with trapped air. At a depth equivalent perhaps to an accumulation of 3 to 5 years of firn, the pore spaces become smaller or may even disappear; the transition into blue glacier ice

made up of interlocking ice crystals is then complete (Fig. 14-4). The process is accompanied by an increase in the specific gravity from about 0.1 in newly fallen snow up to 0.9 in solid ice.

The conversion of snow to ice is an excellent example of present-day rapid metamorphism. Snow falls on the ground and builds up sedimentary layers (Fig. 14-5). As temperature and pressure conditions change, the snow is metamorphosed into ice, which moves downvalley when depths reach several tens of meters, developing fold patterns that resemble those in metamorphic rocks. As the snow becomes ice, much of the glacier remains solid, just as rocks do during metamorphic change. If the formation of glaciers were not a solid-state change, the ice would melt and the glacier would vanish.

## Mechanism of Glacier Movement

Nobody doubts that glaciers flow. A common proof is that the rocks being deposited at the front of alpine glaciers are derived from the cliffs that

**Fig. 14-5.** Sedimentary layers in glacial ice are shown at the edges of this glacier in Vatnajökull, southeast Iceland. (J.D. Ives)

flank the upper end of the glacier. Unusual evidence of such movement came from the disappearance of an early mountaineer in the European Alps. Efforts to find him failed, but his body appeared decades later, at the lower end of the glacier.

**Fig. 14-6. A.** Plan view of an alpine glacier that shows how velocity measurements are made. Stakes are set in the ice in a straight line (A–B–C). Eventually the movement of the ice will cause the line to curve (A′–B′–C′). The distances between the original and present positions of the stakes indicate the amount of displacement; from this the velocity is calculated. **B.** Vertical profile of an alpine glacier, showing the velocity distribution with depth. A vertical pipe extending down to bedrock is placed at X. After several years the pipe will bend (Y–Y′), its top moving from X to Y and its base from X′ to Y′.

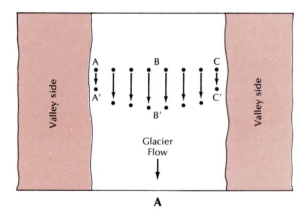

**A**

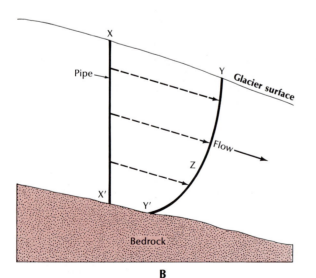

**B**

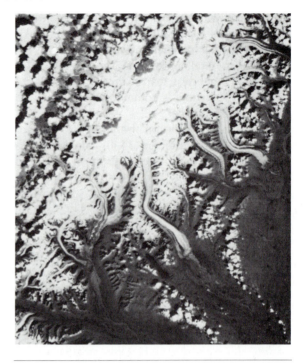

**Fig. 14-7. A.** *(Above)* Satellite imagery detects surging glaciers in the Alaska Range just south of Mount McKinley. Glaciers that have not surged have straight or streamlined moraines (seen as dark lines on the ice), whereas those that have surged have convoluted moraines. (NASA) **B.** *(Opposite)* Vertical view of Jacobshavns Glacier, a surging glacier in Greenland. Moving 23 m (75 ft) per day, this is one of the world's fastest glaciers. The numerous icebergs that choke the fjord result from the rapid retreat of the front of the glacier, which is afloat and calving. (Markhurd Corporation, Minneapolis, Minn., for studies by Terence J. Hughes)

The cross-valley velocity of a glacier in an alpine valley is similar to that of water in a stream. The flow is greatest in the center, least on the sides and bottom (Fig. 14-6). The diminishing velocity toward the valley sides and bottom is the result of friction between the ice and the bedrock. Average velocities vary from glacier to glacier, but most fall between 3 and 300 m (10 and 1000 ft) per year.

Occasionally, a glacier decouples from the rock floor and sides and advances down valley at truly fantastic velocities, some as much as 8 km (5 mi) per year (Fig. 14-7). Such movements are known as **glacial surges** and are usually characteristic of valley glaciers although ice caps have also been known to surge. In some cases a high-velocity wave of ice passes through the glacier without

increasing its length; in other cases the surge results in a longer glacier. Several theories attempt to account for these surges. One is that increasing amounts of water at the base of the glacier speed it on its way. Another theory attributes the unpredictable surge to a block of stagnant ice breaking through in the lower reaches of a glacier.

Ice flows in complex ways. Generally, the base of most glaciers moves by sliding over the rocks as shown in Figure 14-6B, by the displacement of the base of the pipe from X′ to Y′. Evidence for such movement, called **basal slip,** are the **striated rocks** polished and scratched left behind when glaciers melt as well as the horseshoelike inden-

**Fig. 14-9.** Crevasses in a glacier on Mount Rainier, Washington. Irregular bedrock topography sets up complex flow patterns in the plastically deforming ice near the base of the glacier, causing the brittle uppermost ice to crack and form crevasses. (Peter W. Birkeland)

**Fig. 14-8.** Features on bedrock surfaces that are the result of basal slip. **A.** *(Above)* Polished and striated limestone bedrock in Illinois. (Illinois State Geological Survey) **B.** *(Below)* Chattermarks indent Sierra Nevada granite, California. (G.K. Gilbert, USGS)

tations called **chattermarks** (Fig. 14-8). In contrast, the lower part of a glacier (Fig. 14-6B, between Y' and Z) creeps downvalley through **plastic flow** because the ice crystals deform under the pressure of the overlying ice. The upper part of a glacier (above Z in Fig. 14-6B) consists of brittle ice

because there is not enough ice and snow above it to cause it to deform plastically. Such ice rides piggyback on the lower and continually deforming ice. Deep flowage combined with the valleyside friction and the low strength of the brittle ice causes innumerable cracks or **crevasses** to form in the brittle ice (Fig. 14-9). Such crevasses can extend to 30 m (100 ft) or more in depth; at greater depths, plastic flow of the ice seals them off. Crevasses are especially common where glaciers plunge steeply downvalley to form an ice fall (Fig. 14-10).

**Fig. 14-10. A.** *(Opposite above)* View of Mount Everest from the east, and the Khumbu Glacier. **B.** *(Opposite below)* Ice fall on the Khumbu Glacier. Placing a route through this jumble of collapsing ice is one of the main problems mountaineers face when climbing Everest from the Nepal side. (Glen Porzak)

433

**Fig. 14-11.** The snowline on these glaciers ("S") is marked by the transition from clean snow at their upper ends to streaked dirty ice at their lower ends. The accumulation areas lie above the snowline; the wastage areas below it. Tweedsmuir Glacier, St. Elias Mountains, Yukon Territory, Canada. (Austin S. Post, USGS Glaciology, Tacoma, Washington)

This description, however, is not accurate for those glaciers located in polar regions where temperatures are so low that glaciers are likely to be frozen to their bedrock bases. Basal slip, therefore, is not so important in the motion of polar glaciers as it is in the motion of temperate-climate glaciers, which are not frozen to their bases.

## The Glacier Budget

Glaciers continually gain and lose mass; by a series of detailed measurements, we can determine their gains and losses. In short, we can determine a glacier budget. If, for example, a glacier has a surplus year, the front may advance downvalley; in a deficit year, the front may retreat. Unlike the financial condition of humans, the budget of a glacier is exposed to all who care to see it.

In order to understand a glacier's budget, we need to determine where the gains and losses are taking place. Every glacier has a fairly narrow zone on its surface known as the *snowline* (Fig. 14-11). That zone is identified late in the melting season; above it, some snow lingers from year to year and will accumulate; below it, the snow from the previous winter and some of the underlying ice melt and are lost. Thus, there are distinct areas of *accumulation* and areas of wastage, or *ablation,* on glaciers (Fig. 14-12). In a very general way the

area of accumulation makes up about two-thirds of the total surface area of a valley glacier.

With the glacier budget in mind, we now can examine how a glacier advances and retreats. Let us focus our attention on the front of a stable glacier, that is, one that is neither advancing nor retreating because the accumulation of snow is balanced by ablation (Fig. 14-13, position A). The glacier maintains the same form over the years, so that all losses in the ablation area are precisely balanced by the flow of ice into that area from upvalley. To illustrate the point, we might compare a glacier to a side of bacon being fed into a slicer. Although the bacon is continuously being shoved forward, it never advances beyond a fixed point because it is always being cut off by the rotating blade. In contrast, if the snowline were to lower (Fig. 14-13, position B), the area of accumulation would enlarge. The amount of ablation would then be less than the amount gained, and the front would advance. Advance takes place until, for that snowline position, a balance between gains and losses is reached. Similarly, if the snowline rises, the opposite effects are seen—the area of wastage enlarges, more ice melts than can be compensated for by accumulation, and the front retreats until a balance with respect to the new snowline (Fig. 14-13, position C) is reached. If the snowline continues to rise until it no longer intersects the land surface, the glacier melts away.

## KINDS OF GLACIERS

Glaciers can be placed in two broad categories: *alpine glaciers* and *continental ice sheets.* Alpine glaciers have their origin on mountain slopes and summits that rise above the snowline. They advance downslope under the pull of gravity and, more often than not, take the path of least resistance by following a preexisting stream valley. Alpine glaciers are more likely to accentuate irregularities in the landscape, making bold peaks even more jagged and steepening the walls of already deep canyons.

In contrast continental ice sheets are irregular in shape, cover a large land area, and are not necessarily guided in their flow pattern by the underlying topography. They generally flow in the direction that the ice surface slopes even though

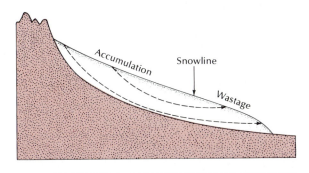

**Fig. 14-12.** Cross section of a valley glacier showing areas of accumulation and wastage. The dashed lines are the approximate flow lines of the glacier particles. Notice how they descend and then rise toward the surface of the glacier in the wastage area. A snow particle in the accumulation area will follow such paths, convert to ice at depth, and melt when it reaches the surface.

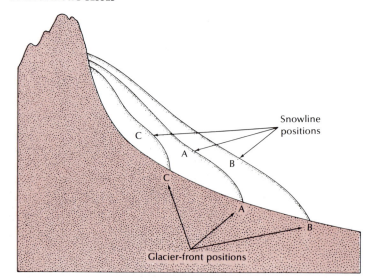

**Fig. 14-13.** Cross section of an alpine glacier showing how its size is governed by the snowline. A change in size comes only through frontal advance or retreat because the upper end of the glacier is always anchored to the headwall.

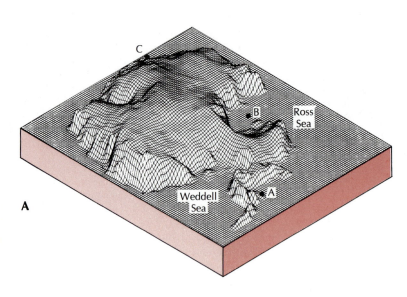

**Fig. 14-14. A.** Computer-generated view of Antarctica. **B.** Cross section of the ice sheet between points A, B, and C. (After D.J. Drewry, 1983, *Antarctica: Glaciological and Geophysical Folio,* Scott Polar Research Institute, Cambridge, Sheet 2)

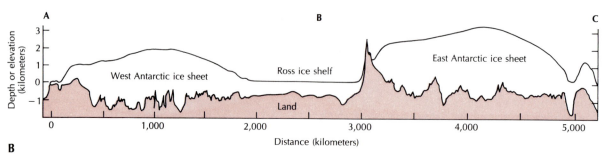

that might be against the slope of the underlying bedrock. Two good examples are the Greenland and Antarctic ice sheets (Fig. 14-14). They have much in common: both have parts of their bases below sea level; both have several kilometers of ice at a maximum; and both flow through mountains on the way to the sea. In fact, it was on two outlet glaciers through the Transantarctic Mountains that teams led by Norwegian Roald Amundsen (1872–1928) and Englishman Robert F. Scott (1868–1912) waged their epic race to the South Pole in the southern summer of 1911–12.

Continental ice sheets also have a central accumulation area that is flanked by a wastage rim, a more symmetrical arrangement than that present in alpine glaciers. Antarctica is extremely cold, so that accumulation of ice is low (parts of the high interior plateau receive less than 5 cm (2 in.) of water-equivalent precipitation per year) and the calving of icebergs to the sea is the main source of wastage. Some entrepreneurs have their eyes on such icebergs as freshwater sources for water-impoverished countries, but they have yet to figure out how to tow the huge hunks of ice there before they melt.

Today, interest in the Antarctic Ice Sheet is high. Because the amount of water locked up in it is so great, small changes in its volume will strongly affect worldwide sea level. Most likely, the changes will take place very slowly, but some earth scientists feel that parts of the sheet could surge, just as alpine glaciers can. If this happened over a short enough time period and involved large masses of calving ice, the effect surely could be catastrophic in many coastal areas.

The worst cast of the latter could involve an area of the Antarctic Ice sheet west of the Transantarctic Mountains. Much of this sector of the ice sheet is below sea level and seems to be prone to surging. A glacial meltdown could add enough water to the ocean to raise sea level some 3 to 6 m (10 to 20 ft), inflicting untold damage on coastal communities. This change could come about in as little as 500 years, according to one estimate.

A final, and unusual, kind of glacier is the **rock glacier.** In some areas with a continental climate, such as in the U.S. Rocky Mountains, ice glaciers are not too common—probably the result of a combination of light winter snowfall, cold tem-

**Fig. 14-15.** Rock glacier on Mount Sopris, Colorado. Rockfalls from the high cliffs continually feed debris to the glacier (see Fig. 11.5A). The slow movement downvalley results in transverse furrows and ridges on its surface. Rock avalanches are common on the oversteepened front of the glacier, which is kept steep by constant pressure from behind. (Edwin E. Larson)

peratures, and relatively high summer snowmelt. In their place we commonly see rock glaciers—tongue-shaped masses of rocky debris and ice slowly moving downvalley (Fig. 14-15). Because of the inherent problems of digging into a rock glacier, we have little data on their interiors. Some seem to consist of essentially clean glacier ice overlain by a thin mantle of rocky debris, whereas others probably contain rocks throughout cemented together by ice. Rock glaciers move downvalley just as ice glaciers do, through basal sliding and plastic flow. As they advance, their fronts are being oversteepened constantly and rock avalanches are common.

Special conditions seem necessary for the formation of rock glaciers. One is that the altitude must be high enough for some snow to linger into the late summer. The other is that the surrounding cliffs must supply rocks rapidly enough to bury the summer snow and protect it from melting. In fact, the surface mantle of rock protects or insulates the underlying ice so well that rock glaciers persist at lower altitudes than those at which some present-day ice glaciers could survive.

Ice and rock glaciers have different sensitivities to climatic change. A warming climate during the last century has caused many ice glaciers to retreat, but many rock glaciers seem unaffected. Some of the larger rock glaciers are still advancing downvalley, knocking down trees in their path. We need not send out a rock-glacier alert, however, as their movement is very slow—about 1 m (3 ft) per year.

## THE EFFECTS OF GLACIERS ON LAND

### Glacial Erosion

There appear to be two major ways in which glaciers shape the rock surface on which they rest: *glacial quarrying* and *glacial abrasion.* In the first process rocks are plucked, or pried, out of place in much the same way that they are in commercial rock quarries—except, of course, at a far slower rate. In temperate-zone glaciers there is ample water along the bed of the glacier, either from melting caused by high overburden pressures—the same process that makes ice-skate blades slide along—or from seepage along the sides or through crevasses. Parts of the bed of a glacier can have water in liquid form; parts can be frozen. When water-filled bedrock joints freeze, blocks can be jacked up and entrained by the moving ice—hence the term *plucking.*

In contrast abrasion takes place when the rock surface on which the ice rests is scoured or worn down by ice containing abundant rock or finger-grained particles—in much the same way that a wood surface may be sandpapered. Some bedrock may even obtain enough polish to glisten in the sun. Abrasion also polishes and scratches clasts carried in the glacier—an important clue in differentiating a glacial deposit from a mudflow as both can be as equally unsorted.

### Landforms Produced by Glacial Erosion

Scoured and smoothed rock surfaces are erosional landforms common to both alpine and continental glaciers (Fig. 14-16). Because of either glacial flow patterns or weaknesses in the rock, such as joints, the result is an undulating bedrock topography. Once the ice melts the irregularities fill with water, thus producing many of the bedrock-enclosed lakes in alpine valleys or in the vast flatlands once burdened by thousands of meters of continental ice. In alpine country these are known as **tarns.**

One major topographic feature of North America—the Great Lakes—resulted from the last major glaciation. In part the lake basins are ice-gouged and their floors lie far below sea level; the bottom of Lake Superior is 213 m (700 ft) below sea level and that of Lake Michigan 104 m (340 ft).

Some of the most spectacular features of glacial erosion are the result of alpine glaciation (Fig. 14-17). At the head of a glaciated valley is a horseshoe-shaped basin enclosed in steep, high walls, called a **cirque** (Fig. 14-18). Its origin is rather obscure. The floor of a cirque appears to be the result of the usual erosional processes and the headwall to be the result of some combination of frost weathering that loosens the blocks and quarrying that removes the latter and maintains a near-vertical headwall. In time the cirque extends headward into the mountain range. In some glaciated mountains glaciers radiate away from the summit area like spokes of a wheel. Should the glaciers continue to quarry actively at the upper cirque ends, then the mountain may be whittled away until only a jagged, saw-toothed pinnacle called a **horn** survives. The Matterhorn is the world's most familiar example of such a glacially accentuated peak (Fig. 14-19).

Downvalley, a primary result of glacial erosion is the overdeepening of valley floors and the steepening of valley walls. The part of an alpine valley occupied by a glacier resembles the channel occupied by a river, but on a far grander scale because the volume of water in the glacier is far

**Fig. 14-16.** Satellite image of northern Northwest Territories, Canada. Glacial scouring of the Precambrian rocks has been such that rock structures, such as folds, joints and dikes show up clearly. (NASA)

greater. The resulting ice-scoured valley not only has a **U-shaped** cross section, but it is more linear than a stream-carved valley, because glaciers selectively erode the lower ends of spurs and ridges. Classic examples are Yosemite Valley (Fig. 14-20) and the valley at Lauterbrunnen, Switzerland (Fig. 14-21). Waterfalls that plunge into space over vertical cliffs are part of the scene, the result of tributary valleys—glaciated or not—that erode at a slower rate than the ice-filled main canyon. Such valleys are called **hanging valleys.**

A journey across Peru clearly demonstrates the difference between glaciated and nonglaciated valleys. The Peruvian Andes are a geologically recent mountain range, and numerous earthquakes indicate that the mountain-building task is not finished. Rivers have incised the rising block, forming spectacular, twisted, **V-shaped valleys** with a steep gradient—a form common in mountainous terrain (Fig. 14-22). In stark contrast once the glaciated upper portion of the valleys is surmounted, they widen and become straighter, with a flatter floor (Fig. 14-23). These changes in valley form are so obvious that the limits of formerly glaciated valleys can be easily marked on satellite images.

Two other glaciated landforms in alpine regions are important, perhaps primarily to mountaineers. One is the **arête** (from the French word for "ridge" or "fishbone"), a slender almost razor-sharp rock ridge formed when glaciers in two adjacent parallel valleys erode the intervening divide until a mere rock screen survives (Fig. 14-24). The other is the **col** (pass) used for transmontane travel, which is formed by continued erosion of an arête or when headward-eroding glaciers in adjacent valleys reduce the rock barrier separating them and forges a pass.

Other spectacular landforms of glaciated terrain are the fjords that flank many high-latitude coasts, as in Alaska, the Canadian Arctic, Norway, Chile, and New Zealand (Fig. 14-25). Narrow troughs, they differ from land-based flat-floored glacial valleys (compare with Fig. 14-21) mainly in that they are submerged by the sea, but also in their truly remarkable depths. One in Norway, for example, is 1300-m (4260-ft) deep; another in Antarctica is more than 1900-m (6230-ft) deep. Not uncommonly there are fairly shallow submerged rock barriers that lie 100 to 200 m (330 to 660 ft) below

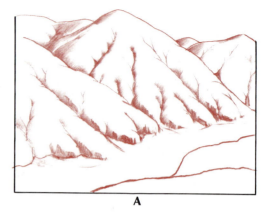

A

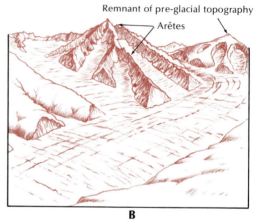

Remnant of pre-glacial topography
Arêtes

B

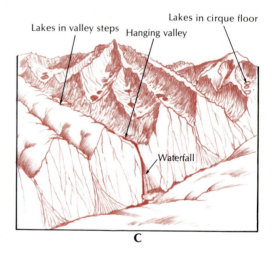

Lakes in cirque floor
Lakes in valley steps    Hanging valley

Waterfall

C

**Fig. 14-17.** The alteration of a stream valley by alpine glaciation. **A.** Terrain before glaciation, with smooth, rounded hillslopes and streams in **V**-shaped valleys. **B.** Maximum extent of glaciation. Ridges and peaks close to the active glaciers are steepened and take on a jagged appearance. Not all the rounded mountain summits are consumed, however. **C.** After the ice melts, the characteristic glaciated landscape appears: cirques, U-shaped valley, hanging valleys, waterfalls, and lakes.

**Fig. 14-18.** Cirques at the heads of glaciated valleys in the Wind River Mountains, Wyoming. The cirques are cut into an old erosion surface, the dissected remnants of which now form rolling plateau-like uplands. Tarns or glacial lakes commonly occur in the cirque basin, but lakes also occur down-valley in bedrock-rimmed basins and are a characteristic feature of glaciated terrain. (Austin S. Post)

**Fig. 14-19.** Clouds collecting on one side of the Matterhorn during high winds. (Karl W. Birkeland)

**Fig. 14-20.** Distant view of Yosemite Valley, California, looking east. Glacial erosion extensively modified its original sloping walls, creating a U-shaped valley. Yosemite's numerous vertical walls challenge rock climbers from around the world. (USGS)

**Fig. 14-21.** Classic U-shaped valley at Lauterbrunnen, Switzerland. (Karl W. Birkeland)

sea level at their mouths. Geologists have long speculated on the origin of fjords; their theories range from one of tectonic origin to one of glacial scouring and overdeepening of former stream valleys. Surely, many fjords owe much of their origin to glacial erosion and stand as testimony to the tremendous work a glacier can perform.

### Glacial Deposits and Depositional Landforms

Much of the debris carried by glaciers comes to rest beneath or along the periphery of the ice, downvalley from the snowline. The sheer volume of debris brought down by ice, which may include extremely large fragments, is more than glacial meltwater can remove—at least not as quickly as it is supplied. Consequently, the sides and lower end of the glacier nearly always carry a rock burden, which is heaped into ridges called *moraines* (Fig. 14-26). Those looped around the snout of a glacier are called *terminal* or *end moraines* (Fig. 14-27).

Ridges of debris that continue upvalley along the sides of the glacier form *lateral moraines;* their crests slope forward with about the same inclination as the glacier surface. Should two glaciers join, the lateral moraines that meet at their intersection unite and continue together down the middle of the ice stream as a dark band of rocky debris known as a *medial moraine.* In fact, there may be as many bands as there are unions of trunk and tributary glaciers, resulting in wonderfully banded strips of white ice interspersed with darker morainal layers (Fig. 14-26).

Lateral and terminal moraines form by several

**Fig. 14-22.** The narrow twisting gorge of the Urubamba River, near Machu Picchu, is typical of many nonglaciated valleys in the high Andes of Peru. (Penny Patterson)

processes. Most of the material in lateral moraines is derived from the valley walls above the ice. Material that avalanches or slowly moves down the slopes is caught in the trough between the ice and the valley wall and forms a ridge as it is dragged downvalley by the moving ice. Terminal moraines have a different origin and form either from the bulldozing of loose material in front of

an advancing glacier or by the melting of debris included in the glacial snout.

Moraines impound the waters of some of the world's most scenic lakes. The best known probably are those bordering the Alps, such as Como and Garda on the Italian side and Geneva and Constance in Switzerland. North America has such moraine-blocked water bodies, too; Lake Mary and Lake MacDonald in Glacier National Park are two of the most striking examples.

In many places morainal material does not form such predictable looping forms. For example, materials deposited beneath the periphery of former continental ice sheets leave an undulating terrain known as **ground moraine** (Fig. 14-28).

Material laid down directly by ice is called **till** and consists of poorly sorted debris in which particles range from the size of clay particles to huge boulders (Fig. 14-29). Till makes up the moraines. Few geological processes leave such a jumble of debris, except perhaps desert and volcanic mudflows. In fact, in places it is difficult to determine if a deposit is the result of glacial or mudflow

deposition. The material can be identified as glacial by examining the boulders it contains for striations and the bedrock on which it rests for polishing. Differentiation of the deposits is not merely an academic exercise, however, because assessment of geological hazards on the high glaciated volcanoes of the Northwest, for example, demands that one be able to tell one deposit from another. One might feel rather foolish in identifying as a mudflow—and thus assigning to an area a high-hazard potential—what was actually a till laid down by a sluggish glacier 20,000 years ago.

Indeed, many of our major northern cities are built on till. Where it is sandy and bouldery, till makes a good base for buildings and freeways, but

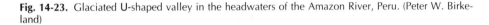

**Fig. 14-23.** Glaciated U-shaped valley in the headwaters of the Amazon River, Peru. (Peter W. Birkeland)

446

**Fig. 14-24.** *(Opposite)* Extensively glaciated terrain in the region around Mount Blanc, France [4807 m (15,771 ft)]. Characteristic of this area are the knifelike ridges called arêtes and the slender needlelike forms of many peaks, features that have challenged mountaineers for more than 200 years. (Courtesy Wild Heerbrugg, Ltd., Heerbrugg, Switzerland)

**Fig. 14-25. A.** *(Right)* Satellite image of Norway clearly displays its many narrow fjords, some of which head near the backbone of the country. (USAF DMSP, CIRES/National Snow and Ice Data Center, Boulder, Colorado.) **B.** *(Below)* High, steep valley walls plunging below sea level are common to many fjords, as shown by this one in Norway. (Russell A. Thompson)

447

**Fig. 14-26.** Lateral and medial moraines of the Yentna Glacier, Alaska Range. (Austin S. Post, USGS Glaciology, Tacoma, Washington)

where it is rich in clay, it makes for very slippery foundations. Tills and associated deposits often change properties over short distances laterally; thus geologists find it difficult to predict the kind of surface materials that might be encountered during a construction project. Extensive field mapping and drill-hole data are essential to builders working in glaciated areas.

In places, blocks of ice may have been entrapped in the till that on melting produced a depression called a *kettle* (Fig. 14-30). It is no wonder then that glaciated areas abound with lakes.

Some glacial forms on ground moraine are not random heaps of till, but show a regular geometry.

Among the shaped features are swarms of curious elliptical, rounded low hills that resemble a whale or the bowl of a teaspoon turned upside down. Such hills are called *drumlins,* from an Irish Gaelic word *druim* (ridge of a hill) (Fig. 14-31). Of those curious features, certainly the most renowned is Bunker Hill. Drumlins vary widely in size and shape, but few are more than 1-km (0.6-mi) long or 30-m (100-ft) high; they are also always aligned parallel to known directions of ice transport. Conjectures on their origin range from erosion of preexisting till or bedrock to a purely depositional landform feature related to the mechanics of till deposition.

Interspersed with the till in some areas of ground

**Fig. 14-27.** Terminal moraine forming along the northeast margin of the Barnes Ice Cap, Baffin Island, Northwest Territories, Canada. (J.D. Ives)

**Fig. 14-28.** Hummocky ground moraine near Cypress Hills, Saskatchewan, Canada. (Geological Survey of Canada)

moraine are deposits laid down by streams that once flowed in ice tunnels at the base of the glacier. Such elongate, narrow, sinuous ridges of stratified sediment are called *eskers.* Generally their crests are rounded, their side slopes are moderately steep, and their longitudinal slope is gentle (Fig. 14-32). Like conventional streams, they may meander; occasionally, they are joined by tributaries, but unlike ordinary streams they may climb up hill slopes, especially where they cross low ridges through passes. Seldom do their crests stand much more than 30 m (100 ft) above their surroundings. Some may be as long as 500 km (310 mi) although most are a great deal shorter.

Although drumlins and eskers can be found in any glacial environment, they are more commonly associated with former continental ice sheets.

## Deposits Closely Associated with Glaciation

Two characteristic deposits are closely associated with proximity to glacier, be they alpine or continental. Streams that drain off the front of the ice are heavily loaded with debris and build up their beds to form vast floodplains, or **outwash plains** (see Fig. 12-21). These featureless plains have been graded and regraded endlessly by ever-shifting streams. Thus the material is fairly well sorted and, when formed, is unvegetated. Because the material is so fresh, so uniformly textured, and in the midcontinent so fine grained, it makes fertile soil. In fact, much of the best farmland is located on glacial outwash plains. But the outwash associated with alpine glaciers is commonly bouldery and difficult to farm.

**Fig. 14-29.** Glacial till in the Sierra Nevada, California. Typical of many alpine areas, the till is very bouldery, with boulders set in a wide range of smaller sizes. (W.C. Putnam)

450

**Fig. 14-30.** Closed depression in till in the Matapedia Valley, Quebec, that probably formed when a block of glacial ice melted. (Geological Survey of Canada)

Because the glacial mill grinds particles so exceedingly fine, the surface of an outwash plain can be covered with **rock flour,** a fine silt that winds can pick up and transport. With strong winds sweeping across an open, unprotected, silt- and sand-covered plain, great clouds of dust are picked up and swept for scores of miles beyond the barren floodplain (Fig. 14-33). Deposited, the fine, wind-transported glacial silt, called loess (see chap. 13), may blanket much of the neighboring countryside, sometimes to depths of 30 m (100 ft). Loess was broadcast over the length and breadth of the Mississippi Valley as well as across.the lowlands of Central Europe and far down into the Danubian plain of Hungary. It is always thickest near its floodplain source, and it systematically thins downwind. The deposits are responsible for some exceptionally fertile soils the world over.

## THE COMPLEX PATTERN OF FORMER GLACIATIONS

When the theory of the Great Ice Age was advanced over a century ago, people had little idea of how complex the story would be. It was the most recent of several worldwide ice ages, and during this time glaciers advanced and retreated many times. Geologists call this last ice age the **Pleistocene Epoch.** This epoch is thought to cover

**Fig. 14-31.** A drumlin field at Strangford Lough, Northern Ireland. The ice moved from left to right. (Aerofilms Ltd.)

**Fig. 14-32.** This esker, near Fort Ripley, Minnesota, consists of gravel and sand probably deposited in a winding tunnel at the base of an ice sheet. When the ice melted, a ridge was left behind. (W.S. Cooper)

the period from about 2 million to 10,000 years ago; yet the Pleistocene is not synonymous with the last glaciation. There is evidence that the last period of glacial advances and retreats began before 2 million years ago in such places as Alaska, Antarctica, and South America.

We find good evidence for multiple glaciation in the deep cuts or quarries in our midcontinent. Tills in the cuts are the evidence for the glacial periods, and soils or nonglacial deposits between the tills attest to a time of ice withdrawal during the interglacial periods (Fig. 14-34). Contained within the sediment might be fossil trees or pollen grains indicating that the vegetation during the interglacials, and therefore the climate, were much like those of the present.

The limits of the last glaciation are fairly well known around the world (Fig. 14-35), for about one-third of the land was glaciated. From the initial staging areas in the northern regions of North America and Europe, the two large ice

sheets grew larger and thickened to several kilometers of ice as they expanded southward. The North American sheet at its maximum pushed into westernmost Canada, where it met an ice sheet covering the Coast Ranges and Rocky Mountains, and southward to the Missouri and Ohio rivers. In the northeastern United States the ice sheet over-ran the Catskills [1280 m (4200 ft)]; the Adirondacks ₁ 615 m (5300 ft)]; Mount Washington [1917 m (6288 ft)], the highest point in the Presidential Range; and Mount Katahdin [1791 m (5268 ft)] (Fig. 14-36). A virtual wall of ice extended 6400 km (3980 mi) across the full width of Canada. In Europe the ice sheet pushed as far south as Hanover, Cracow, and Volgograd. Most alpine areas were also glaciated at this time, wherever falling snowlines intersected mountains. The approximate age of the maximum extent of ice worldwide is about 20,000 years—plus or minus several thousand years, depending on the locality.

Most areas show evidence of more than one

**Fig. 14-33.** Wind lifts clouds of silt from the floodplains of the Knik River and tributaries, Alaska. (W.C. Bradley)

Pleistocene glaciation. The U.S. midcontinent is an example. Glacial geologists were once fixed on the idea of four major glaciations, but it is now becoming apparent that there were more than four. The mapping of these events is incomplete however, and small-scale maps commonly still show four (Fig. 14-37).

The glaciations and interglaciations of the central part of the United States generally were named after the localities or areas in which deposits or soils were well exposed. Names for the glaciations were taken from states, but those for the interglaciations came from a county, a town, and a road junction. The major glaciations and interglaciations are (youngest at top):

| Glaciation | Interglaciation |
|---|---|
| Wisconsin | |
| | Sangamon |
| Illinoian | |
| | Yarmouth |
| Kansan | |
| | Aftonian |
| Nebraskan | |

It is difficult to determine exactly when the various midcontinent glaciations occurred. The Wisconsin episode commonly is considered to have started about 75,000 years ago, lasting until about 10,000 years ago. The only other dated glaciation is the Kansan, which occurred at about the same time as a 600,000-year-old volcanic ashfall deposit from an enormous volcanic explosion in what is now Yellowstone National Park. Estimates on the timing and age of other glaciations can be obtained from the ocean-sediment record (chap. 16), but not all workers agree that the ocean-sediment–glaciation connection is fully clear. In brief, the record suggests a glacial/interglacial cycle every 100,000 years.

Alpine glaciation seems to have occurred more or less in concert with the continental glaciations and interglaciations, as shown by weathering and topographic features as well as by absolute dates on the various tills. Major canyons in the high mountain ranges were glaciated, and bulky moraines commonly are found at their mouths. In a general way the preservation of alpine moraines is related to the age of the till, just as it is in the

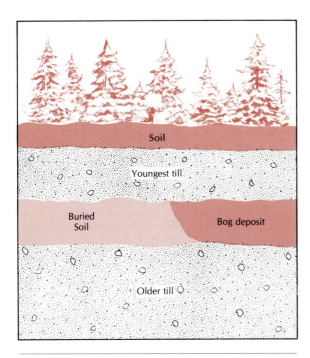

**Fig. 14-34.** The kind of field evidence necessary to prove multiple glaciation. At the surface is the youngest till. Since the retreat of ice some 15,000 years ago, a weak soil has formed on its surface. Beneath it is a soil that formed on an older till. This buried soil, exposed to weathering and soil formation for more than 15,000 years, is deeper and richer in clay than the surface soil. Fossils found in the bog deposit are clues to interglacial vegetation and climate.

midcontinent (Fig. 14-38). Moraines dating from the Wisconsin glaciation are relatively well preserved; Illinoian moraines, less well; and older moraines have been completely worn down through erosion. In addition to the moraines that mark the maximum extent of the ice, smaller moraines are found far upvalley in an area adjacent to the cirques. These smaller moraines record slight climatic changes over the last 5000 years, but these changes were enough to have caused cirque glaciers to advance and retreat several times.

The beginning of the last major glaciation seems to have varied from place to place. The Antarctic Ice Sheet reached a substantial size about 10 million years ago. Extensive glaciation also took place 10 million years ago in Alaska. In contrast the oldest glaciation in lower latitudes is probably less than 3 million years. In like manner the end of the last glaciation varies from place to place. Glaciers have all but disappeared from many of our western mountains outside of the Pacific Northwest, so that we are now in an interglacial period. Such examples are an important lesson for glacial geologists—climatic change does not take place in all areas at the same time. Thus we would not expect climatic change to be synchronous worldwide. In geological terminology the change is **time-transgressive,** occurring in one place first, and in another place later—depending on local conditions.

The Pleistocene ice ages profoundly changed the topographic face of North America. At one time rivers rising in the Canadian Rockies flowed east toward Hudson Bay, as did many of the rivers draining the vast region of the present-day upper Missouri and Mississippi. Glacial erosion and deposition changed all that. Drainage directions were reversed as large areas were added to the upper reaches of the Missouri and Mississippi rivers. Furthermore, glacial moraines trapped water into a string of huge lakes, from the Great Lakes to Great Bear Lake in northwestern Canada. Many of these lakes formed in the peripheral wastage zone of the former ice sheet.

The Great Lakes had a different pattern during the last part of the Ice Age than they have today. For one thing, they were dammed to the north by the retreating wall of the receding ice sheet; for another, their levels were higher then and their outlets were quite different. One outlet was via the Mohawk Valley and the Hudson River; another was down the course of the Illinois River to the Mississippi and the Gulf of Mexico rather than to the Atlantic by way of the Gulf of Saint Lawrence, the present outlet.

## POSTGLACIAL CLIMATIC CHANGES

Enough is now known to tell us that the climate since the end of the Pleistocene has not always been the same. Most of the information comes from nonclimatic sources because the length of time during which climatic observations have been made is far too short. Examples of such historic information are records of drought, floods, crop failure, blocking of alpine passes by long-lasting ice and snow, and successions of unusually cold winters.

The doomed Norse colony in Greenland offers a dramatic example of climatic change in the recent past. The colony was founded in A.D. 984 and perished around 1410. In its early history the Arctic seas were iceless and perhaps calmer, thus Viking ships could make passage where today ice floes and stormy seas bar the way. The colonists raised cattle and hay, built permanent habitations, and the settlement flourished. With a climatic change that brought the Greenland ice southward again, with the pressure of the Eskimos at their gates, with a succession of crop failures, with the rise of permafrost in the ground—so that even such shallow excavations as graves were no longer possible—and with the perils of the ocean crossing too great for the frail vessels of that day, the colony and all its inhabitants perished.

But the geological record can reveal some facts about recent climate changes as well. For ex-

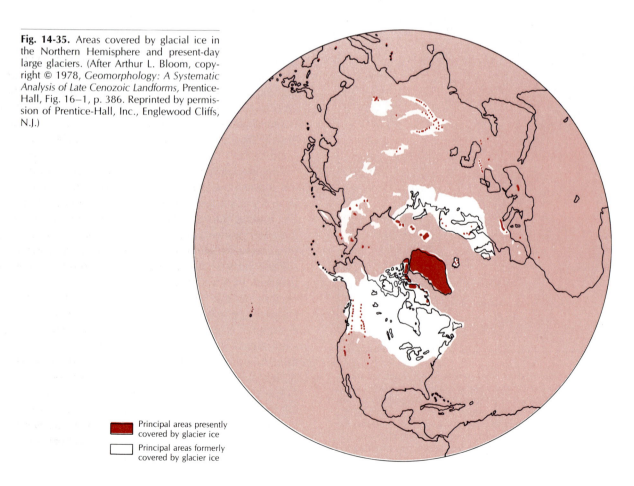

**Fig. 14-35.** Areas covered by glacial ice in the Northern Hemisphere and present-day large glaciers. (After Arthur L. Bloom, copyright © 1978, *Geomorphology: A Systematic Analysis of Late Cenozoic Landforms,* Prentice-Hall, Fig. 16–1, p. 386. Reprinted by permission of Prentice-Hall, Inc., Englewood Cliffs, N.J.)

Principal areas presently covered by glacier ice

Principal areas formerly covered by glacier ice

**Fig. 14-36.** View of Mount Katahdin, Maine, showing well-developed cirques. The history of this mountain is rather unusual in that cirque glaciation took place first. The Wisconsin-age North American ice sheet came next, overwhelming the mountain. In the deglaciation interval following, the cirque glaciers did not reform. (P.T. Davis)

ample, high in many mountain ranges, young moraines front existing glaciers or lie in ice-free cirques. We know that most of these moraines date from the last 5000 years, but the timing of advances and retreats was not synchronous everywhere. In many places the advance that culminated in the last several centuries—dubbed the Little Ice Age by some—was the most extensive. To convert the moraine record to a climatic record requires comparing the altitude of the ancient snowlines with that of the present snowline. Using the rate of temperature cooling with altitude, temperature fluctuations of about 1° C (2° F) for these minor advances and retreats can be obtained.

Ice has disappeared or retreated from the Little Ice Age moraines only recently. Until about 1950 the rate of recession for most Northern Hemisphere glaciers was rapid. Since then, however, slightly cooler and moister climates have slowed the retreat of some glaciers and hastened the advance of others. Retreat has been most extensive in midlatitude glaciers and slight to nonexistent in some Polar areas.

The climate record parallels these glacial recessions. Severely cold weather marked the period from about 1550 to 1850, freezing rivers that were once ice-free in winter and making farming impossible in parts of the European Alps, Norway, and Iceland. Temperatures began to rise thereafter, followed by a cooling from the 1940s to 1960s. By the mid-1970s, there were indications that the North Atlantic region was undergoing yet another

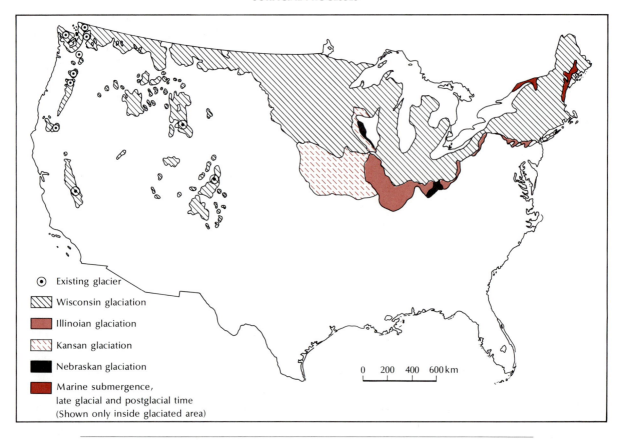

**Fig. 14-37.** Glacial geology of the coterminous United States. (After *National Atlas of the United States of America*, USGS, 1970)

temperature rise. Looking at a larger chunk of time, one estimate is that, of the last 1 million years, only 10 percent of those years have been as warm as the present.

Where the present trend is leading, no one can say. Will the earth's atmosphere continue its warming trend, so that the ice captured in glaciers melts? Or will the air chill once more, starting massive glaciers on the march once again? Understanding the long-range effects of subtle climatic changes is important because they come at a time when the world's population is expanding rapidly and the need for increasing food production is greater than ever. Humans have a short memory. It has been pointed out in *The Climate Mandate* (1979) that the period from the late 1950s to the early 1970s was one of unusually good

weather in the North American grainbelt. But periods like that cannot last forever. The book closes with this thoughtful statement:

The earth is marvelously bountiful. It should have the ability to produce adequate food for a doubled or tripled world population if necessary. But even the job of feeding those who are now alive will require us to apply our scientific and technical knowledge responsibly and with great care and foresight in order to protect and preserve the resources of the planet. Clearly, we must recognize the climate mandate, which dictates that the earth's bounty will rise and fall from time to time and place to place in response to climatic fluctuations. If we heed the climate mandate, and if we accept the fact that the earth's people are bound together by mutual needs and expectations that must transcend our rivalries and contests, humanity should be able not only to survive but to prevail over the hunger and starvation that have threatened so many people for so many centuries.*

## OTHER ICE AGES

A single catastrophic deep freeze of the earth would not fit very well with our ideas of uniformitarianism. When we look back through the geological record, indeed, we do find evidence for other periods of extensive glaciation preceding the

*W.O. Roberts and H. Lansford, 1979, *The Climate Mandate,* Freeman, San Francisco, pp. 188–189.

last one. The evidence consists largely of deposits known as ***tillites,*** which are nothing more than consolidated glacial till. These unsorted deposits are considered of glacial origin only when they rest on striated and polished rock surfaces, and the bounders they contain show striations. Other relatively unusual deposits indicate proximity to glaciers but are marine in origin. In such deposits boulders and cobbles from melting icebergs appear to have dropped into fine-grained bottom sediments, locally deforming their horizontal layers; there is no other way to explain such large clasts in a quiet-water environment.

A variety of evidence points to at least five

**Fig. 14-38.** Moraines of two ages encircle the lower part of Fremont Lake, Wind River Range, Wyoming. Moraines of the last glaciation are well preserved **(A)**, whereas those of the previous glaciation are quite subdued **(B)**. (Dale Johnson)

periods of major global glaciation—and evidence comes to us from many improbable places. For example, the younger Precambrian glaciation has been etched into all continents but Antarctica. The Ordovician–Silurian glaciation is best seen in rocks that stretch some 4000-km (2400-mi) across the Sahara, but to complicate matters, ice movement at the time was northward, from equatorial latitudes into the hottest part of the desert. On the other hand, the extensive Permian glaciation that touched the southern continents seems to have left the northern ones alone, and along many coasts ice was moving from what is now sea to land.

If evidence taken from tills and features of other sedimentary rocks can be used to indicate former climate conditions, global trends in temperature and precipitation can be proposed (Fig. 14-39). A plot of these trends indicates that much of the last 200 million years has been relatively warm; indeed, no glaciation has been found during the Mesozoic, one of the warmer intervals, when dinosaurs surely enjoyed the balmy days.

## CAUSES OF GLACIATION

Many theories have been proposed to explain glaciation—and many have been discarded. For pre-Pleistocene glaciations, it is now well established that the earth enters a glacial period whenever sufficiently large landmasses are situated in polar positions. Glaciers build up, which sets into action a series of events, including an interaction with oceans and atmosphere. The result is a cooling over much of the globe. Thus are explained some of the strange phenomena mentioned earlier (some of these phenomena were discussed in chap. 4). Continental drift makes all of this possible. When landmasses move far enough away from polar positions glaciation is interrupted, and the earth basks under a warmer, more equable climate. Indeed, only when Antarctica moved to its present polar position did the last episode of glaciation begin.

Some scientists believe that glaciation has a **terrestrial cause** and point out the relationship between the growth of high mountains and glaciation. Most of the world's major mountain ranges

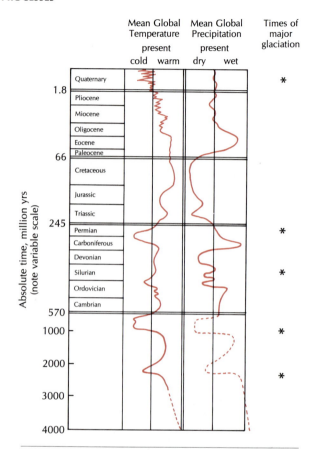

**Fig. 14-39.** Estimated trends in mean global temperature and precipitation through geological time. Notice that the times of major glaciation coincide with low temperatures and with precipitation either close to that of the present or drier. (After L.A. Frakes, 1979, *Climates Throughout Geological Time,* Elsevier, Fig. 9–1)

were elevated to their present heights during or just before the Pleistocene, including such lofty summits as those of the Himalayas, Andes, Caucasus, and the European Alps. In fact, it has been suggested that some ranges missed the earlier Pleistocene glaciations because they were then not high enough to intersect the regional snowline. Others point out that this youngest mountain-building episode resulted in significant north–south trending ranges in the Western Hemisphere. As those ranges lie athwart the general planetary circulation of the atmosphere, they may have had a significant effect on the growth and location of glaciers. Neither of these theories, however, explains multiple glaciation: only a heretic would call for mountain ranges to rise and fall like yo-yos to produce Pleistocene glaciations and interglaciations.

Before attempting to explain Pleistocene glaciations, we should keep in mind the following:

1. A massive accumulation of ice has occurred many times in geological history.
2. Glacial growth seems to have been a slow process, but retreat was so rapid as to verge on the catastrophic. For example, it is estimated that the ice sheet of the Wisconsin glaciation took 20,000 years or more to attain maximum size, but it disappeared in only 7000 years.
3. The glaciations appear to have been synchronous on both sides of the Atlantic; apparently, Northern Hemisphere glaciers advanced or receded at the same time as Southern Hemisphere glaciers. In other words, the entire earth seems to have responded to the same climatic pulses.
4. Times of glacial advance also appear to have been times of lowered temperature, as demonstrated by data from sediment cores from the bottom of the sea and by evidence of a lowering of the regional snowline on the high mountain ranges. Furthermore, in areas such as Arctic Canada, much more precipitation is required to initiate an ice sheet than occurs at present.

## Atmospheric Causes

Some hypotheses cite changes in the composition of the atmosphere to explain the last glaciation. One hypothesis says that climatic variations extensive enough to trigger glacial epochs may be caused by variations in the amount of carbon dioxide ($CO_2$) in the atmosphere. A marked increase of $CO_2$ would produce a so-called **greenhouse effect,** which results in a general rise in temperature. Energy from the sun reaches the earth's surface because our atmosphere is transparent to much of it. Part of this solar energy radiates back into the atmosphere as heat (the infrared portion of the spectrum) where it is absorbed by the $CO_2$ and

returned to the earth to further heat the surface. The result is that the more $CO_2$ in the atmosphere, the warmer the atmosphere.

Another theory involves the dust in the atmosphere. Dispersed particles can partially block the sun's rays from the earth's surface, thus cooling it down. In support of this idea are ash layers in ocean cores, which suggest that volcanism during the last 2 million years has been more active. Objections to such a theory are that major volcanic eruptions, such as that of Krakatoa in 1883, did not affect the weather enough to bring about a major change in the budget of glaciers or to initiate glaciers where none had previously existed. Indeed, some geologists are turning the argument around, suggesting that increased volcanism could result from crustal stresses induced by the loading of ice on land.

One problem with these two hypotheses is that they do not adequately explain the strong cyclic pattern of glacial/interglacial episodes in the Pleistocene. Recent work on dating and on marine sediments suggests that each cycle lasts about 100,000 years.

## Astronomical Causes

One astronomical explanation involves variation in the amount of incoming solar radiation. We know that variation does occur, but is it sufficent to trigger major glaciations as well as produce cyclicity? No one has the answers to these questions.

The astronomical theory favored today was long in being accepted. Its present form owes much to the Yugoslavian mathematician Milutin Milankovitch and is based on three changes in the geometry of the earth's orbit. One is that the path of the earth around the sun forms an ellipse, the shape of which changes over a period of about 100,000 years. The second is that the rotational axis of the earth is tilted with respect to the plane of its orbit; the tilt is now 23.5°, but it has varied several degrees over a period of 41,000 years. The third phenomenon is a wobble in the axis of rotation, an event that seems to recur every 21,000 years. Together, these factors influence the solar radiation received at the top of the earth's atmosphere—and this, of course, influences mean global temperature. Whether or not the earth is in an ice age, these cycles recur.

The Milankovitch theory has gained acceptance primarily because young marine sediments exhibit cycles of 23,000, 42,000, and 100,000 years (see chap. 16)—very close to the cycles Milankovitch calculated. If these calculations can predict past cycles of climate change, we may be able to plan for future change. It is inevitable that another

period of glaciation will arrive—but when? Given the average length of the average cycle, it could be that the present interglacial is about to end.

Keeping in mind the many theories we have considered, our best strategy would be to try to understand the combination of events—continental positioning, mountain building, and orbital cycles—that together produce the glacial and interglacial epochs and their awesome effects across the globe.

## SUMMARY

1. Glaciers form when winter snowfall does not melt entirely but, instead, builds up year by year until the mass, partly converted to ice, flows under its own weight.
2. The mechanisms of glacier movement are (1) slippage along a bedrock base and (2) the internal deformation of the ice crystals.
3. The *glacier budget* is determined by the annual accumulation and the annual loss *(ablation)* of ice. If the two are balanced, the glacier, although flowing, will maintain a stationary front. If accumulation exceeds ablation, the glacier advances. If ablation exceeds accumulation, the glacier retreats.
4. Glacial erosion is responsible for much of the world's scenic mountain terrain. *Cirques* form at the valley heads; the valley lower down is U-shaped. Also, glaciers gouge out bedrock depressions that can fill with water to become lakes—the legacy of glacial erosion either in mountainous terrain or in flat shield areas overrun by continental glaciers.
5. *Till* is the common sediment deposited by glaciers. It is found heaped into *moraines* around the former periphery of the glacier or as *ground moraine* deposited at the base of a glacier. Deposits closely associated with glaciation are glacial *outwash,* along rivers that drained glaciers, and *loess,* a fine silt blown from the glacial-age floodplains.
6. Multiple glaciation took place during the Pleistocene Epoch, with no less than four major glaciations recognized on land. The main mountain glaciations seem to coincide with the major continental glaciations, and the times of glaciations in both the Northern and Southern hemispheres seem to have been synchronous.
7. Glaciation apparently involves several processes. One is continental drift, which allows landmasses to move to polar positions—seemingly an essential factor. Another is mountain building, allowing land to intersect the snowline. Finally, changes in the earth's orbital geometry match proxy data for climatic changes found in marine sediments closely enough to suggest a cause-and-effect relationship.

## QUESTIONS

1. Compare the formation and movement of a glacier in a dry polar area with one in a more temperate region.
2. What major erosional landforms are associated with alpine glaciation and with continental glaciation?
3. How are lateral, medial, and terminal moraines formed?
4. What is the evidence for multiple glaciation, and how can we decipher environmental conditions during interglaciations?
5. Considering the pre-Pleistocene glaciations, how can one be sure the tillite deposits are of glacial origin? How can one determine the direction of flow of one of these ancient glaciers?
6. Why are geologists focusing on the marine-sediment record rather than on the terrestrial record in trying to work out the timing of climatic cycles during the Pleistocene?

## SELECTED REFERENCES

Andrews, J.T., 1975, Glacial systems, Duxbury, North Scituate, Mass.

Covey, C., 1984, The earth's orbit and the ice ages, Scientific American, vol. 250, pp. 58–66.

Denton, G.H., and Hughes, T.J., eds., 1980, The last great ice sheets, Wiley-Interscience, New York.

Flint, R.F., 1971, Glacial and Quaternary geology, Wiley, New York.

Imbrie, J., and Imbrie, K.P., 1979, Ice ages; solving the mystery, Harvard University Press, Cambridge. (Paperback, 1986.)

Roberts, W.O., and Lansford, H., 1979, The climate mandate, W.H. Freeman, San Francisco.

Sugden, D.E., and John, B.S., 1976, Glaciers and landscape, Edward Arnold, London.

Weiner, J., 1986, Planet earth, Bantam, New York.

Wright, H.E., Jr., ed., 1983, Late-Quaternary environments of the United States, vols. 1 and 2, University of Minnesota Press, Minneapolis, Minn.

**Fig. 15-1.** A contestant in the 1983 Sydney-to-Hobart Yacht Race rounds the spectacular wave-eroded cliffs of Tasmania. The cliffs stand vertically because they consist of columnar basaltic sills. (Richard Bennett, Geeveston, Tasmania, Australia)

# 15

# SHORE PROCESSES AND LANDFORMS

One of the most visible and dramatic interfaces on earth occurs where the land meets the ocean. There, energy contained in the water mass that covers 71 percent of the earth is unleashed against the shore, eroding large quantities of rock (Fig. 15-1), or moving vast amounts of sediment. As in most geological systems, an equilibrium is established between the available energy and the movement of sediment. Because coasts are pleasant places to live—two-thirds of the people of the world live in close proximity to them—there is a long history of human interference with coastal processes, usually with disastrous results in a relatively short time.

The position of the shoreline is far from being stable because the amount of water in the oceans is closely linked to the volume of glaciers on land. So, the two move in concert—as glaciers build up, sea level falls. Sea level has changed little during the past 5000 years. The worldwide relative stability in sea level coincides roughly with the appearance of the maritime civilizations around the shores of the Mediterranean Sea and the Persian Gulf; thus the ancient harbors of the Egyptians, Persians, Phoenicians, and Minoans correspond roughly to the sea level of today. The Minoan

inhabitants of the island of Crete disappeared in an instant of archaeological time, perhaps at the hand of a stupendous sea wave (tsunami). The death toll from such waves is often very high.

## WAVES

Waves can be almost hypnotic in their effect. Their rhythm depends not only on the local wind for the shorter, steeper waves, but also on distant fierce winds that start the long, even-spaced ridges of a ground swell moving outward from a storm center half a world away.

Waves have various sizes and velocities. **Wave length** is the horizontal distance separating adjacent and equivalent parts of two waves, such as two crests or two troughs. **Wave height** is the distance from crest to trough. The **wave period** of a wave is the length of time required for two crests or two troughs to pass a fixed point. **Wave frequency** is the number of periods that occur within an interval of time—say a minute. Finally, **wave velocity** is the distance traveled by a wave in a unit of time. A simple equation indicates the relationship between some of these properties:

wave length = period × velocity

For example, if the period is 10 seconds and the velocity is 5 m/sec, the wave length is 50 m (165 ft).

Watching the endless procession of waves, it is difficult to comprehend that it is the form of the wave that moves forward through the water and not the water itself. This statement may not appear to make sense at first, but watch a bottle bobbing on the surface of a bay. Waves pass under it repeatedly, but the bottle moves relatively little other than to drift along with the current. An analogy is the rippling motion wind makes as it blows across a wheat field. Waves follow one another across the stalks of wheat, yet the wheat does not pile up in a heap on the far side of the field. Instead, the motion in the grain results from the nodding of the individual stalks each time a wave passes.

As long ago as 1802 it was known that water particles within a wave do not move forward with the advancing wave itself, but follow a circular orbit (Fig. 15-2). As the wave crest approaches, the water in the preceding trough moves toward the advancing crest. Then, progressively, the particles move upward, forward with the crest, downward, and seaward again in preparation for the passage of the next wave crest. This motion is known as a **wave of oscillation.**

The same cross section (Fig. 15-2) also shows how rapidly wave motion diminishes with depth.

For practical purposes wave motion approaches zero when the water depth is approximately equal to one-half the wave length.

Some of the effects of this orbital motion are familiar to every surfboarder, and others can test the effects by simply wading or swimming into the ocean a short distance. The water first runs strongly seaward toward the oncoming wave. As the wave surges shoreward, however, the water sweeps strongly toward the beach. To avoid being caught in the force of a breaking wave, swimmers dive to the deeper water either in the trough or beneath the next crest, where motion is less. Many are aware of the backward and forward pulse of the sea in the breaker zone, but few probably realize that the motion is only part of the orbital path described by water particles within a wave.

Waves in the open ocean can reach great heights (Fig. 15-3). Among the largest waves observed was one in the North Pacific that rose 34-m (112-ft) high when it was sighted off the stern of the USS *Ramapo* in 1933 during a typhoon with winds registering 68 knots. The ship withstood the waves by keeping them directly astern; approaching from any other direction, they probably would have destroyed the vessel. Wave lengths also can be extensive. One of the longest was almost 800 m (2620 ft) with a period of 22.5 seconds; using the equation given earlier, the wave's velocity would have been 128 km/hr (79 mph). Huge volumes of water are involved in the passage of such waves.

How are such volumes of water set in motion?

**Fig. 15-2.** Cross section of an ocean wave showing the paths the water particles follow. The wave profiles and positions of the water particles are shown at two moments that are one-quarter of a period apart in time. Notice that the orbit of the water particles diminishes with depth. The nearly vertical lines show how blades of grass would be bent as a wave form passes. The stalks are vertical only beneath a crest or a trough.

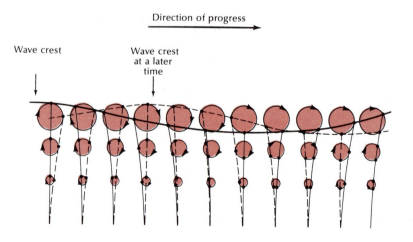

Direction of progress

Wave crest

Wave crest at a later time

**Fig. 15-3.** Wave estimated to be 12-m (40-ft) high rapidly approaching a tanker in the Bay of Biscay off the coast of France. *(Nautica)*

Almost everyone knows that the wind driving across the surface of the sea is the primary cause. How is it then that on completely windless days a tremendous surf may pound some exposed coast? Or that in a violent gale the wind may hammer the sea flat into a turbulent mass of dark water streaked to the horizon with foam?

For large waves to form in deep water, first there must be a strong wind to set large masses of water moving. Second, the wind must be of fairly long duration—more than just a sudden gust is needed. Third, the water must be deep, at least deep enough to round out the full circular pattern—waves 9-m (30-ft) high are not likely to occur in a water basin only 3-m (10-ft) deep. Fourth, the distance over which wind friction can affect waves—called **fetch**—is important. The ripples crossing a pond are a good example. They are small on the upwind side of the pond, yet they grow to fair dimensions by the time they reach the downwind shore. Together these parameters can produce large waves (Table 15-1).

It is not surprising, then, that some of the largest seas are those driven before the strong and persistent westerly winds just north of the margins of Antarctica. There, where the ocean encircles the earth uninhibited by land masses, fetch can be said to be unlimited.

To return to an earlier question: How, if waves are formed by the wind, can they travel shoreward in an endless, rhythmically advancing succession on a dead-calm day? The answer is that such waves, to which the name of **swell** is given, may have originated in gales thousands of miles away.

467

**Table 15-1** Characteristics of a Well-Developed Sea

| Wind Speed (m/sec) | Minimum Fetch (km) | Minimum Duration (hr) | Peak Wave Period | Average Wave Period | Significant Wave Height (m) | Average Wave Height (m) |
|---|---|---|---|---|---|---|
| 5 | 10 | 2 | 4.0 | 2.8 | 0.43 | 0.27 |
| 10 | 100 | 10 | 8.1 | 5.7 | 2.44 | 1.52 |
| 15 | 250 | 22 | 12.1 | 8.5 | 6.58 | 4.11 |
| 20 | 750 | 45 | 16.1 | 11.4 | 13.80 | 8.50 |
| 25 | 1400 | 70 | 20.2 | 14.2 | 23.80 | 14.90 |

After T. Beer, 1983, *Environmental Oceanography: An Introduction to Behaviour of Coastal Waters*, Pergamon, N.Y., Table 3.2, p. 50.

Waves that break on the English coast may have had their start in the distant reaches of the South Atlantic. Waves crashing on the Australian shores or on the west coast of the United States may have been born in the Antarctic, whereas waves the surfers ride in Hawaii may have come from the Arctic.

Waves on the open ocean are more complicated. Not uncommonly, swells come from more than one direction and when they meet complex wave patterns are developed. Local winds also create complex wave patterns. In such seas ships plunge and roll in a dizzying set of motions.

Swells change direction on approach to land. In the South Pacific they bend and are deflected as they reach near-vertical atolls, setting up distinctive wave patterns. Seafarers used these patterns to navigate between islands with charts fash-

**Fig. 15-4.** Stick chart used by Micronesians in navigating between islands in the Pacific. The shells represent islands and the stick pattern, swell orientation and diffraction. (Janet Robertson, Geological Society of America collection)

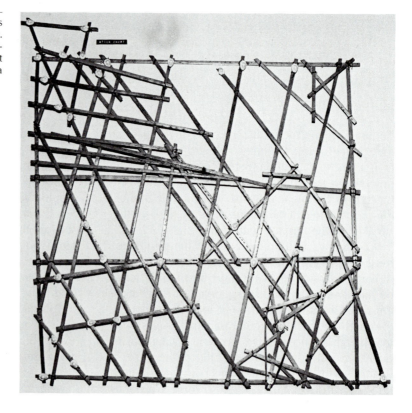

ioned from sticks and shells (Fig. 15-4). By matching local conditions with the chart, they could navigate with great accuracy between islands, which are visible only at very close range.

## Surf Formation

The formation of surf is a complex phenomenon. The endlessly changing pattern of breaking waves—and their variations with the tide, with wind and calm, with storm, and with the lulls between—has inspired generations of painters, photographers, writers, and ordinary daydreamers. Few manifestations of the natural world are more dynamic or make us more conscious of the force of moving water than a strongly running surf.

Surf forms as waves move from deep water shoreward. Waves begin to touch bottom when the depth of water is equivalent to about one-half the wave length. As they move into still shallower water velocity is reduced, the wave length shortens and the height increases (Fig. 15-5). In a general way waves of oscillation break when the stillwater depth is roughly equal to 1.3 times the height of the wave. There appear to be at least two major causes leading to the formation of breakers. The first is an increase in the circular velocity of particles in a wave that has moved into shallowing water, so that the velocity of the particles at the crest exceeds the decreasing velocity of the wave form. The second is that in the shallow depths near shore, there is not enough water to complete the wave form.

Breaking waves are almost always picturesque. Two of the most dramatic are sought by surfers. One is the **spilling breaker:** The crest foams over and cascades down the front of the advancing wave without actually toppling over. The other is the **plunging breaker:** A wave curls over in a beautifully molded half cylinder, topples, and crashes with a thunderous roar; the entrained air is violently compressed and converts the entire roller into a froth of foam-whitened water. The goal of more accomplished and daring surfers is to tuck into the tube and stay ahead of the collapsing portion of the cylinder (Fig. 15-6).

The slope of a beach helps determine the kind of breaker: Spilling breakers form on horizontal beaches; plunging breakers on steep beaches. Very

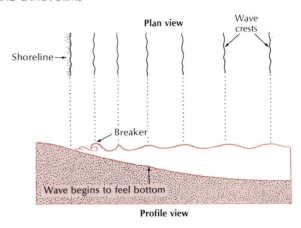

**Fig. 15-5.** The formation of surf as waves approach the shore. Waves shorten, increase in height, steepen, and break as they advance into shallow water.

steep beaches are avoided by surfers because the wave form either collapses rapidly or the water merely surges up and down with little form.

## Tsunamis

One of the most destructive waves on earth is the **tsunami** (Japanese for harbor or bay mouth wave) or **seismic sea wave.** It is usually generated in response to disturbances on the ocean floor, such as fault displacement or earthquake-caused landslides (Fig. 15-7). Other possible causes are related to volcanism. One could be an underwater volcanic blast. Another could be the formation of a caldera at or below sea level. Common to all causes is the instantaneous displacement of large volumes of water, usually to the sea floor. In contrast ordinary waves are formed gradually—by the frictional drag of air masses over water—and involve only a surface skim of the ocean.

A tsunami-creating event sends out several waves at velocities over half that of sound; yet on the open sea they go undetected. Far out in the ocean the tsunami is less than 1-m (3-ft) high, takes 10 to 20 minutes to pass, and has a wave length of up to 200 km (124 mi). Like a gentle swell, it goes unnoticed by people on ships and escapes detection by satellite. A tsunami formed on the coast of Chile in 1960 will give an idea of the velocities attained. The wave traveled the 10,630 km (6600

**Fig. 15-6.** Surfer riding a plunging breaker. (Aaron Chang, *Surfing Magazine*)

mi) to Hawaii in less than 15 hours. Seven hours later it had reached Japan [17,070 km (10,600 mi)]. The maximum velocity of this tsunami was about 775 km/hr (482 mi/hr).

Only when this oceanic ripple approaches a shelving coast is violence unleashed (Fig. 15-8). Like most waves, it feels bottom, increases in height as it piles up, and breaks. Most tsunamis are less than 30-m (100-ft) high, but gigantic ones between 30 to 60 m (100 to 200 ft) have been reported.

Tsunamis have been called the scourge of the Pacific, for they are most common in that basin wracked by earthquakes and volcanism. They have struck the Hawaiian Islands 41 times since record keeping was started there in 1819. Today a tsunami warning system has been set up around the Pacific so that such catastrophes can be avoided.

A well-known passage from the Old Testament is thought by some to describe an ancient tsunami, one that may have taken place during the journey of Moses and the Israelites from Egypt. Note that

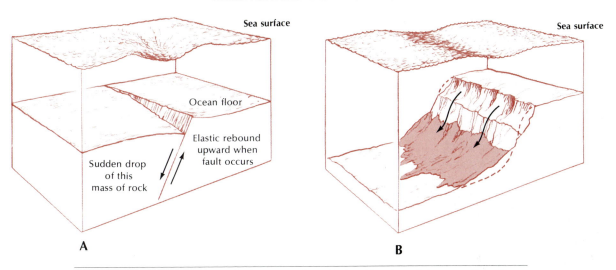

**Fig. 15-7.** Two common ways in which tsunamis form. **A.** A sudden drop of the ocean floor along a fault causes the water surface to drop and a wave is generated. **B.** An earthquake triggers a submarine landslide of loose sediment, which displaces water and sets up a tsunami. (After W. Bascom, 1964, *Waves and Beaches,* Doubleday, Figs. 41, 42)

**Fig. 15-8.** Engraving of tsunamis battering Lisbon, Portugal, during the earthquake of 1755. (Mary Evans Picture Library, London)

withdrawal of the sea prior to the onrushing wave is a precursor to all tsunamis:

21 And Moses stretched out his hand
over the sea;
and the LORD caused the sea to go back
by a strong east wind all that night,
and made the sea dry land,
and the waters were divided.

22 And the children of Israel
went into the midst of the sea
upon the dry ground;
and the waters were a wall unto them
on their right hand,
and on their left.

23 And the Egyptians pursued,
and went in after them
to the midst of the sea,
even all Pharaoh's horses,
his chariots, and his horsemen.

24 And it came to pass,
that in the morning watch
the LORD looked
unto the host of the Egyptians
through the pillar of fire and of the cloud,
and troubled the host of the Egyptians,

25 And took off their chariot wheels,
that they drave them heavily:
so that the Egyptians said,
Let us flee from the face of Israel
for the LORD fighteth for them
against the Egyptians.

26 And the LORD said unto Moses,
Stretch out thine hand over the sea,
that the waters may come again
upon the Egyptians, upon their chariots,
and upon their horsemen.

27 And Moses stretched forth his hand
over the sea,
and the sea returned to his strength
when the morning appeared;
and the Egyptians fled against it;
and the LORD overthrew the Egyptians
in the midst of the sea.

28 And the waters returned,
and covered the chariots, and the horsemen,
and all the host of Pharaoh
that came into the sea after them;
there remained not so much as one of them.*

What could have produced this phenomenon? Some point to the beautiful island of Santorini,

*Exodus 14.

whose fragmented crescent shape rims the collapsed caldera of a volcano. In the fifteenth century B.C. Santorini erupted in a violent explosion. It is thought that a series of huge sea waves followed, sweeping over coasts along the eastern Mediterranean and giving aid, perhaps, to Moses when he most needed it.

## Wave Refraction

As wave fronts approach land they commonly are bent, or refracted, so that they almost always parallel the shoreline closely, no matter what their configuration. Refraction occurs because of variations in the velocity of a wave as it reaches shallower water: Different parts of a wave touch bottom at different times, slowing its forward progress and changing its direction.

Wave refraction explains the variation in the intensity with which waves break on an indented coast (Fig. 15-9). Land promontories are eroded because the wave energy is focused there by refraction. In contrast the energy is diffused through an adjacent bay, as any mariner knows. Because of a diffusion, wave energy decreases in bays, which are commonly the site of deposition of sediments eroded from the promontories.

## WAVE EROSION AND RESULTING LANDFORMS

It is in the shore zone where breakers bring their full force to bear against the land that most of the erosive work of the sea occurs. Indeed, the shore zone is one of the most dynamically active erosional areas on earth. Because there is an upper and lower limit to each wave, the work of waves might be likened to that of an enormous horizontal saw. Their upper effective limit is the maximum height they reach at high tide during a storm. That height will differ among coasts, depending on how the land faces the sea and how strong is the gale driving the waves. Some storm-driven waves have been known to reach heights of 65 m (213 ft) or so.

The lower limit of wave erosion is less certainly known. Estimates are that waves can generate currents capable of moving sand at depths of not

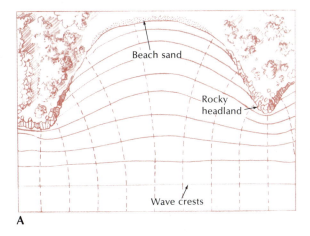

A

**Fig. 15-9. A.** How waves are refracted. Wave crests slow down and crowd together as they move into the shallow water around the rocky headlands; in the deeper waters of the bay, however, they maintain the original velocity longer and, therefore, are more widely spaced. This accounts for the bending of the waves. Wave energy, equally distributed between the parallel dashed orthogonal lines in deep water, is concentrated on the headlands where the lines converge but is diminished in the bay where they diverge. **B.** *(Below)* Wave refraction on a coast with rocky headlands and small islands. Eyre Peninsula, South Australia. (South Australia Lands Department)

much more than 10 m (33 ft) and gravels at about half that depth. The lower effective limit of wave transportation and erosion is called **wave base;** obviously, its depth differs on different coasts, just as the upper limit does. Erosion, therefore, is restricted to a narrow zone of the shore.

Several processes contribute to wave erosion. One is **shock pressure.** When breaking waves hit a cliff, erosion results from the full weight of thousands of tons of water and from the greatly increased air pressures in voids in the rocks. Most waves, however, erode by abrasion, just as streams and glaciers do, with the sediment load carried acting as the tool of erosion. During storms, when the waves are highest and most capable of carrying such abrasive agents as gravel in addition to sand, their erosive power is great (Fig. 15-10). Little erosion occurs on quiet days when sediment is not, or just barely, moving.

Other factors that contribute to the erosion of a coast are salt weathering, as salt spray from breaking waves is ubiquitous; organisms boring into rocks, which can be locally important; and on limestone coasts where there may be little abrasive material for erosion, weathering can rapidly produce a solution platform with a notch several meters deep at the coast's landward edge (Fig. 15-11).

Because the processes of shore erosion have upper and lower limits and work horizontally, they produce landforms that differ from those made by downward-cutting streams and glaciers.

The most noticeable coastal landform is a **sea cliff** (Figs. 15-12, 15-13). Its height depends on several factors, including the energy of the attacking waves, the slope of the land surface, and the resistance of its rocks. Obviously, a granite headland is likely to be much more resistant than one of chalk. The imposing height of the cliffs near Dover results from the erosional weakness of chalk as well as the fury of the sea's attack in a North Atlantic gale.

As a sea cliff recedes before the onslaught of the waves, a planed-off rock bench called an **abrasion platform** is cut at its base (Fig. 15-12). Sometimes it will be bare, abraded rock interrupted (perhaps) by tide pools or unreduced remnants of the cliff, known as **stacks** (Fig. 15-13). Such stacks are part of the rugged beauty of the

**Fig. 15-10.** Storm waves from the Southern Ocean and large boulders of the beach have eroded these coastal cliffs on Banks Peninsula, New Zealand. (Peter W. Birkeland)

Oregon coast. Where the shoreward portion of the platform is mantled with sand or gravel, it is called a **beach.**

The rate of wave erosion varies from place to place. Some coasts show little change in a generation; on other coasts, erosion can be quite rapid, as shown by undercut roads, patios, and steps built down the face of sea cliffs. Rates in excess of 1 m (3 ft) per year are common along the coast of California (Fig. 15-14) and the East coast. The rate is a function of the strength of rock making up the cliff and the energy of the waves. Cliffs along the English Channel have retreated as much as 12 to 27 m (39 to 89 ft) overnight during a single storm. In many places landslides are probably the main agent in cliff retreat; the process continues as waves carry away the slide material and the cliff is again undercut.

## BEACH DEPOSITION AND RESULTING LANDFORMS

Wave and current erosion and sediment transport cause some portions of a coast to retreat and other portions to advance. Material that is eroded from a point or headland may be deposited by a longshore current in a nearby bay or marsh. Eventually, erosion and deposition may attain a rough equi-

**Fig. 15-11.** Weathering notch at sea level forms "mushroom islands" at Belau (formerly Palau), a group of islands east of the Philippines in the western Pacific. (Peter Richards)

librium or one may outpace the other. In Britain, for example, in the 35 years immediately preceding 1911, it was estimated that 1900 hectares (4692 acres) were lost to the sea, whereas 14,345 hectares (35,444 acres) were gained, the latter chiefly as salt marshes, sand spits, beaches, and bars. On the other hand, the island of Heligoland in the North Sea off the German coast in A.D. 800 had a 200-km (124-mi) long shoreline, and was a much-fought-over piece of terrain in the Viking era. By 1900 the one-time independent dukedom

had been reduced to a strongly fortified rock, only 5 km (3 mi) around, enclosed by seawalls, breakwaters, and other defenses.

All beaches are similar in form (Fig. 15-15). The beach face is swept by waves as the water level oscillates between high and low tide. The slope of the face adjusts to the particle size moved: the larger the particles, the steeper the slope. At the upper end of the beach face is the berm, a gently sloping area formed by vertically accreting sediment deposited by waves that swash to the top of

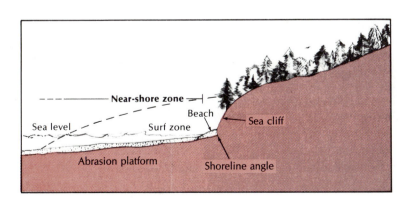

**Fig. 15-12.** Cross section of a steep coast showing characteristic landforms. Dashed line indicates the land profile before erosion.

**Fig. 15-13.** Near-vertical sea cliffs on the Daisei Islands, located off the west coast of Korea, testify to the high rate of beach erosion and shoreline retreat. The stacks of vertical rock remnants detached from the cliffs have not yet been consumed by the sea. Notice the contrast in topography between the sea cliff and the hilly terrain behind it, which was formed by subaerial processes. In places, the upper parts of the valleys have been truncated by sea-cliff retreat. (U.S. Department of Defense)

**Fig. 15-14.** The cliff on this beach near Montara, California, retreated 51 m (167 ft) between 1866 and 1971, an average of about 0.5 m (1.6 ft) per year. Large rocks are put at the base of the cliff in other places along this coast in an attempt to halt the erosion. (K.R. Lajoie, USGS)

the face. If the beach face is not too steep, material is moved offshore and molded into a longshore bar that parallels the beach.

Storms can change the shape of a beach rapidly. Storm waves are especially powerful, eroding the face and berm and depositing sediment in a longshore bar. When waters are relatively calm, however, longshore bars are pushed landward and beach face and berm are rebuilt to their former dimensions.

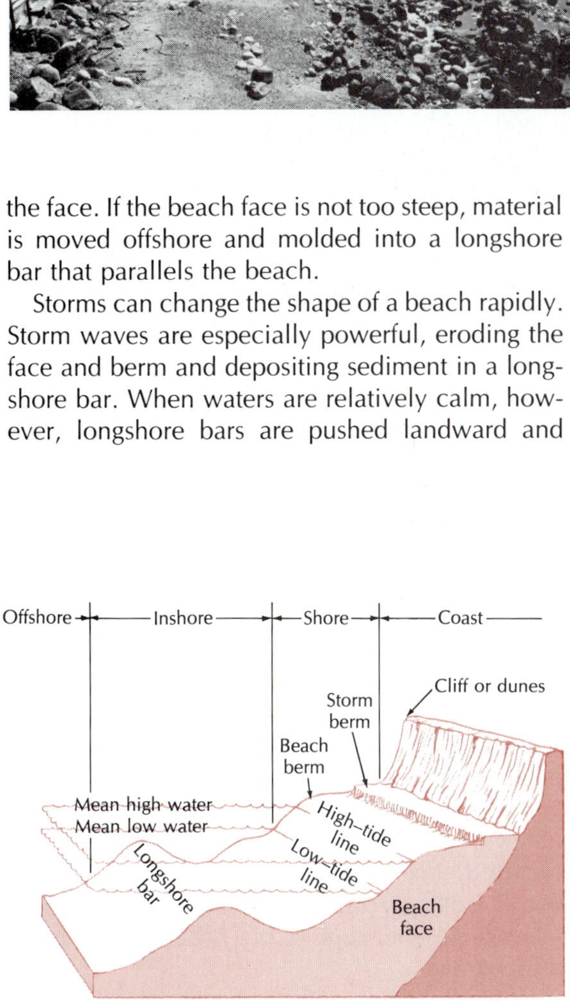

How are beaches formed? Most beaches are composed of sand that has been carried to the coast as the bedload of streams. The transporting power of the stream is checked where it meets the sea and materials like sand and gravel are quickly deposited. Suspended materials like silt and clay are carried to deeper waters offshore by currents. Sediment is also derived from the erosion of the sea cliff or the platform at the cliff's base.

Some sediment remains at the river mouth to form a delta, but most is swept along the shore by the **longshore current.** This current is formed because waves, even though refracted in shoaling water, break against the beach face at a slight angle and recede in a more or less straight path down the face. Water, therefore, moves up and down the face in a zigzag pattern, carrying beach sediment along with it. This sediment is termed **longshore drift** (Fig. 15-16). Beaches have been called "rivers of sand" for, as a whole, sediment moves parallel to the face. One edge of the "river" is that point offshore where waves first touch bottom and sediment begins to move; the other edge is the upper face, or berm, where sediment movement ceases.

The volume of sand moved along beaches is difficult to measure, but estimates have been made

**Fig. 15-15.** Sketch of features common to most beaches. (After T. Beer, 1983, *Environmental Oceanography: An Introduction to the Behaviour of Coastal Waters,* Pergamon, p. 20, Fig. 2-1)

**Fig. 15-16.** Waves striking the shore in southern California. The angle at which the waves meet the coast indicates that longshore transport is to the left. (From *Geology Illustrated* by John S. Shelton. Copyright © 1966 W.H. Freeman and Company. Used by permission)

based on the amount deposited behind artificial barriers (Table 15-2).

Several depositional landforms result from longshore drift, usually along irregular coasts with relatively shallow water (Fig. 15-17). Longshore drift will mimic the shape of the coast, but not in detail. Some sediment is deposited as a **spit** that trails downcurrent from the land. In places spits grow almost long enough to close off a bay, and they are then called **bars.** In other places sand moving along the coast piles up in the low-energy zone behind near-shore islands, eventually connecting them to the mainland by a narrow strip called a **tombolo.** Streams entering bays deposit sediment when their velocity is abruptly checked, building deltas out into the relatively still water.

Beaches are the land's first line of defense against wave erosion, particularly by hurricane and other storm waves. Much of the force of storm waves is expended on the gradual slope of the beach; even when the beach is removed, it will usually be rebuilt by deposition in calmer weather if a source of sediment exists. Sea cliffs do not have such renewal ability; once parts of them are eroded and removed, they are gone forever. There are protective as well as recreational reasons, then, for close observation and study of our beaches in order to preserve them.

Coasts made up of carbonate reefs are also classified as depositional. Because most carbonate reefs are in the oceanic realm, they are described in chapter 16.

**Table 15-2** Typical Rates of Sand Movement at Various U.S. Coastal Localities

| Location | Direction of Drift | Rates (m³ per yr) |
|---|---|---|
| Sandy Hook, N.J. | N | 331,400 |
| Barnegat Inlet, N.J. | S | 190,000 |
| Ocean City, Md. | S | 114,000 |
| Atlantic Beach, N.C. | E | 22,400 |
| Hillsboro Inlet, Fla. | S | 57,000 |
| Vinellas County, Fla. | S | 38,000 |
| Perdido Pass, Ala. | W | 152,000 |
| Galveston, Tex. | E | 332,500 |
| Santa Barbara, Cal. | E | 213,000 |
| Oxnard Plain, Cal. | S | 205,000 |
| Santa Monica, Cal. | S | 760,000 |
| Anaheim Bay, Cal. | E | 114,000 |
| Waikiki Beach, Hawaii | W | 7,600 |

After W. Kaufman and O. Pilkey, 1983, *The Beaches Are Moving: The Drowning of America's Shoreline,* Duke University Press, Durham, N.C., p. 82.

## BEACH EQUILIBRIUM AND ITS DISRUPTION

On beaches, an equilibrium is commonly reached between the amount of sand available and the amount of wave energy expended. This equilibrium has many attributes common to stream equilibrium. Even if some of the sand is deposited in submarine valleys and shunted to deeper parts of the ocean floor, it is replaced by sand brought down to the sea by streams or by sea-cliff erosion;

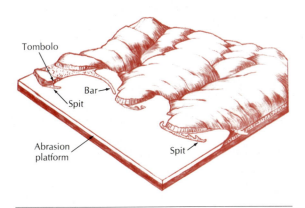

**Fig. 15-17.** Depositional landforms common to some coasts. Longshore drift is to the lower right. (After D.W. Johnson, 1919, *Shore Processes and Shoreline Development,* John Wiley, Fig. 88)

thus equilibrium is maintained. Yet the balance can be disrupted easily and in several ways.

Permanent loss of beaches is becoming increasingly common. Some of this erosion is natural and part of the continuing geological process. However, once the shore is lined with such costly structures as hotels, casinos, condominiums, and vacation houses, beach erosion becomes a serious problem. In order to arrest erosion or to stabilize the beach position, *jetties* are built at harbor or river entrances and *groins* are built along coasts. Both are short walls erected at right angles from the shore; both trap sand and other sediments on the upcurrent side. However, the undernourished longshore current rapidly carries off beach sands downcurrent from the groin. Property owners thus deprived may even erect groins on their own property to restore their beach fronts. Thus, groin building has a natural tendency to proliferate and, in fact, some shores bristle with them (Fig. 15-18).

Breakwaters also retard beach erosion, but they have detrimental effects farther down the beach. One type of breakwater is anchored to the shore at one end (Fig. 15-19); another is built offshore parallel to the beach (Fig. 15-20). Although both types protect parts of the beach from waves, they act as sand traps, disturbing equilibrium. Beyond the structure wave energy is normal, but because the sand load has been trapped, the beach erodes and the face moves landward.

Concrete seawalls are also built along the inner beaches in an attempt to control erosion. These are seldom effective, however, because they provide a vertical battering surface rather than a gradual slope as a natural beach would. The excessive wave erosion that can result not only prevents accumulation of a new beach, but also quickly undermines the structure itself. Commonly, too, seawalls are not built high enough to exclude damaging storm waves. "Bigger and better" does not seem to be the answer.

Another remedy for beach erosion that has proved useful is an artificial beach, which is made by dumping a large quantity of imported sand. Wave energy, then, will be directed to sediment transport rather than erosion; in a sense equilibrium can be reestablished. The sediment, usually sand, may be dredged from offshore bars or even quarried on land. Before such methods are used, however, specific beach processes should be well researched, so that enough sediment of the correct size is placed in the proper location.

The extent of beach-erosion problems makes it obvious that control cannot long be left to individual landowners. For reasons of economy as well as the large-scale research, planning, and organization required, such operations will increasingly fall to local, state, and federal agencies. The situation may lead to more public ownership of beach property and to more beaches made accessible for recreation.

Artificial changes in the bedload of coastal streams can also disrupt the equilibrium of the beach adjacent to its mouth. If bedload is drastically reduced, say by the construction of a dam and the trapping of sediment in the reservoir, the sediment delivered to the beach is decreased. This

could trigger an episode of erosion because the longshore current is deprived of load. In contrast alterations that increase stream bedload could have the opposite effect—that is, sediment supply exceeds available energy to transport the sediment, and the beach builds out.

## SEA LEVEL FLUCTUATIONS

Along some coasts tectonic instability has caused the land to be uplifted or depressed relative to sea level. In addition there have been actual changes in the sea level itself, and by their very nature these changes have worldwide effects.

Sea level has fluctuated over the past several million years (Fig. 15-21). For about the last 6000 years it has been close to the present level. Before

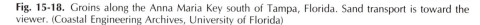

**Fig. 15-18.** Groins along the Anna Maria Key south of Tampa, Florida. Sand transport is toward the viewer. (Coastal Engineering Archives, University of Florida)

**Fig. 15-19.** Santa Barbara, California: A breakwater constructed to protect small boats not only cut off the supply of sand to the coast beyond, but interrupted longshore currents. This resulted in the formation of a spit (just beyond the moored boats) that threatened to close off the harbor. The resulting lack of sand on the far side of the harbor caused that portion to erode. Equilibrium has been restored, but at the cost of dredging the spit and transporting the sand to the far side of the harbor. (John S. Shelton)

**Fig. 15-20.** Artificial rock barriers placed offshore protect the sandy beach from erosion at Tel Aviv, Israel. (Peter W. Birkeland)

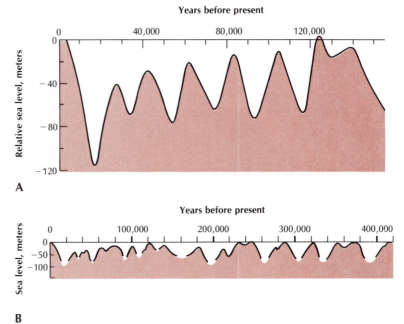

**Fig. 15-21. A.** Worldwide sea level fluctuations based on studies of marine terraces in New Guinea and elsewhere. (After A.L. Bloom et al., 1974, "Quaternary Sea-Level Fluctuations on a Tectonic Coast, New $^{230}$Th/$^{234}$U dates from the Huon Peninsula, New Guinea," *Quaternary Research*, vol. 4, pp. 185–205, Fig. 5.) **B.** Approximate sea level curve for last 400,000 years, based on studies of marine terraces in New Guinea. (After J. Chappell, 1974, "Geology of Coral Terraces, Huon Peninsula, New Guinea: A Study of Quaternary Tectonic Movements and Sea-Level Changes," *Geological Society of America Bulletin*, vol. 85, pp. 553–570, Fig. 19)

that time, sea level was lower and fluctuating; the only time it was close to its present position was about 120,000 years ago. If we look at the position of sea level for the last 500,000 years, it has been lower than the present level for fully 90 percent of the time.

Present-day sea level is far from stationary. Beginning around 1750, or even earlier, it started to rise, as shown by the tide gauges in seaports all over the world. Cuchlaine A.M. King has compiled data from worldwide ports on the annual rate of present-day rise—that for the Atlantic coast of North America is 4 to 7 mm (0.16 to 0.28 in.), Italy 1.7 mm (0.07 in.), Holland 1.5 mm (0.06 in.), and western North Africa 0.6 mm (0.02 in.). There are various reasons for the data varying from port to port, but there is no doubt that they all show that the sea level is rising.

The important question is what accounts for the fluctuations in sea level. The current rise is probably related to the worldwide recession of glaciers and the return of their waters to the sea. Thus, over the last 120,000 years (Fig. 15-21), periods of relatively low sea level correlate with major advances of continental glaciers and periods of

relatively high sea level with major recessions, most during the Wisconsin glaciation (chap. 14). The similarity between Figures 15-21 and 16-26 suggests a very strong glacial/interglacial influence on sea level even farther back in time. Such changes are called **eustatic.** The amount of sea level fluctuation can also allow us to estimate the extent of glaciation, global ice volume, and the rate of ice-sheet growth during the Pleistocene.

Fluctuations in sea level and their repetitive nature complicate the interpretation of coastal landforms. Surely some of the features have formed during the last 6000 years of relative sea level stability. However, as sea level continues to return to about the same position on stable coasts (Fig. 15-21), some features could be relict—formed at least 120,000 years ago. To further complicate matters, wind patterns and therefore longshore drift could have been much different in the past.

## COASTAL DEVELOPMENT AND EVOLUTION

The formation of coasts is complex, for not only do present-day processes operate on coasts, but many coastal landforms carry a legacy of the past.

There are three major kinds of coasts: emergent, submergent, and plains.

## Emergent Coasts

The land is rising along some coastlines. The best evidence for this change is found in areas that were once covered by the Pleistocene glaciers. The sheer weight of the ice pushed the earth's crust down and when the ice melted, the crust rebounded. Thus, in areas that were under or near ice caps, prehistoric shoreline features are found far above present sea level (Fig. 15-22). The pat-terns of shoreline uplift are related to the size of the ice cap; the larger the ice cap, the greater the uplift. These glaciated areas are still emerging. For example, the Gulf of Bothnia is rising at the rate of 0.9 cm (0.35 in.); Hudson Bay is rising at the rate of 2 cm (0.8 in.) per year. It is estimated that these areas still have a considerable amount of rebounding to do, so that we can expect them to become increasingly shallow or even to partially dry up.

Other coasts far from centers of glaciation are also rising. Again, the evidence is found in old beaches and abrasion platforms—collectively called

**Fig. 15-22.** Elevated beach deposits near Fort Severn, Canada, on the southwest shore of Hudson Bay clearly show that the land has been rising since the disappearance of Pleistocene ice. (Department of Energy, Mines, and Resources, Canada)

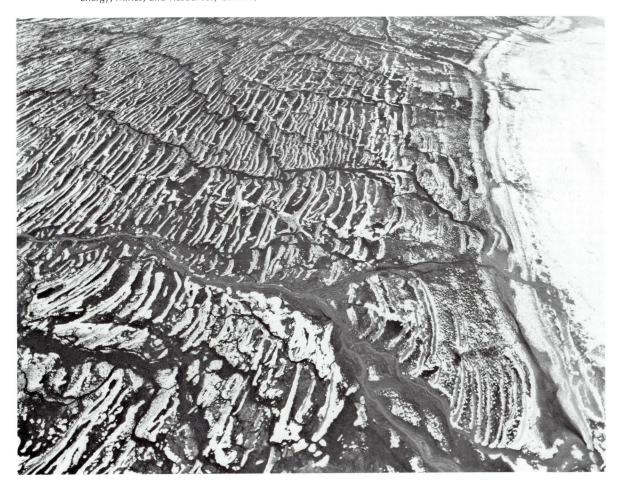

*marine terraces*—and sea cliffs standing far above sea level. Such landforms almost certainly were pushed up by tectonic movements because there is no evidence that sea level has risen much above its present level for the past million years. Marine terraces are well expressed along the coasts of California, Peru, and New Zealand (Fig. 15-23). Perhaps the most spectacular flight of terraces are the coral reef terraces of New Guinea (Fig. 15-24). Also, part of the New Guinea coast is the most rapidly rising one in the world—some 3 m (10 ft) in 1000 years. This means that beach deposits of the last interglaciation (about 120,000 years ago) are now 360 m (1180 ft) above sea level.

### Submergent Coasts

Coasts in other areas show distinct signs of submergence. A familiar example is the coast of the Netherlands, which is in a deltaic area of rapid subsidence and sinking at a rate of nearly 10 cm (4 in.) per century. No wonder that with a total rate of sinking of around 20 cm (8 in.) per century—about one-half from subsidence and one-half from rising sea level—that, beginning with the great floods of medieval times, the Dutch have been compelled to construct an extraordinary complex of dikes and coastal defenses on which their very survival depends.

Typical submergent coasts are those in which the sea extends inland, sometimes for long distances, in embayments. If the indentations were shaped by stream erosion before their invasion by saltwater, the coast is then known as a *ria coast*, from the name applied to the southern shore of the Bay of Biscay. The Gulf of Maine; the north coast of South Island, New Zealand (Fig. 15-25); and the southern coast of Brazil also are ria coasts.

The way in which a particular coast becomes embayed may be difficult to determine. Perhaps the land subsided, in which case the sea invaded or drowned preexisting river valleys. Or it could be that the land remained stationary, and the postglacial sea level rise flooded low-lying areas. In either case the effect is the same.

---

**Fig. 15-23.** Multiple marine terraces preserved on the west side of San Clemente Island, southern California. (John S. Shelton)

**Fig. 15-24.** Marine coral reef terraces near Finschhafen, Papua New Guinea. The highest terrace, about 260 m (853 ft) above sea level, was formed about 140,000 years ago. This coast, therefore, has been uplifted that amount because the highest terrace was formed when sea level was only a few meters above its present position (Fig. 15-21.). (U.S. Department of Defense)

Some of the modifications a ria coast is likely to undergo as a result of wave and current action are shown in Fig. 15-26. In Figure 15-26A the sea has drowned a landmass shaped by subaerial erosion. Former ridges now extend seaward as headlands and the sea may reach inland as an embayment, or **estuary,** perhaps much as Chesapeake Bay does (Fig. 15-27).

In the later history of such a coast (Fig. 15-26B), sea cliffs form on exposed headlands and islands. As the cliffs retreat, an abrasion platform is left behind (Fig. 15-26C). Finally, the coast has been smoothed and an equilibrium may be reached between sea erosion and deposition of sediments from terrestrial erosion. The coast will have lost its original indented character and will be cliffed along much of its length (Fig. 15-26D). Any irregularities will largely reflect the relative resistance of the rocks cropping out along the cliff face.

## Plains Coasts

Some coasts with a low gradient and an adequate sand supply are fringed by **barrier islands:** long narrow islands of sand separated from the coast by narrow bodies of waters called **lagoons.** Some 282 barrier islands extend, like linked sausages, from Long Island, New York, around Florida and the Gulf of Mexico to the southern end of Texas (Fig. 15-28). Cities, such as Atlantic City and Galveston, and many resorts have been built on these islands, which lie from 3 to 30 km (1.9 to 19 mi)

**Fig. 15-25.** Drowned stream-carved valleys form the Marlborough Sounds, South Island, New Zealand, and provide a classic example of a ria coast. (Lloyd Homer, New Zealand Geological Survey)

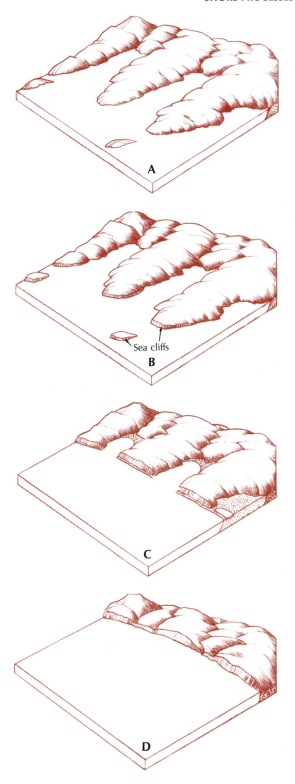

**Fig. 15-26.** The possible successive stages in the development of a submergent or submerging coastline. (After D.W. Johnson, 1919, *Shore Processes and Shoreline Development,* John Wiley, Fig. 88)

**Fig. 15-27.** Chesapeake Bay is a fine example of a landscape shaped mainly by subaerial processes, but later drowned by the rising sea. (NASA)

**Fig. 15-28.** Barrier islands fringing the outer coast of North Carolina, photographed on the *Apollo 9* space mission, 1969. (NASA)

offshore and range from 2 to 5 km (1.2 to 3 mi) in width and 10 to 100 km (6.2 to 62 mi) in length. Most of them rise no more than 6 m (20 ft) above sea level. A comparable chain of sandy barriers stretches along the low coasts of the Netherlands (particularly Friesland) and Denmark around the southeastern margin of the North Sea.

Barrier islands form in two different ways. One is the growth and ultimate breaching of a spit (Fig. 15-29). The other, which is probably the way most barrier islands originate, involves the slow submergence of a coast (or a rise in sea level), so that the beach or sand dune ridges become an island

and the low area behind them becomes a brackish lagoon (Fig. 15-30). Some barrier islands have a complex history. The seaward parts may be of recent origin, whereas landward parts are relics of islands formed 120,000 years ago when sea level was last at its present height (Fig. 15-21).

The stability of any barrier islands depends on several factors. Abundant sand supply is crucial. Waves carry sand in from the ocean floor and longshore currents sweep it along the coast. During storms large waves wash over parts of the island, depositing additional sand. In time winds may pile up sand dunes, which further increase

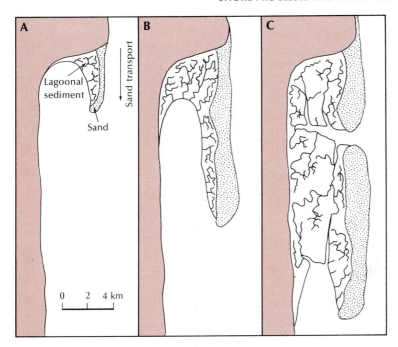

**Fig. 15-29.** Formation of a barrier island from a spit. In **A** and **B** the sandy spit grows and fine-grained lagoonal sediments fill in behind it. In **C** the spit has been breached by a tidal channel and the lower end of the spit has become a barrier island. (After J.H. Hoyt, 1967, "Barrier Island Formation," *Geological Society of America Bulletin,* vol. 78, pp. 1125–36. By permission of the Geological Society of America and J.H. Hoyt)

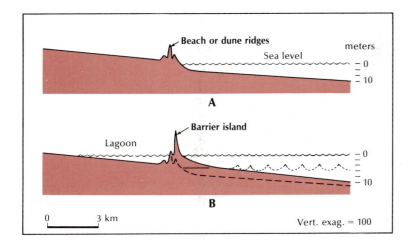

**Fig. 15-30.** Formation of barrier islands by submergence. **A.** Waves and wind build a rampart of beach and dune ridges. **B.** With a submerging coast or rising sea level, the ridges are separated from the land and form a barrier island and a lagoon. Under appropriate conditions they can maintain themselves by growing upward and seaward. (After J.H. Hoyt, 1967, "Barrier Island Formation," *Geological Society of America Bulletin,* vol. 78, pp. 1125–36. By permission of the Geological Society of America and J.H. Hoyt)

**Fig. 15-31.** Condominiums built at the water's edge, Galt Ocean Mile, Fort Lauderdale, Florida. (Charles W. Finkl, Jr., Coastal Education and Research Foundation)

the height of the island. Storms can carve inlets to the lagoons; although currents may keep some of the inlets open, others will disappear as sediment fills them in. Depending on the rate of submergence, the amount of wave energy, and the rate of sediment supply, a barrier island can maintain its general shape and position, be pushed landward, or grow seaward.

With the development of barrier islands for recreational and other uses, more has been learned about their dynamic nature, but not without cost in lives and dollars. About one-fourth of the barrier islands along the Atlantic and Gulf coasts of the United States are developed and urbanized and about another one-third have been put aside for recreational use or as wildlife refuges. The rest— more than 100—are privately owned and undeveloped. Throughout recent geological time, these islands have acted as buffer zones, absorbing the impact of winter storms and hurricanes, shifting and changing shape with time. Yet people often build houses, hotels, and other buildings virtually at the water's edge with little thought for protection against exceptional storms (Figs. 15-31, 15-32). They also try to control the flow of sediment or to retain the naturally shifting sands by building

**Fig. 15-32.** The mid-Atlantic coast was struck by the most severe winter storm of record on March 7, 1962. Most of the damage occurred on low-lying barrier islands, this one being Fire Island. The storm caused over $500 million in damages between Long Island and Cape Lookout and left 32 dead. (UPI)

groins, but this can erode away other parts of the shore. The sea level is still rising—25 cm (1ft) from 1900 to 1974 along the east coast of the United States—bringing the edge of the ocean still closer to the already precariously placed buildings. A study of a section of the mid-Atlantic coast indicates an average rate of erosion of 1.5 m/yr (5 ft/yr).

What can be done to prevent further erosion and to compensate for the effects of erosion in developed areas? One answer would be to add sand to the beaches, perhaps by dredging offshore deposits. This maneuver would be expensive, possibly never ending, and the offshore changes might lead to some unforeseen damage. Another answer to save life and to prevent property damage would be to move threatened buildings back from the shore onto the more protected landward side of the barrier island.

To prevent human-induced erosion in areas being developed, better geological information concerning cause and effect could be incorporated into development schemes. Thus, restricted development could be limited to inland areas and more fragile shorelines protected, perhaps through public ownership as wild areas.

## SUMMARY

1. Waves cut away at the edges of land and the depth of their most effective erosion is only one-half the wave length. Water particles move in an orbital fashion as waves pass by, but the orbital motion is not completed in shallow waters; there the waves break in the surf zone.

2. *Tsunamis* commonly are earthquake-generated waves that reach exceptional heights as they approach land; hence they are among the most destructive waves on earth.

3. Waves are refracted as they approach coasts and most of their erosive energy is concentrated on headlands. Bays between headlands commonly are sites of deposition of material derived from the headlands. On steep coasts wave erosion will form a *sea cliff* and an *abrasion platform*. Beach materials (sand and gravel) cover the platform.

4. Wave refraction produces a *longshore current* that moves beach material along the coast. In places an equilibrium is set up between the wave energy and the amount of material moving along the coast. Interference by humans has caused problems by altering coastal erosion and deposition.

5. Sea level during the Pleistocene was low during times of glaciation and high during times of the interglaciations. At no time during the Pleistocene was it much higher than it is at present.

6. Three major types of coasts are recognized: *emergent* coasts, *submergent* coasts, and *plains* coasts.

## QUESTIONS

1. Explain the movement of sand along a beach. What would be the consequence of building a jetty into deep water at the mouth of a river?

2. What recommendations would you make for the development of a barrier island?

3. If sea level remains constant, what factors would influence erosion of the coastline?

4. What combination of landforms and deposits are used to distinguish an emergent coast from a submergent coast?

5. How can it be shown that a particular coast is subsiding tectonically rather than being submerged by a rise in sea level related to the melting of glaciers?

## SELECTED REFERENCES

Bascom, W., 1980, Waves and beaches: the dynamics of the ocean surface, Doubleday, Anchor Books, New York.

Beer T., 1983, Environmental oceanography: an introduction to the behaviour of coastal waters, Pergamon, Oxford.

Bird, E.C., 1985, Coastline changes: a global review, Wiley, New York.

Bloom, A.L., 1978, Geomorphology: a systematic analysis of late Cenozoic landforms, Prentice-Hall, Englewood Cliffs, N.J.

Dolan, R., Lins, H., and Steward, J., 1980, Geographical analysis of Fenwick Island, Maryland, a Middle Atlantic coast barrier island, U.S. Geological Survey Professional Paper 1177–A.

Kaufman, W., and Pilkey, O., 1983, The beaches are moving: the drowning of America's shoreline, Duke University Press, Durham, N.C.

King, C.A.M., 1972, Beaches and coasts, Edward Arnold, London.

Komar, P.D., 1976, Beach processes and sedimentation, Prentice-Hall, Englewood Cliffs, N.J.

Myles, D., 1985, The great waves, McGraw-Hill, New York.

Shephard, F.P., 1973, Submarine geology, Harper & Row, New York.

Trenhaile, A.S., 1987, The geomorphology of rock coasts, Oxford University Press, Oxford.

**Fig. 16-1.** Nineteenth-century engraving of H.M.S. *Challenger* passing an iceberg.

# 16

# THE OCEAN FLOOR

The oceans, covering 72 percent of the earth's surface, have been the last great scientific frontier on earth. Yet scientific exploration started well over 200 years ago. Through the voyages of Captain James Cook (1728–79) between 1768 and 1779 and that of the H.M.S. *Beagle* from 1831 to 1836 the geography of the oceans began to come into focus. Charles Darwin was on the voyage of the *Beagle* and put forth a perceptive theory to explain the variety of coral reefs he saw. But it was the H.M.S. *Challenger* expedition, from 1872 to 1876, that was truly oceanographic in its epic study of the Atlantic, Pacific, and Southern oceans (Fig. 16-1).

Our present understanding of the oceans began during World War II. Because of the many naval engagements, research into all phases of oceanography was essential. Something as insignificant as the tide became important not only to the timing of a particular landing, but also to the beach conditions that could be expected. Ocean research was stimulated further with the development of the plate-tectonic theory (see chap. 4), which is based largely on oceanic features. The ocean floors are being mapped and rifts, plateaus, and other topographic features under the sea are being investigated. The study of marine organisms sensitive to variations in temperature and the chemical composition of seawater has given us a fairly accurate picture of the timing of Pleistocene glaciations and interglaciations. Recent deep-sea dives have revealed life forms previously unknown (Fig. 16-2) and have even located the sunken remains of famous shipwrecks such as the *Titanic*.

Economic reasons also help explain our present interest in the ocean or, more accurately, in ocean sediments. Several billion dollars worth of oil and gas are extracted annually from shelf sediments, with about one-quarter coming from shelves off the United States. Although environmental problems caused by offshore drilling are numerous, the solutions to many of them are in sight, and deposits on the oceans' shelves probably will increase in importance as a fuel source. More industries may look to the seafloor for concentrations in ancient submerged beaches of such valuable heavy minerals as tin, gold, or diamonds; but exploration costs are high, and profitable concentrations may not be common.

Earlier (chap. 4), we looked at evidence indicating that both the continents and, therefore, the ocean basins have shifted position with time. For

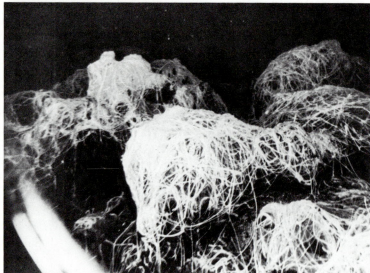

**Fig. 16-2.** Scientists aboard the research submersible *Alvin* were astonished at the density and the size of the animals found at the hot-water vents. Although relatives of the ''spaghetti'' had been seen in shallow water, the tube worms had never been seen before. Because there is generally little food in the deep ocean, it is usually sparsely settled, by very small organisms. **A.** *(Above)* Tube worms. (Kathleen Crane. Woods Hole Oceanographic Institution) **B.** *(Below)* Worms nicknamed ''spaghetti,'' possibly acorn worms. (James Childress, University of California, Santa Barbara/Woods Hole Oceanographic Institution)

example, 61 percent of the Northern Hemisphere is water, compared with 81 percent of the Southern Hemisphere. But about 475 million years ago the bulk of the landmasses were south of the equator. The positions of the continents affect directly or indirectly the movement of ocean currents, global climate, and the timing of glaciations.

## FEATURES OF THE OCEAN FLOOR

Until recently the only means of determining the depth of the ocean or the form of ocean-floor features was to drop heavy weights on cables to the bottom; it is no wonder that progress was slow and ocean-floor topography was little known. In the 1930s, however, German scientists developed the electronic echo sounder to measure ocean depths. The device operates by creating a loud noise ("ping") under water; the sound travels to the bottom and is reflected back to the ship. Because the speed of sound in water is almost constant, the time it takes a sound wave to journey to the seafloor and back is directly related to the depth. The echo sounder provides a continuous visible recording of the bottom profile as a ship cruises along. In 1969 the American oceanographer Henry W. Menard described it as follows:

We cruised over hills and low mountains a mile and a half below and visible only on the echo-sounder. However, after a while the distance between the echo-sounder and the bottom is forgotten. As the marine geologist surveys a new range of undersea mountains, he senses them around him. I was once surveying with a captain new to the game. He remarked, as we headed again toward a peak on the map we were making, that he could not suppress a captain's feeling that we would hit it, even though he knew perfectly well it was a mile below the ship.*

Modern instruments are so refined that changes as little as 1 m (3 ft) can be detected in 5000 m (16,400 ft) of water. Of course, the finer features are revealed by photographs taken by people in submersibles or by cameras lowered on cables.

The ocean floor, far from being featureless (Fig. 16-3), has deep canyons and high mountains that rival any found on land. Yet the general configuration is more regular and predictable than the land surface (Fig. 16-4). The origin of many oceanic features will be discussed here, but those related to plate tectonics are given in chapter 4.

A simple diagram shows differences in elevation distribution between the continents and oceans (Fig. 16-5). For example, the average elevation of the landmasses is much less than the average depth of the oceans. Continents and oceans are alike in that few areas reach far above or below sea level—0.5 percent of the world is above 4000 m (13,120 ft), and 1 percent is at depths greater than 6000 m (19,700 ft).

*H.W. Menard, 1969, *Anatomy of an Expedition*, McGraw-Hill, N.Y., p. 103.

## Continental Shelf

A relatively shallow platform called the **continental shelf** surrounds almost all the continents; it makes up 7.5 percent of the ocean floor (see Fig. 16-4). The shelf has an average width of 80 km (50 mi) but in places extends out as far as 1500 km (900 mi). It approximates a smooth, gradually sloping platform with an average slope of 0°07'. Often there are depressions, hills, and terraces, showing as much as 100 m (330 ft) of relief. In formerly glaciated regions, glacial erosion and deposition give the shelf pronounced relief. The outer edge of the shelf, called the **shelf break,** is characterized by a sudden steepening of the slope. The depth of the shelf break ranges from 20 to 550 m (66 to 1800 ft) and averages 133 m (440 ft). This latter figure is close to that for the maximum ice-age lowering of sea level (see chap. 15).

Most continental shelves are composed of sediments and sedimentary rocks averaging 2-km (1.2-mi) thick. The form of many shelves is related to the position and origin of various natural dams (Fig. 16-6). Common around the Pacific Ocean are tectonic dams, formed by long blocks of rocks that have been faulted, or folded, with some blocks pushed upward. The basins formed behind the dams have been filled by sediments from the continents. In some places parts of the dam rise above the surface of the sea as islands, such as the Farallons near San Francisco. At other places, chiefly off the east coast of the United States, the dams are ancient; sediments have not only filled the basins behind them, but have covered the dams as well. In these cases and where there are no dams, the shelf sediments form a great wedge, often thickening in their seaward direction.

Off the coast in the western Gulf of Mexico, the dams are giant domes of salt, called **diapirs,** which have pushed up through overlying sedimentary rocks. In tropical waters dams are often formed by outlying reefs of coral and marine microscopic plants (algae). This type of dam is also seen on shelves off the eastern Gulf coast and southeastern United States.

During the Pleistocene, when part of the ocean's water lay in great ice sheets on the land, extensive areas of the continental shelves were exposed to

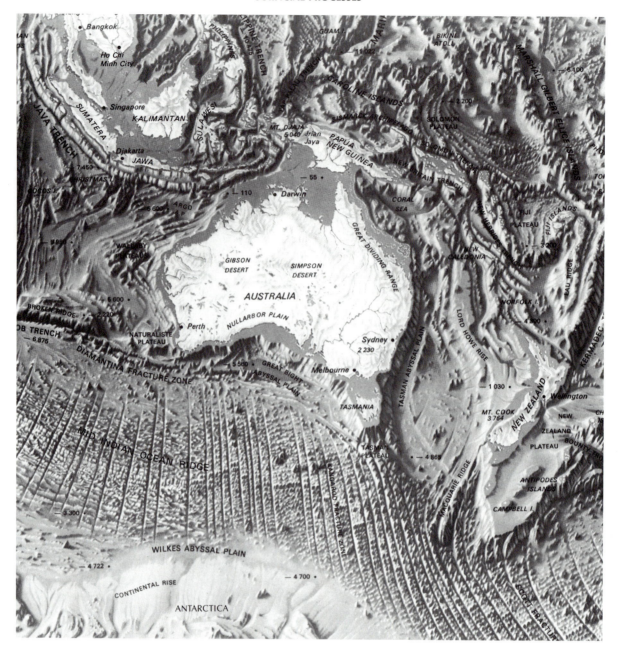

**Fig. 16-3.** Relief map of the rugged ocean floor in the vicinity of Australia and Antarctica. A major ocean ridge lies between the two continents. Elevations are in meters. (From B.C. Heezen and M. Tharpe, 1976, *The Floor of the Ocean*. Map painted by Tangy de Remur, American Geographical Society. Reprinted by permission of B.C. Heezen and M. Tharpe)

# THE OCEAN FLOOR

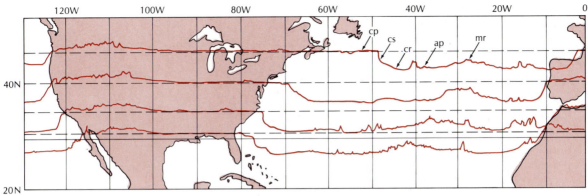

cp = Continental platform
cs = Continental slope
cr = Continental rise
ap = Abyssal plain
mr = Mid–ocean ridge

**Fig. 16-4.** Computer-drawn topographic profiles (solid lines) from the eastern Atlantic Ocean shores to the Pacific Ocean. The profiles represent the topography along the dashed lines; the dashed lines represent sea level for each profile. (After computer graphic by Dr. Peter W. Sloss, National Geophysical Data Center, Boulder, Colorado)

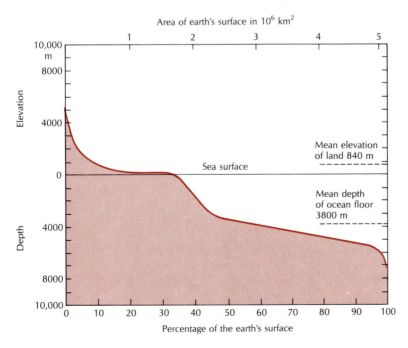

**Fig. 16-5.** The percentage distribution of various elevations on the earth's surface. (After Sverdrup, Johnson and Fleming, © 1942, renewed 1970, *The Oceans*, p. 19. Adapted by permission of Prentice-Hall, Inc., Englewood Cliffs, N. J.)

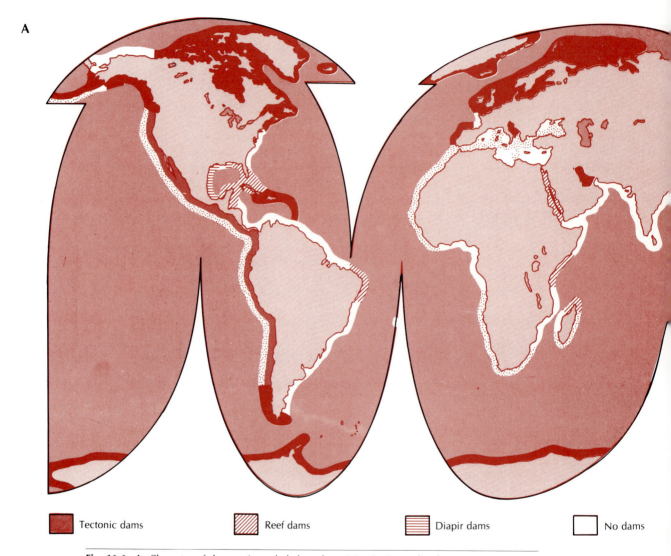

A

| | | | |
|---|---|---|---|
| ▨ Tectonic dams | ▨ Reef dams | ▥ Diapir dams | □ No dams |

**Fig. 16-6. A.** Character of the continental shelves, by origin. **B.** Generalized cross sections across continental shelves of different origin. (After K.O. Emery, 1969, *The Continental Shelves in the Ocean,* W.H. Freeman, pp. 44–45)

subaerial and glacial erosion. Natural causeways, **land bridges,** then connected Tasmania and Papua New Guinea with Australia (see Fig. 16-3), some of the islands of Indonesia with one another, India with Sri Lanka, and Siberia with Alaska. People migrating from Asia to the Americas came across the Alaska land bridge. As they moved south their path was blocked by a massive ice barrier. But in

time the glaciers retreated, forming an ice-free corridor to the south, perhaps explaining the sudden appearance of humans in southern North America about 13,000 years ago. Earlier migrations under similar conditions may have taken place, but firmly dated evidence of human occupation has not yet been found.

Fluctuations in sea level during the Pleistocene

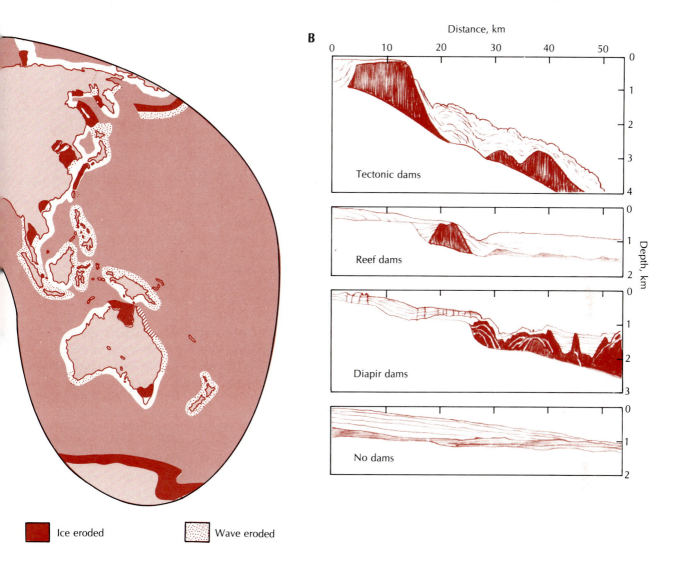

Ice eroded    Wave eroded

resulted in complex sediment and erosion patterns on the shelves. Whenever sea level stabilized for a time, shoreline erosional features formed and beach and finer-grained offshore sediments were deposited. Because fluctuations in sea level during the Pleistocene were numerous, the pattern of deposits and erosional features on any shelf can be quite complex. In the coming years as terrestrial sources of gravel for building materials become more scarce, we may have to start quarrying Ice Age sand and gravel beach deposits located offshore.

## Continental Slope

The *continental slope,* making up about 14 percent of the ocean floor, extends downward from the shelf break to the continental rise (see Fig. 16-4). The grade of the continental slope is gradual—averaging 4°17′ for the first 1825 m (4000 ft). Like the shelves the slope is not a flat expanse, but is marked by basins and hills and is cut by valleys and canyons. In general the slope is a transition zone, about 16- to 32-km (10- to 20-mi) wide, which connects the two main levels of the earth's surface—sea level (or close to it) and ocean floor, 3660 m (12,000 ft) below sea level.

It should be noted that the boundary between the continental and ocean crust is close to the base of the continental slope. In contrast present-day sea level is primarily the result of the volume of glacial ice on land.

## Continental Rise

Between the continental slope and the ocean basins is the **continental rise** (see Fig. 16-4). About 5 percent of the ocean floor is made up of this gently sloping (about 1°) low-relief feature. The rise is a depositional landform consisting of sediments derived from the continents, which are several kilometers thick in places. Some of these sediments form enormous fans, called **submarine fans,** at the mouths of submarine canyons (Fig. 16-7).

## Submarine Canyons

One curious feature of the continental slope are the numerous valleys, or canyons, that have been cut into it. These valleys are among the most impressive features of the ocean floor and, until recently, were the most difficult to explain.

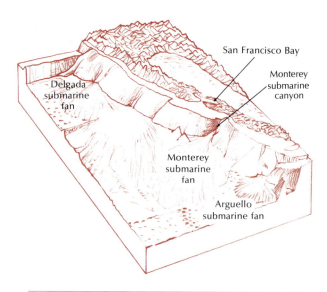

San Francisco Bay

Monterey submarine canyon

Delgada submarine fan

Monterey submarine fan

Arguello submarine fan

**Fig. 16-7.** Diagram of the coast and seafloor off central California, showing several submarine fans.

Submarine canyons are found throughout the world. Probably the best known is La Jolla Submarine Canyon and Fan Valley, with its major tributary, Scripps Canyon. These canyons are located just off the coast of La Jolla, California, the site of Scripps Institution of Oceanography, and within a few miles of the Naval Undersea Research and Development Center.

From the beach a flat and gently sloping sand-covered terrace extends seaward. About 213 m (700 ft) from shore, at a depth of 12 m (40 ft), the bottom suddenly drops away in the precipitous 24-m (80-ft) headwall of La Jolla Canyon. At the base of the cliff is the wide, bowl-shaped head of the canyon. Seaward, the valley widens and then narrows to a steep, rock-walled gorge. A series of giant steps lead down the canyon, caused by slumping in the loose bottom sediments.

Partway down, Scripps Canyon joins La Jolla Canyon, and through the submarine gloom it can be seen that walls overhang as much as 6 m (20 ft). The height of the walls decreases gradually and the valley widens until it is a little more than 1-km (0.6-mi) across, with an entrenched channel wandering through it. Natural levees line either side. This is the La Jolla Fan Valley. Eventually, the valley evens out and merges with the ocean floor. Such submarine fans are formed by sediment channeled down the canyon.

Larger canyons exist. One off the California coast—known as the Monterey Canyon—rivals the Grand Canyon of the Colorado River in its vertical dimension (Fig. 16-8). Off the east coast, the most famous canyon is the Hudson Canyon. It heads off Long Island, New York, and trends southeastward and reaches a depth of about 915 m (3000 ft). For world-record length we look to the deep canyon off the mouth of the Zaïre River, extending 240 km (150 mi) offshore.

The origin of submarine canyons has been debated by geologists for decades. Current thought seems to favor subaerial erosion as a starting point for many canyons. Such erosion would have started with the lowering of sea level during Pleistocene glaciations. Yet subaerial erosion could not carve out seafloor canyons as continuous features, because sea level has never dropped far enough to expose entire canyons.

The one exception is the Mediterranean Sea,

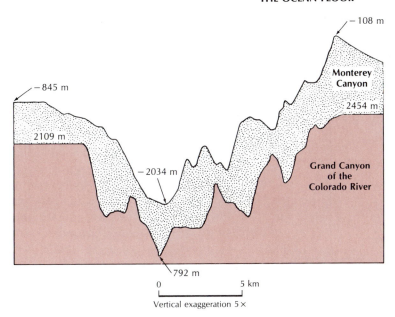

−108 m

Monterey
Canyon

2454 m

−845 m

2109 m

−2034 m

Grand Canyon
of the
Colorado River

792 m

0        5 km

Vertical exaggeration 5×

**Fig. 16-8.** Transverse profile across Monterey Canyon off the California coast compared to that of the Grand Canyon of the Colorado River. (From R.P. Shepard, 1963, *Submarine Geology, 2nd ed.,* Harper & Row, Fig. 150)

where the submarine canyons appear to be subaerial. Six million years ago the Mediterranean was isolated from the Atlantic Ocean. The sea quickly dried up, leaving a desert basin some 3000 m (9840 ft) below sea level. Rivers carved gorges into the sides of the basin and these gorges became submarine canyons when the ocean waters returned.

The consensus, then, is that submarine processes form most of the canyons. Strong bottom currents can transport sand-size sediment, and marine organisms, no doubt, play a part in the mechanical and chemical disintegration of canyon walls. Density currents of debris-laden water, set into motion by the slumping of oversteepened and unconsolidated material near the heads of canyons, are another erosional agent; they are called **turbidity currents.**

Turbidity currents, as their name implies, are clouded or muddy streams of moving water. They were first described by a Swiss engineer who noticed that the muddy waters of the Rhone disappeared where the river flowed into the clear waters of Lake Geneva. He reasoned, quite correctly, that the colder water of the Rhone, laden with glacial silt, was heavier than the water of the lake. Therefore, the river sank into the lake and

flowed as an undercurrent along its bottom. Much the same thing happens where the muddy Colorado River flows into Lake Mead.

That turbidity currents may exist in the ocean was suggested strongly on the afternoon of November 18, 1929, when a severe earthquake shook the Grand Banks off the Newfoundland coast. Apparently the shock was violent enough to trigger what may have been a submarine landslide, or slump, which was soon converted into a cloud of suspended sediment that swirled down the continental slope. Such an event normally would go unnoticed were it not for the unique circumstance that the current's path was directly across the submarine cable network linking North America with Europe.

One after another the cables were broken and, because the time as well as the point of rupture could be determined electrically, the pattern of failure was determined (Fig. 16-9). The cables had snapped progressively downslope, and when they were fished up for splicing, many of them were chaotically snarled and jumbled.

From the distance separating the individual breaks and from the time of failure, it was estimated that early on the current had a velocity of about 20 m/sec (66 ft/sec) or more but that the velocity grad-

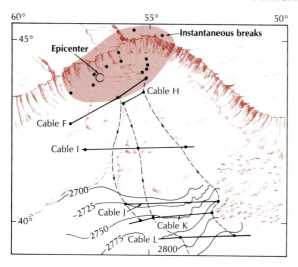

**Fig. 16-9.** The setting of the Grand Banks earthquake of 1929 showing the area where the cables broke simultaneously and the positions of parts of other cables that broke as the turbidity current progressed downslope. The dark circles are points of breakage; the dashed lines are paths that parts of the turbidity current might have taken. The numbers on the contour lines in the lower part of the diagram represent water depth in fathoms (a fathom = 1.8 m or 6 ft).

ually decreased with time until the last cable, located about 600 km (375 mi) from the epicenter, broke some 13 hours after the earthquake.

Whether or not the cable-breaking pulse traveling across the seafloor was a turbidity current probably never will be known. About all we do know from that and a few similar episodes is that occasionally currents set in motion in the ocean depths for short periods are capable of doing an immense amount of work and of shifting great volumes of sediment.

## Abyssal Plains and Hills

At the base of the continental rise stretches the deep-ocean floor. Some parts of the floor are a dead-level plain and leave an unwavering line on the fathometer, kilometer after kilometer; these are the *abyssal plains* (see Fig. 16-4). Depths range from 3 to 6 km (1.9 to 3.7 mi) and lateral extents from 200 to 2000 km (125 to 1240 mi). Abyssal plains are most common in the Atlantic and Indian oceans. The floor of the Indian Ocean appears to

be the nearly level surface of the vast subocean floods of basalt, perhaps comparable to the Deccan lava plateau on land in nearby India. Other plains are covered with sediments, as is a broad expanse in the northwest Atlantic Ocean south of Newfoundland.

Hills, known as *abyssal hills,* commonly rise less than 1000 m (3280 ft) above the plain. They are either of volcanic or tectonic origin, and composed of basalt. Abyssal hills are more common in the Pacific than in other oceans; a slow rate of sediment deposition in the Pacific basin may account for this difference. Together, abyssal hills and plains make up about 42 percent of the ocean floor.

## Oceanic Ridges

The *midocean ridges* are a major feature of the oceans. They commonly occur in a midocean position, are broad and continuous, and are flanked on both sides by lower-lying areas (Fig. 16-10; see Fig. 16.4). The length of the entire ridge system is 80,000 km (49,600 mi); its width is greater than 1000 km (620 mi); and its crest is 1000 to 3000 m (3280 to 9840 ft) above the ocean floor—surely one of the great mountain ranges on earth. Where the ridge pierces the surface of the sea, its lofty peaks are the foundations of such volcanic islands as Ascension, the Azores, Iceland, Saint Helena, and Tristan da Cunha. The highest of these is Pico in the Azores, which towers 2300 m (7544 ft) above sea level and stands on a base that rises 6100 m (20,000 ft) above the ocean floor.

Along the crests of most of the midocean ridges is a *median valley,* a central depression, or rift, with a nearly level floor and steep sides. The rift scoring the mid-Atlantic ridge is 1000- to 2000-m (3280- to 6560-ft) deep and tens of kilometers wide. In many ways it resembles the rift valleys of eastern Africa. In contrast some ridges, such as the east Pacific ridge have no such prominent central rift valley. In all midocean ridges, however, subsidiary rift valleys parallel the trend of the ridges along their flanks.

Another feature common to midocean ridges are numerous long fracture zones that cut across the ridges at right angles (see Figs. 16-3, 16-10).

**Fig. 16-10.** The Mid-Atlantic Ridge in the North Atlantic Ocean. Notice the elongate valley along the ridge crest and the numerous fracture zones that offset the ridge. (From B.C. Heezen and M. Tharpe, 1976, *The Floor of the Oceans*. Map painted by Tangy de Remur, American Geographical Society. Reprinted by permission of B.C. Heezen and M. Tharpe)

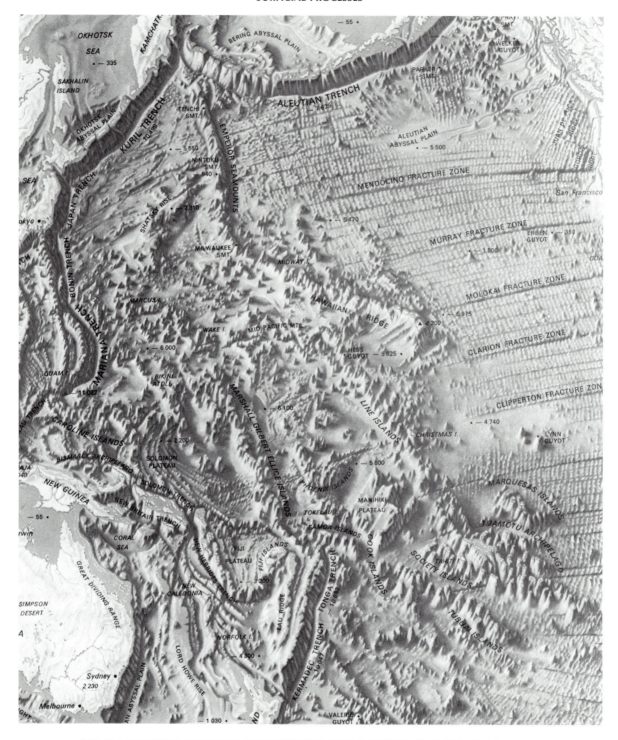

**Fig. 16-11.** Features of the ocean bottom in the western Pacific. Deep trenches are clustered along the Asian margin of the ocean. Many of the hills are volcanoes. (From B.C. Heezen and M.Tharpe, 1976, *The Floor of the Oceans*. Map painted by Tangy de Remur, American Geographical Society. Reprinted by permission of B.C. Heezen and M. Tharpe.)

Some reach a length of 3500 km (2170 mi); local relief ranges from 100 to 4000 m (328 to 13,120 ft). Both the linearity and length of these suggest that a fault probably determined their shape.

## Ocean Trenches

Some oceans—the Atlantic is a prime example—are very symmetrical; the central ridge trends laterally and downward to abyssal plains and hills and, on approaching the continents, one sees, in order, the continental rise, slope, and shelf.

A common feature of some oceans, especially the Pacific, are great trenches countersunk below the level of the ocean floor. Like many mountain ranges they are long narrow features (Fig. 16-11). The trench off the coast of Peru and Chile is the longest, at 5900 km (3660 mi) (Fig. 16-12); most are less than 100-km (62-mi) wide. Except for the Peru–Chile trench, such submarine wrinkles are usually associated with the chains of volcanic islands that festoon the sea between the trenches and the adjacent mainland, roughly making the arc pattern of island arcs.

No deeper spots on earth exist than these trenches. The Mariana Trench off the Mariana Islands at 11,034 m (36,192 ft) and the Tonga Trench north of the Tonga Islands at 10,770 m (35,326 ft) are the two deepest known. It is interesting that adventurers in their efforts to get as far from sea level as possible have always gone up—to the top of Mount Everest—in fact, the points farthest from the sea level are down, deep under the sea.

## Underwater Volcanoes

Not all oceanic volcanoes are associated with trenches, island arcs, and ridges. Many have the familiar shape of land volcanoes of the intermediate type, with steep sides and a small summit area (see Fig. 16-11). Known as **seamounts,** they stand about 1 km (0.6 mi) above the ocean floor.

The number of such submerged volcanoes is almost unbelievable. In 1964 the oceanographer Henry W. Menard estimated that there are 10,000 in the Pacific Basin alone, which makes them one of the ocean floor's most prominent features. Although some are isolated cones, many others oc-

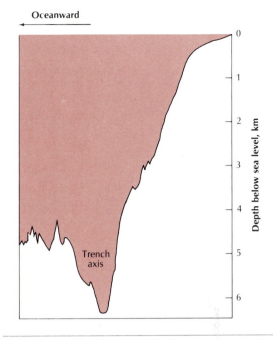

**Fig. 16-12.** Profile of the Peru–Chile Trench, westward off the coast of Peru. The trench is nearly devoid of sediment.

cur in clusters or linear arrangements. All of those investigated thus far are of basaltic composition.

Much more puzzling than seamounts are the features called **guyots** (Figs. 16-11, 16-13). They, too, are volcanic mountains, but their truncated summits form a nearly level submarine plateau. Their strange name honors Arnold Guyot (1807–84), a Swiss associate of Louis Agassiz, who came to the United States more than a century ago to teach at Princeton University. Guyots commonly rise to within 910 to 1520 m (3000 to 5000 ft) of sea level. Their surfaces typically consist of either barren, planed-off rock or of shallow-water sediments, including rounded gravel.

How were the summits of guyots planed off to such remarkable uniformity at such great depths? If we argue that they were beveled by wave action—which seems the most logical solution—then we are caught in the dilemma of believing either (1) that sea level has risen in the recent geological past or (2) that the guyots were carried down to their present depth by regional subsidence. The latter explanation is the more likely as plate tectonics (see chap. 4) has now adequately

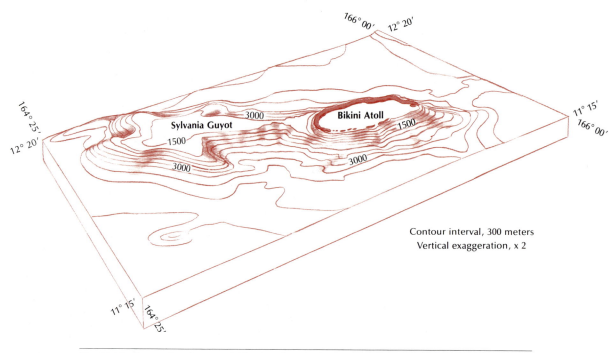

**Fig. 16-13.** Perspective diagram of Bikini Atoll and Sylvania Guyot adjoining it on the west. Only the rim (shaded) of Bikini Atoll is above sea level. Contours are in meters below sea level.

**Fig. 16-14.** The sizes of various volcanoes on land and at sea. (From H.W. Menard, 1964, *Marine Geology of the Pacific*, McGraw-Hill, Fig. 4–8. By permission of McGraw-Hill Book Co.)

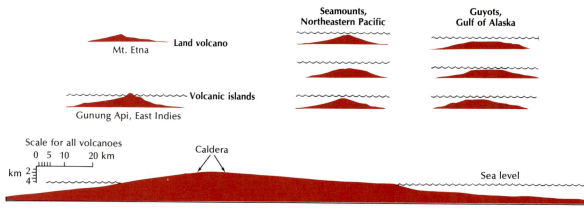

**Fig. 16-15.** Atolls of the Tuamotu Archipelago in the southern Pacific Ocean. (NASA)

shown how great expanses of the ocean floor can sink.

Before we leave the subject of submarine volcanoes, we should note their enormous size. We commonly conceive of terrestrial volcanoes as large features, but they are dwarfed by the size of many of their submarine cousins (Fig. 16-14).

## THE CORAL ATOLL

The origin of coral **atolls,** unique ring-shaped oceanic islands made up of the skeletons of marine animals, has been another subject of intense discussion among geologists (Figs. 16-13, 16-15). Their appeal reaches back across the centuries when the first Western European seafarers sailed distant reaches of the topical seas. This was the world made famous by Herman Melville (1819–91), by Robert Louis Stevenson (1850–94), by the

events arising from the contentious voyage of H.M.S. *Bounty,* and by the naval conflict waged in the Pacific during the years from 1942 to 1945. The modern world of those islands is effectively described in the novels and essays of Charles B. Nordhoff (1883–1947), James Norman Hall (1887–1951), and James Michener:

Scattered over a thousand miles of ocean in the eastern tropical Pacific, below the Equator, lies a vast collection of coral islands extending in a general northwesterly, southeasterly direction across ten degrees of latitude. Seventy-eight atolls, surf-battered dikes of coral, enclosing lagoons, make up this barrier to the steady westward roll of the sea. Some of the lagoons are scarcely more than saltwater ponds; others, like those of Rangiroa and Fakarava, are as much as fifty miles long by twenty or thirty across. The *motu,* or islets, composing the land, are threaded at wide intervals on the encircling reef. The smaller ones are frequented by sea fowl which nest in the pandanus trees and among the fronds of scattered coconut palms. Others, enchantingly green and restful to sea-weary eyes, follow the curve of the reef for many miles, sloping away over the arc of the world until they are lost to view. But whatever their extent, one feature is common to all: they are mere fringes of land seldom more than a quarter of a mile in width, and rising only a few feet above the sea which seems always on the point of overwhelming them.*

A coral atoll, circular in form, subtended a shallow lagoon. On the outer edge giant green combers of the Pacific thundered in majestic fury. Inside, the water was blue and calm. Along the shore of the lagoon palm trees bent their towering heads as the wind directed, and after a thousand more years brown men in frail canoes came to the atoll and decided it should be their home.

The world contains certain patterns of beauty that impress the mind forever. They might be termed the sovereign sights and most men will agree as to what they are: the Pyramids at dawn, the Grand Tetons at dusk, a Rembrandt self-portrait, the Arctic wastes. The list need not be long, but to be inclusive it must contain a coral atoll with its placid lagoon, the terrifyingly brilliant sands and the outer reef shooting great spires of spindrift a hundred feet into the air. Such a sight is one of the incomparable visual images of the world.†

For the origin of these islands, we turn to the work of Charles Darwin, the naturalist aboard the H.M.S. *Beagle* when it circumnavigated the world. As most people know, Darwin used both his geological and zoological observations from that voyage in formulating his theory of evolution published in his *On the Origin of Species* (1859). But few are aware that Darwin's training had been in geology as much as in any other branch of science and that on the voyage of the *Beagle* he made a great many observations on rocks, volcanic features, and fossils. Among the many wonders he beheld, few aroused his interest more than the world below the sea created by the corals. Although he actually saw only a few reefs, Darwin recognized that there were three major kinds—(1) fringing reefs, (2) barrier reefs, and (3) atolls—and that the three were related to each other in a logical and gradational sequence.

Darwin believed that the succession from one reef type to another could be achieved by the upgrowth of coral from a sinking foundation, such as a subsiding volcano. As long as the rate of coral growth was more rapid than the rate of sinking, Darwin argued, the progression would be from a fringing reef through the barrier stage (see Fig. 17-24) and, with the disappearance through subsidence and erosion of the central island, only a reef-enclosed lagoon or atoll would survive (Figs. 16-16, 16-17). Plate tectonics (see chap. 4) now provides a mechanism that explains this well.

Some atolls are very large—Kwajelein in the Marshall Islands of the South Pacific is 120-km (75-mi) long and averages as much as 24-km (15-mi) across. Most are far smaller. Nearly all true atolls are low—few have a freeboard of more than 3 m (10 ft). The reef ring consists of solid coral and most of the time may be just barely awash. During high tide the sea may sweep over it freely, with telling effect during storms. On the ring are the reef islands, where the life-giving pandanus and coco palms take root.

In that two-dimensional world, one is always aware of the encircling sea. No place on earth could be more vulnerable to the savagery of a storm (Fig. 16-18); its impact on such a defenseless world is vividly described by Nordhoff and Hall in their novel *Hurricane* (1936):

The very land itself had all but disappeared beneath the seas that swept the islet from the north. In the swiftly changing light, now bright, now dim, only the trees could be seen, and the church with its gleaming white walls. It gave the impression of sinking slowly, as the seas, a fathom and more deep, swept around it, meeting beyond in great bursts of spray. Then it would appear to rise a little, as though buoyant, to meet the onset of the next wave. . . .

By the middle of the afternoon, we knew that we had seen the worst, and at nightfall the hurricane left us, moving southward like the monster it was, in search of new lands to lay waste. The stars came out in a cloudless sky, and when the moon, one night past the full, rose, its mild light revealed as pitiable a scene of desolation as could have been found the world over. One might well have said that Manukura had ceased to exist. The village islet, certainly, had been destroyed as a habitation for man. From our mound of rocks we looked down upon . . . what shall I call that moon-blanched corpse of an island? It bore no resemblance to the place we had known. Nothing remained to show where the vil-

*Charles B. Nordhoff and James Norman Hall, 1936, *Hurricane,* Little, Brown, Boston, Mass., p. 1.
†James A. Michener, 1950, *Return to Paradise,* Random House, N.Y., p. 10.

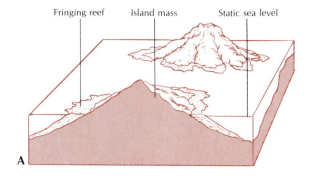

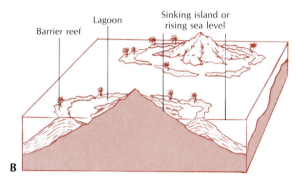

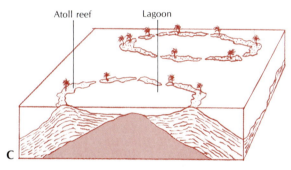

**Fig. 16-16.** Types of coral reefs and the steps in the formation of a coral atoll: **A.** fringing reef, **B.** barrier reef, and **C.** atoll reef. (After Alyn C. and Alison B. Duxbury, 1984, *Oceanography*, Addison-Wesley Publishing Company. By permission of William C. Brown Publishers, Dubuque, Iowa)

lage had been. The sea had half devoured the land itself, and what had been one islet was now two, divided by a channel swept clean to the reef bed, and a full fifty yards wide.*

Darwin knew that if his theory was correct, atolls should be underlain by hundreds or thou-

*Charles B. Nordhoff and James Norman Hall, 1936, *Hurricane*, Little, Brown, Boston, Mass., pp. 156, 168.

sands of meters of coral, all of which grew in shallow water. As part of the environmental studies made in connection with atom bomb testing during the 1940s the U.S. Navy drilled some deep holes in the Marshall Islands' reefs. On Bikini Island drill cores revealed at least 767 m (2520 ft) of coralline material; on Eniwetok the drills cored 1200 m (3940 ft) of shallow-water reef limestone and bottomed on basalt. The basal limestone at Eniwetok is Eocene, meaning that Eniwetok atoll is at the top of a coralline accumulation that has grown upward during the last 60 million years. The rate of subsidence does not appear to have been constant, but ranged between perhaps 15 m (50 ft) and 51 m (170 ft) per million years, slowing down later in the history of the reef. Thus, some of the events ushering in the nuclear age, a little more than a century after Darwin saw his first atoll, serve to vindicate his remarkable insight.

The same nuclear age, however, has brought difficult times to the atolls and their inhabitants. Because of their isolation and sparse populations, atolls were favored locations for testing atomic weapons. Some also have been studied as possible depositories of nuclear wastes. One of nature's most delicate and beautiful ecosystems is being threatened again.

Another complicating factor in atoll formation—or for that matter in the formation of reefs in general—is that of changing sea level with glaciations. During a period of glaciation when the sea level was at least 100 m (330 ft) below the present level, part of the reef would be exposed to subaerial weathering and karst topography would form (see chap. 18). In an interglacial period, however, sea level rises and reefs build upward. If that cycle recurs many times, as has happened over at least the past million years, the formation of these islands becomes quite complicated, especially the formation of the top 100 m (330 ft) or so.

## OCEANIC CIRCULATION

The waters of the ocean are constantly in motion, driven by the prevailing winds. This movement creates a slow-moving ocean drift that is affected by atmospheric patterns, the earth's rotation, and the position of continents. The results are large, rotating current cells (Fig. 16-19). In one of these

**Fig. 16-17.** Barrier reef encircling a volcanic island, an intermediate stage in the formation of an atoll. (Office of Naval Research)

cells is the Gulf Stream, an enormous current that moves warm tropical waters northward along the eastern coast of North America and crosses the Atlantic to the United Kingdom and Norway. In this way the latter two countries are kept warmer than one might suspect from their latitude.

There also are currents that produce vertical mixing in the oceans. These are driven by density differences in the water column. In the polar areas the surface waters are cold, relatively salty, therefore, relatively dense; in contrast equatorial waters are warm and relatively light. This density difference sets in motion a large-scale vertical circulation in which the denser polar waters flow along the ocean bottom toward the equator. There, they replace the warmer water, which rises and flows toward the pole.

The oceanic circulation patterns influence some sedimentation patterns, which we will discuss in the next section. Their most pronounced effect is in the production of oozes. In contrast, their effect on the depositional pattern of land-derived sediments is minimal.

## OCEAN SEDIMENTS

A great variety of sediments blankets the ocean floor. Most people are familiar with the sand and

**Fig. 16-18.** A storm in the late 1970s sent water and fragments of the coral reef across Majuro Atoll, Marshall Islands, Micronesia. Damage was severe, but there were no deaths. (Judy Knape)

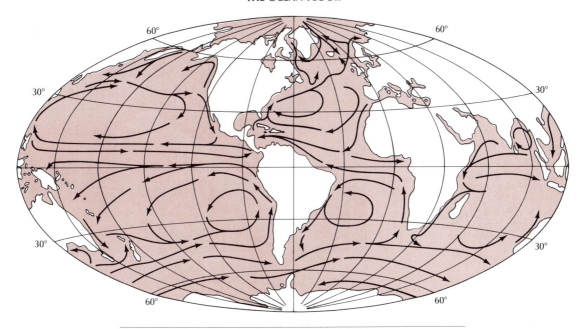

**Fig. 16-19.** Generalized world pattern of surface oceanic currents and gyres.

gravel of beaches. Seaward of the beach, silt and clay mantle the continental shelf. But what of the material that covers the deep-ocean floor?

Many of the original descriptions of sediment on the floor of the abyss are taken from the reports of the H.M.S. *Challenger* expedition of more than a century ago (see Fig. 16-1). Scientists aboard the *Challenger* gathered data to better understand the physical, chemical, and biological properties of ocean water and the ocean sediments. Of course, only surface samples could be taken at the time. Today we know that the pattern of the ocean floor, although it may resemble the one worked out by the men of the *Challenger* in broad outline, is far more complex.

Technological advances have now made it possible to look far below the ocean floor, and to estimate rock thickness and structure. The instrument used is an extension of the echo sounder. If the frequency of the sonic "ping" is increased, some of the sound waves penetrate the ocean floor and are reflected back from the rocks and sediment layers beneath it (Fig. 16-20). Marine geol-

ogists are able to "see" not only the shape of the bottom, but also the nature of the rocks below the surface. The process is called ***subbottom profiling.*** With every track of the ship, another continuous cross section of the rocks below the ocean bottom—to some 10-km (6.2-mi) depth—is obtained.

What we know today of ocean sediments, their distribution, and the age of ocean crust beneath them is the result of an ambitious drilling program. Several major problems, however, had first to be overcome; for example, how to secure the drilling platform in waters too deep for anchors or how to relocate and enter the same holes in waters up to 6 km (4 mi) deep. Modern technology solved these problems. The research vessel *Glomar Challenger*, run by Scripps Institution of Oceanography and funded by the National Science Foundation, went into operation in 1968. Scientists from many countries participated in the program, called the ***Deep Sea Drilling Project (DSDP).*** By the end of its mission in 1983, the vessel had logged over 375,000 nautical miles at sea, drilled 1105 holes, and penetrated an accumulated sediment thickness of

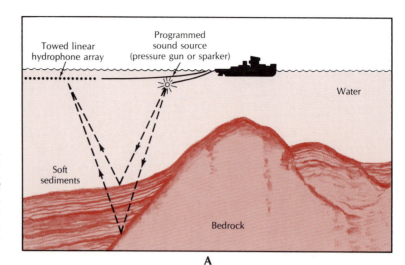

**Fig. 16-20. A.** The method of subbottom profiling, with the profile shown. **B.** Sedimentary layers beneath the ocean floor off the west coast of Africa, as indicated by subbottom profiling. Some of the layers have been offset along faults, shown by the dashed lines. (After R.G. Todd and R.M. Mitchum, Jr., 1977, AAPG Memoir 26, p. 157, Fig. 9)

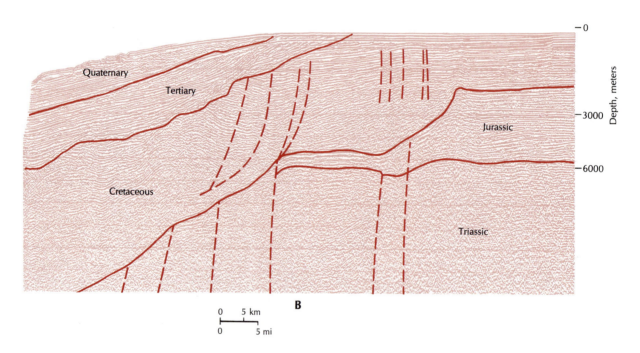

over 325 km (200 mi). In 1985 a more modern research vessel *JOIDES Resolution* began an international effort, called the ***Ocean Drilling Program*** (Fig. 16-21).

Three major kinds of sediment are found along the deep-ocean floor. ***Terrigenous (marine) sediments*** are derived from land, ***biogenic sediments*** are formed from marine organisms, and ***authigenic sediments*** are formed from precipitation of ions in seawater. Mixtures of all three kinds of sediment are found on most parts of the ocean floor, with one or another predominating (Fig. 16-22).

### Terrigenous Sediments

We can identify several types of terrigenous sediment. One is ***glacial-marine*** sediment, found only in high latitudes in close proximity to present glaciers and sea ice. Sediment distribution off Antarctica is a good example. Close to the ice are glacial tills, then silts and clays carried offshore by currents. Glacial-marine sediments have a curious component: land-derived boulders so far from shore that no ocean currents could have transported them. The best explanation is that they were rafted offshore on icebergs and sank to the bottom when the ice melted (Fig. 16-23). Such sediment is identified as far as 1000 km (620 mi) from Antarctica.

Along other contintental margins are coarse-grained beach sediments that grade offshore to silts and clays on the continental shelf, slope, and rise. In places, relatively coarse sands are found where only clays would normally exist; turbidity currents moving down the slope or in canyons are considered a logical mode of transport for such sediments.

In the deepest parts of the ocean [4000 m (13,120 ft)] are found the **brown clays.** The color comes from the weathering of iron-bearing minerals that coat the grains with iron oxides. Because of the depth of the water and the fine grain size of the clay, years are required for the grains to reach the ocean floor. The sedimentation rate is extremely slow, less than 1 mm (0.04 in.) in 1000 years; at such a rate it would require almost all of earth history to fill up an ocean 4000-m (13,120-ft) deep! As we discovered in chapter 4, this could never happen because the oceans are continuously being recycled on a grand scale, so that most ocean basins are fairly young.

We have identified several sources for the brown clay. By determining the clay mineralogy of the sediments in some locales, we have discovered that they were transported offshore from the mouths of river systems. For example, plumes of kaolinite-rich sediments, the main clay mineral produced in the soils of warm and humid regions, carpet the ocean floor far offshore from the mouth of the Amazon River and of the rivers of equatorial Africa.

Grains of clay travel by another means. With the wind as an agent, clay has been carried away from the world's deserts. Eventually it is deposited on the ocean surface. Plumes of clay particles from the Sahara have been identified all across the Atlantic Ocean floor; we find them being transported westward into the Pacific from the southwestern United States, and eastward from Australia, where they are rained out of the atmosphere over the Tasman Sea.

The pyroclastic material (ash) erupted from volcanoes is another source of brown clay. These clays are most common near major ash-producing sources—for example, the andesitic volcanoes rimming the Pacific Ocean.

### Biogenic Sediments

An important discovery of the *Challenger* expedition was an abundant sediment called **ooze.** *Challenger* scientists found that ooze was most prevalent in the area between the lower part of the continental slope and a depth of around 4 km (2.5 mi). Ooze, which resembles flour, forms an ivory-colored blanket, and the slightest disturbance sends a dustlike cloud swirling up through the dark water.

Ooze does not form in place on the ocean floor, but accumulates as the result of a gentle, unceasing "snowfall" of the remains of microscopic, free-floating near-surface organisms. When they die, they sift down from the sunlit surface to the lightless floor of the sea. It is only because the terrestrial supply of sediment is so slight relative to organic productivity that such organic deposits can build up.

Not all oozes are the same, however; their composition varies systematically across the ocean floor. Much of the abyssal plain beneath tropical and warm-temperate seas, in depths of less than about 4500 m (14,760 ft), is carpeted with microscopic shells of **foraminifera,** a single-celled creature that secretes calcite (Fig. 16-24A). Like so many single-celled creatures, they do not die but reproduce themselves by division. That is, one organism divides to make two, each of which grows a new shell, and the original shell, now vacated, sinks to the bottom of the sea to form **calcareous oozes.** These are most abundant in the Atlantic, Indian, and southern Pacific oceans (see Fig. 16-22). In other parts of the ocean, **siliceous oozes,** which consist mainly of **radiolarian** (Fig. 16-24B) and **diatom** remains, predominate (see Fig. 16-22). They are minute animals and plants, respectively, that secrete a silica shell. When these

**Fig. 16-21. A.** *(Opposite)* The research vessel *JOIDES Resolution,* of the Ocean Drilling Program. **B.** *(Left)* Roughnecks working with the drilling equipment on board the ship. (Ocean Drilling Program, Texas A&M University)

organisms die, their shells begin the long descent to the ocean floor.

Siliceous oozes, that consist mainly of diatoms are abundant in a broad band in the cooler seas along the margins of the Arctic and Antarctic. In a sense these plants compose the true pastures of the sea. It is curious that the largest creatures on earth, the Arctic whales, should depend for sustenance on the humble diatom, among the smallest of living things. Radiolarian oozes, on the other hand, are most abundant in the band along the equator in the Pacific Ocean.

Several factors account for the distribution of oozes. One is dilution by other sediments; for example, if biological productivity is low relative to terrestrial input, as in the north Pacific, the clays dominate. Another factor is supply. For example, in the equatorial Pacific upwelling of nutrient-rich waters results in such high productivity of siliceous organisms that siliceous oozes dominate. Dissolution also plays a part. Below a mean depth of about 4500 m (14,760 ft), the water temperatures and chemistry are such that calcareous forms dissolve, and disappear before reaching the ocean floor.

Balancing out these factors, ooze sedimentation

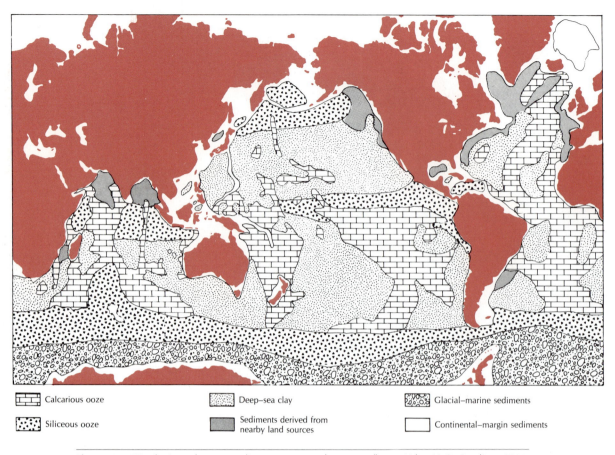

| | Calcarious ooze | | Deep–sea clay | | Glacial–marine sediments |
|---|---|---|---|---|---|
| | Siliceous ooze | | Sediments derived from nearby land sources | | Continental–margin sediments |

**Fig. 16-22.** Distribution of recent sediment types on the ocean floors. (After H.G. Reading, 1986, *Sedimentary Environments and Facies, 2nd ed.,* Blackwell Scientific Publications, Fig. 11.3, p. 347)

**Fig. 16-23.** Iceberg with boulders carried piggyback fashion. As the iceberg rotates or melts, the boulders and sands rain down on the deep-sea floor, which can be several thousand meters below.

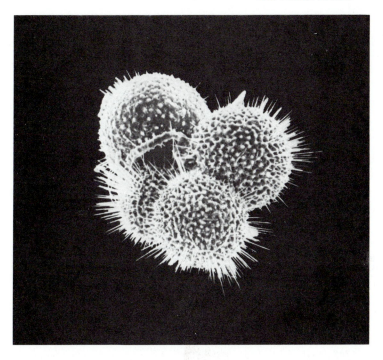

**Fig. 16-24. A.** *(Above)* Scanning electron photomicrograph of the foraminifer *Globergina* as it appears floating in the ocean waters. Once the shell is shed, the spines are lost, and as a fossil it looks similar to other foraminifera. (American Museum of Natural History) **B.** *(Below)* Middle Eocene ooze (45 million years old) from a DSDP core in the western Indian Ocean. Most of the fossils are radiolarian, but a few are foraminifera (F) and sponge spicules (S). (Scripps Institution of Oceanography)

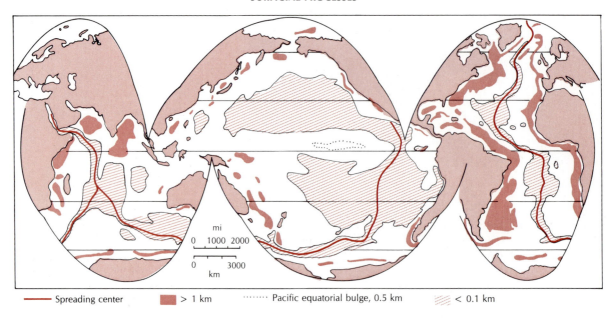

| | | | |
|---|---|---|---|
| —— Spreading center | ▨ > 1 km | ⋯⋯ Pacific equatorial bulge, 0.5 km | ▨ < 0.1 km |

**Fig. 16-25.** Thickness of sediment over the basaltic crust of the ocean floor. (After W.H. Berger, 1974, "Deep-Sea Sedimentation," in *Geology of Continental Margins,* C.A. Burk, and C.D. Drake, eds., Springer-Verlag New York, Inc., pp. 213–241)

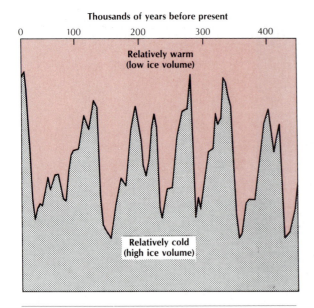

**Fig. 16-26.** Isotopic data on oxygen in fossil shells of foraminifera can be used to help date the time of glacials and interglacials in the past. Here the data from a core brought up from the floor of the Caribbean Sea are interpreted. (After W.S. Broecker and J. van Donk, "Insolation Changes, Ice Volumes, and the $0^{18}$ Record in Deep-Sea Cores," *Reviews of Geophysics and Space Physics,* vol. 8, no. 1, Fig. 3)

rates are fairly high. Those for calcareous ooze reach a maximum near 10 to 20 cm (4 to 8 in.) per 1000 years, whereas those for siliceous oozes are 4 to 5 cm (1.6 in. to 2 in.) per 1000 years.

## Authigenic Sediments

Sediments formed by precipitation from seawater are of minor importance. Perhaps the material that has attracted the most attention are the manganese nodules. In appearance they resemble cannonballs strewn across the ocean floor. More will be said of them in chapter 18.

## Ocean-Sediment Thickness

If we think of the ocean basins as static features impressed long ago into the surface of the earth,

**Fig. 16.27. A.** Modern sea-surface temperatures (°C) for August. **B.** Estimated sea-surface temperatures for August 18,000 years ago, during the last glacial maximum. Note the different position of shoreline due to the lowering of sea level; this created many land bridges. (A. McIntyre, ed., CLIMAP 1981)

**A**

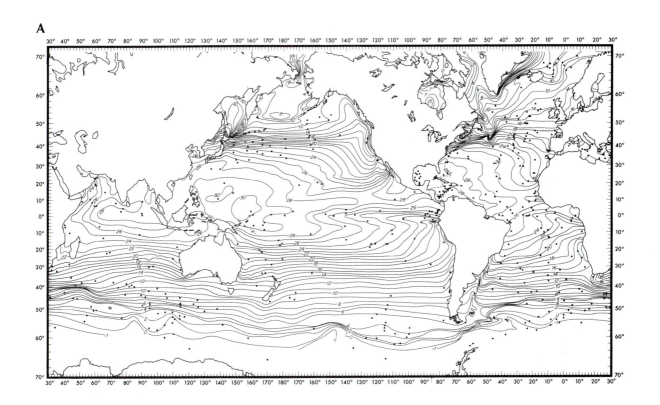

**B**

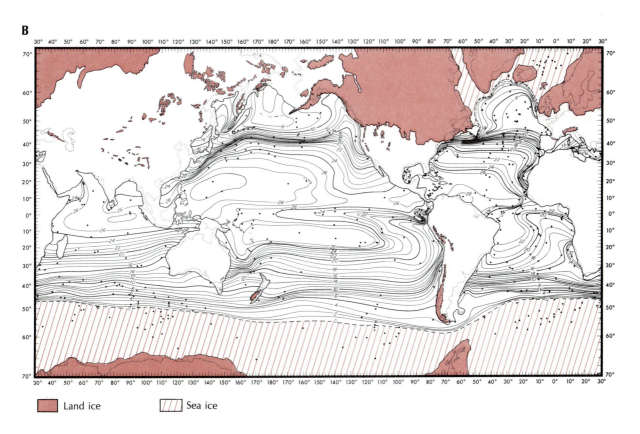

Land ice      Sea ice

sediment thickness would be a reflection of influx rate from various sources. The picture is much more complex, however, and many former anomalies are now adequately explained by the plate tectonics theory (chap. 4).

With numerous seismic studies probing the sediment column to the basaltic floor of the ocean, it was the thinness of the sediments rather than the thickness that surprised workers the most (Fig. 16-25). As expected, sediments are thickest near the continents where large rivers deposit their loads. Sediments are also thick in areas farthest from the midocean ridges—these are the oldest parts of the ocean floor and so have been sediment repositories for the longest time. In contrast the midocean ridges, born in recent, geological time, have accumulated relatively little sediment. Biological productivity is another important factor, and it surely helps explain the presence of thick sediments in the circum–Antarctic region and the equatorial Pacific.

## GLACIATIONS/INTERGLACIATIONS AND PAST GLOBAL CLIMATE—THE STORY FROM THE OCEAN DEEP

Recent intensive study of cores from the ocean floor, many from DSDP, has caused scientists to alter their ideas on the numbers and timing of Pleistocene glaciations. How was their thinking changed? A sample core is a historical record of events in earth history, with the youngest events recorded near the top of the core and the older ones at the base. Furthermore, many scientists think that the deep oceans were repositories of sediment throughout the Pleistocene, and even farther back in time and that little erosion occurred in such submarine areas. Thus, the record preserved by the sea is nearly complete.

What do ocean sediments tell us about glacial periods? For one thing some cores show glacial-marine deposits interbedded with normal marine deposits. In other cores the foraminifera population changes with depth in number, in species, and in isotopic composition of oxygen in shells. Such marine life seems to be sensitive to either

changes in water temperature or variations in water composition that accompany climatic change and the stockpiling on land of water as ice.

When data are carefully analyzed, a procession of alternating warm and cold intervals can be seen that somehow must be related to glaciations and interglaciations on land (Fig. 16-26). These data compare well with that from the marine terraces (see Fig. 15-21). The revelation of more glacial periods than had hitherto been recognized on land was revolutionary. From Figure 16-26, for example, we can postulate six major glaciations in little more than the past 400,000 years. How are the six to be correlated to the traditional four recognized in the land record? An answer for many earth scientists is that some of the land record has been lost through erosion. The other interesting aspect of the curve in Figure 16-26 is its sawtooth shape, which can be interpreted to mean that ice sheets take a long time to build up—some 50,000 years or more—but that their disappearance by melting verges on the catastrophic—it is all over in as little as 10,000 years.

Recently, a group of scientists decided to reconstruct ocean-water temperatures by examining one moment in time in selected cores. The project, called Climate: Long-range Investigation, Mapping, and Prediction (CLIMAP), chose the month of August 18,000 years ago as that "moment." Combined with what is known of the land record, we now have a fairly detailed glimpse of those August days in the past, and the contrast with the present is marked (Fig. 16-27).

Much work still needs to be done. We have to learn to read the deep-sea sediment record better. We also need to correlate more carefully the ocean record with the glacial record on land, which will require more thorough fieldwork. Also, from studying the relationship between warm and cold intervals and periods of glaciation, we might well be able to predict future glaciations, the climatic conditions signaling their onset, or the catastrophic warming conditions that accompany their disappearance. Figure 16-26 also reveals that the interglacials are of relatively short duration. Could we be nearing the end of a warm period?

## SUMMARY

1. The ocean floor consists of several major topographic features. The continental shelf forms the gently sloping platform off the coast of most continents. At depths near 133 m (440 ft), the shelf grades to a steeper landform, called the *continental slope*. At the base of the slope is a gently sloping depositional surface, the *continental rise*. Submarine canyons, some larger than any on land, are cut into the above features in places. Continuing seaward, the extensive and flat *abyssal plain* is reached. Finally, in the middle of the ocean is a *midocean ridge,* the most continuous mountain range on earth. In some oceans these features are symmetrically arranged on both sides of the midocean ridge; in others a deep linear *trench* runs either adjacent to land or on the convex side of volcanic island arcs.

2. Volcanoes are a prominent feature of the ocean. Some rise to the surface as islands, others are submerged and are called *seamounts,* and some submerged ones have flat tops and are called *guyots.* The flat tops of the guyots are generally ascribed to erosion when the guyot was closer to sea level. *Atolls* consist mainly of the buildup of coral limestone deposits on slowly subsiding volcanoes. Plate tectonics (see chap. 4) explains most of the major topographic features of the oceans.

3. Deep-ocean sediment consists mainly of the fine-grained *oozes* of calcareous or siliceous shell grains (the remains of minute animals and plants) and clay. In some places manganese nodules are present. In other places gravel clasts rest upon the fine-grained ocean bed far from shore. They probably were carried far offshore on icebergs and deposited when the iceberg melted.

4. The chemical and biological properties of the oozes vary with depth in ocean cores, and the variations provide important clues as to the number of glaciations and interglaciations in the Pleistocene. There are more such periods recorded in the ocean sediments than in the record of terrestrial glacial tills, as the latter are presently interpreted. Data from the ocean floor also can be used to reconstruct global ocean-water temperatures at specific times in the past.

## QUESTIONS

1. Describe the processes that helped shape the continental shelves.
2. What are the arguments against submarine fans being submerged subaerial features?
3. What are several processes that would reduce the height of the volcanic portion of an island, relative to sea level, in progressively going from an island with a fringing reef to an atoll?
4. What are the major factors influencing the rate of deposition of terrestrially derived sediments in a deep-ocean basin?
5. Why are we looking to the oceans for data on past glaciations? What kinds of data are used?

## SELECTED REFERENCES

Bloom, A.L., 1978, Geomorphology; a systematic analysis of late Cenozoic landforms, Prentice-Hall, Englewood Cliffs, N.J.

Boggs, S., Jr., 1987, Principles of sedimentology and stratigraphy, Merrill, Columbia, Ohio.

Duxbury, A.C., and Duxbury, A., 1984, An introduction to the world's oceans, Addison-Wesley, Reading, Mass.

Heezen, B.C., and Hollister, C.D., 1971, The face of the deep, Oxford University Press, New York.

Kennett, J.P., 1982, Marine geology, Prentice-Hall, Englewood Cliffs, N.J.

Menard, H.W., 1964, Marine geology of the Pacific, McGraw-Hill, New York.

van Andel, T., 1981, Science at sea: tales of an old ocean, W.H. Freeman, San Francisco.

# 17

# SEDIMENTARY ROCKS

Widely spread over the surface of the earth, a relatively thin blanket of sediment has been consolidated into rock through slow-acting processes that are relatively simple to understand (Fig. 17-1). Sedimentary rocks form in land or sea environments much more familiar to us than the deep crustal realm, where metamorphic and igneous rocks originate. Sedimentary rocks constitute 66 percent of the area of the continents and, considering both continents and oceans, their average thickness is about 2 km (1.2 mi).

Some of the world's most picturesque landscapes are carved in sedimentary rocks. The Grand Canyon of the Colorado cuts through layer upon layer of varicolored rocks in probably the most famous display of sedimentary rocks in the world. Another colorful display is in the Atlas Mountains of North Africa.

Ancient civilizations built their monuments and temples and carved statuary from sedimentary rocks.

The buildings of the old city of Jerusalem and its massive encircling walls are constructed almost entirely of limestone (Fig. 17-2). In Egypt the Great Sphinx, probably the best-known monument in the world, was carved in a spur of limestone from the same quarries that supplied the gleaming stone that once faced the pyramids at Giza.

For the most part, sedimentary rocks are secondary, or derived, rocks. The major category of sediments consists of layers made up of clay, sand, or gravel particles derived from the disintegration or decomposition of preexisting rocks.

A second category is organic sediments, which are important to human beings. Coal is in this category, as are oil shales which in recent years have been studied as a source of oil. Another familiar kind of organic sedimentary rock is limestone; of its many forms, most represent the slow accumulation over centuries of deposits made by plants and animals.

A third large and economically important category of sedimentary rocks is chemically precipitated in bodies of water such as evaporating lakes or shallow embayments of the sea. Well-known examples of this category are rock salt and gypsum.

**Fig. 17-1.** This monastery at Meteora, Greece, surmounts bluffs of layered, cemented gravel deposits (conglomerate) of Tertiary age. The area has undergone tilting since deposition, as shown by the layers that are no longer horizontal. (Greek National Tourist Organization)

**Fig. 17-2.** Wall of the Old City of Jerusalem near the Jaffa Gate, built by the Turks in the late 1530s. At the foot of this wall there are ruins of an older wall. (Peter W. Birkeland)

In this chapter we will describe the main properties of sedimentary rocks and comment on how these properties can be related to the environment in which the rocks were formed. Although such a relationship may not seem particularly significant, it can be of the utmost importance to a geologist searching for oil, minerals, or even groundwater. For example, a typical property, or set of properties, possessed by a sedimentary rock formation may indicate the presence of an important deposit in that formation or in similar formations nearby.

## ORIGIN OF SEDIMENTARY MATERIALS

Many of the materials that make up sedimentary rocks are derived from weathering reactions (see chap. 10). A much simplified example is:

$$
\begin{array}{cc}
\text{Rocks} & \text{Rainwater and acids} \\
\text{and} & + \text{from } CO_2 \text{ and} \qquad \xrightarrow{\text{weathering}} \\
\text{minerals} & \text{organic matter} \\
\\
& \text{Smaller} \quad \text{Clay} \\
& \text{solid} \quad + \text{minerals} + \text{Ions} \\
& \text{particles} \\
\end{array}
$$

The smaller solid particles—gravel, sand, and silt—are derived directly from the source rock.

The clay minerals form in soils and in other near-surface, low-temperature environments. Each of the many clay minerals recognized reflects the environment in which it forms; however, if it is transferred to an appreciably different environment, it can change to another clay mineral. Thus the interpretation of clay minerals in sedimentary rocks is complex and depends in part on the rocks and climate of the source area and the chemical conditions underground where sediment was transformed into rock.

Any weathering reaction produces ions or salts that are dissolved in water. Some common ions are calcium ($Ca^{2+}$), sodium ($Na^+$), carbonate ($CO_3^{2-}$), and chloride ($Cl^-$); they and others are responsible for the mineral taste of certain waters. Under appropriate conditions, commonly in areas of evaporation, these and other ions precipitate out of solution. The minerals that they form are complex and are determined in large part by the chemical composition of the parent waters.

Still other kinds of sediments are made up of the largely insoluble shells of organisms. Such shells are formed in part by ions, which the organisms extract from their watery environment.

## DETERMINING THE SOURCE AREA

In interpreting sedimentary rock, we want to know its source area, the area from which most of it was derived. The larger particles in the rock—gravel, sand, and silt—should mimic the rocks and minerals of the source area. For example, a source area in which basalts predominate will produce sands high in olivine, calcic-plagioclase, and augite, whereas a granitic area will produce sands high in quartz, potassium-feldspar, and biotite. Weathering on hillslopes in the source area before material is transported to its deposition site, however, can alter the mineralogy of the deposited sands. In climates where chemical weathering is intense, more quartz is seen in granite-derived sediments than in the granite itself because quartz is more resistant to weathering than other minerals in granite.

More difficult to determine are the source areas of sedimentary rocks consisting of the clay minerals and precipitates. Although some clay minerals reflect the rocks of their source area, others are altered on burial and are not good indicators. Precipitated sediments reflect only the chemistry of the water in which they were dissolved and offer no clues to the source area from which the ions were derived.

## FEATURES OF SEDIMENTARY ROCKS

Several key features distinguish sedimentary rocks from the other major rock types. Some of these features have even survived metamorphism and thus provide clues to the past history of some metamorphic rocks.

### Sorting

Sorting refers to the degree of similarity in particle size in a sediment. In a well-sorted sediment, for example, most particles are about the same size. In contrast a poorly sorted sediment contains a wide assortment of particle sizes. Moderate sorting refers to an intermediate stage between the two.

Important to sorting is the settling velocity in the transporting medium. Consider water. If particles of different size are dropped into water, the larger ones, being heavier, sink to the bottom first. Along shorelines exposed to constant wave action, beach sediments are always in motion. Heavy gravels remain close to the point of highest wave energy (Fig. 17-3); sands remain on the beach or are moved seaward; and silts and clay, which remain in suspension for some time, are carried far offshore. This continual restirring results in a well-sorted sediment.

### Stratification

Most sedimentary rocks are made up of layers—ranging from those measurable in millimeters to those measurable in meters (Fig. 17-4). Such depositional layers are called *strata;* an individual layer is a stratum.

In "quiet water" depositional environments, such as the bottom of lakes or the ocean floor, strata are laid down almost horizontally, in layers of more or less uniform thickness. In contrast the strata of river deposits tend to vary in thickness when the deposits are traced laterally; they are thickest where river channels occur (Fig. 17-5).

Layering can occur through variations in the energy of the transporting medium or through

**Fig. 17-3.** Well-sorted clasts that are characteristic of many high-energy beach environments. (William Estavillo)

variations in the size, amount, mineral composition, or color of the sediment being delivered to a particular depositional site. A high-energy, turbulent flood, for example, could wash large particles out onto a lake bottom. Following the flood, silt and clay—the usual sediment carried to the lake—might be deposited on top of the flood debris. The result is two layers of contrasting particle size. Discontinuous deposition also can cause layering because the size of the material deposited in a new cycle is not likely to be exactly the same as the sediment that came before it.

Some sediments and sedimentary rocks are characterized by the repetition of distinctive layers. Well-known examples are *varves,* which seem to form best in glacier-fed lakes. Each varve consists of two layers—a coarser silty layer overlaid by a finer-grained layer of silt and clay (Fig. 17-6). The varve is thought to represent a seasonal event. The coarser fraction in the lower layer is thought to be laid down during the spring and summer, the seasons of active snow and ice melt. The finer-grained upper layer, however, represents quiet sedimentation during the fall and winter, when the ice stops melting and the surface of the lake is frozen over.

Some sediments show a very different sort of stratification. The particles of the individual layers, instead of being distributed uniformly throughout, are graded—the larger particles concentrated at the bottom, the smaller at the top. Such layers are said to have *graded bedding*—a feature that occurs when a mass of sediment is discharged suddenly into a relatively quiet body of water. The largest

**Fig. 17-4.** Sedimentary rocks commonly are characterized by horizontal layering, which is usually most visible in arid regions. The layering in the rocks in this vertical aerial photograph closely simulates the contour lines of a topographic map. (William A. Garnett)

**Fig. 17-5.** A conglomerate channel fill of an ancient stream in Wyoming. The maximum thickness of the fill is about 6 m (20 ft.). The hammer in the center of the photograph indicates scale. (Emmett Evanoff)

particles drop out quickly, followed by those of medium size, and finally the finest particles. The change from one dominant particle to another is gradational, and the whole makes up the graded bed. An excellent example of how graded beds form is found in the description of the turbidity current that swept across part of the Atlantic Ocean in 1929 (chap. 16).

Finally, it should be emphasized that in layered sequences of rocks, the younger layers rest on top of the older layers, rather like a layered birthday cake. This is a very important point to keep in mind when trying to determine the relative ages of rocks in a sedimentary rock sequence.

## Roundness of Grains

An important property of clastic sediments is the roundness of the grains. How are they rounded off? Assume that a rock tumbles from a cliff into a riverbed. Initially, the rock has sharp corners, but as it slowly moves downstream it collides with, and scrapes against, neighboring rocks and the corners become more rounded Fig. 17-7. Some of the smoother rounded gravels are found on beaches; and good-to-moderate rounding characterizes rivers and streams. In contrast glaciers and mudflows do not allow much grain-to-grain contact; consequently, rounding of the coarse materials is generally poor.

## Color

Igneous rocks, unaltered by exposure to the atmosphere, typically are shades of gray or black because these shades prevail in the most abundant constituents of the rocks, feldspar and ferromagnesian minerals. Sedimentary rocks are often more

**Fig. 17-6.** Well-expressed varves exposed along the shore of Yellowstone Lake, Yellowstone National Park. These were deposited in a glacial-age precursor to the present lake. (Don Rodbell)

colorful, however, with the materials that color them either filling the voids between the grains or coating them like stain or paint.

An important pigment in sediments is iron released by the weathering of iron-bearing minerals. If the pigment contains hematite (iron oxide, $Fe_2O_3$), the resulting rock is likely to be red. Hematite is the source of most of the rich red coloring in the walls of the canyons of Utah and Arizona. Other forms of iron may stain a rock brown or even shades of pink and yellow. Iron may even be partly responsible for the purple, green, or black shades of some sedimentary rocks, but often the origin of coloring matter is poorly known. Pigments may have been carried to the site of deposition along with the sediments or produced later by chemical weathering of the original sedimentary grains throughout eons of deep burial.

Many of the dark sedimentary rocks owe their color to the organic material they contain. Coal, an excellent example, is of organic derivation, and its very name is a synonym for black. Depending on the amount of organic material, sedimentary rocks may range from light gray to black.

## Mud Cracks

When wet clayey mud is exposed to the air it dries, shrinks, and then cracks—generally in a

**Fig. 17-7.** Rounded river boulders of basalt stacked up to form a wall on a farm in southern Idaho. This amount of rounding occurred in 13 km (8 mi) of stream transport. (H.E. Malde, USGS)

**Fig. 17-8.** Mud cracks, Colorado River, head of Lake Mead. (Tad Nichols)

nearly uniform pattern of polygons (often four-sided) that in some ways resemble the tops of lava columns (Fig. 17-8). In lava, cracking is caused by contraction on cooling; in wet muds, by dehydration. As drying continues the mud layers on the surface of the polygons may curl up at the edges, so much so that at times they form complete rolls, much like cardboard tubes or fancy pastries.

Mud cracks indicate that the original sediment was alternately wet and dry; thus such features are typical of mud-bottomed shallow lakes that on occasion dry up. They can also form on muddy tidal flats if the time of exposure at low tide allows drying to occur.

### Ripple Marks

Nearly everyone has noticed the characteristic corrugated surface made by currents flowing across the sandy bottom of a lake or stream (Fig. 17-9) or has seen photographs of virtually the same pattern produced by the wind blowing across a desert sand dune. Such ripples, called **current ripples,** develop at right angles to the current and are likely to have steep slopes on the downcurrent side but gentle ones upcurrent. The asymmetrical form results when an air or water current rolls sand gains upslope and gravity pulls them down the opposite side, or *slip face.* This slope stands at an inclination known as the **angle of repose** (the maximum slope at which sand grains will remain stationary without sliding down the slope). Current ripples in solid rock can be used to establish the direction of ancient currents in the atmosphere or under water. In the past it was thought that such ripples in water-laid sediments indicated only shallow depths, but in recent years underwater photography has shown ripple patterns on the seafloor at depths of several thousand meters.

Another type of ripple has symmetrical sides, sharper crests, and more gently rounded troughs. Such corrugations are called **oscillation ripples** and presumably are the result of surface waves (called *waves of oscillation*) stirring up the sandy bottom of a shallow body of water.

## Cross-bedding

Earlier in the chapter, the point was made that sedimentary rocks usually are deposited in essentially parallel horizontal layers. However, in several varieties of stratification, the layers are inclined at steep angles to the horizontal plane. Such layering is known as ***cross-bedding.***

One kind of cross-bedding can be seen in sand dunes. Each layer of the dune at some time past was part of the surface and, as the dune's configuration was established largely through a balance of wind transport upslope on the windward side of the dune and gravity-sliding downslope on the leeward side, most of the layers are sweeping curves—more often than not they are concave upward. Because sand dunes are ephemeral landforms, which change in position and orientation with the inconstant wind, it is not surprising that the sweeping, shingled layers may intersect one another in complex patterns, such as those in the walls on Canyon de Chelly, Arizona, or in Zion National Park, Utah (Fig. 17-10).

Another kind of large-scale cross-bedding occurs in deltas. Streams carrying a fairly large load of moderately coarse debris deposit their sediment rapidly when they reach a body of water such as a lake or gulf. There, the sediment deposited by the stream constructs a leading edge out into the water, much as a highway-bridge fill is built out into a canyon by end-dumping from gravel trucks. The outer slope of the delta, like the slip face of a sand dune or a current ripple, also stands at the angle of repose. When such sediments are consolidated into rock, three distinctive layers may result. At the top and bottom of a deltaic deposit

**Fig. 17-9.** Ripple marks on a bedding plane in the Dakota Sandstone, Colorado. (J.R. Stacy, USGS)

are horizontal strata, known as **topset** and **bottomset beds,** respectively; the steeply inclined layers in the middle that were once the advancing delta front are called **foreset** beds (Fig. 17-11).

River deposits also can be cross-bedded, but usually on a much smaller scale than those discussed earlier.

## Fossils

No other property is so characteristic of sedimentary rocks as the presence of fossils (Fig. 17-12). They are the remains of once-living things that on their death were buried in sand, silt, lime, or mud. Over the centuries much of their organic matter was gradually replaced by inorganic matter until, to use petrified wood as an example, many of the woody fibers and cellulose were replaced by silica. Many species of plants and animals have been preserved as fossils. Even such improbable creatures as jellyfish (whose composition must be more than 95 percent water), or such fragile structures as the compound eyes of flies, or the delicate tracery of dragonfly wings have been preserved. Those creatures are the exception, however, because the organisms most commonly preserved as fossils are those with such durable elements as shells, bones, and teeth. In fact, most fossils are the remains of shells or skeletons. Some rocks may consist entirely of organic matter. Coal is made up of plant fragments; some limestones are the remains of coral or calcareous algae or may be a mass of seashells. In addition to the remains of organisms, footprints, tracks, trails, and burrows also are considered fossils.

Fossils are extremely important to geologists in dating rocks and in correlating rocks from area to area. They also provide vital clues to the depositional environments of the rocks in which they are found. Shell collectors are well aware of this—if they are to have a varied collection, they must search many environments for their specimens.

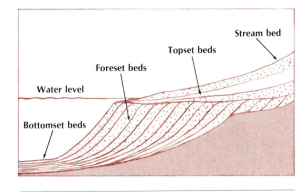

**Fig. 17-11.** Cross section through a delta illustrating the major sets of beds.

## Concretions

Round (or almost round) solid bodies are sometimes found within sedimentary rocks. **Concretions,** as most such bodies are called (Fig. 17-13), are composed of sedimentary material that solidifies around a small hard nucleus, which is usually not sedimentary in nature. Any small particle—a grain of sand, a piece of shell, even a small insect—can act as a nucleus. A cement, which can be limy, collects around the nucleus, which eventually binds all the particles together and gradually enlarges the concretion. Some may reach 1 m (3 ft) or more in diameter. Most are much smaller, however.

## CONVERSION OF SEDIMENTS TO SEDIMENTARY ROCKS

If rock is to form from loose clastic sediments, something has to bind the materials together. Could it be the high pressure exerted by the weight of overlying rocks? The answer is no. If enough pressure were applied to cause sand grains to adhere to one another, it would simply crush them into smaller particles. Pressure does play a role, however, in the process of **compaction,** which is the squeezing together of the particles in a sediment. If, for instance, enough pressure is applied to fine-grained muds, such as clay or silt, most of the interstitial water is squeezed out and the sediment shrinks markedly. If clay is a dominant constitu-

**Fig. 17-10.** Giant cross-bedding in ancient sand dune deposits at Checkerboard Mesa, Zion National Park, Utah. (Ray Atkeson)

**Fig. 17-12. A.** *(Opposite, above)* Brachiopod fossils in the Trenton Limestone of Ordovician age, Watertown, New York. (Smithsonian Institution) **B.** *(Opposite, below)* Dinosaur bones excavated at Como Bluff, Wyoming, in the late 1800s. (Western Research Center, University of Wyoming) **C.** *(Left, above)* Fossilized tree stump on Axel Heiberg Island, in the northern Canadian Arctic, about 1120 km (700 mi) from the North Pole. The stump is part of a forest that grew about 45 million years ago when global temperatures were much warmer. (Courtesy of James Basinger. Photo by E. McIver) **D.** *(Left, below)* Dome-shaped stromatolite exposed in the North Pole region of Western Australia. Built of algae and bacteria in shallow waters about 3.5 billion years ago, this is among the oldest direct evidence of life on earth. The fossil is 31 cm (1.2 ft) high and 43 cm (1.7 ft) wide. (Smithsonian Institution, photo by Victor E. Krantz)

**Fig. 17-13.** Concretions weathered out of limestone are left scattered on the desert floor in the Khārga Oasis, Egypt. (Tad Nichols)

ent, the particles will then tend to adhere to one another.

*Cementation* is the most significant process involved in the conversion of sediments into sedimentary rock (Fig. 17-14). Fundamentally the process involves the deposition from solution of a soluble substance, such as calcite ($CaCO_3$), and its building up as a layer or film on the surface of sand grains, silt particles, or clay flakes (as the case may be) until much of the pore space separating them is filled. Such a limy cement is precipitated in much the same way, although at a lower temperature, as scale is deposited inside a teakettle.

Calcite is one of the most abundant natural cements because it is among the more soluble of the common substances that may be dissolved in groundwater. Another important natural cement is silica ($SiO_2$), which is also soluble, although less

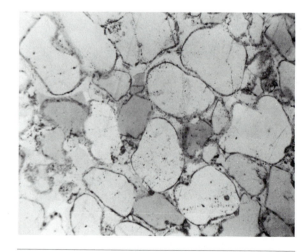

**Fig. 17-14.** Microscope photograph of a thin section of the Potsdam Sandstone from New York. The rounded quartz grains are cemented together by fine-grained silica that fills the void between the grains. The dark material along the grain boundaries is hematite and other impurities. (Edwin E. Larson)

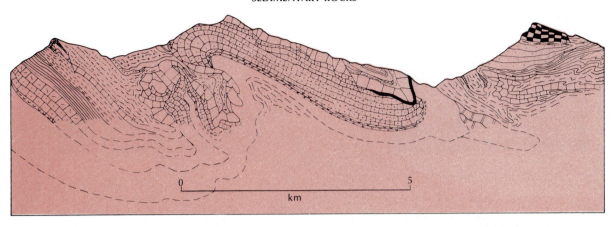

**Fig. 17-15.** Tightly folded and overturned sedimentary rocks in the European Alps. (From R. Trümpy, 1960, "Paleotectonic Evolution of the Central and Western Alps," *Geological Society of American Bulletin,* vol. 71, pp. 843–908, p. 2)

so than calcite. Iron oxide ($Fe_2O_3$), too, is a common cementing agent.

## WHICH WAY IS UP?

The deformation of many sedimentary rock layers is so slight that it is easy to determine if the sequence is right-side up or up-side down (Fig. 17-1). But pity the poor geologist who tries, for example, to unravel the complex structural geology of some of our western mountain ranges or the European Alps. In places the rocks have been so intensely folded that they are now upside down (Fig. 17-15). Some sedimentary rock features can help determine which way is up (Fig. 17-16). Baked zones associated with volcanic rocks also are an indicator of the up direction. Interpretation of up versus down is considerably more difficult in metamorphic rocks (especially if intense metamorphism has taken place) and impossible in plutonic rocks.

## TYPES OF SEDIMENTARY ROCKS

The point was made earlier that there are three major categories of sedimentary rocks: fragmented, or *clastic* rocks; *organic deposits;* and *chemical precipitates.* However, not only are there gradational types, there are also varieties that might just as logically be placed in one category as another.

### Clastic Sedimentary Rocks

The clastic rocks are truly secondary rocks because they are fragments of preexisting rocks. These "fragments" range in size from blocks the size of boxcars to particles so fine that they remain in suspension almost indefinitely. The range of composition is so wide that in classifying them, the first property to be considered is the size of the particles that are cemented together rather than the material of which they are made.

Sand is an example. Sand can consist of almost any material of sufficient durability. Many beach sands contain mostly feldspar grains as well as a liberal sprinkling of other sand-size rock particles and mineral grains. Along some U.S. rivers in the Atlantic states, there are sand bars of coal fragments. On some of the beaches of Hawaii the sands are coal black, too, but they are ground-up basalt. In the islands of the South Pacific the straw-colored sands of their fabled shores consist of fragmented coral heads, pieces of shell, and other organic debris. It would be confusing, therefore, if all sands of slightly different composition were given different names.

Clastic sedimentary rocks, then, are classified according to the size of the particles making up the rock (Table 17-1). This system has the advantage that almost all the terms used are everyday words and that the size ranges are close to those in common usage. The term *clay* as used in the

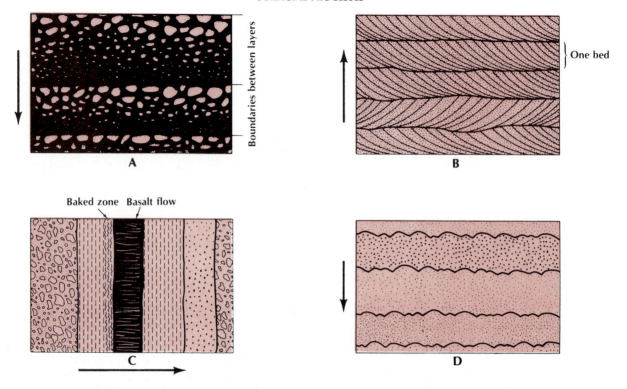

**Fig. 17-16.** Various ways to tell which way is up. For a true sense of direction tilt the book one way or the other. In each case the arrow points in the original "up" direction. **A.** Graded beds, with pebbles and sand shown in white; silt and clay in black. The sequence is upside down because each graded-bed layer is coarse at the bottom, gradually becoming finer at the top. **B.** These cross-bedded sediments are right-side up because the base of each younger bed truncates the cross-beds, which are concave upward. **C.** Basalt flow interbedded with sediments. Because the flow bakes only the sediments beneath it, the up direction is to the right. If the basalt were a sill rather than a flow, could you determine up from down? **D.** Beds of sand with oscillation ripple marks. Because such marks are concave upward when formed, the sequence is upside down.

**Table 17-1** Classification of Clastic Sediments and Sedimentary Rocks

| Sediment | Particle Term | Smallest Grain Size in Class (in mm) | Rock |
|---|---|---|---|
| Gravel | Boulder | 256 | |
| | Cobble | 64 | Conglomerate |
| | Pebble | 4 | |
| | Granule | 2 | |
| Sand | Very coarse | 1 | |
| | Coarse | 0.5 | |
| | Medium | 0.25 | Sandstone |
| | Fine | 0.125 | |
| | Very fine | 0.0625 | |
| Mud | Silt | 0.0039 | Shale or mudstone |
| | Clay | | |

classification refers only to particle size. It should not be confused with the term *clay mineral,* which refers to very small minerals with micalike structures (some of which are described in chap. 3), most of which are of clay size.

Particle size can be related to the energy of the transporting medium. Large gravels, for example, are moved by glaciers, swift rivers, and debris-laden mudflows. Less energy is required to move sand; so sand is common to sand dunes and to some beaches and rivers. Because silts and clays settle so slowly, they commonly settle out only in quiet water, thus their association with lake and marine environments.

**Conglomerates** *Conglomerates* are cemented, rounded gravels. The larger fragments may range in size from boulders several meters in diameter to particles the size of sand. More often than not, the interstices, or pore spaces, between the larger boulders, cobbles, or gravel are filled with sand or mud; the whole mass of sediment is then cemented together to form a single rock (Fig. 17-17).

*Breccia,* a variety of conglomerate, is made up of angular rather than rounded fragments. The same word is used for pyroclastic volcanic rocks (see chap. 8). The adjective *sedimentary* or *vol-* *canic* is usually added to indicate its origin. The fragments in breccias either have been transported only a short distance or have been carried in a medium that does not result in rounded clasts, such as a mudflow.

**Sandstones** The sedimentary rocks called *sandstones* consist of cemented particles with a diameter between 2 mm and 0.0625 mm. Commonly, they include shale layers or lenses of conglomerate.

Pure, well-sorted sandstones are often used as

**Fig. 17-17.** Conglomerate forms this hanging rock at the foot of Echo Canyon in Utah, photographed by Andrew Joseph Russell about 1868. (Beinecke Rare Book and Manuscript Library, Yale University)

a building material. Many college campuses are adorned with examples of academic Gothic hewn from sandstone blocks. The White House and the Capitol are both built of sandstone quarried a short distance down the Potomac from Washington, D.C. In the Victorian Era—especially the General Grant period—one of the favorite construction materials was the so-called brownstone—a brownish-red sandstone. Many of Europe's celebrated landmarks are made of clastic sedimentary rocks—the castles at Heidelberg and Salzburg and most of the great ducal palaces of Florence are but a few. Sedimentary rocks were greatly preferred over granite by builders in the past because such stratified rocks split more readily along their bedding planes and thus could be worked far more easily with the hand tools of the time.

The cement determines the degree of induration, or hardness, of sandstones and, therefore, how well they perform as building stone. In some sandstones the cement is weak and individual grains separate readily from their neighbors; in others the cement may actually be tougher than the material it holds together, thus when the rock breaks, it breaks across the grains.

Compositional differences affect the appearance of sandstone. Among the many kinds of sandstones, two common varieties are **arkose** and **graywacke.** Arkoses are made up largely of quartz and feldspar grains and are commonly reddish to pale gray or buff. They result, typically, from the erosion of granitic rocks. They can grade into a quartz sandstone if the more durable quartz grains are all that survive weathering and transportation to the site of deposition.

Graywackes (from the German *Grauwacke,* meaning gray stone) were originally named for distinctive sandstones in the Harz Mountains of Germany. They are darker than arkoses, and although they commonly contain quartz and feldspar minerals, they have a much higher content of rock fragments—chiefly of the darker varieties of igneous and metamorphic rocks. The sand-size particles in many instances are set in a clayey or silty matrix, which was essentially a muddy or clayey paste at the time of deposition. Some graywackes appear to have been deposited in the sea close to a steep mountain range. Muddy water carrying a large volume of sediments, including sand, was moved but a short distance from its source and deposited so rapidly that sorting is not well expressed.

**Shale**   The original constituents of **shale,** a fine-grained rock, were clay and silt particles that form a typically laminated rock that splits readily into thin layers. Because shales are composed of clay grains and of individual mineral grains or rock particles less than 0.0625 mm in diameter, few of the constituents can be distinguished by the unaided eye. Under the microscope they can be resolved; there we can see that most shales are made of minute grains of quartz, feldspar, and mica and of larger rock fragments along with the ubiquitous clay. Nearly one-half of all sedimentary rocks are shale.

Many shales are shades of dark gray or even black, especially if they contain organic matter. Others are dark red or green or particolored, depending on how much iron or other kinds of pigment they contain.

Although fissility, or the ability to split along well-developed and closely spaced planes, is an important property of shales, it is by no means characteristic of all shales. Some varieties with comparable composition and grain size are not fissile at all but break in massive chunks or small compact blocks (Fig. 17-18). They are best given the descriptive name of mudstone.

### Precipitated Sedimentary Rocks

In addition to the clastic rocks consisting of fragments and mineral grains derived from preexisting rocks, there is a second large group of sedimentary rocks, of chemically precipitated materials. Chemically formed rocks are discussed according to their composition and their mode of origin. Such an approach can be confusing because some varieties of rocks—specifically the carbonates—may be similar in composition but not in origin. Thus, of necessity, the same term appears more than once in the classification.

**Evaporites**   Rocks or minerals that result primarily from the evaporation of water containing dissolved

**Fig. 17-18.** Factory Butte on the Colorado Plateau in Utah is capped by cretaceous sandstone but underlain by massive bedded shales. (Eldon L. Byland)

solids are known as *evaporites.* As the water becomes saturated, the ions precipitate out of solution to form a crystalline residue.

Halite, or common table salt (sodium chloride, NaCl), is the most familiar evaporite mineral. Commonly, it is formed when evaporation in an arm of the sea occurs faster than the inflow of water.

From earliest times salt has been a highly prized commodity. Today we take it for granted, but in ancient times people gave their lives in battle to win control over salt deposits or to seize the trade routes over which salt was transported. Famous among historic deposits were those of northern India—the center of a flourishing trade before the time of Alexander—and those of ancient Palmyra in Syria, from whence salt moved by caravan to the Persian Gulf. The salt mines of Austria are also famous, and in the Salzkammergut region around Salzburg they were in operation at least as early as 2000 B.C.

*Gypsum* ($CaSO_4 \cdot 2H_2O$) is closely related to halite in origin, for it, too, is a product of the evaporation of seawater. Along with it is found an anhydrous (water-lacking) salt, calcium sulfate ($CaSO_4$), called *anhydrite.*

As a body of water evaporates, the salt minerals precipitate out in a specific order. The order is determined by the solubility of the minerals in water—the less soluble first, the more soluble, which stay in solution the longest, last.

Gypsum is less soluble than halite and thus precipitates first when seawater is evaporated. However, great quantities of water must evaporate before either mineral forms. Both gypsum and anhydrite come out of solution after about 80 percent of the seawater has evaporated; halite appears after 90 percent has evaporated.

There are many other kinds of evaporites, of minor significance in amount, but highly significant economically. Among them are borax ($Na_2B_4O_7 \cdot 10H_2O$) and potash ($KCl$)—both of which are found in lakes or ancient lake deposits—of arid regions such as the Mojave Desert in California. Years ago picturesque 20-mule teams hauled those salts across the rough desert floor in large wooden boxcars (Fig. 17-19).

Some salt deposits are extremely thick—in places several hundred meters. There is a vexing problem concerning such deposits: The evaporation of 300 m (984 ft) of seawater will produce a salt bed about 4.5-m (15-ft) thick, of which 0.1 m (0.3 ft)

**Fig. 17-19.** Twenty-mule team on its way to the Lila–C mine in 1892. (U.S. Borax)

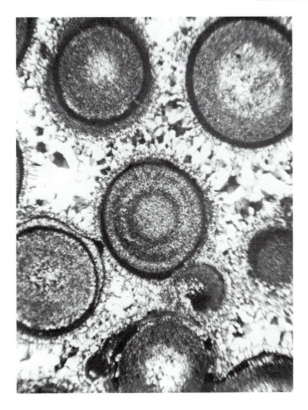

**Fig. 17-20.** Microscope photograph of a thin section of oolite from Cambrian sediments in Pennsylvania. Although silica has replaced the original carbonate, the original layering of the ooliths is well preserved. (N.C. Nielson)

should be gypsum and anhydrite, 3.5 m (11.5 ft) halite, and the remaining 1 m (3 ft) potassium- and magnesium-bearing salts. Do several hundred meters of salt deposits represent the evaporation of thousands of meters of water without a fresh supply being added? Hardly, for that would require the evaporation of the deeper oceans, an event that certainly has not taken place.

How then, do we explain the great thicknesses of evaporite deposits? Consider the fact that gypsum and anhydrite make up strata many hundreds of meters thick in west Texas and New Mexico. A great amount of seawater must have been evaporated there in the geological past. But what took place was not simple evaporation because the extensive bodies of sodium chloride that should be associated with the gypsum beds are absent.

An explanation advanced by the American geologist Philip B. King is that water in a shallow, sun-warmed lagoon reached the temperature and concentration at which calcium sulfate precipitated out and settled on the floor of the lagoon. As seawater was continually replenished, the precipitation process continued. Were such a basin a subsiding one, a thick layer of evaporites could accumulate without the water necessarily being deep at any time.

Salt beds have been studied as possible repository sites of high-level radioactive wastes in the United States. Initially, they were thought to be highly impermeable, but recent studies suggest that the beds are relatively mobile and that wastes could leak from the repository sites and contaminate groundwater supplies (see chap. 18).

**Carbonate rocks**   Rocks known as *carbonates* are made up of calcium or magnesium compounded with carbonate, generally in the form of calcite ($CaCO_3$) or dolomite ($CaMg(CO_3)_2$). If calcite predominates, the rock is called *limestone;* if dolomite predominates the rock is called *dolomite.* The latter term is virtually the only one used to designate both a mineral and a rock. Limestones are for the most part organic in origin, but a few are true chemical deposits.

Seawater is very nearly saturated with calcium carbonate ($CaCO_3$). This means that very slight changes in the temperature of the water or in its chemical composition can precipitate calcite. Usually, the initial mineral precipitated is *aragonite,* an unstable form of $CaCO_3$, which in time converts to calcite.

There seems to be little doubt about the inorganic origin of one curious type of limestone known as *oolite.* It is made of minute spherical grains of calcium carbonate the size of fish roe, called *ooliths,* from the Greek words *oo* ("egg") and *lithos* ("stone") (Fig. 17-20). The grains form as layers of calcium carbonate around a nucleus—perhaps in much the same way that layers of pearl shell are built up. A well-formed oolith is the result of grains rotating again and again in a bottom current, and a logical place for such tumbling would be in a tidal area. Common environments of present-day formation are the coasts of Florida and the Bahamas.

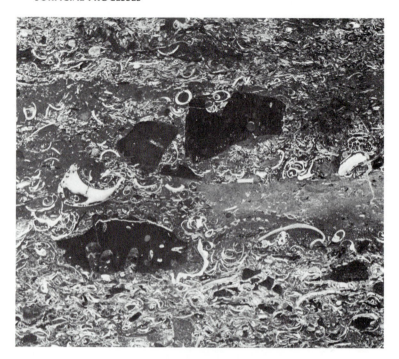

**Fig. 17-21.** Polished slab of shelly limestone probably deposited as a submarine mudflow generated by an earthquake several million years ago in the Los Angeles Basin, California. (Janet Robertson, Geological Society of America Collection)

*Travertine* is a good example of a limy deposit that appears to have been deposited by spring waters saturated with calcium carbonate. Travertine has been greatly favored as an architectural material because it is soft and readily worked, with an interesting array of colors—generally pale yellow or cream, if pure; brown and darker yellow if it contains impurities. *Tufa* also forms in springs and lime-saturated lakes, and its deposition seems to be fostered to some degree by lime-secreting algae. Tufa and travertine when cut and polished were until recent times much favored for the lobbies of banks and the large railway terminals of a past era. Much of monumental Rome is built of tufa, including Bernini's columns that nearly encircle the piazza in front of Saint Peter's.

The origin of dolomite has been much debated. The argument centers around whether dolomite forms as a primary precipitate from seawater or whether it forms from calcite or aragonite. In some places, dolomite masses cut across limestone layers, evidence for the partial replacement of calcite in the limestone by magnesia-bearing solutions. In other places, however, dolomite occurs as widespread layers interlayered with limestone strata.

The origin of such deposits is less clear cut. Some geologists believe that the dolomite was precipitated directly on the seafloor. Others take the view that the dolomite layers represent selectively replaced layers of limestone.

The favored theory at present states that dolomite is not precipitated directly from water. In support of this theory, it should be noted that dolomite is very rare in modern sediments. Also, studies of a desert lake have shown that calcite forms first, probably being converted to dolomite over the decades or centuries it rests on the lake bottom.

### Organic Sedimentary Rocks

Rocks made of the remains of organisms, both animals and plants, are called organic sedimentary rocks.

**Limestone**   The most abundant organic sedimentary rock is limestone; probably most limestone is truly organic and not precipitated. In a strict sense limestones made up of shell debris that has been reworked by currents should be classified as clastic sedimentary rocks (Fig. 17-21). Of limestones from large fossils and fossil fragments, perhaps the most impressive are the reefs.

Modern reefs are made up largely of corals and carbonate-secreting algae (Fig. 17-22). In order to grow so profusely, they require an environment free of most land-derived sediment and shallow, warm water that is both agitated by wave action and high in nutrients. These conditions are met today off many coasts within about 30° north or south of the equator (Fig. 17-23). The reefs most familiar to most North Americans form the Florida Keys, the Bahama Islands, and part of the Hawaiian Islands. But the most impressive reef is Australia's Great Barrier Reef (Fig. 17-24). That great carbonate barrier lies just off the coast, stretches a distance of 2300 km (1425 mi), contains more than 2100 individual reefs, and was the scourge of early sea captains. Witness the experiences of Captain James Cook and his crew:

June, 1770, found him cautiously sailing northwards in his tiny ship, the 70-foot *Endeavour Bark*. The hazardous journey ultimately led to parts close to those discovered by the Spanish explorer, Luis Vaes de Torres (1605). Going finally ashore on a sea-girt spot of land near the tip of Cape York, Cook named this Possession Island, and there took formal possession of the east coast of Australia for Britain.

Never before had a ship sailed so dangerous and unknown a sea as that skirting the Queensland mainland. Cook found the waters dotted with islands, shoals and coral banks. His way was through twisting passages and shallows into a strange world of mystery and beauty. He was for a long time unaware of a great barrier to seawards that was closing in upon his track. At a spot near the present site of Cooktown, the coral banks crowded in on his ship in baffling array. She finally ran

**Fig. 17-22.** This patch reef community at Mosquito Bank, Florida, is characterized by a wide variety of coral and harbors a diverse and beautiful assemblage of reef fish. (H.G. Multer)

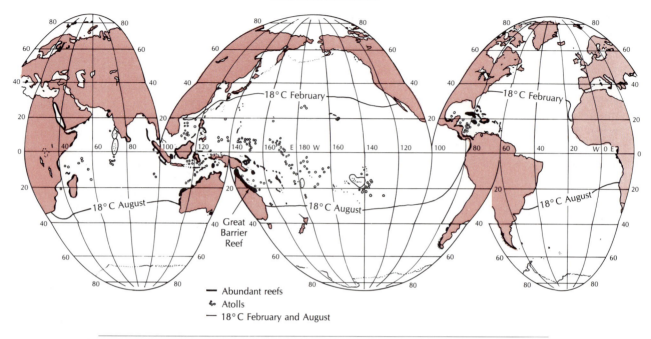

**Fig. 17-23.** Present coral reefs and atolls around the world. The contour lines bound the waters that are never colder than about 18°C (64° F). (After Gross, 1977, *Oceanography: A View of the Earth, 2nd Ed.,* Prentice-Hall, p. 74, Fig. 3–14)

aground and was all but lost on a treacherous patch. The stirring story of that mishap and the masterful saving of ship and crew is one of the highlights of Australia's early history.*

Other limestone deposits consist of the heaped remains of microscopic limestone fossils—commonly minute, free-floating, single-celled animals—the **foraminifera.** A striking example is the chalk deposits that make up the cliffs near Dover, England.

**Coal and Oil Shale** Coal is an excellent example of an organic sediment that consists of plant remains that accumulate in a low-oxygen swampy area. It can be thought of as a somewhat metamorphosed sediment. In such an environment vegetation accumulates to form **peat,** a brown, soft organic deposit in which plant remains are easily recognized. When peat is buried under younger sediments, pressure and heat start the process of

*From K. Gillett and F. McNeill, 1959, *The Great Barrier Reef and Adjacent Isles,* Coral Press, Sydney, Australia.

coal formation. In the first step a brownish material called **lignite** is formed. Subsequently, the carbon content of lignite increases, and black **bituminous** coal and **anthracite** coal (the highest quality product) form. A good example of a present-day swamp that will eventually become a coal deposit is the Dismal Swamp of Virginia and North Carolina.

Most of the world's coal deposits were laid down during the Carboniferous Period beginning about 350 million years ago and lasting about 100 million years. In the eastern United States individual coal seams can be traced for hundreds of kilometers, indicating the extent of the swamps of that period. Many of the eastern coal deposits are closely associated with sediments that reveal at least 50 cycles of marine and nonmarine sedimentation, cycles that may record worldwide changes in sea level because there is evidence for widespread glaciation at that time. It seems paradoxical that alternating cooling and warming cycles in the Carboniferous Period set the stage for the production of the material that keeps us warm today!

In recent years oil shale has attracted wide

attention as a new source of oil in the United States. In parts of the Rocky Mountains there are extensive deposits of oil shales laid down some 50 million years ago. In that ancient environment plant and animal life flourished. Subsequently, it was incorporated into sediments; in time the organic matter in the sediments was converted to **kerogen,** which on distillation and refining yields oil. The deposits are well bedded and contain a wide variety of fossils, the most noted of which are well-preserved fossil fishes.

**Siliceous Rocks**   The most widely occurring siliceous rock is **chert,** a name used to cover a host of varieties of very dense, hard, nonclastic rocks made of microcrystalline silica. One familiar form is **flint,** which is dark-colored because it contains organic matter. Because flint is uniformly textured, has a conchoidal fracture much like obsidian, and is easy to chip and at the same time retains a sharp edge, it proved to be the ideal strategic material for arrowheads and spear points in the Stone Age of Europe and the United States. Red varieties of the same rock commonly are called **jasper,** the color being derived from the iron oxide ($Fe_2O_3$) they contain.

By volume the most important bodies of chert are thick and well bedded (Fig. 17-25). Although some of them may have precipitated directly from the sea, thick cherts are most likely to be organic. Microscopic marine animals, such as the *radiolaria* and the **sponges,** and plants, such as the *diatoms,* build their shells from silica extracted from seawater (Fig. 17-26). In an environment in which both carbonate shells and clastic material are excluded, a thick deposit of chert can build up in time.

Two other sources of siliceous deposits, although minor, should be mentioned. One is the direct precipitation of silica from hot springs, for example, the pedestals at the bases of active geysers in Yellowstone National Park. Such deposits, called **sinter,** are spongy and porous. Another source has to be sought for the irregularly shaped nodules of chert contained in limestones. Because such deposits cut across the bedding planes, they clearly replaced part of the limestone at a later date. The process is similar to that in which silica replaces woody fibers to form petrified wood.

## SEDIMENTARY FACIES

Now that the major sedimentary rocks have been described, we should issue a warning—the basic properties of any rock unit probably will change as the unit is traced laterally. For example, suppose we are examining a thick layer of conglomerate in a canyon wall. Over a distance of several kilometers, the layer might change gradually to a sandstone and then to shale. The reasons are fairly simple: In any depositional basin such conditions as sediment supply, current velocity, and biological by-products can vary laterally. The lateral differences in rock types within a single unit are called **facies changes;** hence, as depositional conditions vary, so do the resulting sediments (Fig. 17-27).

The concept of facies was exceedingly important to the early geologists; without it they would not have been aware that a limestone sequence in one place was the same age as a sandstone sequence 100 km (62 mi) or so away. It is hard enough to follow facies changes where outcrops are clear and rocks are not deformed. Consider the plight of the geologist working with strongly deformed sediments in forested terrain with few rock outcrops. Only with hard work and a good imagination can such a rock record be unscrambled. In order to understand facies more fully, a brief description of depositional environments is in order.

## DEPOSITIONAL ENVIRONMENTS

Sedimentary rocks accumulate in a wide variety of environments—about as many as there are different kinds of landscapes or climates. Most such environments occupy the lower parts of the landscape—the parts to which material moves from higher areas. The two major realms of sedimentation are the sea (marine) and the land (continental).

Chapters 11 through 16 describe these environments in some detail. Here we will concentrate on the sedimentary rocks characteristic of each.

A major problem in determining depositional environments is that some rocks can be formed in any of several environments, for example, silts and clay could indicate a river floodplain, but they

**Fig. 17-24. A.** The Great Barrier Reef of Australia is one of the finest examples of a barrier reef.

A

could also indicate marine surroundings. Often, a field geologist must examine miles of valley walls or mountainsides before the identity of an environment can be established. Further, when we consider how widespread the processes of erosion are, it is not unreasonable to suppose that the rocks providing the best evidence of a depositional environment might have been eroded away. It is an excellent example of multiple working hypotheses to keep one's mind open and to keep seeking more clues.

## Continental Deposition

The ways in which sediments may be trapped on land are familiar to many of us. In the passages

**Fig. 17-24. B.** Satellite image of the Great Reef near Cook's Passage. (NASA)

**Fig. 17-25.** Well-bedded chert in the Cook Inlet region, Alaska. Originally horizontal, the chert has been subsequently folded and a fault trends diagonally across the outcrop from the man's right foot. (G.K. Gilbert, USGS)

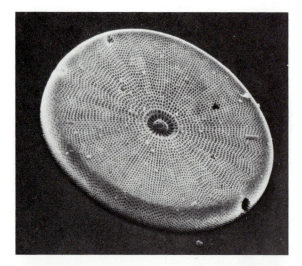

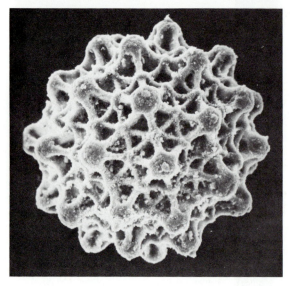

**Fig. 17-26.** Scanning electron micrographs (SEM) of fossil marine organisms from sediments from the California coast. Top fossil is a diatom (AIME). Bottom fossil is a radiolaria. (Emile A. Pessagno, Jr.)

that follow the main kinds of environments and deposits will be discussed, starting with mountainous terrain and progressing toward the seacoast.

**Glacial Deposits** Glaciers, most of which today are confined to higher mountains and to the Arctic and Antarctic, were once more widespread; *gla-cial deposits* blanket much of North America and northern Europe. A good example of a typical continental glacial deposit is ***boulder clay,*** which is literally that—rock the size of boulders set in a clayey matrix.

Poor sorting is a major characteristic of glacial deposits, but such deposits can be confused with other poorly sorted deposits. Ideally, the bedrock on which a glacial deposit rests will be polished and striated.

**Flood Plains** Large amounts of material usually are deposited along the margins of rivers—the flood plain—and in the river channel. In some places, especially in the channels of braided streams (Fig. 17-28), much of the material is gravel. In contrast the deposits of meandering streams grade laterally from sand in the channel to clay and silt on the flood plain. If such a river is aggrading (building up deposits) one sees isolated channel deposits at various levels set in the finer-grained overbank deposits (Fig. 17-29). Generally, the clasts in the river deposits can be of any size, and sorting and rounding vary from moderate to good.

**Lakebeds** Lakes are prime sediment traps. Glaciers, rivers, and wind action all move sediments into lakes. However, lakes are a relatively temporary feature by the geological time scale, being formed mainly by glaciers, landslides, volcanic activity, and faulting. Eventually, they fill in with sediment. Lake sediments generally are well sorted and layered.

**Deltas** *Deltas* form where sediment-laden streams enter bodies of relatively still water, be they fresh or marine. Sediment sorting is good, and ideally, the topset-foreset-bottomset configuration (see Fig. 17-11) will indicate a deltaic environment. Commonly, these deposits grade laterally to beach deposits and to fine-grained sediments of the deeper waters offshore.

**Desert Basins** Although not restricted to arid environments, alluvial fans literally flourish in them. Clastic sediment of all sizes is flushed from the mountain valleys and deposited as a fan where the stream enters the basin (Fig. 17-30). Poorly sorted and angular gravel is deposited at the upper

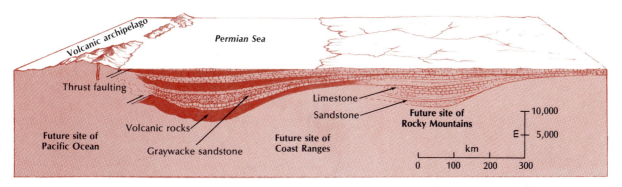

**Fig. 17-27.** Reconstruction of facies relationship of Paleozoic rocks in southeastern Alaska and British Columbia at the close of the Permian Period. Notice that the thickness of the sediments varies. Also, the dominant volcanic rocks and graywacke sandstones to the west grade eastward to limestones and sandstones with distance from the volcanic archipelago. (After A.J. Eardley, 1951, *Structural Geology of North America*, Harper & Bros., Fig. 21)

**Fig. 17-28.** The present flood plain of the Rakaia River, New Zealand, is the lighter-colored gravel strip adjacent to the river. Overall the flood plain makes up one-third to one-half of the valley floor, the rest being a slightly higher terrace—the former flood plain. (Peter W. Birkeland)

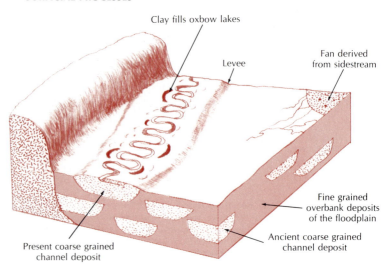

**Fig. 17-29.** Drawing showing facies relationship between channel and overbank deposits of an aggrading meandering stream system. (After A.T. Allen, 1965, "A Review of the Origin and Characteristics of Recent Alluvial Deposits," *Sedimentology*, Vol. 5, p. 89+)

end of the fan and the smallest material at the outer margins of the fan.

Temporary lakes are common in **desert basins;** their deposits range from silts and clays to salts left behind as the last waters evaporate. In time a basin accumulates a thick deposit: a coarse alluvial fan near the mountains, grading to lake silts and clays and finally to salts in the center of the basin (Fig. 17-31).

Sand dunes are also common in many desert environments. These deposits are characterized by excellent sorting and unique, sweeping bedding features (see Fig. 17-10). As in the other sedimentary environments, deposits of one kind can grade laterally into others, for example, dune sands can grade laterally into the deposits of desert lakes.

### Marine Deposition

Some of the factors affecting the distribution of sediment in the sea are (1) distance of the deposit from land, (2) depth of the water, and (3) the physical and chemical properties of the water and the types of plant and animal life.

Consider moving seaward from a beach. On some shelves gravels occur in the "high-energy" area where the waves are hitting the shore, but farther offshore on the shelf where the deeper and quieter waters move smaller particles, sands and, eventually, silts and clays are found. At all localities sediments are well sorted and the sands and larger materials are well rounded.

Other shelves, however, are the sites of deposition of calcium carbonate ($CaCO_3$) mud. Still others, including many ocean islands, are characterized by coral reefs. Three main reef zones are commonly recognized (Fig. 17-32). One is the reef itself, which is formed of organisms building upward from the shallow seafloor to sea level. Such creatures can flourish in the face of the prevailing winds, for these winds drive currents that bring the nutrients needed for growth. Waves constantly crash against the reef, breaking off pieces of the delicate organisms and forming an apron of steeply dipping organic debris—a sort of submarine talus of reef material on the windward side. Landward of the reef is a quiet lagoon in which carbonate and clastic muds accumulate.

Finally, far from shore on the deep-ocean floor is the repository of deep-sea clay and biological oozes. Only under special circumstances can rel-

**Fig. 17-30.** Alluvial fans flank the ranges in the Mojave Desert, California, and grade to the white saline deposits of temporary lakes in the intermontane basins. (USAF)

557

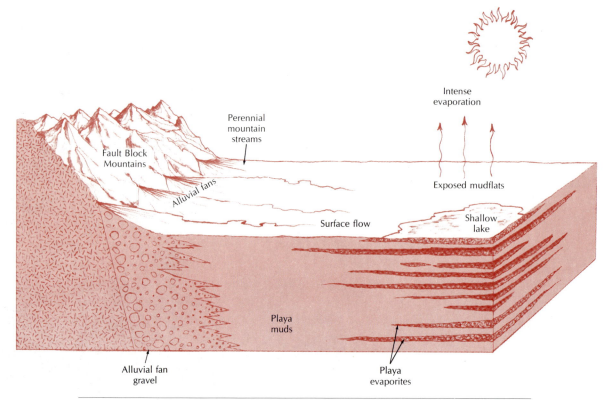

**Fig. 17-31.** Drawing showing facies relationship between alluvial fan deposits and deposits of temporary lakes in a desert basin. Evaporite deposits form when the lakes dry up completely or are quite small. Under appropriate conditions sand dunes could move across the exposed mud flats and some dune deposits could be interbedded with the playa-lake deposits. (After H.P. Eugster and L.A. Hardie, 1975, "Sedimentation in an Ancient Playa-lake Complex: The Wilkins Peak Member of the Green River Formation of Wyoming," Geological Society of America Bulletin, Vol. 86, p. 319+)

**Fig. 17-32.** Diagram of the zones commonly associated with an offshore reef. (After C.O. Dunbar, 1960, *Historical Geology*, John Wiley, Fig. 215)

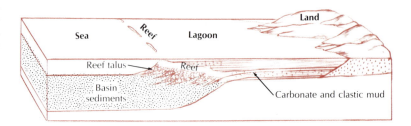

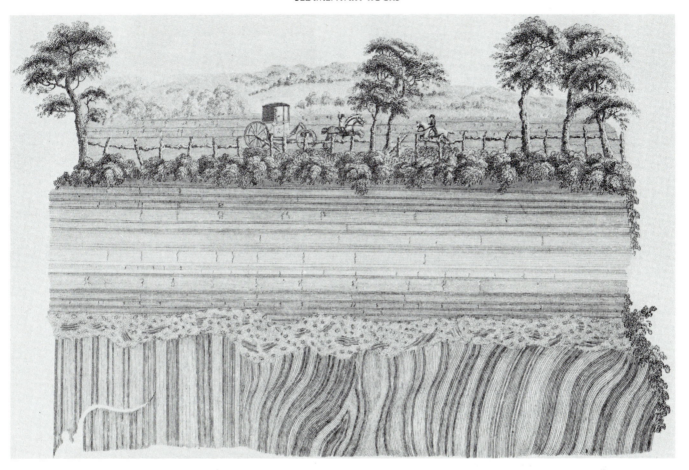

**Fig. 17-33.** An eighteenth-century engraving showing a distinct angular unconformity in which horizontal beds rest on vertical and near-vertical beds. The sequence of events is as follows: (1) horizontal deposition of lower beds, (2) tilting of beds during an episode of mountain building, (3) erosion of the beds to form a relatively flat surface, and finally (4) deposition of younger, flat-lying beds. The unconformity shown here could represent millions of years.

atively coarse-grained, land-derived sediment intrude this realm. One is when turbidity currents traveling from the continental slope deposit material; the other follows release of sediments by melting icebergs.

## UNCONFORMITIES

Sedimentation is not a continuous process—commonly, it is interrupted by other environmental changes. We would expect the oceans, for example, to be areas of continuous sedimentation, but the seas shift and change—what was once an area of marine deposition can be raised above sea level for a time and erode, to be followed at a later date by another cycle of marine deposition. Fairly long-term gaps in the sedimentation record of rocks, called **unconformities,** usually occur at a time of erosion, which is often associated with uplift in a particular region.

Several kinds of unconformities are recognized. By far the easiest to identify is an **angular unconformity,** in which the beds above and below the unconformity are not parallel but meet at an angle (Fig. 17-33). Obviously, to arrive at this geometric pattern, the older beds must themselves have been deformed during a previous upheaval, subsequently eroded, and finally buried by younger sediments. In a **disconformity** the rock layers are parallel above and below the unconformity, but the upper surface of the lower beds can have considerable relief (Fig. 17-34). Finally, a **nonconformity** is characterized by sedimentary rocks resting on igneous or metamorphic rocks (Fig. 17-34); obviously, much time, erosion, and geological history are represented where such contrasting rocks are seen together.

Unconformities are usually localized. The three types can occur alone or blend laterally with one another, even laterally to areas where sedimentation occurred without any signs of interruption. Being of regional extent, they can aid in the correlation of widely separated rock units.

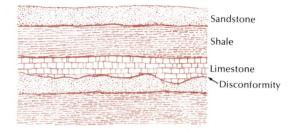

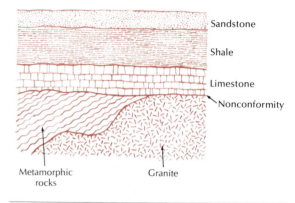

**Fig. 17-34.** Examples of a disconformity and a nonconformity.

## SUMMARY

1. Sedimentary rocks result from the weathering, erosion, transport, and deposition of any preexisting rocks. Rock and mineral particles that result from weathering make up the *clastic* sedimentary rocks; the ions from weathering recombine to form the various sedimentary rocks that are known as *chemical precipitates. Organic* deposits form the third major subdivision of sedimentary rocks.

2. The source areas for sediments are usually mountains or hills, whereas the desert basins, river flood plains, lakes, and oceans are the common depositional sites.

3. The sediments have a set of field characteristics that can be used to identify the environment in which they formed. Such characteristics include stratification, sorting and roundness of grains, cross-bedding fossils, mud cracks, and ripple marks. In many environments the major stratification layers are horizontal.

4. Once sediments are deposited, *calcite* or *silica* may precipitate from the circulating groundwater and cement the grains together to form sedimentary rock.

5. In classifying rocks the primary criterion for clastic sediments is grain size; for chemically precipitated rocks, it is composition. *Organic sediments* are the products of the accumulation of plants or animals, and the common organic sediments are limestone, coal, and oil shale.

6. The concept of sedimentary facies is essential to interpreting the origin of sedimentary rocks. In short, it means that the characteristics of sediments laid down in a specific time period will change laterally because the environments of deposition change laterally. Hence, stream deposits may grade laterally to clastic marine deposits and then to carbonate reef deposits.

7. Unconformities record a time when no sediments are deposited; most indicate long periods of erosion.

SEDIMENTARY ROCKS

# QUESTIONS

1. What are the arguments against using the mineral content of sandstones or the rock type of conglomerates as the major criteria for classifying the clastic sedimentary rocks?
2. List all properties useful in differentiating river deposits from sand dune deposits, lake deposits, and mudflows.
3. List the features of sedimentary, plutonic, and igneous rocks that can be used to determine the up direction in a sequence of rocks.
4. Do extremely thick sequences of rocks—for example, shales or evaporites—mean that the sea in which they were deposited was deep?
5. Examine all the ways in which a limestone rock can form from a reef complex offshore to the deep ocean.
6. What sedimentary facies can you recognize in Figures 12-26 (the Nile), 13-11 (Death Valley), and 14-1 (a glacier margin)?

# SELECTED REFERENCES

Blatt, H., Middleton, G., and Murray, R., 1980, Origin of sedimentary rocks, 2nd ed.: Prentice-Hall, Englewood Cliffs, N.J.

Eicher, D.L., and McAlester, A.L., 1980, History of the earth: Prentice-Hall, Englewood Cliffs, N.J.

Gillett, K., and McNeill, F., 1959, The Great Barrier Reef and adjacent isles: Coral Press, Sydney, Australia.

LaPorte, L.F., 1968, Ancient environments, Prentice-Hall, Englewood Cliffs, N.J.

Reader's Digest, 1984, Reader's Digest book of the Great Barrier Reef: Readers Digest services, Sydney, Australia.

Reading, H.G., Sedimentary environments and facies, 2nd ed., 1986, Blackwell Scientific Publications.

**Fig. 18-1.** Roman aqueduct on the Mediterranean Sea coast, Caesarea, Israel. (Peter W. Birkeland)

# 18

# GROUNDWATER

In Xanadu did Kubla Khan
A stately pleasure-dome decree:
Where Alph, the sacred river, ran
Through caverns measureless to man
    Down to a sunless sea.

Samuel Taylor Coleridge (1772–1834) here reflects the remarkable image that many people have of water within the earth. Some of us are prone to speak glibly of underground rivers flowing for miles beneath the parched surface of deserts and to some of us springs are nearly as mysterious as they were to people long ago.

Springs played a leading role in Greek and Roman mythology. An example is the spring of Arethusa, which appears on the island of Sicily in the ancient harbor of Syracuse. The river god Alpheus, in pursuit of the wood nymph Arethusa, flowed as an underground river all the way from Greece to Sicily, where he finally caught the elusive spirit and changed her into a spring.

Comparable beliefs were held in those early days. Generally, there were two leading schools of thought. One held that springs drew their water from the sea. The other belief was that springs and streams had their own origin within subterranean caverns, large enough perhaps to have atmospheres of their own from which water condensed as a sort of rain to feed them. Aristotle was content with that idea because rainfall was obviously inadequate:

The air surrounding the earth is turned into water by the cold of the heavens and falls as rain, [so] the air which penetrates and passes into the crust of the earth also becomes transformed into water owing to the cold which it encounters there. The water coming out of the earth unites with the rain water to produce rivers. The rainfall alone is quite insufficient to supply the rivers of the world with water. The ocean into which the rivers run does not overflow because, while some of the water is evaporated, the rest of it changes back into the air or into one of the other elements.

It is now well accepted that the primary source of groundwater is rainfall and that groundwater feeds springs, but it was not until the seventeenth century that the connection was established by the work of two Frenchmen, Pierre Perrault (1611?–80) and Edme Mariotte (1620–84). The hydrologist Oscar Meinzer tells of their discoveries:

Perrault made measurements of the rainfall during three years, and he roughly estimated the area of the drainage basin of the Seine River above a point in Burgundy and of the run-off from this same basin. Thus he computed that the quantity of water that fell on the basin as rain or snow was about six times the quantity discharged by the river. Crude as was his work, he nevertheless demonstrated the fallacy of the age-old assumption of the inadequacy of the rainfall to account for the discharge of springs and streams. . . .

Mariotte computed the discharge of the Seine at Paris. . . . He essentially verified Perrault's results. . . . He maintained that the water derived from rain and snow penetrates into the pores of the earth and accumulates in wells; that this water percolates downward till it reaches impenetrable rock and thence percolates laterally; and that it is sufficient in quantity to supply the springs. . . . He also showed that the flow of springs increases in rainy weather and diminishes in time of drought, and explained that the more constant springs are supplied from larger underground reservoirs.*

In the past, considerable use was made of water from the ground. Wells were the center of village life for centuries and were absolutely essential to the survival of a walled city or castle.

The demand for water created by large concentrations of people in urban centers has resulted in the extensive development of water-transport systems. Almost everyone knows that the Romans brought water to their cities by building imposing, valley-spanning aqueducts (Fig. 18-1). Oddly enough, the Romans knew little about water in the ground. They placed their chief dependence on springs and streams, going prodigious distances to the Apennines for water when they had a perfectly adequate supply almost directly underfoot—had they dug for it.

In the Mideast, people dug remarkable burrow-like excavations in their pursuit of water. The chief examples are the *kanats* of Iran (Fig. 18-2). They center largely around Teheran and for the most part are dug in the gravels of alluvial fans at the base of the Elburz Mountains. The Kanats, which serve as collection galleries, are long passages that follow a water-bearing layer of porous sand or gravel within the fan. The longest known passage is 80-km (50-mi) long and individual tunnels may be as deep as 300 m (984 ft) below the surface.

Some of these early wells were surprisingly deep. An outstanding achievement was the one dug at

Orvieto, Italy, which was sunk to a depth of 61 m (200 ft) in 1540. Two spiral staircases lined the walls, one above the other, with one used by descending and one by ascending water-bearing donkeys. Deep wells drilled at Artois, France, in the twelfth century and at Modena, in the Po Valley of Italy, resulted in water flowing out at the surface; this excited great interest because they were the first true artesian wells of medieval times. In China there were wells as deep as 1500 m (4920 ft).

In this chapter we will discuss the origin and occurrence of groundwater, ways to prospect for it, and landscape features that groundwater helps to create in limestone areas (Fig. 18-3). Hot springs and geysers—groundwater phenomenon as well as energy sources—and contamination of underground reservoirs will also be covered. Contamination of water supplies worldwide is increasing rapidly. The World Health Organization estimates that about fifty thousand people die each day from diseases associated with impure water, some of which is contaminated groundwater. And in the United States and Europe the siting of underground repositories for nuclear wastes has become a highly controversial issue, partly because of fear of radioactive pollution of groundwater.

## ORIGIN OF GROUNDWATER

Nearly all of the water in the ground comes from precipitation that has soaked into the earth. In addition, some water is included with marine sediments when they are deposited and some reaches the upper levels of the crust when it is carried there by igneous intrusions, volcanoes, and hot springs. However, the latter sources provide only a minor part of the total budget of usable water in the ground.

Many things happen to water that falls as rain or snow, as we saw in the discussion of the hydrologic cycle (see Fig. 12-2). For the United States, earth scientists Luna Leopold and Gordon Wolman have estimated that water is divided up as follows: There is an average of 76 cm (30 in.) of precipitation each year. Of that amount, approximately 53 cm (21 in.) is returned directly to the

*O.E. Meinzer, 1949, "Introduction" in O.E. Meinzer, ed., Hydrology, Dover, N.Y. pp. 14–15.

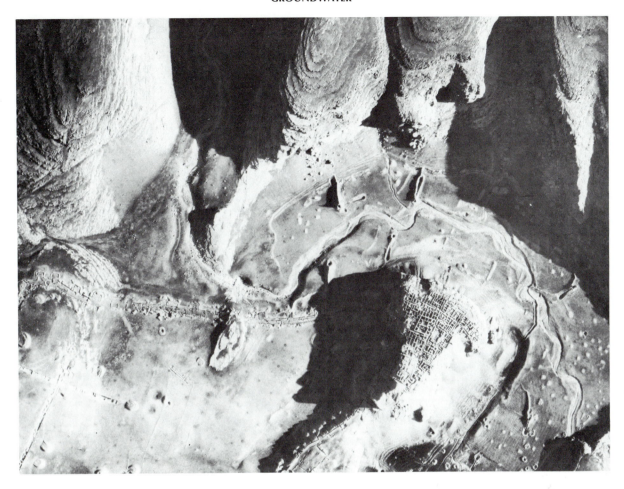

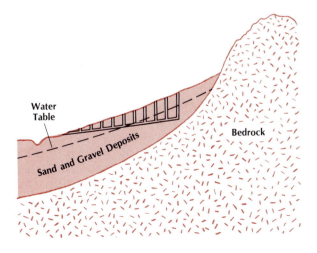

**Fig. 18-2. A.** *(Above)* Ruins of an ancient city along a dry riverbed in Iran. The circular mounds mark the tops of the vertical shafts of *kanats*, ancient underground water-transportation systems. (Oriental Institute, University of Chicago) **B.** *(Left)* Cross-sectional diagram of a kanat system. The vertical shafts and near-horizontal tunnels intersect the water table in the alluvium of the fans flanking the mountains and the tunnels guide the water to the surface. (After M. Bvbordi, 1974, diagram by G. Olson, Department of Agronomy, Cornell University)

**Fig. 18-3.** Chinese painting showing the beauty of the landscape in limestone terrain. (The Metropolitan Museum of Art, Purchase, The Dillon Fund Gift, 1979)

atmosphere by evaporation and transpiration. Only 23 cm (9 in.) runs off in streams directly to the sea; of the total runoff, nearly 40 percent escapes by the Mississippi River—an impressive fraction of the continental supply.

Where does groundwater come from, then, if the budget balances as closely as these figures indicate? The answer is that although the amount of water entering the ground by infiltration is slight—perhaps as little as 0.25 mm (0.01 in.) per year in some places, more in others—with the passage of many millennia, great quantities of water slowly accumulate in the ground. It is that vast reservoir built up gradually over thousands of years that we draw on today—unfortunately, in some areas more

rapidly than it is replenished. Where the latter situation exists, for example, in the Tucson–Phoenix area in Arizona and in parts of the Great Plains and California's Great Valley, the groundwater can be considered a nonrenewable resource that will be depleted some time in the future.

The total amount of groundwater in the world has been estimated at some 8.4 million km$^3$ (1.8 million mi$^3$). A glance back at Figure 12-2 indicates that the amount is more than that in the rivers and lakes at any one time but less than that in the oceans or glaciers. It is that great readily available abundance that makes groundwater so important to the development of farms, communities, and industries.

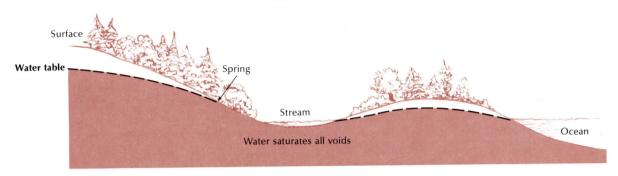

**Fig. 18-4.** Position of the water table in a humid region.

## OCCURRENCE AND MOVEMENT OF GROUNDWATER

Probably many of us have seen groundwater in shallow wells. If we were to determine the level of that water and then to compare it with the level in nearby wells, we would find that in many regions the water surface stands at about the same level. The surface at which water stands in wells is called the **water table.** All the voids, or openings, in unconsolidated sediments and rocks below the water table are filled with water or are saturated; this area is called the saturated zone. Above the water table, in the **zone of aeration,** or **vadose zone** (from the Latin, meaning "shallow"), the pore spaces in the ground may range from completely empty to partially full. In many places this water moves slowly downward through the zone of aeration to the water table.

Rarely is the water table dead level. Instead, in humid regions it is more likely to be a blurred replica of the ground surface, rising under hills and sinking under valleys (Fig. 18-4). It intersects the surface at lakes, streams and springs. Where groundwater contributes to the discharge of a stream, it is called an **effluent stream** (Fig. 18-5). However, if the stream flows above the water table and thus adds to the supply of water in the ground, it is an **influent stream** (Fig. 18-5). The latter are common in arid regions. The position of the water table related to an effluent stream is more or less stable. That beneath an influent stream, however,

**Fig. 18-5.** Position of the water table with respect to effluent and influent streams. Arrows depict the flow of water beneath the water table.

Effluent stream

Influent stream

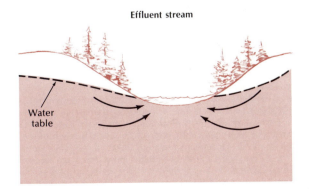

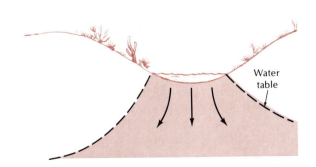

is apt to fluctuate. It will intersect the surface when the rivers are flowing, but it will drop below the surface if they run dry.

A closer look at the zone of aeration shows that it is actually made up of three zones: (1) the *zone of soil moisture,* (2) the *intermediate zone,* and (3) the zone encompassing the *capillary fringe.*

The zone of soil moisture is the portion of the profile most familiar to us. It is the ground layer that becomes wet after a rain or when a lawn is watered; at other times, it may be bone dry and dusty.

The lower margin of the zone of soil moisture may be only a few centimeters or several meters down. When we dig into the ground, it generally becomes drier until the soil is no longer moist. Typically, in this belt water percolates slowly and intermittently downward through openings until it reaches the water table.

Extending a short distance upward—4 m (13 ft) or less from the water table, the capillary fringe is comprised of water that has migrated upward in the minute passageways between individual grains. This movement occurs in about the same way that kerosene climbs the wick of a kerosene lamp.

The zone of high water content under the water table does not continue indefinitely downward. In other words drilling a well to greater depths will not necessarily increase the flow of water. With increasing depth, the pore spaces in the rocks close up and their water-bearing capacity diminishes until the water is in low supply and virtually immobile. As an example, the upper levels of deep mines may require constant pumping to avoid flooding, whereas the lower levels may be so dry that water has to be piped down for use in drilling.

### Porosity and Permeability

If water is to move underground, rock and unconsolidated sediment must be both porous and permeable. These two properties influence not only the movement of groundwater, but also the rate at which rainwater reaches the water table.

*Porosity* is a major factor in controlling the movement of water in the ground. We are familiar with the general meaning of the word when we think of a porous substance as one that contains many holes. Porosity is expressed as the percent-

age of the total volume of the rock that is occupied by such openings. If one-half the volume of a rock is taken up by openings, the material has a porosity of 50 percent, and so on.

Many factors determine the porosity of a rock (Fig. 18-6). In clastic sedimentary rocks, the packing arrangement is important, but so, too, is the degree of sorting, the porosity of the clasts themselves, and the amount of cementing material in the void spaces. Hand specimens of limestone may be quite dense, but large masses of it are porous because dissolution of the rock can result in large connected cavities and thus in high porosity. The porosity of igneous and metamorphic rocks is largely determined by joint frequency because the rock itself is so dense (1 percent or less porosity).

It is important to note that grain size does not influence porosity in clastic sediments. BB's or basketballs, if packed in the same manner, would have identical porosities. In fact, relatively fine-grained materials, such as silt, may have higher porosities than such seemingly open material as gravel.

*Permeability* is a measure of the capability of a rock or sediment to transmit a liquid through it. Therefore the actual size of the openings in a rock is much more important than the percentage of open space. Large void spaces obviously are more

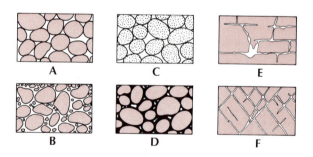

**Fig. 18-6.** Degrees of rock porosity. **A.** Highly porous, well-sorted sandstone. **B.** Poorly sorted sandstone of low porosity. **C.** Highly porous, well-sorted pebble conglomerate made up of porous sandstone pebbles. **D.** Well-sorted sandstone of low porosity because cementing material (black) fills much of the void space. **E.** Limestone rendered porous through solution. Some of the voids could eventually grow into large underground caverns. **F.** Dense igneous rock rendered porous by joints.

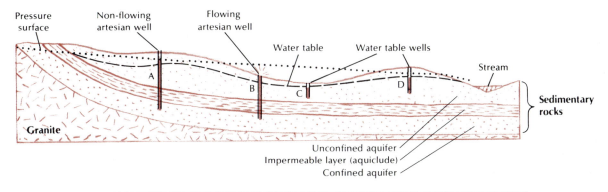

**Fig. 18-7.** The two main kinds of aquifers, the geological setting necessary for artesian wells, and an aquiclude. The dark pattern depicts the water level in each well.

permeable than small ones. For example, a silt or clay may have a higher porosity than a gravel, but since the void spaces are so small the permeability is less. Also important are connections between the openings. If the pore spaces in a rock are not interconnected, the water will not flow, no matter how large those spaces may be.

## Aquifers

Not all rocks are equally permeable, nor do they all have the same capacity to hold water. A permeable, highly porous sandstone layer, for example, may not only be able to hold much more water than its enclosing rocks, but it may provide a route along which groundwater moves with relative ease. Such a layer, one that readily yields water to a well, is called an *aquifer.* In contrast a layer that is too impermeable to accept water and allow it to flow, such as one high in silt and clay, is appropriately called an *aquiclude.*

There are two common kinds of aquifers (Fig. 18-7). One is an **unconfined aquifer,** which may be nothing more than a surficial layer of permeable sand or gravel. The other is a **confined aquifer,** a layer of permeable material, such as sandstone, between layers of impermeable material, such as shale. Typically, a sandstone aquifer may crop out in a band paralleling a mountain front, dip below the adjacent plain, and flatten as it extends away from the mountains. Such an aquifer, called Dakota Sandstone, dips east of the

Rocky Mountains and under the Great Plains in the Dakotas and Colorado (Fig. 18-8). The first wells were drilled into that sandstone in the 1880s; since then it has yielded a prodigious quantity of water.

Water enters an aquifer by surface-water infiltration. The area in which this takes place is the *recharge area.*

**Springs**  Where a valley wall intersects either a confined or unconfined aquifer, water seeps to the surface, giving rise to a spring. Springs have been the focus of settlements since ancient times. In the United States virtually every state has a town named after a spring. A resort area grew up around Silver Springs, Florida, and both the Current River in Missouri and the Snake River Canyon are noted for their springs.

## Water Wells

Wells are constructed to tap underground supplies of water. In an unconfined aquifer, the water level in the well lies at the water table (Fig. 18-7, wells C and D). The level can vary, however, depending on precipitation patterns and pumping practices. Those who have lived on a ranch dependent on a well for irrigation water, for example, are quite aware that with pumping, the water level drops. The water will rise again, but not always to the same level if the amount of water removed is exceptionally large. The time required for water levels to be restored is a function of the permeability of the aquifer.

How much does pumping from a single well affect the water table of the surrounding area? Do the water levels in adjacent wells rise or fall in concert? The answers to such questions have been established through observation in many localities over many years. If a well is pumped heavily and water is taken out of the ground faster than it can be replenished, then the water table is pulled down in the form of an inverted cone centering on the well; this is known as a **cone of depression.** Obviously, the water level in nearby wells will be affected more markedly than that in more distant ones. Studies show that the effect of an individual well may be seen in others over distances of as much as 0.4 km (0.25 mi). Hard pumping in many wells will cause the rims of individual cones to overlap; this may result in the lowering of the water level of an entire basin.

Some wells that tap confined aquifers flow at the surface of the ground. They are called **artesian wells,** from the Roman province of Artesium (now Artois) in southern France. To almost everyone the term artesian means a well that flows freely. Yet in practice the word has a more restricted use and is applied to a well in which the water is under pressure because a confined aquifer has been penetrated.

Whether or not water reaches the ground surface depends on the relationship of the **pressure surface** and the shape of the terrain (Fig. 18-7). The pressure surface is the level to which water rises in a confined or unconfined aquifer. Theoretically, in a confined system it equals the highest point in the aquifer. However, the pressure surface does not coincide with that point because energy is lost through friction as the water moves through the aquifer; hence, the pressure surface lies at a lower level and declines in altitude away from the recharge area.

Artesian wells can be flowing or nonflowing. They flow where the pressure surface is above ground (Fig. 18-7, well B). In contrast, if the pressure surface is below ground, water must be pumped (Fig. 18-7, well A). The pressure in a confined aquifer, however, may cause the water to rise to a higher level than that of the water table in an adjacent unconfined aquifer.

## EXCESSIVE WITHDRAWAL OF GROUNDWATER

If supply and demand were perfectly balanced, the volume of groundwater withdrawn would be balanced by water entering the water table from the recharge area; the result would be a relatively constant position of the water table.

However, replenishment takes place slowly, whereas pumping commonly goes on at a rapid pace. The result is that the water table is lowered and wells must be drilled to greater depths at greater cost. Another problem is that in some places the water from progressively greater depths may contain brines or hydrocarbons.

The problem of overdraft is especially critical in the American Southwest, where groundwater is crucial to the billion-dollar agricultural industry as well as to other industries. In parts of California and Arizona, for example, maximum water table decline approaches or exceeds 5 m (16 ft) per year. Obviously, this drawing off of water cannot go on forever and other sources of water are being sought. One proposed solution is to transfer water from northwest Canada not only to the American Southwest but to the Great Lakes, the Great Plains, and Mexico. This scheme would involve enormous engineering, environmental, and political problems. Perhaps there are better solutions, including conservation on a massive scale.

The history of the Dakota Sandstone and other aquifers associated with it (Fig. 18-8) is another example of overdraft. Earlier in this century water issued from the ground under pressure high enough in some places to operate waterwheels. Today, however, after the drilling of thousands of wells, pressure has dropped to the point where many wells must be pumped and the yield has greatly diminished in flowing wells. This will have an enormous impact on our agriculture in the near future.

Another well-known aquifer in the Great Plains is in great trouble. Back in the Pliocene rivers flowing eastward from the Rocky Mountains constructed an enormous alluvial fan system. Erosion has isolated most of the deposits from the mountain fronts; known as the Ogallala Formation, these deposits form much of the high Great Plains from

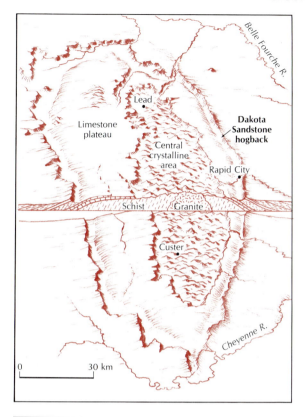

**Fig. 18-8.** Diagram of the Black Hills and the surrounding plains. Water that falls in the mountains enters the Dakota Sandstone, a major aquifer, and slowly travels along that confined aquifer to great depths beneath the dry plains. (After A.N. Strahler, 1969, *Physical Geography*, John Wiley, Fig. 32–29)

Texas to South Dakota (Fig. 18-9). Water drawn from this unconfined aquifer is responsible for the great productivity of the region: half our beef cattle, and significant amounts of sorghum, cotton, wheat, and corn. In places groundwater is being drawn off in amounts fifteen or more times greater than it is being replaced. Experts predict that shortages will become evident in the early part of the twenty-first century as water from the Ogallala Formation becomes increasingly scarce. Among the proposed solutions are to convert the region to dryland farming or to import water in canals from distant sources; the former would increase the costs of agricultural products, whereas the latter has a price tag of billions of dollars.

Coastal areas have their own peculiar problems

when it comes to pumping overdrafts. The groundwater beneath the land is fresh but that beneath the ocean is salty. Freshwater, because it is less dense than saltwater, rests on the latter, and the contact between the two dips beneath the land. If fresh groundwater is removed faster than it is replenished, salty groundwater moves inland.

A case in point is an area on Long Island, New York, where saltwater contamination of local wells has been a problem for many years. A good freshwater aquifer underlies the island, and before extensive groundwater development took place, salty groundwater did not enter the aquifer (Fig. 18-10). In earlier days much of the fresh groundwater was returned to the aquifer via cesspools. But as the population and the number of cesspools increased, contamination became such a problem that sewage was dumped into the sea. By that time

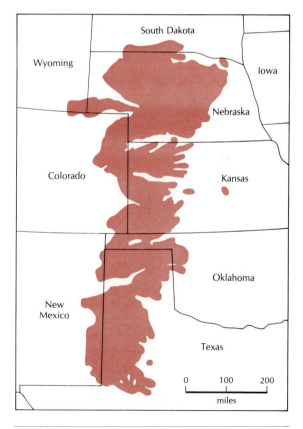

**Fig. 18-9.** The Ogallala Formation in the central United States. (From the *Colorado Daily*, University of Colorado)

571

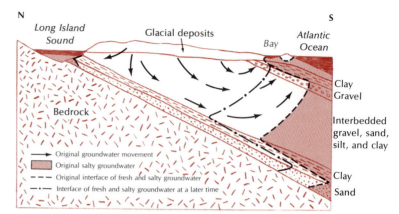

**Fig. 18-10.** Groundwater conditions on Long Island, New York. (From R.C. Heath et al., 1966, *The Changing Pattern of Ground-Water Development on Long Island, New York,* USGS Circular no. 524)

water was being removed from the system more quickly than it was replenished, the interface between freshwater and saltwater moved inland, and deep wells that once tapped freshwater became brackish.

If the situation is allowed to continue, a very important freshwater aquifer will be destroyed. What can be done? Perhaps one of the simplest solutions would be to reclaim the sewage and to pump water back underground. In that way the freshwater–saltwater interface would be forced seaward, away from domestic water wells.

Efficient water management in many areas may require similar artificial recharge by pumping. Another way to aid recharge, especially in arid areas, is to build barriers across the floodplains of rivers and streams in order to impound the waters of flash floods. The trapped water will then percolate to the water table rather than be carried off by surface flow.

Extensive withdrawal of groundwater can have other deleterious effects; for example, the land may subside. Sinking occurs because unconsolidated sediments compact when groundwater contained in the sedimentary pores is removed. Large ground cracks can accompany the subsidence (Fig. 8-11). In some areas the amount of subsidence is large indeed: up to 2.7 m (8.9 ft) in the Houston–Galveston area, 1 m (3 ft) in Las Vegas, 8.5 m (28 ft) in the southern portion of the Great Valley of California, and 9 m (30 ft) in the Long Beach–Los Angeles area. The areas involved measure from hundreds to thousands of square kilometers.

In Italy, too, are some noted examples of subsidence caused by groundwater withdrawal. One is Venice. On certain days one has to wade into Saint Mark's (Fig. 18-12). Subsidence [14 cm (5.5 in.)], sea level rise, and high waters during storms combine to create damaging floods. The other is the Leaning Tower of Pisa (Fig. 18-13). Although it began to tilt as it was being constructed because of weak foundation materials, the rate of tilting increased in the 1960s, coincident with groundwater withdrawals.

Subsidence creates enormous problems. Coastal areas suffer from the encroachment by the sea, and levees have to be built to keep it back. Perhaps a larger problem is that water-transport systems that require gravity flow, such as canals and sewers, can be disrupted by the changes in gradient that accompany subsidence or they may be broken entirely. One way to counteract subsidence is to recharge the aquifer with water.

## CONTAMINATION OF GROUNDWATER

Pollutants are a by-product of our society that are not easily disposed of. Just as we can no longer consider our streams, lakes, and oceans as unlimited reservoirs for dumping wastes, we must guard against contaminating our reservoirs. Recharge areas, especially, should be kept from contamination.

A common practice today is to dispose of refuse in sanitary landfills (garbage dumps). What makes the operation "sanitary" is that periodically the

**Fig. 18-11.** Large crack developed in alluvial fan deposits after groundwater withdrawal, near Phoenix, Arizona. The crack opened to a depth of at least 15 m (50 ft). (T.L. Péwé, Photograph no. 3357, July 25, 1972)

refuse is bulldozed over and buried beneath compacted earth. Such landfills can become unsanitary, however, if the water draining through them picks up undesirable substances in solution and subsequently contaminates the water supply. Communities should site their dumps only after the geology and the groundwater conditions of their area have been thoroughly studied. Where sufficient information on the subsurface is available, regional maps have been prepared to aid planners in siting landfills (Fig. 18-14).

In places it is feasible to dispose of liquid waste by pumping it to great depths. Obviously, the geology of an area must be studied with great care to avoid any possibility of contaminating present or future aquifers or disturbing areas in which earthquakes are a possibility. In one case (see chap. 5), deep pumping of radioactive wastes temporarily made Denver, Colorado, a seismically active region.

The underground isolation of radioactive wastes from nuclear reactor and nuclear weapons' man-

**Fig. 18-12.** High water blocks the entrance to St. Mark's Basilica, Venice, following a storm. (Peter W. Birkeland)

ufacture is a major problem in countries with nuclear capabilities. In the United States several disposal sites have been proposed. A major factor in considering any one of them is the nature and location of the aquifer in the area. The slightest amount of contamination could result in illness and possibly loss of lives.

## "PROSPECTING" FOR GROUNDWATER

"Prospecting" for water is not so strange as it may sound because groundwater is essentially a mine-

able resource. Radiometric dating has shown that thousands of years may be required for water to accumulate in underground reservoirs. Furthermore, continued pumping is permanently lowering the water table, and annual rainfall does not

**Fig. 18-13.** The Leaning Tower of Pisa tilts partly because of subsidence associated with groundwater withdrawal. The tilting began during construction of the tower in the twelfth century. Notice that the upper part of the tower leans less than the lower part; this is an adjustment made in its design during construction. (Paulina Franz)

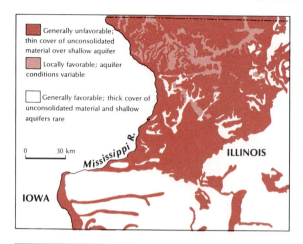

**Fig. 18-14.** Map of northern Illinois indicating the chances of success in locating a suitable site for sanitary landfill. (After K. Cartwright and F.B. Sherman, 1969, *Evaluating Sanitary Landfill Sites in Illinois*, Illinois State Geological Survey Environmental Notes no. 27)

Prospecting for groundwater in sedimentary rocks is not too difficult because the positions of the rocks at depth are predictable. Prospecting in the glaciated part of the midcontinent, however, is much more difficult. In places glacial tills of low permeability are interlayered with permeable stream deposits consisting of sands and gravels. These materials are good groundwater sources, but as they vary in thickness over short distances and do not form extensive blanket deposits, it is very difficult to predict where water will occur. In places the gravels of preglacial valleys buried beneath the glacial deposits are sought as groundwater reservoirs.

Once aquifers are located we can call on computers to help us decide how to develop the re-

replenish the water in most places. Because the demand for water is increasing, the need to prospect for new supplies is becoming more and more evident to urban planners, irrigation scientists, hydrologists, and others. Of course, this need would be lessened if more water-conservation measures were practiced.

The search for groundwater has produced its colorful prospectors. Some use dowsing or water witching to determine the presence of water. The dowser, or water witch, walks back and forth over the land holding two ends of a forked stick, or divining rod, keeping it horizontal (Fig. 18-15). When the stick, guided by some magical power, dips sharply downward, the dowser announces that this is the place to dig.

More accurate methods, of course, are now used to find water. The developing countries especially—many of which are in arid regions where water must be found before agricultural production can be increased—require reliable and accurate prospecting. One method is to map the rock units in an area, incorporating records of existing wells, to help determine potential aquifers. Geophysical methods used in petroleum exploration can also be used to help locate aquifers (e.g., the seismic methods described in chap. 5).

**Fig. 18-15.** Seventeenth-century water witch at work. (From Pierre de Vallemont, *La Physique occulte*, Paris, 1663.) (Rare Book Division, New York Public Library)

source most efficiently. Computer models for an entire groundwater system can simulate the flow of the water, the conditions at each well, the amount of water being removed, and the actual watertable level at any time. Models can also predict the effects of future withdrawals and help to plan the distribution of wells and the timing of water removals.

## GEOLOGICAL ROLE OF GROUNDWATER

### A Cementing Agent

Among the more significant accomplishments of groundwater is that of providing the means by which the various natural cements, such as calcite

$(CaCO_3)$, silica $(SiO_2)$, and iron oxide $(Fe_2O_3)$ are introduced into the pore spaces of unconsolidated sediments. These cements are derived from weathering and dissolved in water when it starts its journey underground. Later, when concentration, temperature, and the pressure are right, these substances precipitate out of solution and are deposited on the surface of the individual grains, binding them together.

### Underground Caverns and Dripstone

Groundwater plays a unique role in areas underlain by rocks that are readily soluble in water. Limestone, dolomite, marble, and such evaporite deposits as gypsum and salt are examples of these

**Fig. 18-16.** Cleaveland Avenue, a tube-shaped passage at Mammoth Cave, Kentucky. Solution along a bedding plane is responsible for the elliptical cross section. (William B. White)

**Fig. 18-17.** Stalactites hang from the roof, whereas stalagmites grow up from the floor of a cavern at Lehman Caves National Monument, eastern Nevada. Where the two join, as in the background left, a column forms. (Hal Roth)

**Fig. 18-18.** Karst topography creates a surrealistic landscape in Guangxi (formerly Kwangsi) Autonomous Region, China. Compare with Figure 18-3. (H.E. Malde, USGS)

rocks, and when they dissolve and slowly waste away, large underground caves are formed. The Carlsbad Caverns in New Mexico, Mammoth Cave in Kentucky (Fig. 18-16), and the caves decorated by Stone Age peoples near Lascaux in southern France are renowned.

Bedrock removal by solution seems to account for much of the formation of caverns. In a humid environment above the water table, waters that quickly descend to the water table are undersaturated with respect to $CaCO_3$—that is, they are still capable of dissolving more limestone and in doing so increase the amount of calcium and bicarbonate ions in the water. Dissolution of the bedrock is usually concentrated along bedding planes and joints. For a variety of reasons dissolution increases in a narrow zone just below the water table, thus producing the largest caverns. The approximate lower limit of the dissolution process is just below that zone where the waters are saturated with respect to $CaCO_3$. This means that the waters are already carrying all the dissolved calcium and bicarbonate ions they are capable of carrying, and that solution weathering no longer takes place.

In areas where cavern systems are extensive vertically, a falling water table seems to have been operative. As mentioned earlier, if the water table remains stable for a period of time, a system of caverns related to it can develop. If the water table

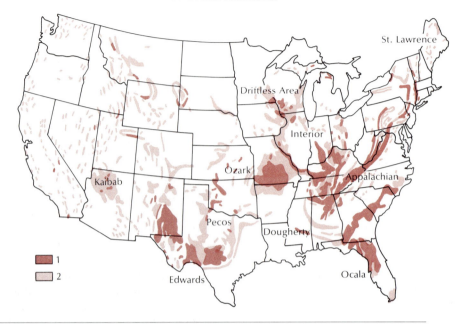

**Fig. 18-19.** Distribution of karst areas in the United States. 1 = karst areas; 2 = carbonate and sulphate rocks at or near the surface with few or no karst features. (Copyright © William E. Davies)

is then lowered through river downcutting, another set of caverns forms with passageways connecting the two sets. This process can be repeated many times until a vertical series of caverns tens or hundreds of meters in extent is formed.

Once caverns are isolated above the water table, percolating waters drip from the roof and form the unusual deposits that are the key tourist attraction of most caverns. Appropriately called **dripstone,** they consist of deposits of calcium carbonate. Two major kinds are recognized. **Stalactites** are the iciclelike pendants that hang down from the cave roof (Fig. 18-17). When dripping water is exposed to the air of the cave, some of the carbon dioxide contained in solution escapes and calcium carbonate is precipitated. In time a long pendant forms, customarily with a narrow tube extending through its full length. Seldom, though, is such perfection achieved. The tube may become plugged, the amount of water may vary, or new holes may break out along the sides of the stalactite rather than at the tip. All such changes lead to a great variation in form.

**Stalagmites** (Fig. 18-17) are deposits built upward from the cave floor. They usually form below stalactites, from saturated waters dripping from the latter, and they are usually the thicker and more diversified in shape of the two structures. Stalagmites do not contain a central tube. Stalactite and stalagmite may eventually meet and fuse to form a column.

Other cave deposits may take on a wide variety of shapes—fluted, columnar, or sheetlike masses—that are often enhanced by indirect lighting in commercially developed caves.

### Karst

The landscape that develops in a region underlain by limestone or other soluble rock has certain pronounced characteristics and is given a special name—**karst.** Parts of China display such a landscape well (Fig. 18-18). In the United States there are many karst areas—the major ones being in Pennsylvania, Maryland, Virginia, Indiana, Kentucky, Tennessee, Florida, Missouri, New Mexico, and Texas (Fig. 18-19). Puerto Rico is also known for well-expressed karst topography (Fig. 18-20).

One of the most extensive landscapes of that type is the Karst, the Dalmation coast portion of

Yugoslavia bordering the Adriatic. It is one of the most picturesque coasts in the world, with the sea penetrating far inland in long, fjordlike inlets, separated by barren, whitish limestone ridges and islands.

The region has one of the heavier rainfalls of Europe. Limestone—if joints and other fractures abound—is so permeable that rainwater sinks rapidly into the ground. A stream will flow for short distances, disappear underground, and then reappear several kilometers away from a giant spring. Such a limestone terrain is pocked with large numbers of closed depressions—some large, some small—called **sinkholes.** Commonly, the depressions are floored with red clay; this accumulation of reddish soil is likely to be all that is available for agriculture. In Yugoslavia the larger depressions may be several kilometers across—large enough to shelter a village and its surrounding patchwork of fields.

Sometimes sinkholes serve as natural wells. Their steep sides extend downward for scores or even hundreds of meters until they intercept the water table, which stands as a pool at the bottom. Re-nowned examples of such formation are the cenotes of Yucatán. Mexico's Yucatán Peninsula is a nearly level limestone plain without surface streams because rainwater sinks almost immediately into the ground. When the peninsula was the site of the Mayan Empire, the dense agglomerations of people at such cities as Chichén Itzá depended on so slender a supply of water as the dank fluid at the bottom of a limestone sink. It is no wonder that humans were ceremoniously hurled into he well in a sacrificial rite to assure a continued supply of water.

Most sinkholes form when soil is transported through solution channels in limestone. However, some form when the roof of a shallow underground cavern collapses.

Sinkhole formation is still an active process, as some unfortunate landowners have discovered. Florida and Texas especially are areas where soil-transport sinkholes are a common geological hazard. They cause damage to houses, building foundation, and roads, and they have even been known to gobble up cars if they form quickly enough (Fig. 18-21). Current research is aimed at pinpointing

**Fig. 18-20.** Cone karst in the northern karst belt of Puerto Rico. The conical hills are underlain by the massive Lares Limestone. The low-relief foreground is underlain by the clay-rich Cibao Formation. (William B. White)

**Fig. 18-21.** "Wink Sink," sinkhole in northwestern Texas, is the result of the collapse of underlying rock into a cavern formed by solution of a subsurface salt layer. The depression, which formed in only 48 hours, is 110-m (361-ft) across and 34-m (112-ft) deep. (R.W. Baumgardner, Jr.)

areas prone to sinkhole formation. One method is to study thermal and vegetation patterns on photographs obtained from remote sensing satellites, which could indicate areas likely to develop sinkholes.

In regions with temperate climates sinkholes are the most common features of karst terrain (Fig. 18-22). In rainy tropical areas with thick limestone deposits, the most common karst landforms are conical tower-shaped hills separated by narrow, irregular ravines or flat plains. Most of the karst on islands in the Caribbean and the South Pacific are of the cone and tower type. The celebrated karst of China is largely of the tower type, with many isolated towers separated by flat plains. (Fig. 18-18). They form by the progressive downward and lateral solution of the limestone, guided perhaps by joint patterns in the rock.

### Hot Springs and Geysers

By far the most spectacular manifestation of groundwater is its appearance at the surface in the form of hots springs and geysers. Hot springs are

widely distributed over the face of the earth. More than one thousand are found in the United States, most of them in the montane parts of the West. In Yellowstone National Park geysers are the leading attraction; few have not heard of Old Faithful, and millions have seen it run through its repertoire (Fig. 18-23). A geyser basin in Iceland gives its name, *geysir*, to this sort of aqueous outburst. Another area of hot springs and geysers is the Rotorua region of North Island in New Zealand, part of which has been developed for thermal power.

Hot springs are the result of descending groundwater coming into contact with heated rocks or magma beneath the earth's surface. The heated water rises to the surface and emerges as a warm spring. If the water is very hot, it owes its high temperature to superheated steam present in volcanically active areas.

Geysers occur when a column of steam and hot water is explosively discharged at intervals. Vertical underground caverns connected by passages seem to be required for their formation. The plumbing can be thought of as a vertical irregular tube with many side caverns. Water in the column is heated from below, and it approaches the boiling point. Once the boiling point is reached in a particular part of the column, the water flashes to steam and pushes out the overlying water. Should enough water drain off, pressure throughout the water column is reduced, with the result that the superheated water at greater depths also flashes into steam. This action is enough to propel the whole column of water upward; because a similar pressure reduction and a near-instantaneous conversion to steam occurs throughout its length, a mixture of hot water and steam is hurled sky-ward—in Old Faithful for about 46 m (150 ft). In time water again fills the underground caverns and passageways and the eruptive process repeats itself. The filling and emptying of Old Faithful is so predictable that the time of the next geyser eruption (about 1 hour later) is posted at the site.

The castellated rims, platforms, and particolored deposits surrounding the geysers and hot springs of Yellowstone are especially interesting features to park visitors (Fig. 18-24). In general there are two kinds of hot-water deposits, those composed of silica and those composed of $CaCO_3$.

**Fig. 18-22.** Sinkholes in limestone form karst topography in Monroe County, West Virginia. (William K. Jones)

**Fig. 18-23.** Steam rising from Old Faithful Geyser in winter, Yellowstone National Park. (Janet Robertson)

**Fig. 18-24.** Travertine terraces at Mammoth Hot Springs, Yellowstone National Park. Photographed by W.H. Jackson in 1864. (Metropolitan Museum of Art, Rogers Fund, 1974)

## SUMMARY

1. Rainwater percolates beneath the surface of the ground, eventually filling the pores to become *groundwater*. The latter is an extremely valuable source of water.
2. The top of the groundwater zone is the *water table,* and its configuration generally parallels that of the land surface, Where the water table intersects the land surface, water flows; springs are a common example.
3. *Permeability* of rock is the main property controlling the rate at which groundwater flows from one point to another. Rock units that are permeable and hold sufficient quantities of water are called *aquifers.* Po-rous sandstone beds make some of the best aquifers.
4. Before aquifers are extensively used, they should be studied to avoid certain problems: too-rapid lowering of the water table, land subsidence accompanying excessive withdrawal, encroachment of salty groundwater in coastal areas, and pollution of groundwater.
5. The geological role of groundwater ranges from the cementation of sand deposits into sandstone to the formation of underground caverns and the stalactites and stalagmites, for which many are so famous, to the formation of the peculiar topography called *karst.*

# QUESTIONS

1. How do porosity and permeability influence the flow of groundwater?
2. Compare the suitability of the following rock types for groundwater reservoirs: sandstone, shale, limestone, granite, and firmly cemented conglomerate.
3. What factors must be considered in storing toxic chemical wastes in ponds so that the groundwater is not contaminated?
4. How does karst topography form?
5. Are there any similarities between the eruption of a geyser and the eruption of a rhyolitic pyroclastic flow?

# SELECTED REFERENCES

Davis, S.N., and De Wiest, R.J.M., 1966, Hydrogeology, Wiley, New York.

Dolan, R., and Goodell, H.G., 1986, Sinking cities, American Scientist, vol. 74, pp. 38–47.

Freeze, R.A., and Cherry, J.A., 1979, Groundwater, Prentice-Hall, Englewood Cliffs, N.J.

Holzer, T.L., ed., 1984, Man-induced land subsidence, Geological Society of America, Reviews in Engineering Geology 6.

Jennings, J.N., 1971, Karst, Australian National University Press, Canberra.

Leopold, L.B., 1974, Water, a primer, W.H. Freeman, San Francisco.

Monroe, W.H., 1976, The karst landforms of Puerto Rico, U.S. Geological Survey Professional Paper 899.

Rinehart, J.S., 1980, Geysers and geothermal energy, Springer-Verlag, New York.

# PART IV

# Resources
# and Social Issues

# 19

# RESOURCES AND ENERGY

In recent years resources have become a vital issue as we have become increasingly aware that the energy sources and the raw materials on which our society depends are not infinite. Western nations, in particular, and industrial nations, in general, must buy many resources from other nations—which are often developing countries—and this adds a political dimension to the problem of supply and demand. The role geologists must play as energy sources and metal and mineral deposits are depleted is an important one, for their task is to find and evaluate new supplies. Only in the past several decades have earth scientists and others quantitatively estimated the remaining resources of this planet. If their estimates are accurate or even close to being accurate, it is evident that it may be necessary to change our way of life considerably and without too much delay.

All the options open to us are not without some cost or risk. For energy supplies many have been looking to nuclear power (Fig. 19-1) as oil reserves dwindle. However, recent accidents at nuclear

power plants at Three Mile Island in Pennsylvania and at Chernobyl in the USSR may limit this option. The Chernobyl accident was by far the worse, killing 31 persons and sending radioactive fallout across Western Europe. Since the Chernobyl incident there have been many calls for the elimination of such plants. However, one risk many take every day with little thought is a trip in the automobile; yet as one writer put it the loss of fifty thousand lives per year in automobile accidents has not threatened the automobile industry!

In this chapter, we will discuss basic resources, their geological occurrences, and the problem of dwindling supplies relative to their geological setting as well as to the social issues involved.

## BACKGROUND INFORMATION

By late 1987, Al Bartlett, a professor of physics at the University of Colorado, had given a speech on growth and resources eight hundred times and published the same information in two scientific journals. Bartlett has devoted much time and energy to explaining effectively how economic growth affects resource usage, population, and lifespans. Throughout this chapter we will refer to the con-

Fig. 19-1. Big Rock Point nuclear power plant on the north-eastern shore of Lake Michigan. (Consumers Power Company)

cepts he expounds and the examples he used along with statements and reports from scientists and others in government and industry.

## Rate of Growth

When anything grows or increases at a fixed percentage rate, its growth is said to be *exponential,* and a curve plotted to show its growth rate will steepen with time. This idea should not be new to the reader. In chapter 2 we discussed the decay of radioactive elements in relation to carbon–14 dating—a case in which change is exponential although in the opposite direction (decrease) from that of growth. A good example of exponential growth is a savings account yielding 5 percent interest yearly. If the interest is added to the account, the interest rate allows the principal to double every 14 years. If the interest rate is increased to 10 percent, the principal will double in 7 years, or one-half the time. Thus, a deposit of $1000 would grow to $2000 in 7 years, to $4000 in 14 years, to $8000 in 21 years, and to more than $1,000,000 in 70 years. One can determine doubling time by using the equation

$$T = 70/P$$

where $T$ is the doubling time in years and $P$ the percentage growth per year.

Another example of exponential growth is population growth (Fig. 19-2). The annual world rate is estimated to be 1.7 percent, apparently a low rate, but one that translates to a doubling of the world's population every 41 years. Western Europe is a region of slower growth, at 0.2 percent, whereas the Middle East and Africa are two of the faster growth regions at 2.9 percent. At the average world rate, however, population density would reach one person per square meter—excluding the Antarctic continent—in 550 years, if life span remains the same.

It is important to understand that while population is increasing, so is the per capita demand for resources. Add to this the demand from countries hoping to develop industries and one of the major problems of the future comes into focus: one exponential growth process (population) feeding another (resource use). Predicting when a population crisis will arise is difficult. Bartlett, as an

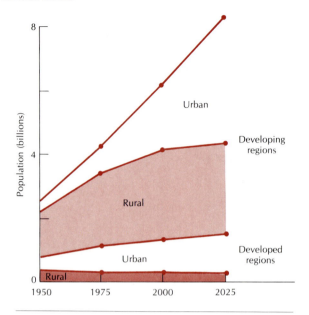

**Fig. 19-2.** Estimated world population trends, 1950–2025. The pattern of exponential growth is very subtle because only 75 years are shown, the upper line flattens to the left.

illustration, hypothesizes a strain of bacteria that doubles its population every minute, and an empty bottle, which represents a nonrenewable resource. A bacterium placed in the bottle multiplies. When the bottle is full, the resource is gone (Fig. 19-3). If the bacterium is placed in the bottle at 11 A.M., the bottle will be filled to the brim by 12 Noon. Only in the last few minutes before noon, however, does it become apparent that only a small amount of the resource remains. At 11:58 A.M., three more bottles are provided—a fourfold increase in resources in a little less than an hour. Yet, by 12:02 P.M., after only four more doubling times, all the bottles are full of bacteria—again the resource is gone! The point to be made: Given exponential consumption of a resource, that resource can be consumed in a very short time.

Bartlett's illustration makes several other points. First, it emphasizes our position in relation to resources so that we may plan for the future. Second, it cautions us to view announcements of newly found or greatly increased resources in perspective, for in reality they will not last long at an exponential consumption rate. For example, if the Alaskan oilfields were our only source of oil, the

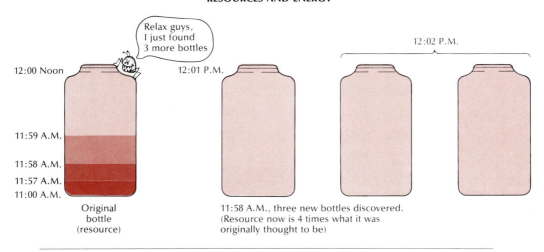

**Fig. 19-3.** Bartlett's example of steady growth: bacteria as the population (or demand) and bottles as the resource.

proved reserves would last little more than a year at the current rate of consumption, whereas the estimated amount of oil present might last only 6 or 7 years. Finally, Bartlett's illustration points up the fallacy of calculating resource use from current consumption without taking growth into consideration.

A few more figures may help to underline the problem. Since 1970 U.S. oil production has declined, yet present annual growth in demand is about 3 percent. To meet our increased demands, we imported more oil—one estimate is that by 2000 we will be importing 75 percent of our demand.

## RESOURCES

### Definitions of Basic Terms

By **resource** we mean a commodity useful or commonly essential to people and thus to industry and to our life style. **Reserve** refers to the estimated amount of an identified resource recoverable with present-day technology (Fig. 19-4). However, reserves are usually a small fraction of the total resource. It is possible that undiscovered resources may be a sizable proportion of the total. A **paramarginal resource** is expensive to recover; the profit margin of its recovery is marginal, but re-

covery is technologically feasible. Recovery of a **submarginal resource** is not profitable; it involves financial loss unless the technology can be improved or unless the value of the resource increases greatly. The reworking of dumps from the placer mining of gold illustrates the latter type of resource: an improved technology and a huge jump in the price of gold. The incentive to recover almost any resource is economic.

### Renewable and Nonrenewable Resources

Resources are divided into two main groups: renewable and nonrenewable. **Renewable resources** may be replenished, and if carefully managed, continue to be available. Examples are farm produce and forest products. **Nonrenewable resources**—the common ore minerals and fossil fuels, for example—once used, they are gone forever. Within the latter group some resources, such as metals, can sometimes be recycled, whereas others, such as oil, cannot be retrieved after use. Groundwater is an important renewable resource, but only if it is replenished at the same or higher rate at which it is pumped from the earth. In many dry areas of the western United States, however, groundwater is being withdrawn faster than it is being renewed.

Similarly, farmland and forests can be exploited

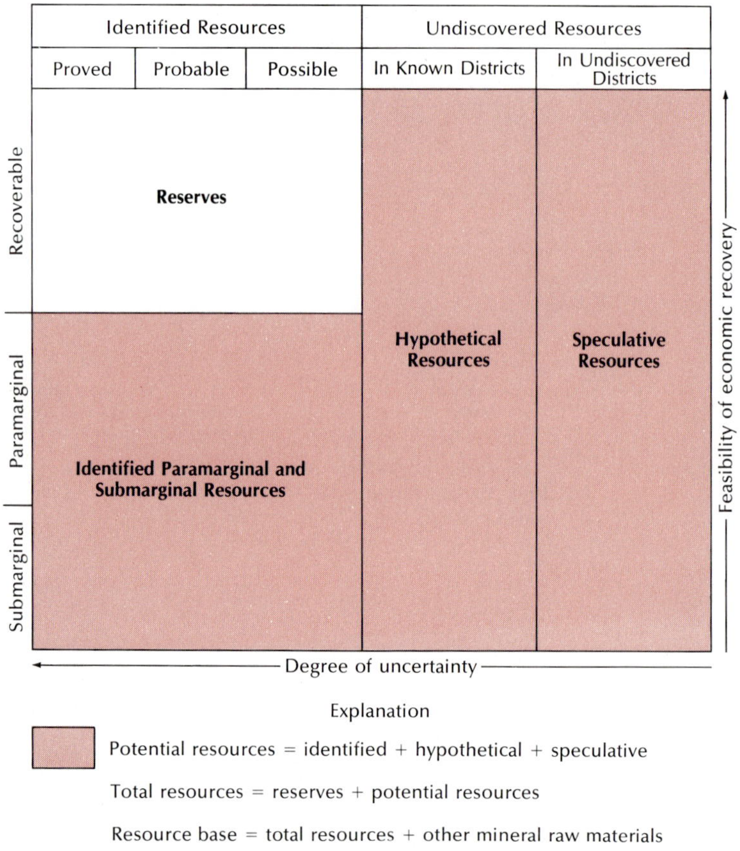

| Identified Resources | | | Undiscovered Resources | |
|---|---|---|---|---|
| Proved | Probable | Possible | In Known Districts | In Undiscovered Districts |

Recoverable

**Reserves**

**Hypothetical Resources**

**Speculative Resources**

Paramarginal

**Identified Paramarginal and Submarginal Resources**

Submarginal

← Degree of uncertainty →

Feasibility of economic recovery →

Explanation

Potential resources = identified + hypothetical + speculative

Total resources = reserves + potential resources

Resource base = total resources + other mineral raw materials

**Fig. 19-4.** Diagram showing categories of a resource. The area of each box is proportional to the amount of the classified resource. (U.S. Geological Survey)

or they can be used judiciously so that they can renew themselves. If we allow our soils to erode, or our valuable croplands to be swallowed up in shopping centers or suburban development, or if we demolish our forests indiscriminately, the damage is irreversible. For example, the area paved over with asphalt since 1945 is about equal to the area of Ohio.

## What Is a Resource?

Most resources can be found in ordinary rock. Gold and phosphorus, and even nuclear energy, are locked up in common basalt, for example, but only in very small quantities. A resource represents a profitable concentration of a particular material. Some resource elements (e.g., iron, aluminum) are found in concentrations only several times average crustal abundance; others may be a hundred times

the abundance (e.g., copper, nickel), several thousand times (e.g., uranium, silver, lead, gold), or as high as a hundred thousand times (e.g., mercury) (Table 19-1). The largest deposits as well as the number of deposits of, say, a particular ore, correlate well with its crustal abundance (Fig. 19-5).

Each resource reflects specific geological or biological processes and, thus, are unevenly distributed worldwide. This explains why, for example, Saudi Arabia has an abundance of oil, yet nearby Egypt and Israel have little; or why, for example some tiny Pacific islands supply vast quantities of phosphate for fertilizers (Fig. 19-6) from bird droppings, yet nearby islands with large bird populations supply none.

Locating economically profitable deposits of minerals and other resources is sometimes easy, sometimes not. The holes that dot part of the Nevada landscape, for example, suggest that there

**Table 19-1** Crustal Abundance of Some Common Ore-forming Elements and the Concentration Factor That Indicates Whether Their Mining Would Be Economical

| Element | Crustal Abundance (percentage weight) | Concentration Factor |
|---|---|---|
| Aluminum | 8.00 | 3–4 |
| Iron | 5.8 | 4–5 |
| Copper | 0.0058 | 80–100 |
| Nickel | 0.0072 | 150 |
| Zinc | 0.0082 | 300 |
| Uranium | 0.00016 | 1,200 |
| Gold | 0.0000002 | 1,800 |
| Lead | 0.00010 | 4,000 |
| Mercury | 0.000002 | 100,000 |

Data compiled from B.J. Skinner, 1986, *Earth Resources*, 3rd ed., Prentice-Hall, Englewood Cliffs, N.J.; D.A. Brobst and W.P. Pratt, eds., 1973, *United States Mineral Resources*, USGS Professional Paper no. 820, Washington, D.C.; and A.D. Howard and I. Remson, 1978, *Geology in Environmental Planning*, McGraw-Hill, N.Y.

are simple ways to search for an ore deposit. Yet exploration can also be difficult, involving complex models. Economic factors affect the decision to exploit a particular deposit. Extracting oil from oil shale, for example, is much more expensive and uses more energy and water than offshore oil drilling because the yield is so low. Another factor is the expense of transporting material to a processing plant or to market.

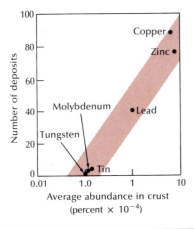

**Fig. 19-5.** Crustal abundance of scarce metals compared with the number of deposits. (After B.J. Skinner, 1986, *Earth Resources*, 3rd ed., Prentice-Hall)

So many variables are involved in determining resource reserves—rate of use, economic value, new technology—that estimates should be used only as guides in planning for the future.

## Rise and Fall in Nonrenewable Resource Production

For those resources that are nonrenewable, we can predict how long a particular resource will last (Fig. 19-7). Curve A represents exponential growth up to the day the resource is exhausted; so far as we know, such growth cannot be sustained. Taking fossil fuel oil as an example, a prominent geologist (M. King Hubbert) argues that a more or less bell-shaped curve (curve B) is more realistic. It has three parts: (1) total production to date, (2) proved reserves, and (3) hoped-for discoveries of further reserves. The area under curve B must equal the final total production of the resource being considered. The curve falls toward zero even though material is still being extracted because fewer deposits are being discovered, or because the deposits are of low-grade, or are more difficult to reach. However, major discoveries can alter the shape of the curve, but only slightly because of population growth and the increase in per capita demand. These curves have been used to predict peak production and the date at which the resource will be exhausted.

## Metallic Resources—Origins and Availability

Metals have changed the course of history, so much so that the major prehistoric periods have been named for the metals that were developed and widely used during those periods. After the Stone Age came the Bronze Age, then the Iron Age. It has been observed, however, that we might still be in the Stone Age: Based on tonnage of materials used in the United States, stone and gravel may well be our main resource. Of the many minerals that make up the earth, only about 1 in 20 is used.

There are many complex ore-containing minerals. Most elements can combine with oxygen to form oxides (e.g., manganese, tin, uranium) or with sulfur to form sulfides (e.g., zinc, molybdenum, lead). Some elements, such as iron, can form

**Fig. 19-6.** Per capita income of Nauru, an independent island-state in the South Pacific, is one of the world's highest. The source of the island's wealth, phosphate fertilizer, is shown being mined; as the mining proceeds, large pillarlike carbonate structures are left behind. The fertilizer supply will last only a few more years, so Nauru is investing part of its income in real estate in Australia and in other ventures that will yield money for the future. (Peter W. Birkeland)

both an oxide (e.g., magnetite) and sulfides (e.g., pyrite) and some usually do not combine with other elements, remaining in what is called their *native state* (e.g., gold, platinum).

Metallic ore deposits can be classified on the basis of abundance of metals in the earth's crust. Two categories are recognized: *abundant* if they are over 0.1 percent of the total; *scarce* if they are less than 0.1 percent. Iron, aluminum, manganese, magnesium, and titanium are the abundant metals;

all the others are scarce. Deposits of scarce metals are relatively small and few.

## GEOLOGICAL ORIGINS OF RESOURCES

### Igneous and Metamorphic Process

Igneous and metamorphic processes are important in the formation of many of the world's ore deposits (Fig. 19-8). Within a magma several pro-

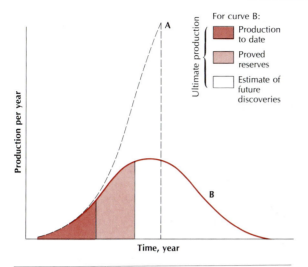

For curve B:

▉ Production to date

▨ Proved reserves

☐ Estimate of future discoveries

Fig. 19-7. Two curves showing predicted production. The area under curve A is approximately equal to the area under curve B. (Modified from A.A. Bartlett, 1980, "Forgotten Fundamentals of the Energy Crisis," *Journal of Geological Education*, vol. 28, pp. 4–35)

cesses might occur to produce an ore deposit. Perhaps the simplest one is the gravity separation of heavy minerals, some of which form early in the crystallization of the magma to form a layer at the bottom of the magma chamber. Another important differentiating mechanism is the separation of liquids within the magma followed by crystallization. Deposits so formed contain minerals rich in iron, chromium, titanium, nickel, and sulfur. Pegmatites can crystallize from the more volatile constituents of magma or from a very fluid magma, and they can mimic their granitic parent in their mineral content. They may contain such rare elements as lithium, boron, fluorine, or uranium. The rock is very coarse-grained, some single crystals are up to 15-m (49-ft) long.

Ores also may be formed by contact metamorphism, which occurs when a body of magma, commonly intermediate in composition, intrudes country rock. The alteration commonly involves circulating hot solutions and can range from the rearrangement of existing materials to complete replacement by the addition of materials from the igneous source; the zone can be 1-meter or several kilometers wide. Limestone is a common host to

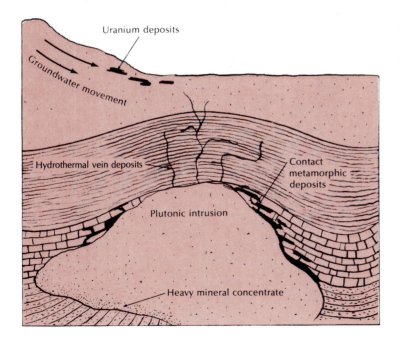

Fig. 19-8. The realms of various kinds of ore deposits. The diagram is schematic only, because not all ore-forming processes are associated with a single intrusion.

ore deposits, and the former limestone is removed by the hot solutions. The ores commonly found at contact zones are simple oxides and sulfides of iron, zinc, copper, and molybdenum.

Contact features are well displayed in the Iron Springs District of southwestern Utah. Magma of intermediate composition intruded a body of limestone, cooled, and forced iron-enriched solutions into the limestone to form iron ore bodies up to 70-m (230-ft) thick. For every million tons of iron introduced to the limestone, 40,000 tons of silicon, 20,000 tons of magnesium, and 10,000 tons of aluminum were also introduced.

Ores may also be precipitated from fluids, in which case they are called *hydrothermal deposits* (see Fig. 19-8). Hot solutions contain high concentrations of dissolved materials; as chemical conditions and temperatures change, ore is deposited either in veins, because the fluids follow joints or faults, or as irregularly shaped bodies or

**Fig. 19-9.** The open-pit copper mine at Bingham Canyon, Utah. (Salt Lake Area Chamber of Commerce)

very small particles disseminated through the host rock. Discrete concentrations of ore are often easy to find, but disseminated ores, if they are fine grained, are more difficult to find.

*Porphyry copper,* made famous by the huge mine at Bingham Canyon, Utah, is an example of a hydrothermal ore deposit (Fig. 19-9). In the Bingham deposit, ore is disseminated throughout a porphyritic plutonic rock. The magma moved to quite shallow depths [about 3 km (1.8 mi)], where late in the intrusive stage shattering occurred, thus creating extensive fracture systems in which ore minerals were later precipitated. Although the amount of copper in these deposits is usually less than 1 percent, relatively inexpensive mining techniques have made the extraction of such ores profitable.

In recent years unusual hydrothermal deposits have been found in the seas and oceans. In the mid-1960s newly discovered pools of hot brine [56° C (133° F)] at the bottom of the Red Sea, whose salinity is second only to that of the Dead Sea, were found to be rich in many elements. The sediment on the seafloor is described as a black powdery substance, too hot to touch, containing the oxide and sulfide ore minerals of iron, manganese, zinc, and copper. Even more recently scientists in the submersible *Alvin* have observed vents at a depth of 2622 m (8600 ft) on the East Pacific Rise off the coast of South America (Fig. 19-10). The superheated water [over 300° C (600° F)] melted part of the *Alvin*'s temperature sensor. The ocean floor was multicolored, carpeted with sulfide minerals rich in copper, iron, zinc, cobalt, lead, silver, and cadmium.

The plate-tectonics theory provides one explanation for such concentrations. Where plates are diverging, as at the midocean ridges, magma ascends to fill the voids thus created. The cool oceanic water penetrates the cracked, newly formed crust some distance from the spreading center and migrates toward it; heating occurs and the water rises into the overlying rocks (Fig. 19-11). As it flows through the crust, the water dissolves quantities of metals from the host rock. The dissolution process is aided by the high chloride content of seawater. On rising, the ore is precipitated either as veins, or as masses in the rock, or directly on the ocean floor.

**Fig. 19-10.** Black smoker vent, East Pacific Rise, 1979. (Dudley Foster, WHOI)

That some ores can originate at diverging plate boundaries is difficult to prove conclusively. If, as is thought, the plates are later consumed in a subduction trench, all evidence of such an origin would be lost. Yet earth scientists think that they have found on the Mediterranean island of Cyprus a slab of ancient ocean floor laced with the original ore deposit. Cyprus, which has a long history of copper mining that extends back to about 3000 B.C., lies along a continental collision zone, but the rock containing the copper ore seems originally to have been a spreading center and its cross section has similarities to that shown in Figure 19-11. During subsequent continental convergence, a slab of ocean floor was upthrust to its present position on the island.

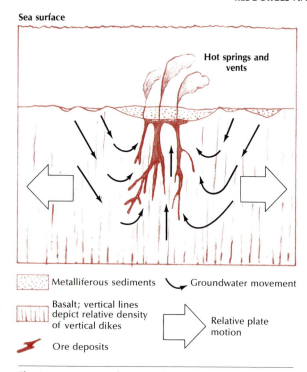

Sea surface

Hot springs and vents

Metalliferous sediments

Basalt; vertical lines depict relative density of vertical dikes

Ore deposits

Groundwater movement

Relative plate motion

**Fig. 19-11.** Diagram depicting deposits, groundwater motion, and hydrothermal activity at a midocean ridge. Because the lavas were extruded beneath water, pillow structures are common. (After J.B. Corliss, 1973, "The Sea as Alchemist," *Oceanus*, vol. 17, p. 39)

The plate-tectonics theory can also help to explain the occurrence of major ore deposits found in relatively young folded mountain belts (Fig. 19-12). It is thought that the ores were first concentrated at a midocean ridge, then transported as part of an oceanic slab toward a subduction zone, carried downward, and melted. Eventually, the molten ores worked their way back toward the surface as hydrothermal deposits.

Island arcs are also areas with an ore plate-tectonic connection. In Japan the sedimentary deposits enclosing ore bodies are of shallow-water marine origin and the ores themselves are hydrothermal and associated with nearby arc volcanism.

## Groundwater Processes

Some ore deposits are precipitated from fairly cool groundwater solutions. Common are the uranium and vanadium deposits found in the sedimentary rocks of the Colorado Plateau, which are important future sources of fuel for atomic reactors (see Fig. 19-8). Organic matter in the sediments helps to create chemical conditions conducive to the precipitation of these minerals. Some prospectors have found buried logs almost completely replaced by uranium ore.

## Sedimentary Processes

Minerals and metals can be concentrated by mechanical and chemical sedimentary processes to form ore deposits. The familiar **placer deposits** are formed by a concentration of heavy mineral particles from weathered rocks, such as gold, which may be deposited in pockets along a river channel or in the sand of a beach. The forty-niners who panned gold during the California gold rush washed sand and gravel in a pan to concentrate the metal—a process similar to that occurring in a stream. Diamonds found in beach placer deposits were first brought to the surface in narrow kimberlite pipes (see chap. 7).

An unusual sedimentary deposit found in places on the deep-ocean floor takes the form of large nodules of manganese dioxide ($MnO_2$)—some of them the size of cannonballs (Fig. 19-13). Their origin is somewhat obscure, but precipitation was probably brought about by a combination of inorganic and organic processes. Growth took place by accretion of ion upon ion on a simple substrate, such as a shark's tooth, over millions of years. Estimates of their rate of growth vary from 1 to 100 mm (0.04 to 4 in.) per million years. The nodules also contain appreciable amounts of iron, cobalt, nickel, and copper and are so abundant in some places that methods for their recovery are being developed even though problems of legal ownership and the environmental impact of mining the ocean floor have not been resolved. It is estimated that billions of tons of nodules lie on the ocean floor—amounting to the largest mineral deposit on earth.

Other chemically precipitated sedimentary ores are the extensively banded iron deposits, some hundreds of kilometers across, that formed in Precambian seas.

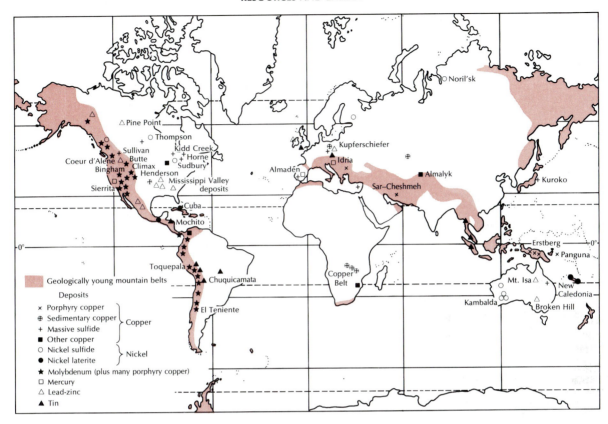

**Fig. 19-12.** The association between mountain belts formed during the last 200 million years and major deposits of the scarce metals. (After S.E. Kesler, 1976, *Our Finite Mineral Resources*, McGraw-Hill, Fig. 8–2)

**Fig. 19-13.** Manganese nodules of cannonball size at a depth of 5417 m (17,768 ft) in the southwest Pacific. (Smithsonian Oceanographic Sorting Center)

**Table 19-2** Reserve Base, Production (1980), and Reserve Base/Production Index for Important Mineral Commodities of the World

| | Reserve Base (thousands of tons)[a] | Production (thousands of tons)[a] | Reserve Base/ Production Index (RB/P) |
|---|---|---|---|
| Salt | ∞ | 181,600 | ∞ |
| Magnesium | ∞ | 351 | ∞ |
| Cement | Very large | 978,000 | Very large |
| Lime | Very large | 131,623 | Very large |
| Gypsum | Very large | 78,290 | Very large |
| Clays (common)[b] | Very large | 449,000 | Very large |
| Sodium carbonate | >43,200,000 | 8,459 | >5,107 |
| Sodium sulfate | 5,100,000 | 2,169 | 2,351 |
| Chromite | 7,540,000 | 10,725 | 703 |
| Potash | 18,739,000 | 30,722 | 610 |
| Vanadium | 18,250 | 40 | 456 |
| Manganese ore | 12,000,000 | 29,000 | 414 |
| Cobalt | 9,200 | 33 | 279 |
| Feldspar | >1,000,000 | 3,782 | >264 |
| Boron minerals | >300,000 | 1,175 | >255 |
| Phosphate rock | 38,580,000 | 151,000 | 255 |
| Bauxite | 24,581,000 | 99,165 | 248 |
| Ilmenite | 905,000 | 3,979 | 227 |
| Iron ore | 231,056,000 | 1,090,432 | 212 |
| Nickel | 111,000 | 850 | 131 |
| Fluorspar | 645,000 | 5,000 | 129 |
| Molybdenum | 12,975 | 120 | 108 |
| Antimony | 5,175 | 74 | 70 |
| Copper | 562,000 | 8,421 | 67 |
| Tungsten | 3,813 | 60 | 63 |
| Sulfur | 2,976,000 | 56,900 | 52 |
| Zinc | 319,670 | 6,340 | 50 |
| Talc and pyrophyllite | 330,000 | 7,366 | 45 |
| Mercury[c] | 7,200 | 191 | 38 |
| Lead | 148,810 | 3,885 | 38 |
| Barite | 244,000 | 8,114 | 30 |
| Asbestos | 114,600 | 5,314 | 22 |
| Tin | 3,307 | 272 | 12 |

[a] Data from U.S. Bureau of Mines, *Mineral Commodity Summaries,* 1982 (production) and 1985 (reserve base).
[b] There are no data for world reserves of special types of clays.
[c] The standard unit for mercury is the flask, a flask being 76 lb. Numbers given are thousands of flasks.
After E.N. Cameron, 1986, *At the Crossroads—the Mineral Problems of the United States,* Wiley, N.Y., p. 146.

## Weathering Processes

Weathering can help to produce soil, a major resource, but under special conditions it can also form an ore body. In humid tropical areas soluble elements, such as potassium, calcium, and magnesium, are rapidly leached from the soil, as is the less-soluble silicon. What is left is a residue of insoluble materials containing sufficient quantities of nickel or iron and aluminum oxides to be designated ores.

In other places weathering has solubilized metals, such as silver and copper, that have been carried downward—sometimes to great depths—to where the chemical conditions at the water table cause them to precipitate. This process is important because it increases ore proportions in the deposit.

## MINERAL RESOURCES AND THE FUTURE

It is difficult to determine the extent of the world's mineral reserves and potential resources and even more difficult to predict how long they will last. The life span of a resource depends on geological factors, advances in mining and extraction, new technological applications, the economic and political climate of the times, demand, and substitute materials. The figures in Table 19-2 provide a base for roughly estimating the life span of a known resource. The minimum number of years is the reserve base/production index (RB/P). Some reserves will be exhausted in the next 100 years. Gold and silver are already in short supply and

there seems to be sufficient uranium for only about a generation.

How does the United States fare with respect to the minerals needed to run its industries? Using for the United States the same analysis as for the world, we see in Table 19-3 that reserves of many important minerals will be exhausted in less than 100 years. The United States relies on many imports at present (Fig. 19-14), and this has both positive and negative aspects. On the positive side, imports extend the life of our own resources and provide income for the exporting nations. On the negative side are the problems of the trade deficit, and the political reality that once friendly countries may not always remain friendly and may at

**Table 19-3** Reserve Base, Production, and Reserve Base/Production Index for the United States for 1984 and 2005

| | Reserve Base[a] | Average Annual Production[a,b] | Reserve Base/ Production Index 1984 (RB/P) | Production[a,b,c] 1985–2005[c] | Remaining Reserve Base[a] 2005 | Reserve Base/ Production Index 2005 (RB/P) |
|---|---|---|---|---|---|---|
| Magnesium | ∞ | 132 | ∞ | 2,772 | ∞ | ∞ |
| Vanadium | 2,400 | 5 | 480 | 105 | 2,295 | 459 |
| Silicon | Very large | 481 | Very large | 10,101 | Very large | Very large |
| Iron ore | 24,800,000 | 74,684 | 332 | 1,568,364 | 23,231,636 | 311 |
| Nickel[d] | 2,800 | 12 | 233 | 252 | 2,548 | 212 |
| Zinc | 53,000 | 384 | 138 | 8,064 | 44,936 | 117 |
| Tungsten | 320 | 3 | 107 | 63 | 257 | 86 |
| Molybdenum | 5,900 | 57 | 104 | 1,197 | 4,703 | 83 |
| Lithium | 460 | 6 | 77 | 126 | 334 | 56 |
| Copper[e] | 90,000 | 1,433 | 63 | 30,093 | 59,907 | 42 |
| Lead | 27,000 | 559 | 48 | 11,739 | 15,261 | 27 |
| Aluminum | 12,000 | 356 | 34 | 7,476 | 4,524 | 13 |
| Tin | 55 | small | ~50 | 50? | 5 | 5? |
| Mercury | 200[f] | 26[f] | 8 | 546[f] | 0 | 0 |
| Antimony | 100 | 13 | 8 | 273 | 0 | 0 |
| Chromium | 0 | 0 | 0 | 0 | 0 | 0 |
| Manganese | 0 | 0 | 0 | 0 | 0 | 0 |
| Cobalt[g] | — | 0 | — | 0 | — | — |

[a] Thousands of tons, except mercury.

[b] Primary metal production.

[c] Assuming average rate of production is equal to the 1974–84 average. Exception mercury, for which the 1979–84 average is used.

[d] Only 150,000 tons of the reserve base is currently ore.

[e] Reserve base includes a large amount of high-cost copper, currently uneconomic.

[f] Thousands of 76-pound flasks.

[g] The reserve base for cobalt is given by the U.S. Bureau of Mines as 950,000 tons. None is economic at present, and "most domestic resources are in subeconomic concentrations that will not be economically usable in the foreseeable future."

Data for 1984 from U.S. Bureau of Mines, Mineral Commodity Summaries. After E.N. Cameron, 1986, *At the Crossroads—the Mineral Problems of the United States*, Wiley, N.Y., p. 182.

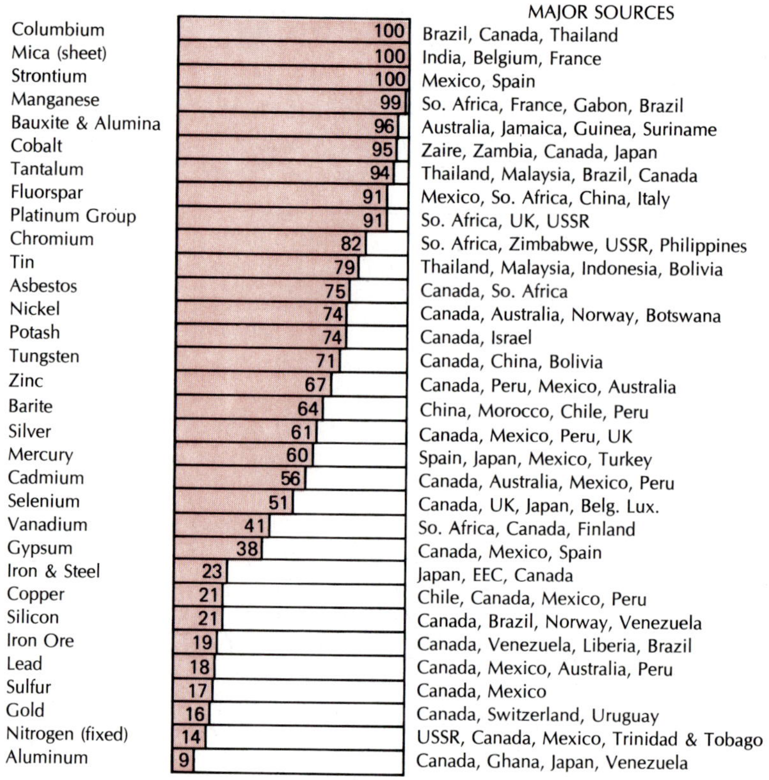

MAJOR SOURCES

| Commodity | Value | Major Sources |
|---|---|---|
| Columbium | 100 | Brazil, Canada, Thailand |
| Mica (sheet) | 100 | India, Belgium, France |
| Strontium | 100 | Mexico, Spain |
| Manganese | 99 | So. Africa, France, Gabon, Brazil |
| Bauxite & Alumina | 96 | Australia, Jamaica, Guinea, Suriname |
| Cobalt | 95 | Zaire, Zambia, Canada, Japan |
| Tantalum | 94 | Thailand, Malaysia, Brazil, Canada |
| Fluorspar | 91 | Mexico, So. Africa, China, Italy |
| Platinum Group | 91 | So. Africa, UK, USSR |
| Chromium | 82 | So. Africa, Zimbabwe, USSR, Philippines |
| Tin | 79 | Thailand, Malaysia, Indonesia, Bolivia |
| Asbestos | 75 | Canada, So. Africa |
| Nickel | 74 | Canada, Australia, Norway, Botswana |
| Potash | 74 | Canada, Israel |
| Tungsten | 71 | Canada, China, Bolivia |
| Zinc | 67 | Canada, Peru, Mexico, Australia |
| Barite | 64 | China, Morocco, Chile, Peru |
| Silver | 61 | Canada, Mexico, Peru, UK |
| Mercury | 60 | Spain, Japan, Mexico, Turkey |
| Cadmium | 56 | Canada, Australia, Mexico, Peru |
| Selenium | 51 | Canada, UK, Japan, Belg. Lux. |
| Vanadium | 41 | So. Africa, Canada, Finland |
| Gypsum | 38 | Canada, Mexico, Spain |
| Iron & Steel | 23 | Japan, EEC, Canada |
| Copper | 21 | Chile, Canada, Mexico, Peru |
| Silicon | 21 | Canada, Brazil, Norway, Venezuela |
| Iron Ore | 19 | Canada, Venezuela, Liberia, Brazil |
| Lead | 18 | Canada, Mexico, Australia, Peru |
| Sulfur | 17 | Canada, Mexico |
| Gold | 16 | Canada, Switzerland, Uruguay |
| Nitrogen (fixed) | 14 | USSR, Canada, Mexico, Trinidad & Tobago |
| Aluminum | 9 | Canada, Ghana, Japan, Venezuela |

**Fig. 19-14.** Import reliance (imports minus exports) of the United States for various mineral commodities in 1984. (U.S. Bureau of Mines)

some future date withhold resources. At present the United States is restricting trade with South Africa because of the latter's policy of apartheid; yet South Africa has been providing us with several key mineral commodities (see Fig. 19-14).

Greater exploration and exploitation of our own resources have been urged by some people to remedy this situation. The minerals, it is said, are there—they simply have not been located. Others hold the opinion that valuable resources are not available because they are protected within national parks or wilderness areas. Still others advocate increased production from lower-grade ore bodies and suggest that there is virtually no limit to the technology for extracting ore from such bodies, provided the financial incentive is there.

Before we start wholesale exploitation of our own resources, however, we should realize that our actions might severely damage the economies of developing countries that now supply us with minerals and other resources, undermining even further our relations with those countries. Or we might ask ourselves what the consequences would be if national parks—visited annually by millions of vacationers from the United States and abroad—and wilderness areas—protectors of delicate ecosystems—were opened to commercial mining. It becomes apparent that the solution is extremely complex, involving sharing of resources among nations and increasing efforts to conserve and to recycle materials. Some warn that the energy necessary for extraction could be a major limiting factor for the future of some ores, many vital to our civilization.

## ENERGY

Not too long ago, firewood was the principal source of energy in most places on earth and animals and slaves did much of the hard, less-

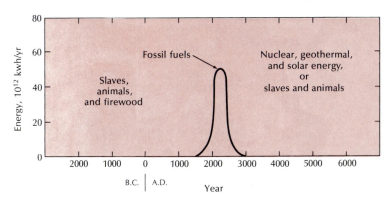

**Fig. 19-15.** Diagram illustrating the energy sources used by people and showing the very brief time span over which fossil fuels will be important. (After M.K. Hubbert, 1962 "Energy Resources: A Report to the Committee on Natural Resources," National Academy of Sciences—National Research Council Publ. 1000–D, Washington, D.C.)

desirable work. The principal energy sources of the developed countries are the fossil fuels, which came into their own only a little more than 100 years ago. Few people realize how short their life span will be—both on a geological and a human time scale (Fig. 19-15). It is predicted that we will exhaust our supply of fossil fuel in 1000 years or less. Hence, it seems clear that nuclear and solar energy do have a future and that we also will have to modify our standard of living and lifestyle. It is possible that wood might be used in countries where the yield can be sustained without harmful effects on the environment. However, it is predicted that many developing countries will experience fuel-wood shortages before the year 2000.

## Formation of Fossil Fuels

Fossil fuels—coal and oil—are derived from organic remains. The ultimate source of energy for their formation is the sun, which is used by plants to convert water and carbon dioxide into various organic materials by **photosynthesis.** The sun's energy, therefore, is stored in living plants, but lost through oxidation when the plant decomposes. Fossil fuels form when some portion of the organic matter, however small, does not decompose and thus accumulates. Low-oxygen, or **reducing,** environments are necessary for this. You can well appreciate how slowly fossil fuels accumulate—millions to hundreds of millions of years—in contrast to the extremely rapid rate at which we are using them up.

The formation of coal and oil shale have already been discussed (chap. 17); here we will concentrate on oil and gas. Unlike coal, most oil and gas (petroleum) usually are not formed in the rock in which they are finally found. They form, instead, through complex chemical reactions and migrate to their final resting places in other rocks. It is the geologist's job to find these reservoir rocks.

Marine sedimentary rocks are the most common source of oil and gas (Fig. 19-16). During their deposition in offshore basins, a substantial amount of organic matter—mostly in the form of microscopic plant remains—was incorporated into the sediment. The organic matter was preserved and a series of complex reactions slowly converted it into oil and gas. Finally, the light, mobile oil worked its way toward the surface and eventually accumulated in permeable **reservoir rocks.** Such accumulations are called **oil pools.** An **oilfield** is usually defined as a group of pools or an isolated pool. The latter term is a misnomer because oil does not collect in pool-like basins. Rather, it fills voids in the rock, such as the spaces between sandstone grains or cracks and hollow solution features in limestone and dolomite. Impermeable **roof rocks** or **cap rocks,** most commonly shale, prevent the hydrocarbons from migrating to the surface and being dissipated. Carbonates are important reservoir rocks; about 20 percent of the hydrocarbons in North America and 50 percent worldwide are found in such rocks.

Several kinds of geological structures can effectively trap hydrocarbons, provided that a cap rock

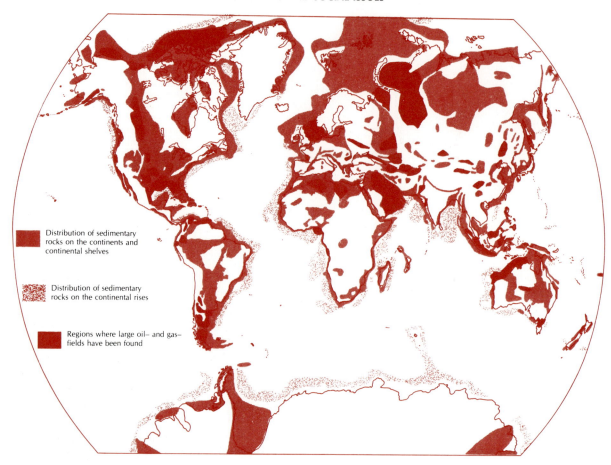

**Fig. 19-16.** World map showing areas of sedimentary rocks and regions with large oil- and gasfields. Notice the large potential areas offshore on the continental shelf and slope. (After B.J. Skinner, 1986, *Earth Resources, 3rd ed.,* Prentice-Hall)

is present (Fig. 19-17). The most simple trap is an anticline in which hydrocarbons migrate to the arch at the top of the structure. More complicated are salt domes, which are formed by the intrusion of the salt through such permeable beds as sandstones. Faulting can also place impermeable beds against permeable beds and so provide favorable environments for the entrapment of oil. Still more complicated traps, largely because they are so difficult to locate, are **stratigraphic traps.** Usually these occur at unconformities where dipping reservoir beds beneath the unconformity are sealed

by an overlying cap rock. Although oil and gas commonly are found together, gas can occur by itself.

Another oil-bearing deposit, the curious **tar sands,** contains oil so viscous that it cements the grains of sand together. The consistency of tar sands normally prevents their removal by pumping; to be recovered from depth they must first be heated to reduce the viscosity. Fortunately, the world's largest tar sand deposit, located in the Athabasca region of northern Alberta, lies close enough to the surface to be easily mined—the present ex-

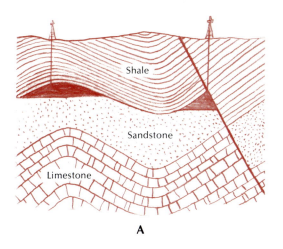

A

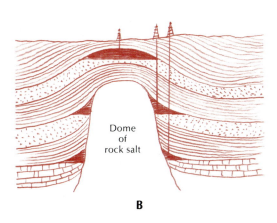

B

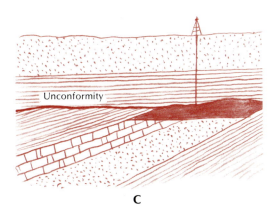

C

traction rate is over 100,000 tons per day. Large deposits are also found in Trinidad, Venezuela, and the USSR.

## FOSSIL FUEL RESOURCES AND THE FUTURE

### Coal

Worldwide, coal is fairly abundant, with the USSR and the United States holding the greatest share. The technologically and economically recoverable amount approaches 1 trillion tons (Fig. 19-18). At the 1982 rate of production of slightly over 4 billion tons, the supply should last about 250 years. However, if estimates of the ultimate production of 4 to 7 trillion tons are correct, coal could last for about 1000 years. Even if this is true, the production curve should follow the form depicted in Figure 19-7, curve B, which predicts that maximum worldwide production will take place between 2100 and 2150, and thereafter decline.

In the United States the distribution of coal is widespread (Fig. 19-19). Production grew between 6 and 7 percent per year until about 1910 when the switch to oil and gas began (Fig. 19-20). It then leveled off to a constant rate until 1972 when production again began to rise. How long our reserves—estimated at about 480 billion tons—will last will be determined by how much we increase the annual rate of production over the 1972 figure of 0.5 billion tons. Simple calculations show that with no increase in production rate, coal will last a long time. Recent administrations have called for a 5 to 10 percent increase in coal production (Fig. 19-20), but few people know what such growth will mean to the life expectancy of the coal industry. Al Bartlett in his lectures

**Fig. 19-17.** Various geological settings for the occurrence of oil and gas (black). In each case gas would rest on the oil; any water present would underlie the oil, which is less dense than water. **A.** Oil trapped at the crest of an anticline capped by shale. To the right, downfaulted shale seals a sandstone unit. **B.** Rocks domed by a salt intrusion form an anticlinal trap as in **A;** traps are also formed along the margins of the dome by an impermeable salt cap. **C.** Shales above a major unconformity form the cap for dipping reservoir rocks beneath the unconformity.

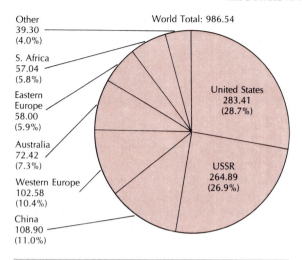

**Fig. 19-18.** Estimated recoverable coal reserves of the world, 1980, in billions of tons. (Energy Information Administration, 1985)

appropriately points out that supplies of coal are plentiful today because for 60 years no growth took place in the coal industry. Had we continued at the pre-1910 growth rate, we would be out of coal by the end of this century.

## Oil

The outlook for oil is worse than that for coal. Distribution is uneven and the resources and proved reserves lie mainly outside the Western Hemisphere (Fig. 19-21). Even so, the demand for oil in the United States as well as in other countries continues to grow, and what we do not produce ourselves we import at sometimes wildly fluctuating prices. In the United States the time of peak production has passed. The discovery of large reserves in Alaska is important, but, put in per-

**Fig. 19-19.** Various grades of coal deposits in the United States. (U.S. Department of the Interior)

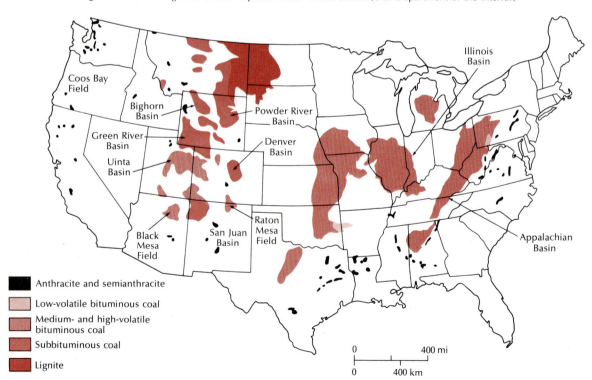

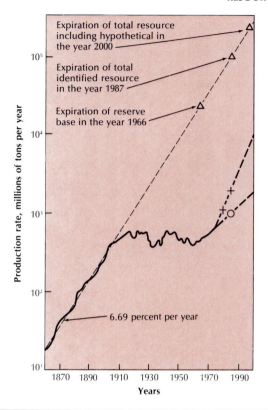

**Fig. 19-20.** History of coal production in the United States. For future production, the goals of the Ford administration (+) and the Carter administration (o) are indicated. (From A.A. Bartlett, 1980, ''Forgotten Fundamentals of the Energy Crisis,'' *Journal of Geological Education,* vol. 28, pp. 4–35, after Hubbert, 1968)

cessful exploration and greatly improved oilfield technology would be needed to ensure that every last drop of oil is squeezed from the reservoir rocks. M. King Hubbert, a geologist and an energy expert who has worked both with Shell Oil Company and the USGS, has long been a leader in predicting the future of fossil fuel production. In 1969 he predicted that by the year 2000, if not sooner, maximum world oil production would be attained. Subsequently, production would decline. In 60 years, he estimates, 80 percent of the available oil will have been produced, and beyond the year 2075 there will be little production. These figures still seem to be good estimates.

Finding new oilfields alters the estimates just quoted only slightly. For example, a relatively new field in Mexico (reserves of 46 billion barrels, total production of about 200 billion barrels) would satisfy the world's needs for only 2 to 3 years. Because of dwindling traditional supplies, there is increased interest in offshore drilling, but such drilling is not without great risk, both financial and environmental.

A continuing dependence on a finite resource such as oil is no longer feasible in a world with a growing population, increasing per capita demands for fuel, a yearning for an adequate standard of living by people in developing countries, and changing politics. Yet the search for new fields continues in the face of an increasing number of dry holes and escalating drilling costs.

Finally, a quote from Earl T. Hayes, former chief scientist of the U.S. Bureau of Mines, is appropriate:

It must be recognized that the United States never had and never will have the petroleum resources to sustain indefinitely the production levels of the last 25 years. In effect, we have been living off our capital all this time and cannot postpone the day of reckoning indefinitely. Talk of rising petroleum (and gas) production for long periods is both immoral and nonsensical. Whatever slight gain might be achieved for a very few years will be at the expense of the youth of today. Predictions of sustained increased production deny the records of 50 years of experience with the exploration, development, and extraction cycle of liquid hydrocarbons. There is a finite amount of easily recovered petroleum in this country, and no act of Congress or false

spective, if the Alaskan fields were our sole source of oil, they would supply the needs of the United States for only about 6 years. Estimates are that we have 20 to 40 more years of oil production at the 1979 rate of extraction.

We can forecast what is in store for the world by comparing oil production with the estimated total resource. From 1890 to 1970 world production grew at about 7 percent per year, which means that the amount produced and used doubled every decade. Put even more dramatically, at this rate of growth the amount used during any one decade is equal to the total consumption of all previous time (Fig. 19.22). Clearly, more suc-

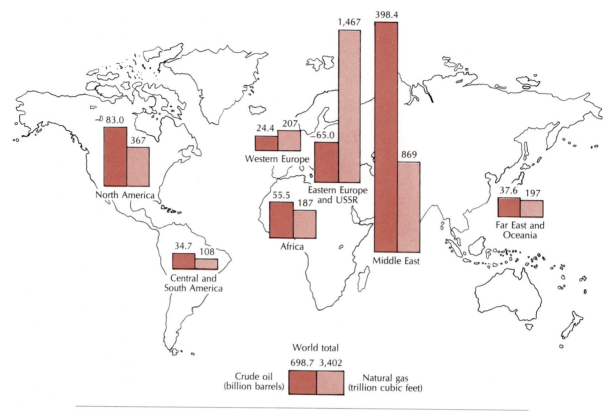

**Fig. 19-21.** Proved crude oil and natural gas reserves, 1984, for regions of the world. One billion barrels of oil are equal to approximately 5.3 trillion ft.³ of natural gas. (Energy Information Administration, 1985)

optimism of government, industrial, or academic planners can add to our natural resource base.*

Tar sands and oil shales as oil sources are largely untapped. The estimated resource oil in tar sands is large—some 2.5 to 6 trillion barrels, which is considerably more than the Middle East resource. Oil shales in the United States yield between 10 and 65 gallons per ton of rock; in some thin layers the yield may be as high as 140 gallons per ton (42 U.S. gallons = 1 barrel). On average, the yield is close to 1 barrel of oil from 2 tons of rock. Present-day reserves in the United States are about 80 billion barrels; the potential resource could be 2 trillion barrels or more. For the world the known resource of the higher grade rocks is about 910

*E.T. Hayes, 1979, Energy resources available to the United States, 1985 to 2000, *Science*, vol. 203, pp. 233–34.

billion barrels; this, however, is only 4 percent of the estimated total resource of higher-grade rocks, and the latter is only 1 percent of the estimated total resource of all grades of oil shale. The figures are staggering; for example, the estimated amount of oil in the highest grade rocks is eight times the estimated world oil resources. Low world oil prices at present have not made this petroleum resource, which is expensive to extract, very attractive.

### Natural Gas

World proved natural gas reserves are large (see Fig. 19-21), amounting to about 63 times the 1982 production of 1.5 trillion m³ (54 trillion ft³). In the United States production peaked in 1972–73 and has declined since. Proved reserves and undiscovered recoverable resources are estimated at 23 to

25 trillion m$^3$ (800 to 900 trillion ft$^3$), or 44 to 50 times the 1982 production of one–half trillion m$^3$ (18 trillion ft$^3$).

## Fossil Fuels in Agriculture

Today the agricultural industry consumes more petroleum than any other industry in this country—a fact that is probably not well known. Oil is used for running machinery on the farm, pumping water for irrigation, manufacturing and distributing fertilizers and pesticides, and transporting food-stuffs. The energy requirements of high-technology agriculture, or agribusiness, are so high that there is some question as to how long it can continue.

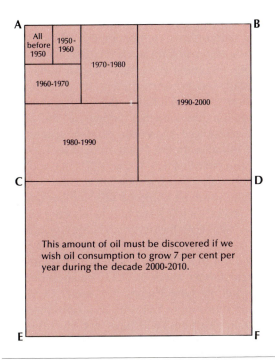

Fig. 19-22. Relative amounts of oil used each decade as a result of a 7 percent annual increase in consumption. At this rate the time required to double consumption is a decade; the diagram clearly shows that the quantity consumed in any one decade is equal to the total of all preceding years. For example, the amount represented by rectangle CDFE is equal to that represented by ABDC. Hence production must double each decade to match consumption. The area of ABDC is an approximate representation of the known world oil reserve. (After Bartlett, 1980, "Forgotten Fundamentals of the Energy Crisis," *Journal of Geological Education,* vol. 28, pp. 4–35, from M. Iona, 1977)

For example, it was recently estimated that the equivalent of about 112 gallons of gasoline was required to grow the food for the average American. If the world were to eat the way Americans eat and use our technology to grow their crops and produce their livestock, the world oil reserve would be depleted in less than 50 years. Or, put in a different way, by the end of the next population doubling (41 years), the percentage of people starving would remain the same, but the actual number would have doubled. Data presented by Worldwatch Institute show that the annual growth in energy use for world agriculture is over 3 percent.

## The Environmental Cost of Fossil-Fuel Use

The use of fossil fuels has been at the cost of some environmental degradation and, as their consumption increases, we must be watchful that the environment does not degrade further. The problems have to do with both the land surface and the atmosphere.

Most people are aware of the unsightly pits and spoil-piles left by the surface mining of coal. Large sums are being spent to landscape and revegetate the sites, with some success.

If the United States begins to extract large amounts of oil from its extensive oil shale deposits, the environmental impact could be massive. When the oil company Exxon announced plans for an oil shale plant with an output of 47,000 barrels per day, it was estimated that processing that amount would require roughly 94,000 tons of rock per day. President Carter had hoped for an industry producing 400,000 barrels per day (about 800,000 tons of rock per day) by 1990. Going one step further, to obtain all of our oil from shale means the processing of 13 billion tons of rock per year or 130 million freight-train cars full. Bartlett makes another calculation of the volume of rock that must be handled: to extract 1 million barrels of oil per day requires the processing each nine months of an amount of rock equivalent to that removed in building the Panama Canal. To put this into proper focus, the United States currently uses about 16 million barrels of oil each day.

The problems involved in processing oil shales are formidable. Considerable amounts of both water and energy are needed: The estimates are 3 to 5 barrels of water for each barrel of oil produced and the energy equivalent of 0.25 barrel of oil to process 1 ton of shale. Stream and air pollution must be avoided and mining areas revegetated. Our high-grade deposits occur in semiarid parts of Wyoming, Utah, and Colorado where pollution can have a marked effect and where water supplies are not overly abundant. Yet another problem is socioeconomic—creating communities in sparsely populated areas for the workers and their families. Dislocations have already occurred in Colorado when companies abandoned oil shale operations because they were unable to compete on the world market.

*Acid rain* is much in the news these days. Examples of damage attributed to acid rain are decimated or wholly eradicated fish populations in the lakes of Scandinavia and the northeastern United States, forest and crop damage in Western Europe, and an alarming increase in the weathering of buildings and outdoor art (see chap. 10). The damage is spread over large areas and knows no boundaries. For example, pollution from the United Kingdom and Western Europe could be the source of damage in Scandinavia; that in the northeastern United States has been cited as affecting parts of Canada, and that in the southwestern United States could affect the Rocky Mountain states.

The oxides of sulfer and nitrogen are the main source of acid. Coal-fired power plants, industrial facilities, and motor vehicles spew tons of these oxides into the atmosphere. Carried downwind, they are deposited as dry matter or as acid rain. It is the combination of these oxides with water and oxygen that creates the acids—sulfuric and nitric—that do the damage.

Some geological processes and materials can absorb or neutralize this acidity: For example, most weathering reactions are enhanced by acid waters and, in the process, the acid is used up. However, in many reactions involving silicate minerals, the rate of weathering may be so slow that little acid is consumed. In contrast to this, the weathering of calcite is rapid and here, too, acid is used up. Many organic- and clay-rich soils can absorb much of the newly introduced acid. Watersheds with different rock types and soils will not be affected to the same degree by acid deposition. For example, the thin soils of glacially scoured granite basins have little capacity to absorb acids and therefore lakes in these basins rapidly become acid. On the other hand, lakes in limestone terrain or in areas with thick soils with the appropriate properties may be little affected. One way to decrease the acidification of lakes is to add crushed limestone dust to them from airplanes. This is only a short-term solution, however, and for the Scandinavian lakes alone it could cost $60 to $100 million.

The burning of fossil fuels also increases the level of $CO_2$ in the atmosphere. Instruments on Mauna Loa volcano in Hawaii are far from local sources of pollution and so may give the best measure of global increases in $CO_2$. Between 1958 and 1984 the increase amounted to about 9 percent.

Increases in atmospheric $CO_2$ could increase global atmospheric temperatures. When atmospheric carbon dioxide levels increase, the balance between incoming and outgoing solar radiation is disrupted; this prevents the escape of long-wave radiation. Subsequently, the atmosphere is warmed—a phenomenon known as the *greenhouse effect* (see chap. 14). Models suggest that mean global warming could amount to a temperature rise between 1.5° C and 4.5° C (3 and 8° F) by the middle of the twenty-first century; in polar areas the warming could be two to three times as high. Several factors keep $CO_2$ levels from becoming too high: both the ocean and vegetation absorb some of the excess $CO_2$. However, it has been pointed out that clearcutting and burning in tropical rain forests alters the $CO_2$ balance because there are no longer trees to absorb $CO_2$, and burning sends more $CO_2$ into the atmosphere.

A warming trend will have two important effects. One is its influence on weather patterns and therefore on agriculture. The second is the melting of glaciers and a rise in sea level. It is difficult to estimate, but one report suggests maximum rises of as much as 1–2 m (39–79 in.) per 100 years over the next 200 to 500 years. Obviously, coastal areas—where a large number of the world's people dwell—would be most affected.

Both the greenhouse effect and acid rain would occur on a wider scale if we increased our use of coal in an effort to cut down on our use of the scarcer fossil fuels. There are many unknowns, however. One is that the impact of greenhouse warming could be offset by the gradual cooling of the earth whenever it moves into the next ice age.

## GEOTHERMAL ENERGY

Areas in which hot water and steam occur naturally are known as **geothermal regions.** The development of such areas is in its infancy, but geothermal energy has a high potential for driving turbines to produce electricity as well as for other uses.

Producing geothermal areas as well as developing those with potential are all in regions of recent volcanism. Magma or masses of hot rock at depth are the primary sources of heat that is transferred to the surface by rising hot water and steam. Of the water involved in such a system, over 95 percent originates at the surface and is circulating; the rest originates at the heat source in the form of magmatic steam.

The first commercial geothermal venture took place in Italy in 1904. Since that time producing areas have been developed in Japan, the Kamchatka Peninsula (USSR), China, New Zealand, Turkey, Mexico, El Salvador, and Iceland. Now there are about 20 countries using geothermal energy. Geothermal energy is particularly appealing to such nations as Japan, where oil and coal are almost nonexistent, hydroelectric power is limited, and about seventy percent of the fuels consumed must be imported. In Iceland volcanic steam is used to heat buildings and steam pipes that run through fields to warm the soil, so that crops that ordinarily would not survive in that severe climate can be grown.

In the United States the first and the only successful large-scale exploitation of geothermal energy is at The Geysers in northern California, about 145 km (90 mi) north of San Francisco. In spite of its name, the area has no geysers, but steam rises from hot springs, wells, and fumaroles (vents from which gases and vapor rise). Steam from wells several thousand meters deep is piped to turbines that produce sufficient electricity to serve the electrical needs of over half a million homes.

Like most resources, the production of geothermal energy presents problems. The hot waters carry materials in solution that can precipitate out as solids and clog pipes or pollute local water supplies. Thermal pollution or surface subsidence accompanying water withdrawal also can occur. Finally, if reserve hot water and steam are used before more can be generated from the depths of the earth—an all too familiar story—then geothermal energy, too, will be in short supply.

Two new sources of geothermal energy are now being investigated. One involves heating water by using underground hot rocks recently crystallized from magma. Such rocks are not in contact with underground water, thus they differ from the geothermal systems just described. What nature has not provided, however, geoscientists have. In Los Alamos, New Mexico, water under high pressure was forced down one of two holes drilled to a depth of 10 km (6.2 mi) where the temperatures are near 204° C (400° F) (Fig. 19-23). This created a system of cracks in the rock that previously was not porous enough to allow the transfer of water. Water was then pumped into the artificially cracked rocks where it heated and then was returned to the surface by way of the other hole, ultimately to drive a steam turbine. Results thus far have been encouraging.

The other new source is the hot saline water contained in sedimentary rocks beneath the Gulf coasts of Texas and Louisiana. When tapped with a drill hole, the water, which is under very high pressure, rises rapidly to the surface. The sources of energy are several: the force of the water itself, the heat contained in the water, and the natural gas dissolved in the water.

Because of the expanding technology of geothermal energy, it is difficult to assess this resource. More than likely it will contribute only a small percentage of the world's energy supply in the foreseeable future. One estimate is that geothermal energy, hydroelectric power, and solar energy will make up about 6 percent of the total energy supply by 2000. Locally, however, and for some countries, it is a valuable energy resource.

## NUCLEAR ENERGY

In 1974 the USGS was optimistic that nuclear power, which had provided 3.5 percent of the nation's electrical energy in 1972, would by 1980

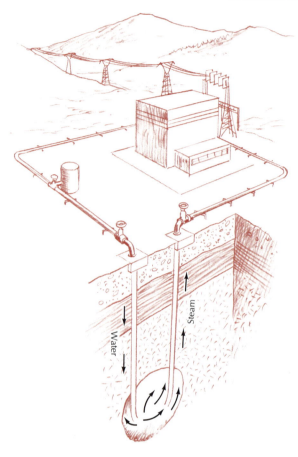

Water

Steam

**Fig. 19-23.** Drawing that shows the concept of extracting geothermal energy from dry hot rock. The arrows indicate directions of the water and steam flow. (From the University of California, Los Alamos Scientific Laboratory, 1979)

provide 21 percent of the total and by 2000 as much as 60 percent of the total. These predictions, however, have not been met, for in 1986 only about 16 percent of our electricity was generated by nuclear power. In some Western European countries, however, nuclear power accounts for up to 65 percent of the electricity generated. Here we will investigate briefly how nuclear power is generated and the problems that the use of nuclear energy poses.

Nuclear energy derives from two reactions, fission and fusion. Fission in a nuclear reactor is

achieved by bombarding uranium atoms with neutrons (Fig. 19-24), which produces lighter fission fragments, or different atoms, and at least two neutrons. The neutrons, in turn, encounter other uranium atoms, and the fission process is repeated. If continued, a chain reaction, such as that in an atomic bomb, occurs. Fusion, on the other hand, is just the opposite, that is, two light particles fuse, or join, to form a heavier element (Fig. 19-24). This same process fuels the sun, where hydrogen atoms fuse to form helium atoms, a reaction similar to that in the hydrogen bomb. The fusion reaction is considered by some the best possible energy source because hydrogen is so abundant, but the technology of controlling the reaction so that it can be used safely has not been worked out. However, fusion has the distinct advantage over other nuclear reactions of leaving very little radioactive waste as a by-product.

In both fission and fusion reactions, energy is released. This energy can be used to heat water, produce steam, and generate electricity, but only if the reaction is controlled. Because fission of uranium can be controlled, it has become the basis for our present nuclear power plants. The advantage of using uranium is that only small amounts of it are needed to produce the same amount of energy produced by large amounts of fossil fuels.

Of the three different isotopes used in nuclear reactors, only uranium–235 occurs naturally. Uranium ores, however, contain only 0.7 percent of uranium–235. If we continue to fuel U.S. power plants with uranium–235, our supply will probably run out before the end of this century—even though the element is plentiful in the United States.

One way to supplement our limited supplies of naturally occurring nuclear fuels is to create fissionable material artificially. This can be done by neutron bombardment of uranium–238 and thorium–232—both of which are more abundant in nature than uranium–235—to create the fissionable products plutonium–239 and uranium–233, respectively. Because the reaction produces more fissionable material than it starts with, it is said to "breed" new material, hence reactors based on this reaction are called **breeder reactors.** Many technological problems have to be overcome, however, before breeder reactors can be used to

FISSION

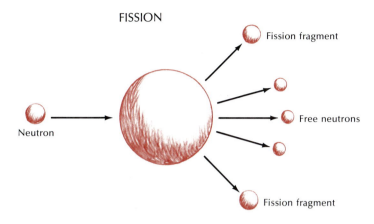

Neutron

Fission fragment

Free neutrons

Fission fragment

FUSION

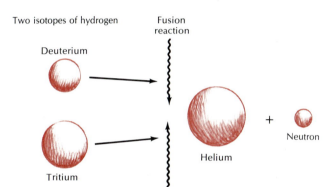

Two isotopes of hydrogen

Deuterium

Tritium

Fusion reaction

Helium

+ Neutron

**Fig. 19-24.** Diagrams illustrating fission and fusion.

generate electricity on a large scale. And another problem, of international import, is that the plutonium produced can be used to produce nuclear weapons.

The number of new nuclear power plants built has not fulfilled earlier predictions (Fig. 19-25). In the mid-1980s there were about 281 power reactors in the 10 leading nuclear power countries, ranging from 84 in the United States to 6 in Belgium. The reactors generated 13 percent of the electricity worldwide and accounted for 3 percent of the total energy production. If countries are to continue to rely on nuclear power, careful planning is required: The average life span of a plant is 30 years and decommissioning old plants is an expensive proposition—costing at least $50 million per plant.

The major problems facing the nuclear power industry are safety and the safe disposal, for all time, of radioactive waste. All facilities must be located in tectonically stable, earthquake-free areas. However, the stability of many sites in California has been questioned. Safety also has to do with the operation of the plant itself, and there is virtually no room for error. The accidents at the Three Mile Island plant near Harrisburg, Pennsylvania, in 1979, and at Chernobyl, USSR, in 1986 have led to questions on the effectiveness of safety systems in nuclear power plants and whether the risks involved in providing energy in this way are

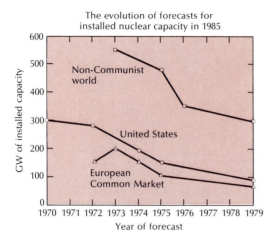

The evolution of forecasts for installed nuclear capacity in 1985

**Fig. 19-25.** Diagram showing the changes in predicted installed nuclear capacity for 1985. The vertical scale is in GW; 1 GW is equal to 1 million KW, the equivalent of 42,000 barrels of oil per day in an oil plant operating full time. All the curves show that, with time, the predicted capacity for the year 1985 has been drastically reduced.

not altogether too great. Only Sweden has reversed its course regarding nuclear power. By 2010 the country hopes to phase it out—a decision made before the accident at Chernobyl.

Disposing of high-level radioactive wastes is a problem that has not been solved. Far-fetched proposals include propelling the wastes into space or burying them in Antarctic ice or ocean-floor sediments. Criteria for land-based sites are (1) isolation from the biosphere, from soil, and from surface and groundwater; (2) a stable geological setting (no nearby volcanism or earthquakes and associated tectonism); and (3) rocks with very low permeabilities.

In the United States three kinds of sites have been studied. Salt beds in the semiarid western United States were once considered especially safe, but they have not proved to be as impermeable and stable as originally thought. Near Hanford, Washington, basaltic terrain is being considered, but there is fear of leakage of contaminants into the Columbia River along the joints and fractures common to most basalt. At present welded tuffs in the arid environment of the Nevada Test Site meet most of the three criteria mentioned, and

these are being studied intensely. Thus far, a final site has not been chosen.

## CONCLUSION

The statistics for resources and reserves can be misleading because they do not reveal that some resources cannot be profitably extracted and that the time gap between discovery of a reserve and the production of energy or materials from it could be years away. The environmental cost may be more than we care to pay now or to pass on to future generations. If we opt for rapid growth rates and all that they entail, how will other countries view our action, let alone history? And by how much more will we despoil our planet and destroy other, possibly more important natural resources? The words of Earl Hayes on the subject are sobering:

The facts point to the inescapable conclusion that exponential growth of energy supply is coming to an end in the United States. Energy and gross national product have risen 3 to 3½ percent a year since 1940, and a decrease in the energy growth rate to less than 1 percent a year by 2000 will occasion some fundamental national problems for which we have no precedent. The involuntary conservation brought on by higher prices and decreased supplies will be exceedingly painful for an unprepared American public.

We have sufficient energy resources to supply our basic needs for many decades, but the costs will rise continually. The country still does not understand the problem. The layman wants to believe in inexhaustible, cheap gasoline and in this has been supported by many unsubstantiated claims. The time has come to realize that no miracle is imminent and we must make do with what we have. We will never again have as much oil or gas as we have today, nor will it be as cheap. Nuclear energy has been a major disappointment. Solar energy will be slow in developing and, contrary to popular opinion, quite expensive. Coal is the only salvation for the next few decades.

In the last analysis, we have entered into a massive experiment to determine what effect energy growth has on economic growth, or how much we can slow the machine down and still maintain a democratic, capitalistic form of government.*

*E.T. Hayes, 1979, Energy resources available to the United States, 1985 to 2000, *Science,* vol. 203, p. 239.

The United States plays a significant role in the future of world energy. With only 6 percent of the world's population, we use one-third of total current energy production. We could still enjoy life, as do many other people in many other countries, and use less energy. Many industrial nations with excellent standards of living have a per capita consumption of energy less than half ours. Surely, it would be possible to moderate our gluttonous use of world resources. By implementing some of the following ideas as outlined by Bartlett, this might be achieved:

1. Curtail and eventually stop population growth (achieve zero population growth) and curtail consumption. Even if we in the United States were able to heat one-half the buildings in which we live and work with solar energy, the resulting 10 percent savings in fossil fuels could be wiped out by 2 year's growth at 5 percent.

2. Conserve and recycle resources and materials to the fullest extent possible. We might attempt to change some of our energy-consuming habits that do not contribute much to the quality of our way of life.

3. Intensify research in all forms of energy to determine how they best can be utilized.

More and more people are concluding that conservation is the key to the energy problem. A quote from *Energy Future,* a report from the energy project at the Harvard Business School seems an appropriate conclusion:

The United States might use 30 or 40 percent less energy than it does, with virtually no penalty for the way Americans live—save that billions of dollars will be spared, save that the environment will be less strained, the air less polluted, the dollar under less pressure, save the growing and alarming dependence on OPEC oil will be reduced, and Western society will be less likely to suffer internal and international tension. These are benefits Americans should be only too happy to accept.*

*R. Stobaugh and D. Yergin, eds., 1979, *Energy Future,* Ballantine, N.Y., p. 229.

## SUMMARY

1. Resources are commodities that are useful or essential to us or our way of life. Resources are classed as *renewable,* those that can be replenished, and *nonrenewable,* those that cannot be replenished. A *paramarginal* resource is technologically possible to recover; recovery of a submarginal resource is not profitable. A *reserve* is the estimated amount of a resource that can be recovered with present-day technology.

2. Population growth and consumption of resources are increasing exponentially or at a fixed percentage rate.

3. Metallic ore deposits can be classified as *abundant* or *scarce*. The abundant metals are iron, aluminum, manganese, magnesium, and titanium. All other metals are scarce, and many are in short supply.

4. Igneous, metamorphic, sedimentary, and weathering processes are responsible for the formation of the ore deposits.

5. Much of our energy is derived from coal and petroleum, both of which are nonrenewable resources, and could be depleted in a relatively short period of time.

6. Oil and gas collect in reservoir rocks, where they accumulate to form *oil pools.* An *oil field* is one or more such pools. Several different geological structures can trap oil and gas, if *roof rocks (cap rocks)* are present.

7. Geothermal energy is mainly available in areas of fairly recent igneous activity, but areas that can be exploited are limited in number.

8. Nuclear energy is considered a leading potential source of energy, but major problems, such as the possibility of catastrophic nuclear accidents, the safe disposal of nuclear wastes, and dwindling supplies of uranium fuel, must be solved before its potential can be realized.

## QUESTIONS

1. What will be some of the environmental effects of attempts to increase production of fossil fuels and other minerals as the world's population increases exponentially?
2. How does the plate tectonics theory explain ore deposits?
3. What hypotheses can you advance to explain the lack of oil and oil shale in a folded mountain belt, such as the European Alps, when other belts have an abundance of these resources?
4. Study the geological map of a particular region, state, or country (all geology departments have them on their walls), and outline prospective areas for mineral and oil exploration.

## SELECTED REFERENCES

Abelson, P.H., and Hammond, A.L., eds. 1976, Materials: renewable and nonrenewable resources, American Association for the Advancement of Science, Washington, D.C.

Bartlett, A.A., 1980, Forgotten fundamentals of the energy crisis, Journal of Geological Education, vol. 28, pp. 4–35.

Bricker, O.P., ed., 1984, Geological aspects of acid deposition, Butterworth, Boston.

Brown, L.R., et al., 1986, State of the world—1986, W.W. Norton, New York.

Brown, L.R., et al, 1987, State of the world—1987, W.W. Norton, New York.

Cameron, E.N., 1986, At the crossroads—the mineral problems of the United States, Wiley, New York.

Committee on Nuclear and Alternative Energy Systems, National Research Council, 1980, Energy in transition, 1985–2010, W.H. Freeman, San Francisco.

Guilbert, J.M., and Park, C.F., Jr., 1986, The geology of ore deposits, W.H. Freeman, San Francisco.

Hayes, E.T., 1979, Energy resources available to the United States, 1985 to 2000, Science, vol. 203, pp. 233–39.

Howard, A.D., and Remson, I., 1978, Geology in environmental planning, McGraw-Hill, New York.

Hubbert, M.K., 1971, The energy resources of the earth, Scientific American, vol. 225, no. 3, pp. 60–70.

Lindholm, R.C., 1980, The oil shortage—a story geologists should tell, Journal of Geological Education, vol. 28, pp. 36–45.

Muffler, L.J.P., ed., 1979, Assessment of geothermal resources of the United States—1978, U.S. Geological Survey Circular 790.

Park, C.F., Jr., 1978, Critical mineral resources, Annual Review of Earth and Planetary Sciences, vol. 6, 305–24.

Pimental, D., et al., 1973, Food production and the energy crisis, Science, vol. 182, pp. 443–49.

Ruedisili, L.C., and Firebaugh, M.W., eds., 1982, Perspectives on energy: issues, ideas, and environmental dilemmas, 3rd ed., Oxford University Press, New York.

Skinner, B.J., 1986, Earth resources, 3rd edition, Prentice-Hall, Englewood Cliffs, N.J.

Stobaugh, R., and Yergin, D., eds. 1979, Energy future, Ballantine, New York.

Weiner, J., Planet earth, Bantam, New York.

# CONVERSION TABLE

## Length

| | | | |
|---|---|---|---|
| Inches | × | 2.54 | = Centimeters |
| | × | 0.0254 | = Meters |
| | × | 0.0833 | = Feet |
| | × | 0.0277 | = Yards |
| Feet | × | 0.3048 | = Meters |
| | × | 12. | = Inches |
| | × | 0.3333 | = Yards |
| | × | 0.00019 | = Miles |
| Miles | × | 1609.34 | = Meters |
| | × | 1.609 | = Kilometers |
| | × | 63360. | = Inches |
| | × | 5280. | = Feet |
| | × | 1760. | = Yards |
| Centimeters | × | 0.3937 | = Inches |
| | × | 0.0328 | = Feet |
| | × | 0.0109 | = Yards |
| | × | 0.0100 | = Meters |
| Meters | × | 39.370 | = Inches |
| | × | 3.2808 | = Feet |
| | × | 1.1936 | = Yards |
| | × | 0.0006 | = Miles |
| Kilometers | × | 3280.83 | = Feet |
| | × | 1093.61 | = Yards |
| | × | 0.6214 | = Miles |
| | × | 1000. | = Meters |

## Mass

| | | | |
|---|---|---|---|
| Ounces | × | 28.3495 | = Grams |
| | × | 0.0625 | = Pounds |
| Pounds | × | 453.5923 | = Grams |
| | × | 0.4536 | = Kilograms |
| Tons (long) | × | 1016.05 | = Kilograms |
| | × | 1.106 | = Tons (metric) |
| | × | 2240. | = Pounds |
| | × | 1.12 | = Tons (short) |
| Tons (short) | × | 907.185 | = Kilograms |
| | × | 0.9072 | = Tons (metric) |
| | × | 2000. | = Pounds |
| | × | 0.8928 | = Tons (long) |
| Grams | × | 0.0352 | = Ounces |
| | × | 0.0022 | = Pounds |
| | × | 0.001 | = Kilograms |
| Kilograms | × | 35.2739 | = Ounces |
| | × | 2.2046 | = Pounds |
| Tons (metric) | × | 2204.62 | = Pounds |
| | × | 0.9842 | = Tons (long) |
| | × | 1.1023 | = Tons (short) |
| | × | 1000. | = Kilograms |

## Area

| | | | |
|---|---|---|---|
| Square inches | × | 6.4516 | = Square centimeters |
| | × | 0.000645 | = Square meters |
| | × | 0.006944 | = Square feet |
| Square feet | × | 929.03 | = Square centimeters |
| | × | 0.0929 | = Square meters |
| | × | 144. | = Square inches |
| | × | 0.1111 | = Square yards |
| | × | $3.587 \times 10^{-8}$ | = Square miles |
| Square miles | × | 2589988. | = Square meters |
| | × | 2.589988 | = Square kilometers |
| | × | $2.7878 \times 10^{7}$ | = Square feet |
| | × | $3.0976 \times 10^{6}$ | = Square yards |
| Square centimeters | × | 0.1550 | = Square inches |
| | × | 0.0011 | = Square feet |
| | × | 0.00012 | = Square yards |
| | × | 0.0001 | = Square meters |
| Square meters | × | 1550.0031 | = Square inches |
| | × | 10.7639 | = Square feet |
| | × | 1.1960 | = Square yards |
| Square kilometers | × | $1.0763 \times 10^{7}$ | = Square feet |
| | × | $1.1960 \times 10^{6}$ | = Square yards |
| | × | $1. \times 10^{6}$ | = Square meters |

## Volume

| | | | |
|---|---|---|---|
| Cubic inches | × | 16.387 | = Cubic centimeters |
| | × | $1.6387 \times 10^{-5}$ | = Cubic meters |
| | × | 0.00058 | = Cubic feet |
| | × | $2.1433 \times 10^{-5}$ | = Cubic yards |
| Cubic feet | × | 28316. | = Cubic centimeters |
| | × | 0.0283 | = Cubic meters |
| Cubic centimeters | × | 0.0610 | = Cubic inches |
| | × | $3.5314 \times 10^{-5}$ | = Cubic feet |
| | × | $1.308 \times 10^{-6}$ | = Cubic yards |
| | × | $1. \times 10^{-6}$ | = Cubic meters |
| Cubic meters | × | 61023.74 | = Cubic inches |
| | × | 35.3146 | = Cubic feet |
| | × | 1.3079 | = Cubic yards |
| Cubic kilometers | × | 0.240 | = Cubic miles |

## Velocity

| | | | |
|---|---|---|---|
| Miles per hour | × | 1.60934 | = Kilometers per hour |
| Kilometers per hour | × | 0.62137 | = Miles per hour |

## Temperature

To convert from °Fahrenheit to °Celsius, subtract 32 and multiply by 5/9
To convert from °Celsius to °Fahrenheit, multiply by 9/5 and add 32

# GLOSSARY

**aa**  A lava flow with a rough, blocky appearance.

**ablation**  The processes by which snow or ice is lost from a glacier.

**abrasion**  The mechanical wearing, grinding, or scraping of a rock surface by the friction and impact of moving rock particles.

**absolute time**  The best estimate in years of the actual age of an object or event. Generally determined by the decay of radioactive elements.

**abyss**  The deep ocean floor.

**abyssal hills**  Hills that rise up to 1000 m (3280 ft) above the deep ocean floor.

**abyssal plains**  The relatively extensive low-relief plains of the deep-ocean floor.

**accreted terrane**  "Foreign" bodies of oceanic rock suites and sediments plastered against a continental margin. Accreted bodies that are thought to have moved great distances before "docking" are called *suspect terrane*.

**acid rain**  Rainfall that has become more acid because of pollution of the atmosphere.

**adobe**  A clayey deposit mixed with silt. Originally used to make sun-dried bricks in the Southwest.

**aftershock**  A generally smaller earthquake that occurs after a larger one. These earthquakes are closely related in origin.

**agglomerate**  See *volcanic agglomerate*.

**aggradation (stream)**  Deposition of a stream's excess load on the channel bottom.

**A horizon**  The dark-colored surface layer, or topsoil, of the soil profile.

**air-fall ash**  Volcanic ash that settles out of the air and can be deposited thousands of kilometers from the volcano's vent.

**alluvial fan**  A fan-shaped deposit of sand and gravel deposited by a stream at the base of a mountain, usually in an area of arid or semiarid climate.

**alluvium**  Any clastic material deposited by a stream along its course.

**alpine chain**  A complex linear mountain belt made up of batholithic igneous rocks, dynamothermal metamorphic rocks, and complexly folded and faulted sedimentary rocks.

**alpine glacier**  A mountain glacier that moves downslope under gravity, often following a preexisting stream valley. Also called a valley glacier.

**amphibole group**  A large group of ferromagnesian silicates with similar physical properties. Their most distinctive property is the possession of two good cleavages intersecting at 124° and 56°.

**amphiboles**  A group of dark, rock-forming ferromagnesian silicate minerals characterized by a double-chain arrangement of the $SiO_4$ tetrahedra. Hornblende, a black mineral, is one of the most common varieties.

**andesite**  A gray to grayish-black volcanic rock composed mostly of intermediate plagioclase, augite, hornblende, and biotite. Volcanic equivalent of a diorite.

**angle of repose**  The maximum slope at which relatively loose material will remain stationary without sliding downslope.

**angular unconformity**  An unconformity in which the beds above and below the unconformity meet at an angle.

**anion**  A negatively charged ion.

**anorthosite**  A variety of gabbro consisting almost entirely of coarse crystals of calcic-plagioclase.

**antecedent stream**  A stream that has maintained its original course despite the occurrence of local uplift or movement along its course.

**anthracite**  The most highly metamorphosed form of coal; it contains a higher percentage of carbon than bituminous coal.

**anticline**  A fold in which the limbs dip away from the hingeline; older beds are found toward the hingeline.

**aphanitic texture**  A crystalline rock texture in which most of the crystals are too small to be seen by the unaided eye.

**apparent polar wander curve**  The curve created by plotting on a map the succession of paleomagnetic poles for rocks of different ages from any one continent or lithospheric plate.

**aquiclude**  A subsurface layer too impermeable or too tight to accept or transmit water.

**aquifer**  A subsurface rock layer that readily yields water.

**arête**  A jagged, thin rock wall or ridge separating two glaciated valleys, generally near their heads.

**arkose**  A sandstone composed mainly of quartz and feldspar.

**arroyo**  The steep-sided, flat-floored channel of an ephemeral or intermittent stream.

**artesian well**  A well that taps a confined aquifer.

**ash**  Volcanic fragments the size of dust blown into the air during an eruption.

**ash-fall tuff**  Consolidated air-fall ash material.

**ash flow**  See *glowing avalanche.*

**asteroid belt**  A zone between Mars and Jupiter in which quantities of relatively small rocklike particles orbit the sun. Some meteorite showers originate in this zone.

**asthenosphere**  The mechanically soft zone extending from the base of the lithosphere to about 250 km (155 mi). In some places it corresponds in position to the *low-velocity zone.*

**asymmetrical fold**  One in which the two limbs make dissimilar angles with the axial plane.

**atmosphere**  The gaseous envelope that surrounds the earth. It is composed largely of nitrogen (78 percent by volume) and oxygen (21 percent).

**atoll**  A ring-shaped island with an interior lagoon and made up of the calcareous skeletons of marine animals.

**atom**  The smallest unit of an element, consisting of protons, electrons, and (in most cases) neutrons.

**atomic number**  A number characteristic of a chemical element and equal to the number of protons in an atom of that element.

**atomic ordering**  See *long-range ordering.*

**atomic weight**  Of an element a number that indicates, on the average, how heavy an atom of that element is compared to an atom of hydrogen.

**augen**  Shear-resistant minerals in cataclastic metamorphic rocks; they appear as relatively large, rounded clots or eyes.

**augite**  A dark ferromagnesian mineral of the pyroxene group with two good cleavages that intersect at 87° and 93°.

**aureole**  A halo of hydrothermally metamorphosed rocks that surrounds most batholiths.

**authigenic (marine) sediment**  Fine-grained sediment formed by the precipitation of ions in seawater.

**axial plane**  A plane connecting the hingelines in successive beds in a fold.

**back-arc upwelling**  A secondary zone of divergence of lithospheric plates that commonly forms between a continent and an island-arc system.

**backswamp**  A section of low-lying ground between a natural levee and the river bluff.

**backwasting**  A process by which valley slopes retreat essentially parallel to themselves.

**badlands**  Rough, steeply gullied terrain usually found in dry areas.

**bajada**  An apron of a desert mountain range formed from lateral coalescing of alluvial fans.

**bar**  A long linear accumulation of marine clastic sediment, aligned parallel to the coast.

**barchan**  A symmetrical, crescent-shaped dune that forms in deserts in which the wind direction is nearly constant, sand supply is limited, and the ground surface (bedrock) is hard.

**barrier islands**  Long narrow sand islands that are roughly parallel to the shore and are separated from it by a narrow body of water.

**barrier reef**  A long, narrow coral reef roughly parallel to the shore and separated from it by a fairly deep and wide lagoon.

**basal slip (glaciers)**  The sliding of glaciers over underlying material.

**basalt**  A dark volcanic rock, fine grained to aphanitic in texture and composed largely of pyroxene, calcic-plagioclase, and olivine. It is the most abundant volcanic rock.

**base level (stream)**  The low point to which most streams flow, utimately the ocean.

**basin (structure)**  A saucer-shaped synclinal structure that lacks an axis and in which the beds dip toward the center.

**batholith**  A pluton with an exposed surface area of at least 100 km² (40 mi²); generally granitic in composition.

**bauxite**  A chief ore of aluminum, formed as an end product of weathering in tropical climates.

**bedding plane**  The surface that parallels the layers in sedimentary rock.

**bed load**  The material moved along the bottom of a stream.

**bedrock**  The rock that underlies soil or other unconsolidated surface material.

**Benioff zone**  The planar dipping zone of shallow, intermediate, and deep earthquakes that is common to the Pacific Ocean margins.

**B horizon**  The subsurface horizon below the A horizon in the soil profile. Contains the most clay and is usually red to brown in color.

**biogenic (marine) sediment**  Fine-grained seediment formed from marine organisms.

**Biotite**  A common ferromagnesian mineral of the mica group. Sometimes called black mica.

**bird'sfoot delta**  A delta formed by many distributaries of a river and resembling a bird's talons.

**bituminous coal**  A soft black coal formed by the metamorphism of lignite.

**Bk horizon**  In arid-region soils, a layer of calcium carbonate beneath the B horizon.

**block faulting**  A type of faulting in which the crust is broken into a number of subparallel blocks that are displaced with respect to each other, primarily by dip-slip motion.

**body waves**  Seismic waves (*P* and *S* waves) that travel through the interior of the earth.

**bonding**  The joining together of atoms by electrical attraction.

**bottomset bed**  The horizontal strata at the bottom of a delta that is deposited progressively in front of the advancing delta front.

**boulder clay**  A glacial deposit consisting of rocks the size of boulders set in a clayey matrix.

**Bowen's reaction series**  The general sequence of crystallization in cooling magmas, going from minerals stable at higher temperatures to those stable at lower temperatures.

**braided stream**  A stream that flows in a network of wide, anastomosing channels separated by low bars or islands.

**breccia**  A clastic sedimentary rock composed of angular fragments cemented together.

**breeder reactor**  A nuclear reactor that creates fissionable material artificially. It is called a breeder reactor because the reaction produces, or "breeds," more material than it begins with.

**calcareous ooze** Fine-grained sediment that covers much of the ocean floor in tropical and warm-temperate seas; consists of microscopic calcareous shells.

**calcite** A calcium-carbonate mineral with three directions of cleavage; effervesces in acid.

**calc-silicate rocks** Metamorphic rocks formed from sedimentary rocks rich in both calcite (or dolomite) and silicate clasts (predominantly quartz).

**caldera** A large, basin-shaped depression formed by the inward collapse of a volcano after an eruption.

**caliche** See *K horizon.*

**calving** The process by which a chunk of ice breaks away from an ice sheet into water.

**capacity** The potential load that a stream can carry.

**capillary fringe** A relatively thin zone of water that migrates upward from the water table.

**cap rock** Impermeable rock, commonly shale, that prevents oil and gas from escaping upward from the reservoir rock.

**carbonate mineral** One of a group of minerals composed of the carbonate anion and metal cations, for example calcite ($CaCO_3$) and dolomite ($CaMg (CO_3)_2$). These minerals are common in some sedimentary rocks.

**carbonates** Sedimentary rocks made up of calcium or magnesium carbonate.

**carbon-14 dating (radiocarbon dating)** A method of radiometric dating in which decay of the isotope carbon–14 into nitrogen–14 is measured. The level of carbon–14 is continually readjusted in all organisms to an equilibrium level. Once they die, decay is not compensated for and steadily decreases, reaching an unmeasurable level in about 50,000 years.

**cataclastic metamorphism** The process by which rocks are formed by shearing and granulation during fault movement.

**catastrophism** The belief that geological history occurs as a sequence of sudden catastrophic events.

**catenary** A curve produced by a wire or chain suspended from two points. The cross section of a U-shaped valley approximates a catenary.

**cation** A positively charged ion.

**cementation** The process by which sediments consolidate: a cement (usually $CaCO_3$ or $SiO_2$) is deposited in the pore spaces separting the sedimentary particles.

**cenote** In Yucatán, Mexico, a sinkhole formed by the collapse of the roof of an underlying cave.

**Cenozic Era** The major division of geological time following the Mesozoic Era and beginning about 65 million years ago. Characterized by the rapid evolution of mammals.

**change of state** Changes in crystalline form of rock material in a zone between 250 and 700 km (155 and 435 mi) in the earth as a result of increased confining pressure.

**channeled scabland** The basalt plateau in eastern Washington State that is crossed by deep, dry channels that formed during a huge prehistoric flood.

**chemical compound** Two or more elements bonded together in definite proportions.

**chemical remanent magnetization(CRM)** Permanent magnetization that results from the growth of magnetic grains, through chemical alteration, in a rock subsequent to its formation.

**chemical weathering** The weathering of rocks and minerals by chemical reactions.

**chert** A dense, hard sedimentary rock made up of cryptocrystalline silica.

**chlorite** A group of hydrous silicates containing magnesium, aluminum, and other metal ions. Occurs in thinly banded masses and is grass-green to blackish-green in color.

**C horizon** The slightly weathered layer in a soil profile beneath the B horizon.

**chromosphere** The reddish, low-density layer just above the photosphere.

**cinder cone** A relatively small volcanic vent commonly composed almost entirely of pyroclastic material.

**cinders** Pebble-sized, reddened pyroclastic fragments blown from a volcano. Abundant in cinder cones.

**cirque** A horseshoe-shaped, steep-walled glaciated valley head.

**clast** A fragment produced from the breakdown of a preexisting rock.

**clastic sediment** Sediment made up of fragmental material derived from preexisting rocks of any origin.

**clastic texture** A fragmental texture usually associated with sedimentary rocks in which the angular-to-rounded grains (clasts) are broken fragments of preexisting minerals and rocks.

**clay minerals** A group of hydrous aluminosilicate minerals that result from the weathering of rocks.

**clay-size** A term used to designate the particle size of a sediment as very fine and smaller than that of silt.

**cleavage** A planar surface or set of parallel surfaces along which a mineral will tend to split when broken.

**coal** A partially metamorphosed sedimentary rock formed from decomposed and altered plant remains.

**columnar jointing** The joint pattern that forms when lava flows or tabular, shallowly buried magma bodies cool.

**compaction** The squeezing together of particles in a sediment.

**competence (stream)** The ability of a stream to transport sedimentary particles; numerically measured as the diameter of the largest particle transported.

**composite volcano** See *stratovolcano.*

**compositional zonation** A partitioning of melt within the magma chamber such that the melt in the upper part of the chamber is relatively cooler and more felsic than that in the lower part.

**conchoidal fracture** A type of fracture that resembles the markings on a conch shell; the surface has a number of concentric irregularities.

**concordant intrusion** A pluton, typically a sill, that intrudes generally parallel to the layering of a metamorphic or sedimentary rock.

**concretion** The rounded bodies in sedimentary rock formed when cement preferentially collects in abundance around a small particle.

**cone of depression** A depression in the water table in the shape of an inverted cone; it develops around a heavily pumped well.

**confined aquifer** A layer of permeable material, such as sandstone, enclosed between layers of impermeable rock.

**confining pressure** The all-sided pressure created at depth by the load of the overlying rock.

**conglomerate** A clastic sedimentary rock composed of rounded gravels cemented together.

**contact** The surface between two different types or ages of rocks.

**contact/hydrothermal metamorphic rocks** Rocks recrystallized through heating marginal to a pluton.

**continental crust** The relatively low-density ($2.7 \ g/cm^3$) material extending beneath continents to depths up to 70 km (44 mi). Also called *lithostatic pressure.*

**continental drift** The concept that the continents have moved apart relative to one another.

**continental rise** The area of coalescent sedimentary fans between the continental slope and the abyssal plain.

**continental shelf** The shallow, gradually sloping platform that surrounds almost all of the continents. Its width averages 80 km (50 mi) but ranges from 0 to 1500 km.

**continental slope** The slope that extends downward from the shelf break to the abyssal plain.

**contour** An imaginary line connecting points of the same value.

**convection current** The postulated circular movement of mantle material in a cell as a result of local heating at the base of the cell; also called thermal convection.

**coral** Any of a large group of shallow-water, bottom-dwelling marine invertebrates having skeletons consisting of calcium carbonate. Common in warm seas.

**core** That portion of the earth below the mantle at a depth of about 2900 km (1800 mi). Apparently composed largely of iron, the core is divisible into two portions—an *outer core,* down to 5100 km (3170 mi), which is liquid; an inner one, from 5100 to 6300 km (3170 to 3900 mi), which probably is solid.

**core-mantle boundary** The boundary at a depth of about 2900 km (1800 mi) that marks a change from solid silicate rocks above to the fluid core below. The core is composed primarily of iron.

**Coriolis force** The apparent force that causes objects traveling over the earth's surface to deflect toward the right or left into a curved path.

**corona** A zone of rapidly moving ionized gases surrounding the photosphere of the sun and extending far out into space. The corona is visible only when the chromosphere and photosphere are blocked out, as during an eclipse of the sun by the moon.

**country rock** Preexisting rocks into which plutonic rocks intrude.

**covalent bonding** A type of bonding in which the electron is shared by two atoms.

**crater** A funnel-shaped depression at the summit of a volcano marking the conduit through which volcanic products are erupted.

**creep** The slow movement of shallow soil material downslope.

**crevasse** A deep-split fissure, or crack, near the surface of a glacier, caused by stresses during glacial movement.

**cross-bedding** The original layering of sedimentary rocks in which the layers are inclined, sometimes at steep angles to the horizontal.

**crossing (river)** The shallow part of a channel at the bend of a meander.

**crust** The rocks above the Moho (Mohorovičić discontinuity).

**crystal** A solid chemical compound or element having a regularly repeating atomic arrangement (crystal lattice) and commonly bounded by plane surfaces (crystal faces) that parallel prominent lattice planes.

**crystal fractionation** The physical separation of crystals from residual magma, producing one type of magmatic differentiation.

**crystal lattice** See *crystal.*

**crystalline texture** A texture of interlocking mineral grains formed as a result of crystallization from a magma, precipitation of minerals, or metamorphic recrystallization.

**crystallinity** See *long-range ordering.*

**cuesta** A ridge formed by a resistant bed that dips at a low angle; has a gentle slope on one side and a steep slope on the other.

**Curie temperature** The temperature above which a rock loses its permanent magnetism and becomes virtually nonmag-

netic. The Curie temperature of pure magnetite is 580° C (1076° F).

**current ripples** The corrugated surface made by a water current flowing across a sandy base or by the wind blowing across desert sands. The ripples are asymmetric in cross section, with the more gentle slope facing the upcurrent or wind direction.

**cut bank** The outside bank in a meander, cut by lateral erosion of the steam.

**cutoff** A channel eroded through the neck of land between two bends, or meanders, of a river.

**cut terrace** A river terrace consisting of a thin layer of gravel resting on a fairly smooth bedrock surface.

**daughter isotope** The product of decay of a radioactive parent isotope.

**debris flow** A mass movement of high fluidity in which over half of the solid material is greater than sand size.

**décollement** A complexly folded overthrust sheet common to alpine chains, composed mainly of rumpled sediments.

**deep-focus earthquake** Originating in the depth zone between 300 and 700 km (190 and 435 mi).

**deep-sea trench** See *trench.*

**deflation** Erosion of earth materials by the wind.

**degradation (stream)** The downcutting of a stream's channel.

**delta** A low, nearly flat body of sediment deposited near the mouth of a river.

**dendritic drainage** A drainage system in which tributary streams join the main stream in an irregular pattern resembling the branching of a tree.

**dendrochronology** The study of tree-ring patterns for dating the recent past (back to a few thousand years ago).

**density current** A gravity-induced underflow of relatively more-dense water. Density differences may be affected by temperature, salinity, or sediment content.

**depositional remanent magnetization (DRM)** Permanent magnetization resulting from the alignment of magnetic clasts with the earth's field during sedimentation processes.

**desertification** Human misuse of land that results in its becoming more desertlike than it was before.

**desert pavement** See *stone pavement.*

**desert varnish** A thin, shiny bluish-black coating composed largely of iron and manganese oxides formed in desert regions on stones and cliffs.

**detachment fault** Nearly horizontal major faults beneath the basin and range province along which large amounts of horizontal movement may have taken place. Many of the range-bounding faults flatten with depth and merge into detachment faults.

**devitrification** Conversion of volcanic glass to a fine crystalline intergrowth over a long period of time.

**diagenesis** Alterations of minerals in rocks occurring after burial to a shallow depth; merges with low-grade metamorphism.

**diapir** A salt dome that has pushed through overlying sedimentary rock.

**diastrophism** Large-scale deformation of the earth's crust, such as folding, faulting, subsidence, or uplift.

**diatom** A marine plant that secrets silica. The remains form siliceous ooze.

**differential weathering** Variation in the rate of weathering in rocks. Ledges, recesses, and irregular forms are the result.

**differentiation** In cooling magma, the process that involves separation of earlier-formed minerals from the residual magma.

**diffraction grating** A finely engraved glass plate capable of diffracting light into its composite wavelengths.

**dike** A discordant tabular intrusion of magma that cuts across the layering of the country rock.

**dike swarm** A group of dikes intruded at about the same time and possessing a geometrical relationship. They most commonly occur in parallel alignment—also can occur concentrically or radially.

**diorite** A gray coarse- to fine-grained plutonic rock with a composition midway between granite and gabbro. Composed mostly of calcium–plagioclase, amphibole, pyroxene, and biotite.

**dip** The angle between a dipping bed or surface and a horizontal plane; measured perpendicular to the line of strike.

**dipole field** A relatively simple magnetic field configuration consisting of two magnetic poles called the north and south poles. Analogous to a bar magnet.

**dipping bed** A layer of sedimentary rock tilted at an angle to horizontal.

**dip-slip fault** A fault in which movement (slip) is parallel to the dip of a fault.

**directed pressure** Unidirectional pressure in response to differential stress that operates during some types of metamorphism associated with mountain building; foliated and lineated metamorphic textures are indicators of its occurrence.

**discharge (stream)** The quantity of water that passes a designated point in a given interval of time.

**disconformity** An unconformity in which the rock layers are parallel above and below the unconformity and a considerable gap in sedimentation is represented by the unconformity.

**discordant intrusion** A pluton, typically a dike, that cuts across the layering of a metamorphic or sedimentary rock.

**disharmonic fold** A complex fold in which the subsurface geometry has little resemblance to the surface geometry.

**dissolved load (stream)** The material a stream carries in solution.

**distributaries** The branches of a stream into which a river divides when it reaches a delta or alluvial fan.

**divergence zone** see *zone of divergence*.

**dolomite** A carbonate mineral with the composition $CaMg(CO_3)_2$.

**dolostone** A carbonate sedimentary rock in which the mineral dolomite predominates.

**dome** A shield-shaped anticlinal structure that lacks an axis and plunges nearly equally in all directions away from the dome crest.

**doubly plunging fold** A fold in which the axis is either arched upward or bowed downward.

**downwasting** A process in which the slopes adjacent to a stream valley are gradually diminished by rock decomposition, soil creep, and other mass-movement processes.

**drainage basin** The entire area that gathers water and ultimately contributes it to a given river.

**dripstone** Calcium carbonate precipitated from dripping waters in underground caverns.

**drumlin** An elliptical, rounded low hill formed by glaciers and occurring in groups.

**dynamic metamorphism** See *cataclastic metamorphic rocks*.

**earth flow** A relatively slow earth movement usually with a spoon-shaped sliding surface and a crescent-shaped cliff at the upper end. Commonly breaks up internally.

**earthquake magnitude** See *Richter scale*.

**earth reference ellipsoid** The ellipsoid of rotation closely approximating the size and shape of the earth.

**echo sounder** An instrument that records ocean depths by measuring the time required for a sound impulse to travel to the seafloor and back.

**ecliptic plane** The plane of rotation of the moon about the earth.

**effluent stream** A stream that flows at the level of the water table and derives some of its water from the latter.

**E horizon** A whitish layer in a podzol that lies between the A and B horizons and from which most of the iron oxides have been removed by downward-percolating water.

**elastic-rebound theory** The theory that movement along a fault results from an abrupt release of elastic strain energy that has accumulated in the rock masses on either side of a fault as a result of their deformation.

**elastic strain energy** The energy stored in a rock body during deformation; can be regained during faulting.

**electron** A negatively charged particle of low mass outside the nucleus of an atom.

**element** Groups of similar atoms; at present there are 106 known elements; examples are hydrogen, carbon, and sulfur.

**ellipsoid of rotation** The shape that is generated when an ellipse is revolved about one of its axes.

**end moraine** See *terminal moraine*.

**eolian** Pertaining to the wind.

**ephemeral stream** A stream that does not flow continuously. Also called an intermittent stream.

**epicenter** The point on the earth's surface above the focus of an earthquake.

**epoch** A geological time unit and subdivision of a period.

**era** A major division in geological time; eras are divided into lesser units called periods.

**erg** A broad, dune-covered area in a desert.

**erratic (glacial)** A rock or boulder carried from its source by glacier ice or an iceberg and deposited when the ice melts.

**escarpment** A steep cliff formed either directly as a result of erosion or as an erosional modification of a previously existing steplike topographic feature, such as a fault scarp.

**esker** A narrow, sinuous ridge of stratified sediment probably deposited by streams flowing in ice tunnels at the bottom of a glacier.

**estuary** A funnel-shaped mouth of a coastal river valley formed as a result of a rise in sea level or land subsidence.

**eustatic change (sea level)** A change in sea level resulting from changes in volume of water in the ocean.

**evaporite** A sedimentary rock that results primarily from the evaporation of water containing dissolved solids.

**evapotranspiration** The transfer of water to the atmosphere through evaporation and through transpiration from plants and animals.

**exfoliation (sheeting)** The process by which concentric exfoliation joints form near or at the surface of bare rocks.

**fabric** The way in which grains or crystals in a rock fit together. Involves consideration of relative sizes and shapes.

**facies changes (sedimentary)** The lateral variation in sedimentary rock bodies owing to lateral variation in the depositional environment.

**fault** A fracture along which significant movement has occurred.

**fault block** A segment of crust bounded by a fault on one side (tilted fault block) or by faults on both sides.

**fault-block mountains** Mountain ranges bounded at one or

both lateral margins by large, high-angle normal or reverse faults.

**fault gouge**   A clayey, soft material formed when the rocks adjacent to a fault are pulverized during slippage.

**fault scarp**   A low, linear cliff resulting from displacement of the earth's surface during fault movement.

**fault-zone breccia**   Rocks in a fault zone that are broken and sheared as a result of fault movement.

**feldspar**   A silicate mineral group that includes both potassium feldspar and plagioclase (rich in sodium and calcium). Has two good cleavages that intersect at 90°, striations on one of the cleavage faces, and a hardness of 6.

**felsic rocks**   Igneous rocks containing a large proportion of feldspar and silica (quartz).

**fenster (window)**   An erosional window through a thrust sheet that displays the rocks beneath the sheet.

**ferromagnesian mineral**   A silicate mineral that contains relatively abundant iron and magnesium and tends to be dark in color.

**fetch**   The distance or area in an open body of water over which wind fiction can affect waves.

**fill terrace**   A river terrace made up of relatively thick deposits of gravel resting on a bedrock surface.

**firn**   The gritty, granular snow formed when snowflakes melt and refreeze several times.

**first motion**   The character of the earthquake wave at its initial recording on a seismograph; it provides information about the type of movement along the fault that triggered the quake.

**fissility**   The tendency of sedimentary rocks to split along well-developed and closely spaced planes.

**fissure eruption**   An eruption that takes place through a fissure, or large crack, rather than through a localized volcanic vent.

**fissure flows**   See *flood lavas*.

**fissures**   Long cracks in the earth through which mafic lava erupts.

**fjord**   A narrow glacial valley partly submerged by the sea.

**flint**   A variety of chert, usually dark colored.

**flood lavas (flood basalts)**   Flows of basaltic lavas that erupt from innumerable cracks or fissures and commonly cover vast areas to thicknesses of 1000 m (3280 ft) or more. Also called fissure flows.

**flood plain**   The flat surface adjacent to a stream over which streams spread in time of flood.

**flow banding**   In an igneous rock alternating layers of different texture and minerals; it is the result of the flowing of viscous magma.

**fluidity**   See *viscosity*.

**focus**   The initiation point of an earthquake within the earth.

**fold**   Wave-like feature caused by structural processes.

**foliated rocks**   Metamorphic rocks characterized by parallel orientation of tabular minerals and varying degrees of banding (color layering). Common to regionally metamorphosed rocks.

**foliation**   Layering, or banding, in some metamorphic rocks caused by the subparallel alignment of platy minerals.

**footwall**   The face of the block below an inclined fault.

**foraminifera**   Single-celled marine animals that secrete a calcite shell. Some limestones consist almost entirely of foraminifera.

**forceful injection**   Name given to the process by which magma at depth pushes aside existing rocks and forces its way into cracks and fissures.

**foreset bed**   The inclined layers that make up the front of a delta as it advances into a body of water.

**formation**   A lithologically distinctive rock unit or deposit with an upper and lower boundary that is large enough to be mapped.

**fossils**   The preserved remains or traces of prehistoric life. Most fossils consist of the bones or exterior shells of organisms.

**fracture zone**   See *transform fault*.

**free oscillations**   Fundamental, long-period vibrations of the whole earth—similar to the ringing of a bell—excited by very large earthquakes

**fringing reef**   A coral reef that is directly attached to the shore of an island or continent.

**frost heave**   The uneven lifting of surface materials owing to subsurface freezing of water.

**frost wedging**   The mechanism by which jointed rocks are pried apart by ice acting as a wedge.

**gabbro**   A dark, plutonic rock consisting typically of coarse-grained crystals of pyroxene, calcium-plagioclase, and olivine.

**garnet**   A group of silicate minerals that lack cleavage and whose crystals are almost always well formed and equidimensional. They are common in some metamorphic rocks.

**geoid**   A generalized earth shape in which high and low spots are represented but smoothed out and reduced. The attraction of gravity is everywhere perpendicular to the geoidal surface, which corresponds to the surface of mean sea level.

**geological map**   A map which records lithology, aerial extent, and orientation of bedrock and surficial deposits at or immediately below the earth's surface.

**geological time**   The vast period of time (about 4.5 billion years) covering the entire history of the earth.

**geology**   The scientific study of the earth.

**geosyncline**   A huge, elongated trough at the earth's surface in which thousands of meters of sedimentary rocks and volcanic rocks accumulate over many tens to hundreds of millions of years. Thought to represent the site of subsequent alpine mountain building.

**geothermal region**   An area in which hot water and steam occur naturally underground and which is commonly characterized by hot springs and geysers.

**geyser**   A hot spring that erupts jets of hot water and steam, resulting from the heating of ground water by hot rocks and steam.

**glacial abrasion**   The scouring or wearing down of the rocks over which a glacier moves.

**glacial marine (sediment)**   A marine sediment containing stones carried out to sea by icebergs. Found around glaciated areas and areas of pack ice.

**glacial quarrying**   The process by which a glacier prys rocks loose from the surface over which it moves.

**glacial striations**   Long grooves and scratches made by a glacier on the rocks it carries or on the bedrock over which it moves.

**glacial surge**   A rapid movement in which a glacier becomes decoupled from its base and advances downvalley at high velocities, some approaching 6000 m/yr (19,700 ft/yr).

**glacier**   A large mass of ice formed by the compaction and recrystallization of snow, which moves slowly under the stress of its own weight.

**glassy texture**   A texture in which a large part of the rock is composed of volcanic glass, as in obsidian.

**glide (translational slide)**   A movement of a large mass of intact rock downslope along a planar surface, such as a bedding plane.

**globigerina** A single-celled, surface-dwelling marine organism with a shell made of calcium carbonate.

**glowing avalanche (ash flow)** A turbulent mass of pyroclastic fragments and some high-temperature gas that erupted from a volcano.

**gneiss** A higher-grade coarse-grained metamorphic rock in which bands of light-colored minerals (quartz, feldspar) alternate with dark-colored ones (amphibole, biotite).

**Gondwana** One of two hypothetical landmasses consisting of all the continents in the Southern Hemisphere. See *Laurasia*.

**graben** A relatively down-dropped fault block with linear margins and generally lying between two horsts (uplifted blocks).

**graded bedding** A type of stratification in which the particles in each layer are graded—the larger particles at the bottom grade into smaller ones at the top.

**graded stream** A stream in equilibrium: One that has the velocity and channel characteristics required to transport the load from its drainage basin.

**gradient (stream)** The downvalley slope of a stream channel.

**granite** A coarse- to fine-grained plutonic rock made up largely of potassium feldspar, sodium-rich plagioclase, quartz, and micas.

**granitization** The theory that most granitic rocks are formed in place by solid-state crystallization of preexisting rocks under the influence of chemical solutions. Also referred to as metasomatism.

**gravity meter (gravimeter)** An instrument used to determine the acceleration or force of gravity at a given spot on the earth's surface. Generally the gravimeter provides a measure of the acceleration, which, of course, is proportional to the force of gravity.

**graywacke** A sandstone consisting mainly of detritus derived from mafic igneous rocks.

**greenhouse effect** The heating of the atmosphere when increasing amounts of gases, such as carbon dioxide, alter the sun's radiation budget.

**ground ice** Ice in the thin soil layer overlying permafrost.

**groundmass** The fine material in a porphyritic rock that surrounds the phenocrysts.

**ground moraine** Till deposited beneath a glacier.

**groundwater** Subsurface water, generally occurring in the pore spaces of rock and soil.

**gumbotil** Highly weathered glacial till.

**guyot** A submerged volcano with a planed-off, nearly level summit.

**gypsum** A common evaporite mineral composed of hydrous calcium sulfate.

**gyres** Large, rotating current cells in the major ocean basins generated by the deflective effect of the Coriolis force on moving water currents.

**half-life** The time required for a mass of radioactive isotope to decay to one-half of its initial amount.

**halide** A mineral composed of a halogen anion (chlorine, bromine, fluorine, and iodine) in combination with some metal cations of low valence; examples are halite (NaC1), fluorite (CaF$_2$).

**halite** Sodium chloride, or table salt; a common evaporite mineral.

**hanging valley** A tributary valley whose floor is higher than the main valley; produced when erosion deepens the main valley more rapidly than the tributary.

**hanging wall** The face of the block above an inclined fault.

**hardness** The relative ability of a mineral to be scratched. See *Mohs hardness scale*.

**harmonic tremor** A near-continuous release of seismic wave energy associated with the movement, or flow, of magma in fissures and chambers at depth.

**hematite** A relatively nonmagnetic oxide of iron having a red to red-brown streak.

**high-grade metamorphic rocks** Dynamothermal metamorphic rocks, such as gneiss, which are crystallized under high temperatures.

**hingeline** A line drawn along the points of maximum curvature of a fold.

**hogback** A narrow, sharp-crested ridge formed by a resistant, steeply dipping bed.

**horizon** A soil layer with characteristic physical, chemical, and biological properties.

**horn (glacial)** A jagged peak formed in areas where mountain glaciers radiate away from a summit area and have removed most of the latter.

**hornblende** A dark ferromagnesian mineral of the amphibole group that resembles augite in color and luster, with two cleavages that intersect at 56° and 124°.

**hornfels** A dense, nonfoliated contact metamorphic rock.

**horst** A fault block with linear margins, it rises above the blocks on either side of it.

**hot spot** See *plume*.

**hot spring** Natural hot-water spring fed by juvenile water emanating from a magma, or by groundwater heated by its passage several thousand feet below the surface, or by both.

**hydration** The volume expansion of salt minerals that occurs when water is added to their crystal structure. Pressures generated can break up a rock.

**hydrologic cycle** The complete cycling, or transfer process, of the earth's water.

**hydrolysis** A weathering process in which minerals are altered by chemical reaction with water and acids.

**hydrothermal solutions** See *hydrothermal metamorphic rocks*.

**hydrosphere** The waters of the earth.

**hydrothermal metamorphic rocks** Rocks formed by recrystallization through the action of hydrothermal solutions (hot fluids and gases) circulating marginal to large plutons.

**hydrothermal ore** An ore precipitated from a high-temperature fluid.

**iceberg** A large chunk of ice that breaks away from an ice sheet and floats away.

**ice dome** The summit of a continental glacier or ice sheet.

**ice sheet (continental)** Large irregular sheets of ice that cover a large area and are not usually guided in their flow by the underlying topography.

**icing** The freezing of groundwater that reaches the surface in permafrost areas.

**igneous rocks** Rocks that have solidified from magma.

**inclined fold** A fold in which the axial plane is moderately to greatly tipped from the vertical.

**index fossil** An easily identified fossil with a wide geographical range and limited span of time on earth. Useful in correlating and dating rocks.

**influent stream** A stream that flows above the water table and from which water flows in pores to the water table. Common in arid regions.

**inner core** See *core*.

**inselberg** An isolated residual hill that rises abruptly above the surrounding plain in a dry region. Characteristic of the late stage of the erosion cycle.

**intensity scale**  A scale that measures the damage done by an earthquake; the modified Mercalli scale is the one most commonly used.

**interior drainage**  A pattern of streams that drain toward the center of a basin rather than toward the sea.

**intermediate-focus earthquake**  An earthquake originating in the depth zone between 70 and 300 km (44 and 190 mi).

**intermediate rocks**  Igneous rocks intermediate in composition between felsic and mafic. Examples are andesite and diorite.

**intermittent stream**  See *ephemeral stream.*

**intrusive contact**  The surface of contact between a pluton and the surrounding rock (country rock).

**ion**  An atom or group of atoms that carries a positive or negative charge.

**ionic bond**  A bond that results from the electrostatic attraction between positively and negatively charged ions.

**ionic-covalent bonding**  A type of bonding common in minerals; alternates between ionic and covalent.

**iron meteorite**  See *meteorite.*

**island arc**  A volcanically active island archipelago.

**isostasy**  The theory that blocks of the earth's crust are in a floating gravitational equilibrium.

**isotope**  A form of an element. Each isotope of a specific element has the same atomic number (contains the same number of protons and electrons) but contains a different number of neutrons, therefore each has a different atomic weight.

**joint**  A fracture in a rock along which no movement has taken place.

**joint set**  A group of joints with a similar geometry. In most cases a set consists of parallel, concentric, or radial joints.

**juvenile water**  Water derived directly from magma that comes to the surface for the first time.

**karst**  The name given to hummocky landscapes characterized by the features caused by the solution of rocks by groundwater, such as sinkholes and caves.

**kerogen**  The organic material usually found in shales that can be converted to petroleum products.

**kettle**  Depression formed by the melting of a large block of ice trapped in glacial till or outwash.

**K horizon (caliche)**  A hard, thick calcareous crust that forms beneath the B horizon in arid-region soils (i.e., a strongly developed Bk horizon).

**kimberlite**  An ultramafic igneous rock that originated as deep as 250 km (155 mi); commonly contains diamonds and forms pipelike intrusive bodies.

**klippe (pl. klippen)**  An erosional remnant of a thrust sheet, or nappe.

**laccolith**  A concordant igneous body, more or less circular in outline, with a flat base and dome-shaped top.

**lag deposit**  See *stone pavement.*

**lagoon**  The body of water separating a barrier island from the mainland.

**lahar**  A mudflow on the flanks of a volcano.

**laminae**  Sedimentary layers whose thickness is less than 1 cm (0.4 in)

**landslide**  A relatively rapid movement of soil and rock downslope.

**lapilli**  Pyroclastic fragments about 2 cm (0.8 in) in diameter.

**lateral fault**  A fault in which the relative displacement is primarily strike slip. Also called strike-slip fault.

**lateral moraine**  A ridge of debris that continually accumulates along the side of a glacier.

**laterite**  A highly weathered tropical soil rich in oxides of iron and aluminum. Hardens on drying and can be used to make bricks.

**lateritic soil**  A deep, highly weathered reddish soil, widespread in tropical climates. Most soluble elements (calcium, sodium, potassium, and silicon) have been leached out, leaving a residuum enriched in oxides of iron and aluminum.

**Laurasia**  One of two hypothetical landmasses consisting of the continents in the Northern Hemisphere. See *Gondwana.*

**lava**  The molten material that erupts from a volcanic vent or the rock that results from the solidification of the molten material.

**law of cross-cutting relationships**  A geological rule used in establishing relative time such that if one rock body (or fracture) cuts across another body (or fracture), the latter must be older than the former.

**law of inclusions**  A geological rule that states: If fragments of one rock body are contained in another, the rock body from which the inclusions came is older.

**law of original horizontality**  The geological law that states: All sediments are originally deposited horizontally or nearly so.

**law of superposition**  One of the laws upon which relative geological chronology is based: In any sequence of sedimentary rocks that has not been disturbed, any one layer will be older than the layer *above* it and younger than the layer *below* it.

**leaching**  The dissolving out or removal of soluble materials from a rock or soil horizon by percolating water.

**levee**  A low embankment adjacent to, and confining, a river channel.

**lichen dating**  See *lichenometry.*

**lichenometry**  The study of lichen sizes on rocks for dating of very recent surface deposits.

**lignite**  A brownish coal-like material formed when peat is buried under younger sediments.

**limb (flank)**  One of the two sides of a fold.

**limestone**  A sedimentary rock in which calcium carbonate predominates.

**linear ocean-floor magnetic anomalies**  Bilaterally similar, paired stripes of higher-and lower-than average magnetic field paralleling oceanic ridges.

**lit-par-lit**  French term meaning "bed-by-bed." Applied to a mixed igneous-metamorphic rock, such as migmatite, composed of alternating layers of apparently metamorphic and igneous rock material.

**lithosphere**  The mechanically strong 5- to 150-km (3 to 93 mi) thick outer layer of the earth; it is thicker under the continents and thinner under the oceans, being thinnest at the ridges.

**lithostatic pressure**  The all-sided pressure at depth exerted by the rock load above. Also called confining pressure.

**lodestone**  See *magnetite.*

**loess**  Wind-deposited silt originating in glacial outwash plains or in desert regions.

**longitudinal dune**  A dune aligned parallel to the prevailing wind.

**long-range ordering**  An ordering of atoms in a mineral that extends throughout the mineral grain. Also referred to as atomic ordering or crystallinity.

**longshore current**  A coastal current, parallel to the shore, produced by wave refraction.

**longshore drift**  The movement of sediment along the coast with the prevailing current.

**low-grade metamorphic rocks**  Dynamothermal metamorphic rocks, such as slate, crystallized under moderate to low temperatures.

**low-velocity zone**  A discontinuous zone in the earth between depths of about 100 to 250 km (62 to 155 mi) in which *S*-wave velocities are much less than expected, mostly as a result of thermal softening; temperatures in places may be high enough that the rock is partly molten. See *asthenosphere*.

**luster**  The subjectively evaluated character of the light reflected from the surface of a mineral; falls into two groups, metallic and nonmetallic.

*L* **wave (long wave)**  A seismic surface wave of relatively high amplitude and long wavelength. It follows the *P* and *S* waves in arrival at a seismograph station.

**mafic rocks**  Igneous rocks containing a large proportion of minerals rich in magnesium, iron, and calcium.

**magma (melt)**  A liquid composed of molten material at high temperatures; generally rich in silicon and oxygen.

**magma chamber**  A cavity beneath the earth's surface surrounded by solid rock and containing magma.

**magmatic differentiation**  Term used to describe the separation of magma into fractions of differing chemical composition during cooling and crystallization.

**magnetic anomaly**  Any departure from the normal magnetic field of the earth. In the ocean anomalous highs and lows occur in alternating ridges and rises.

**magnetic field**  A region in which magnetic forces are exerted.

**magnetic polarity time sequence**  The dated sequence of changes in magnetic field polarity as determined from paleomagnetism of rocks and ocean-floor anomaly stripes.

**magnetic reversal**  The 180° reversal of the earth's magnetic field that has occurred intermittently throughout geological history.

**magnetite**  A black, strongly magnetic iron oxide mineral, sometimes called lodestone.

**magnetometer**  An instrument for measuring magnetism in rocks.

**mantle**  The rocks below the Moho (Mohorovičić discontinuity) and above the core.

**mantle plume**  See *plume*.

**marble**  A fine- to coarse-grained nonfoliated metamorphic rock recrystallized from limestone.

**marine terrace**  (a) An accumulation of wave- and current-transported materials seaward of a wave-cut platform; (b) the term also applies to old marine beach deposits and wave-cut platforms now found above sea level.

**mass movement**  The movement of rock material downslope through the direct pull of gravity.

**mass wasting**  See *mass movement*.

**M-discontinuity**  See Mohorovičić discontinuity.

**meander**  A large curving bend in a river.

**mechanical weathering**  Weathering by physical forces, such as frost action or absorption of water.

**medial moraine**  A moraine in the middle of a glacier formed by the merging of the lateral moraines of two coalescing valley glaciers.

**median valley**  The keystonelike depression, or rift, in the crest of an oceanic ridge or rise.

**melt**  See *magma*.

**melting point**  Temperature at which a mineral melts, that is, changes from a solid to a liquid.

**meltwater**  Melted ice and snow from a glacier.

**mesosphere**  The mantle zone beneath the asthenosphere.

**Mesozoic Era**  The era of geologic time between the Paleozoic and Cenozoic eras lasting from about 225 to 65 million years ago. Characterized by the dominance of reptiles, especially dinosaurs.

**metallic bonding**  A type of covalent bonding in which there are more metal atoms available than are necessary to satify bond requirements. Also called time-shared covalent bonding.

**metallic luster**  A mineral luster characteristic of metals.

**metamorphic rocks**  Rocks that form at depth through solid-state recrystallization of preexisting rocks as a result of internal heat, pressure, and chemical activity of fluids.

**metamorphism**  An isochemical process in which rocks are crystallized from preexisting rock under the influence of higher temperatures, confining and directed pressure, and interstitial fluids.

**metasomatism**  See *granitization*.

**meteorite**  Extraterrestrial fragment of solid material that lands on the earth's surface after falling through the atmosphere. Iron meteorites consist largely of iron–nickel alloys and sulfides; stony meteorites contain a large abundance of silicon, magnesium, aluminum, calcium, and iron.

**mica**  A mineral consisting of parallel sheets of silica tetrahedra strongly bonded at their bases but less strongly bonded across the sheets; the result is a well-developed cleavage in one direction.

**microplate**  A relatively small lithospheric plate.

**Mid-Atlantic ridge**  A submarine mountain range that stretches the length of the Atlantic Ocean.

**midocean ridge**  See *oceanic ridges and rises*.

**migmatite**  A layered zone of igneous and metamorphic rocks between a batholith and the surrounding rocks. Generally found near the deeper parts of the batholith.

**Milky Way galaxy**  The large, rotating pinwheel-shaped celestial grouping of about 10 billion stars in which the solar system is located.

**mineral**  A naturally occurring, crystalline chemical compound (or element) that possesses a fixed composition or restricted range of composition.

**modified Mercalli scale (earthquake)**  A scale used to measure earthquake intensity based on damage.

**Moho**  See *Mohorovičić*.

**Mohorovičić discontinuity**  A seismic-velocity boundary used to demarcate the earth's crust above from the mantle below. A relatively large increase in *P*- and *S*-wave velocities takes place at the boundary, which is also called the Moho or M-discontinuity.

**Mohs hardness scale**  A scale used by mineralogists to judge the hardness of a mineral; ranges from 1 to 10.

**molecule**  Combination of two or more elements.

**monadnock**  An isolated hill of resistant rock that stands above the level of a peneplain.

**monocline**  A one-limbed fold with horizontal strata on either side.

**moraine**  See *lateral, medial,* or *terminal* moraine.

**mud cracks**  The polygonal pattern produced by the drying and shrinking of wet, clayey mud.

**mudflow**  A mass movement, usually of fine-grained earth materials containing a relatively high water content.

**mudstone**  A shale without distinct bedding.

**multiple working hypotheses**  Several equally plausible explanations that may be advanced to fit scientific observations. The formulation of such explanations is one of the most common approaches in geology.

**muscovite (white mica)**  A mineral of the mica group, usually colorless, gray, or transparent. Common in metamorphic rocks and in many sedimentary rocks, especially sandstone.

**mylonite**   A dark, hard, fine-grained cataclastic rock.

**nappe**   A complex, large-scale recumbent anticlinal fold, generally with metamorphic and igneous rocks in their centers. See *klippe*.

**native element**   An element that naturally occurs in an uncombined state; examples are gold, silver, platinum, diamond, copper, sulfur.

**natural law**   A theory that meets the tests of experiments over a long period of time, for example, the law of gravity.

**neutron**   An uncharged particle in the nucleus of an atom.

**new global tectonics**   The hypothesis that new lithospheric material, formed at the ocean ridges, moves away from the divergent zones and toward convergent zones where the lithosphere bends sharply downward into the asthenosphere and disappears below.

**nonconformity**   An unconformity in which sedimentary rocks rest on plutonic or metamorphic rocks.

**nondipole center**   An area on the earth's surface where the magnetic field is greater or smaller than would be expected. These centers are nondipolar inasmuch as any one center is unique and not paired with a center of the opposite polarity on the other side of the world.

**nonplunging fold**   A fold in which the axis is horizontal.

**nonsilicate**   One of a large group of minerals that do not contain silica; includes, for example, sulfides, native elements, oxides, halides, carbonates, and sulfates.

**normal fault**   A fault in which the footwall moves up relative to the hanging wall.

**normally magnetized**   A rock that exhibits magnetization generally parallel to the modern field, that is, downward in the Northern Hemisphere.

**nucleus**   The central small core of an atom, composed of protons and neutrons; carries more than 99.9 percent of the mass of the atom but only one-trillionth of its volume.

**nuée ardente**   A turbulent gaseous cloud that has erupted from a volcano and that contains primarily ash; accompanies eruption of a glowing avalanche.

**oblique slip**   Fault movement (slip) oblique to the dip of the fault.

**obsidian**   See *volcanic glass*.

**oceanic crust**   The 5- to 6-km-thick (3- to 4 mi-thick) rock layer beneath oceans with a denisty of about 3.0 g/cm³.

**oceanic ridges and rises (midocean ridge)**   Elongate suboceanic mountain ranges. See also *zone of divergence*.

**oceanography**   The study of the ocean, including its chemical, physical, biological, and geological aspects.

**oilfield**   An underground accumulation of oil or a group of oil pools.

**oil shale**   Shale rich in kerogen, which on distillation and refining yields oil.

**olivine**   A ferromagnesian mineral that usually occurs as rounded and glassy green crystals.

**oolite**   Small, round grains of calcium carbonate that form as layers of the mineral buildup around a nucleus.

**ooze**   The remains of microscopic free-floating organisms that drift down and cover the deep-ocean floor.

**ophiolite suite**   An association of mafic and ultramafic rocks largely altered to serpentine that appears to have formed initially at, or close to, an oceanic ridge.

**organic evolution**   The passage of life on earth during the last 3.5 billion years—begin with primitive single-cell bacteria and blue-green algae and progresses through increasing complexity to the present. The fossil record provides its chief basis of documentation.

**oscillation ripples**   The corrugated surface made in the bed of a shallow body of water. The ripples are symmetrical in cross section, in contrast to current ripples.

**outer core**   See *core*.

**outwash plain**   Flood plains formed by streams draining from the front of a glacier.

**overthrust fault**   A low-angle thrust fault in which the dip angle is less than 10°.

**overturned bed**   A sedimentary bed that has been structurally rotated more than 90° from the horizontal.

**overturned fold**   A fold in which one limb has been rotated past 90°.

**oxbow lake**   A crescent-shaped lake formed when a stream bend, or meander, is cut off from the main stream.

**oxidation**   A process of chemical weathering in which minerals or elements within the minerals combine with oxygen.

**oxide**   A mineral consisting of oxygen in combination with one or more metal elements, generally iron and titanium; examples are magnetite ($Fe_3O_4$), hematite ($Fe_2O_3$).

**pahoehoe**   A lava flow with a ropy or corrugated surface and a glassy outer rind.

**paleobotany**   The study of fossil plants.

**paleoclimatology**   The study of past climates and the causes of their variations.

**paleomagnetism**   The study of fossil magnetism in rocks.

**paleontology**   The study of past geological life based on plant and animal fossils.

**Paleozoic Era**   The geological time-period between the Precambrian and the Mesozoic eras, lasting from about 600 to 225 million years ago. Characterized by relatively simple invertebrates and backboned animals.

**Pangaea**   The hypothetical single continent postulated by Wegener that split into fragments and began to drift apart during the Jurassic period.

**parent isotope**   The initial, or starting, radioactive isotope that will decay spontaneously and progressively to a daughter isotope.

**parent material**   The material from which a soil forms.

**patterned ground**   Polygonal patterns formed in surface material subject to intensive frost action.

**pedalfers**   Soils with a fairly high content of organic matter in the A horizon, and with no calcium carbonate accumulation beneath the B horizon. Common in humid temperate regions.

**pediment**   An erosional bedrock surface that slopes away from a desert mountain range.

**pediplane**   An extensive erosion surface in deserts formed by the coalescence of two or more pediments.

**pedocal**   An arid-region soil characterized by a thin A horizon and a layer of calcium carbonate beneath the B horizon known as the Bk horizon or the K horizon.

**peneplain**   A broad, almost featureless plain that has been eroded to nearly sea level by mass wasting and stream erosion.

**period**   The fundamental unit of the geological time scale and the subdivision of an era.

**permafrost**   Ground that remains below 0° C and usually contains small to large quantities of ice.

**permafrost table**   The upper surface of permafrost.

**permeability**   The measure of the capability of a rock to transmit a liquid.

**Phanerozoic Eon**   That portion of geological time since the Precambrian Eon, constituting (from oldest to youngest) the Paleozoic, Mesozoic, and Cenozoic eras.

**phenocryst** The larger, usually well-formed crystals in a porphyritic igneous rock.

**photosphere** That part of the luminous envelope of the sun below the chromosphere; composed largely of noncharged gas atoms at temperatures near 5400° C (9750° F).

**phreatic explosion** A volcanic explosion caused by the heating and expansion of groundwater.

**phyllite** A common low-grade metamorphic rock with a well-developed rock cleavage—between a slate and schist in stage of metamorphic development. Contains abundant aligned crystals of muscovite that produce a lustrous sheen.

**pillow lava** Lobes of lava resembling a bed of pillows. Results from extrusion into water or a water-rich environment.

**pingo** A near circular, turf-covered mound with an ice core; found in permafrost terrain.

**placer deposit** A mineral deposit, such as gold, formed by a concentration of heavy mineral particles in a beach or river environment.

**plain** A broad area of low relief occurring generally at low elevations.

**plateau** A broad relatively flat or rolling region occurring at relatively high elevations.

**plate-tectonics theory** The theory that the earth is divided into rigid blocks, or plates, that move relative to one another.

**plastic flow (glacial)** The flow of the lower part of a glacier, caused by deformation of ice crystals under the pressure of the overlying ice.

**playa** A dried-up playa lake consisting of clay and silt or sand and deposits of soluble salts.

**playa lake** A seasonal lake in the center of a desert basin.

**Pleistocene** A recent epoch of the Cenozoic Era (and part of the Quarternary Period) in which 90 to 100 percent of the fossil shells still exist today. Usually thought of as coincident with the ice age of the Cenozoic.

**plume** A column of heated rock rising from the mantle. It can bring about partial melting in the asthenosphere and the lithosphere. Also called hot spot.

**plunging breaker** A wave formed in shallow depths near the shore and whose crest takes the shape of a half cylinder that curls over and breaks suddenly with a crash.

**plunging fold** One in which the hingeline dips at an angle to horizontal.

**pluton** A body of plutonic rock of any size or shape.

**plutonic rocks** Rocks formed from magma that cools and solidifies underground.

**podzol** A variety of pedalfer that forms in cooler climates toward the northern limit of trees. Podzols are characterized by a whitish layer, the E horizon.

**point bar** The bar on the inside of a meander.

**porosity** The percentage of the total volume of a rock that is occupied by open spaces.

**porphyritic texture** See *porphyry.*

**porphyry** An igneous rock containing two grain sizes: relatively large, well-formed phenocrysts imbedded in a finer-grained crystalline, or glassy, groundmass. Texture is said to be porphyrytic.

**porphyry copper** A hydrothermal copper deposit in which copper-bearing minerals occur in veins throughout a large volume of rock.

**Precambrian** All geological time before the beginning of the Paleozoic Era. Characterized in the record by scant primitive life or no life at all.

**pressure surface** The level to which water rises in a confined or unconfined aquifer.

**proton** A positively charged particle in the nucleus of an atom.

**pull-apart zone** See *zone of divergence.*

**pumice** A frothy, porous volcanic glass that is generally rhyolitic in composition.

**P wave (primary wave)** A relatively low-amplitude seismic body wave, the first to arrive at a seismograph station after an earthquake.

**pyroclastic flow** The modern term applied to the combination of the glowing avalanche below and nuée ardente above.

**pyroclastic rocks** Coarse- to fine-grained rocks formed from material hurled into the air during a volcanic eruption.

**quartz** A silicate mineral with a hardness of 7 that has a vitreous luster and commonly fractures conchoidally. Composed almost exclusively of silicon dioxide.

**quartzite** Unbanded metamorphosed sandstone or unmetamorphosed silica-cemented sandstone consisting of quartz grains. In the former, the rock breaks across the grains.

**Quaternary** The period of geological time that includes the Pleistocene and Holocene epochs and covers the last 2 million years of earth history up to the present.

**quick clay** Clay deposits with a high water content that become fluid when jarred (e.g., by an earthquake).

**radioactivity** The spontaneous decay of an atom of one isotope into an entirely different isotope.

**radiocarbon dating** See *carbon–14 dating.*

**radiolariam** A microscopic marine organism that secretes a shell of silica. The remains form siliceous ooze.

**radiometric date** The age of a material as determined through measurement of radioactive decay.

**recumbent fold** An overturned fold: One in which one limb is overturned and roughly parallel to the normal limb and both limbs are nearly horizontal.

**reducing environment** An environment characterized by a low concentration of oxygen in the water.

**reef** A ridge of layered sedimentary rock built by the secretions and remains of marine organisms, usually coral.

**refraction** The bending of seismic body waves as they encounter materials of different seismic velocities.

**regionally metamorphosed rocks** Generally foliated rocks recrystallized by heat, pressure, and chemical solutions during alpine mountain-building processes.

**relative time sequence** Sequence of geological events as established by their order of occurrence.

**remanent magnetization** The permanent magnetization acquired by rocks. In most cases it parallels the earth's magnetic field lines at the time of the rock's origin.

**reserve** The estimated amount of an identified resource recoverable with present-day technology.

**reservoir rock** A permeable rock in which oil accumulates.

**resource** A commodity useful or often essential to people.

**reverse fault** A fault in which the footwall moves down relative to the hanging wall. Also called thrust fault.

**reversely magnetized** A rock that exhibits magnetization generally opposite to the modern field, that is, upward in the Northern Hemisphere.

**rhyolite** A light-colored fine-grained to glassy volcanic rock similar to granite in composition and commonly characterized by flow banding.

**ria coast** A submerged coastline in which the sea extends inland in stream valleys sometimes for long distances.

**Richter scale** A scale of earthquake magnitudes developed by the seismologist C.F. Richter. The magnitudes can be determined from seismographs and are directly related to the amount of energy released during an earthquake.

**ridge-ridge transform fault** Transform faults along which oceanic ridges are offset.

**rift valley** A large, elongate trough, or graben.

**Ring of Fire** A zone that girdles much of the Pacific Ocean and contains about two-thirds of the world's active volcanoes.

**river terrace** Along a stream, an elevated flat area underlain by river deposits. Such a terrace is an abandoned flood plain.

**rock cleavage** The tendency for a rock to break along relatively smooth, closely spaced parallel surfaces. Slaty cleavage is an example.

**rockfall** The relatively free-falling movement of rock material from a cliff or other steep slope.

**rockfall (debris) avalanche** A large mass of rock that slides very rapidly as a unit [100 km/hr (62 mi/hr)] downhill.

**rock flour** A fine silt that covers much of the surface of an outwash plain.

**rock glacier** A tongue-shaped slow-moving mass of rocks and ice in alpine areas.

**salt** A general chemical term that includes many compounds, the most familiar of which is table salt (sodium chloride).

**saltation** A mode of sediment transport in which the particles bounce along the floor of a stream or along a desert surface.

**salt dome** A structure dome produced by the upward movement of a body of salt through enclosing sediments.

**salt playa** A playa consisting of saline residues.

**sandstone** A clastic sedimentary rock consisting mainly of sand-size grains cemented together.

**scarp** A cliff or line of cliffs produced by faulting, erosion, or landsliding.

**schist** A higher-grade foliated metamorphic rock of intermediate grain size. Individual folia are relatively thin; platy minerals commonly make up one-half of the rock; and color banding is not well developed.

**schistosity** The well-developed wavy or undulatory rock cleavage characteristic of schists.

**scientific method** First, observation of a phenomenon or process ensuing in collection of data; next, formulation of one or more hypotheses to explain the data; last, testing of the hypothesis by natural or designed experiments.

**scoriaceous** Adjective applied to a mafic volcanic rock that is frothy and cellular, that is, filled with many vesicles.

**sea cliff** A cliff or slope produced by wave erosion.

**seafloor spreading** The hypothesis that continental drift occurs through cracking and spreading at the oceanic ridges and rises.

**seamount** A submerged volcano of basaltic composition.

**sedimentary rock** Rocks formed by the accumulation of layers of clastic and organic material or precipitated salts.

**seiche** A wave of oscillation—set up in a lake, harbor, or bay—that is initiated by the motion of an earthquake or by local changes in atmospheric pressure.

**seismic gap** Segment of an active seismic zone that has not undergone recent strain release and is therefore considered a prime region of potential seismic activity.

**seismic-reflection profiling** Study of geological bodies by reflection of seismic waves generated by strong, surface detonations.

**seismic sea wave** See *tsunami*.

**seismic tomography** Analysis of multitudes of earthquake records to provide three-dimensional information about the inside of the earth.

**seismic wave** A wave or a vibration produced by an earthquake.

**seismograph** An instrument used to measure the vibrations, or waves, generated during an earthquake.

**seismology** The study of seismic waves in its broadest sense. Includes studies of wave motion, the events that produce them (primarily earthquakes), and the nature of the materials of the earth through which the waves pass.

**self-exciting dynamo** A mechanism consisting of interrelated moving electronic currents and varying magnetic fields. Once primed, it is able to continually regenerate a magnetic field.

**serpentine group** A group of hydrous rock-forming minerals derived by the alteration of magnesium-rich silicate minerals (e.g., olivine) in water-rich environments at low temperatures.

**serpentinization** The process of forming serpentine from magnesium-rich silicates.

**shadow zone** A region from about 102° to 143° from the epicenter of an earthquake in which there is no reception of direct seismic waves.

**shale** A fine-grained laminated sedimentary rock composed of clay and silt.

**shallow-focus earthquake** An earthquake originating between the earth's surface and 70 km (44 mi) in depth.

**sheet flood** Muddy turbulent water that fills and overflows the banks of an arroyo and eventually spreads out over the desert floor.

**sheeting** See *exfoliation*.

**shelf break** The outer edge of the continental shelf characterized by a sudden steepening of the slope.

**shield** A large area of exposed igneous and metamorphic rocks, usually Precambrian, surrounded by sediment platforms, for example, the Canadian Shield.

**shield volcano** A shieldlike volcanic cone built almost entirely of fluid lava flows.

**short-range ordering** A short-term atomic, molecular, and ionic ordering in liquids.

**silicate** A mineral consisting of silicon, oxygen, and varying proportions of one or more metals. Most common minerals are silicates.

**siliceous rocks** Rocks made up largely of silica.

**siliceous sinter** A hot-spring or geyser deposit composed of silica.

**silicon tetrahedron** The tent-shaped (tetrahedral) arrangement of four oxygen ions around one silicon ion; represents the basic building block of all silicate minerals.

**sill** A concordant tabular intrusion of magma that more or less parallels layers of the country rock.

**sinkhole** A pit in karst topography caused by the solution of surficial limestone or the collapse of a cave roof.

**sinter** A spongy, porous sedimentary rock formed from the chemical precipitation of silica from springs.

**skarn** Rocks formed through combined heart from intrusions and hydrothermal solutions in which the initial rock composition has been greatly altered by the addition of ions from the hydrothermal solutions.

**slate** A common low-grade foliated metamorphic rock, usually derived from fine-grained sedimentary rocks. Possesses slaty cleavage.

**slickensides** A polished and smoothly striated rock surface that results from movement along a fault plane.

**slip** The term used to denote actual relative displacement along a fault.

**slip face** The steep face of an asymmetrical dune.

**slump (rotational slide)** The movement of rock material downslope as a unit along a concave-upward slip plane. Characterized by backward tilting of the mass.

**snow** Frozen water vapor that is crystallized directly from the water vapor in the atmosphere.

**snowline** The line, or altitude, on a glacier separating the area in which snow remains from year to year from the area in which the snow of the previous season melts.

**sodium chloride** Common salt.

**soil profile** The three basic layers, or horizons, of most soils.

**solar day** The time between two consecutive passages of the sun at its zenith past a particular spot on earth. A mean solar day is an arbitrary unit of time, one that is always the same length regardless of the season.

**solar wind** A continuous but spasmodic flow of ionized particles from the sun that moves out into space at speeds of 300 to 600 km/sec (675,000 to 1,350,000 mi/hr). The wind represents the outer portion of the sun's corona.

**solid load** The suspended load and bed load of a stream taken together.

**solid solution** A mixed-crystal mineral composed of varying amounts of certain ions that can substitute for one another in the crystal lattice.

**solifluction** The slow downslope movement of water-saturated surface material occurring in permafrost areas.

**solution** A process of chemical weathering in which soluble minerals are dissolved.

**sonar (Sound Navigation Ranging)** A device that sends and receives sound signals underwater.

**sorting** The property that refers to the degree of similarity in particle size in a sediment or sedimentary rock.

**specific gravity** The weight of a specified volume of a mineral divided by the weight of an equal volume of water at 4° C (39° F).

**spilling breaker** A wave whose crest collapses gradually over a relatively long distance as water spills continuously down the wave front.

**spit** A curved embankment formed by a longshore current that trails downcurrent from the land.

**stack** A pillarlike rocky island or mass near a cliffy shore that is separated from the headland by wave erosion.

**stalactites** Iciclelike pendants of travertine that hang from the roof of a cave.

**stalagmites** Deposits of travertine built upward from a cave floor.

**star dunes** Eolian sand accumulations with a star shape.

**stellar day** The time between two consecutive passages of a distant star past a particular spot on earth.

**steppe** An extensive dry region characterized by grass vegetation.

**stock** A large pluton, generally of granitic rock, less than 100 km² (40 mi²) in exposed surface area.

**stone pavement** A thin veneer of stones several stones thick mantling a desert surface. Also called desert pavement or lag deposit.

**stony meteorite** See *meteorite.*

**stoping** Enlarging of a magma chamber by the prying loose of small blocks or rocks from the roof and walls.

**strata (sing. stratum)** Layers of sedimentary rock.

**stratovolcano** A steep-sided volcano consisting of alternating layers of lava and pyroclastic materials. Also called a composite volcano.

**streak** The true color of a mineral as seen in its powdered form, generally as a result of rubbing the mineral on a streak plate, that is, a small plate of unglazed porcelain.

**strike** The line of intersection made by a dipping bed or surface with an imaginary horizontal plane.

**strike slip** Fault movement (slip) parallel to the strike of the fault.

**strike-slip fault** See *lateral fault.*

**structural geology** The field of geology dealing with the deformation of rocks.

**subaerial erosion** Erosion that takes place in the open air (cf. subterranean and submarine erosion).

**subbottom profile** A profile of the ocean floor and a cross section of the stratigraphy and structure of the rocks below it obtained by a shipboard echo sounder.

**subdelta** A small delta forming a part of a complex of deltas.

**subduction** The pulling down, or sinking, of lithospheric plates into the asthenosphere at the convergent zone.

**subduction zone** See *zone of convergence.*

**sublimation** The process by which a solid substance vaporizes without passing through a liquid stage.

**submarginal resource** A resource that is not profitable to recover.

**submarine canyon** A steep-sided valley or canyon cut into the continental shelf or slope.

**submarine fan** A fan-shaped deposit of sediments located seaward of a submarine canyon.

**submarine plateau** A broad relatively low-relief plateau that rises, usually 200 m (660 ft) or more, above the ocean floor.

**sulfate** One of a group of minerals composed of the sulfate anion in combination with a metal ion; common examples are anhydrite ($CaSO_4$), gypsum ($CaSO_4 \cdot 2H_2O$).

**sulfide** A mineral composed of sulfur in combination with one or more metal elements; examples include galena (PbS), sphalerite (ZnS), pyrite ($FeS_2$).

**superposed river** A river that erodes down across preexisting structures, so that the river pattern does not conform to the structures.

**surface waves** Relatively slow seismic waves that travel at, or close to, the earth's surface. Also called *L* waves.

**suspect terrane** See *accreted terrane.*

**suspended load (stream)** The material a stream carries in suspension that is buoyed up by the moving water.

**S wave (secondary wave)** A seismic body wave, the second to arrive at a seismograph station. Characterized by sudden onset and is normally greater in amplitude than a *P* wave.

**swell** A regular, somewhat flat-crested wave that has traveled far from its generating area; made up of long-period waves.

**symmetrical fold** A fold in which the two limbs make nearly equal angles with the axial plane.

**syncline** A fold in which the limbs dip toward the axis; younger beds are found toward the hingeline.

**talus** A loose pile of angular rocks at the base of a cliff.

**tarn** An alpine rock-basin lake commonly resulting from differential glacial scouring.

**tectonic dam** Refers to the origin of a type of continental shelf in which shelf sediments are deposited behind a geological uplife or lava; both act as dams.

**terminal moraine** A mass of debris that accumulates as a hummocky, rocky ridge around the snout of a glacier. Also called an end moraine.

**terrestrial planet** A planet composed chiefly of dense rocky material, like that of the earth.

**terrigenous (marine) sediment** Sediment derived from land (e.g., wind transported dust settling on the ocean surface and eventually on the ocean bottom).

**Tertiary** A period of the Cenozoic Era covering the timespan between 65 and about 2 million years ago.

**Tethys Sea** The hypothetical oceanic waterway separating the supercontinents, Gondwana and Laurasia.

**texture** The interrelations of the size, shape, and arrangement of the particles in a rock.

**thermal convection** See *convection current*.

**thermoremanent magnetization** (TRM) Permanent magnetization that results from thermal cooling of a rock (generally igneous in type) from high temperatures through the Curie temperatures of contained magnetic crystals.

**thrust fault** See *reverse fault*.

**till** Material laid down directly by glacier ice.

**tillite** A consolidated glacial deposit consisting of sand, gravel, boulders, and clay.

**time-shared covalent bonding** See *metallic bonding*.

**tombolo** A strip of sand connecting a near-shore island to the mainland.

**topographic profile** A cross sectional drawing of the surface of the land along a given line.

**transform fault (zone of lateral movement, fracture zone)** One of the numerous fracture zones in the ocean basin, along which the ridges and rises have been offset and along which lateral movement occurs.

**transverse dune** A dune aligned at right angles to the wind.

**travel-time curve** A plot of the times taken for earthquake waves (*P*, *S*, and *L*) to travel to recording stations at increasing distances from a earthquake.

**travertine** A limy cave or spring deposit formed by chemical precipitation of calcium carbonate from solution in surface and groundwater. Can also be precipitated by calcareous algae.

**tree-ring dating** See *dendrochronology*.

**trench** A narrow elongate depression on the deep seafloor that parallels the trend of an island arc or continental margin.

**tsunami** A destructive wave generated by disturbances on the ocean floor. Also called a seismic sea wave.

**tufa** A spongy or porous sedimentary rock formed by the precipitation of calcium carbonate around the mouth of a hot or cold spring or in a stream or lake.

**tuff** A fine-grained rock composed of pyroclastic fragments, primarily ash.

**tuff breccia** Rock consisting of relatively large pyroclastic fragments in an ashy matrix.

**turbidite** A sedimentary rock deposited by a turbidity current and characterized by graded bedding.

**turbidity current** Density currents of sediment-laden water that are triggered by the slumping of oversteepened and unconsolidated material.

**ultramafic** A term applied to magmas and igneous rocks that contain extremely large amounts of magnesium and iron.

**unconfined aquifer** A water-bearing surficial layer of permeable material, such as sand or gravel.

**unconformity** A surface where the sequence of rock units has been interrupted by either erosion or nondeposition. The time represented by the unconformity is variable.

**uniformitarianism** The doctrine that the geological processes now modifying the earth's surface have acted in essentially the same way throughout geological time although possibly at different rates.

**upright fold** A fold in which the axial plane is close to vertical.

**U-shaped valley** A valley suggesting the shape of the letter U, with steep sides and a flat floor carved by a glacier.

**vadose zone** Zone between the ground surface and the water table in which the pore spaces are not completely filled with water; also called zone of aeration.

**valley glacier** See *alpine glacier*.

**varve** A pair of thin, sedimentary layers made up of a coarse, silty lower layer and a fine-grained upper layer. Thought to represent seasonal variation in deposition.

**velocity (stream)** The direction and magnitude of displacement of a portion of a stream per unit of time.

**ventifact** A rock faceted by wind-driven particles.

**vesicles** Small, rounded cavities in lava formed by trapped gas bubbles.

**viscosity** Resistance of a liquid to flow; the opposite of fluidity.

**vitreous (glassy) luster** A term applied to a mineral that reflects light to about the same degree as glass.

**volcanic agglomerate** An unsorted deposit of volcanic bombs, cinders, lapilli, and ash in crude layers. Sometimes called volcanic breccia.

**volcanic bomb** Irregular- to spindle-shaped airborne blocks of lava hurled from a volcanic vent during eruption.

**volcanic breccia** Any rock composed of angular fragments of volcanic rock. See also *volcanic agglomerate*.

**volcanic cone (volcano)** The accumulated eruptive products around a volcanic vent, generally in a steep- to flat-sided cone.

**volcanic dome** A rounded extrusion of glassy lava sqeezed out from a volcano that forms a dome-shaped mound.

**volcanic glass** Magma that has cooled so quickly that the liquid quenched to glass without crystallization taking place. Most commonly associated with rhyolitic magma. See *obsidian*.

**volcanic-hazards assessment** Determining the likely sites for imminent volcanic activity, establishing reasonable recurrence rates between episodes, and estimating the type and intensity of activity that might occur.

**volcanic neck** A pipelike pluton of solidified lava that once connected a magma reservoir with a volcanic vent at the surface.

**volcanic rocks** Rocks formed from magma that erupts at the surface and cools and solidifies.

**volcano** Surface mound, commonly conical, of material resulting from volcanic eruption.

**V-shaped valley** A narrow valley with steep sloping sides that result from downcutting by a stream and mass movement of material down the side slopes.

**water table** The surface at which water stands in wells, the upper surface of groundwater.

**wave base** The lower effective limit of wave transportation and erosion.

**wave-cut platform** A planed-off rock bench cut by wave erosion at the base of a sea cliff.

**wave length** The horizontal distance separating two equivalent wave phases, such as two crests or two troughs.

**wave of oscillation** A water wave in which the individual particles move in orbits with little or no change in position although the wave form itself advances.

**wave period** The length of time required for two crests or two troughs of a wave to pass a fixed point.

**wave refraction** The bending of a wave as it approaches the shore.

**weathering** The mechanical disintegration and chemical decomposition of rocks.

**weight** The force that gravity exerts on a body.

**welded ash-flow tuff** See *welded tuff*.

**welded tuff** A pyroclastic rock whose particles have been fused together by heat still contained in the deposit after it has come to rest; generally associated with large-scale caldera-forming events.

**xerophyte**   A plant adapted to dry conditions.

**yazoo**   A tributary river that runs parallel to the main stream between a natural levee and the river bluff.

**zone of aeration**   See *vadose zone.*

**zone of convergence**   The zone where plates push together and lithospheric material is subducted, or pulled down, into the asthenosphere.

**zone of divergence**   The zone where plates move apart and new lithospheric material is formed; equivalent to active oceanic ridges and rises. Also called the pull-apart zone.

# INDEX

Page references in boldface indicate the presence of an illustration.